Vibrational Spectroscopy of Biological and Polymeric Materials

Vibrational Spectroscopy of Biological and Polymeric Materials

edited by
Vasilis G. Gregoriou
Mark S. Braiman

A CRC title, part of the Taylor & Francis imprint, a member of the Taylor & Francis Group, the academic division of T&F Informa plc.

Published in 2006 by
CRC Press
Taylor & Francis Group
6000 Broken Sound Parkway NW, Suite 300
Boca Raton, FL 33487-2742

Printed in the United States of America on acid-free paper
10 9 8 7 6 5 4 3 2 1

International Standard Book Number-10: 1-57444-539-1 (Hardcover)
International Standard Book Number-13: 978-1-57444-539-8 (Hardcover)
Library of Congress Card Number 2005050682

Library of Congress Cataloging-in-Publication Data

Vibrational spectroscopy of biological and polymeric materials / [edited by] Mark Braiman, Vasilis G. Gregoriou.
p. cm.
Includes bibliographical references and index.
ISBN 1-57444-539-1
1. Polymers--Spectra. 2. Biopolymers--Spectra. 3. Vibrational spectra. 4. Infrared spectroscopy. 5. Raman spectroscopy. I. Braiman, Mark Stephen. II. Gregoriou, Vasilis G.

QC463.P5V53 2005
547'.84--dc22 2005050682

Visit the Taylor & Francis Web site at
http://www.taylorandfrancis.com

and the CRC Press Web site at
http://www.crcpress.com

Dedication

To the little ones: Michaela, Gregory, and Alexandra

Vasilis G. Gregoriou

Preface

Infrared spectroscopy was discovered by William Herschel in the early 19th century, with the discovery of the Raman effect to follow about a century later. Both techniques have enjoyed their ups and downs, a trend that is always associated with specific technical advancements in their subcomponents. Nowadays, thousands of instruments in both academic and industrial laboratories are used primarily for characterizing polymers and biological systems and the number of applications continues to grow monotonically.

For this book, we have asked experts in the field to address selected areas of broad interest in an effort to provide the latest developments in the application of advanced infrared and Raman techniques to the above mentioned materials. We also wanted to give thorough experimental detail, and not just to document recent instrumentation advances, so that experienced vibrational spectroscopists might be able to design and execute their own experiments similar to the ones described in the book. Additionally, we wished to examine methods that are particularly amenable to cross-fertilization between the related fields of polymer chemistry and biochemistry. These two fields tend to have distinct groups of practitioners, each of which is likely to benefit from seeing what is developing in the other group. This book should be a useful addition to the number of volumes on infrared and Raman spectroscopies that have appeared in the last few years, such as *Infrared and Raman Spectroscopy of Biological Materials* by Gremlich and Yan (Marcel Dekker), *Handbook of Vibrational Spectroscopy* by Griffiths and Chalmers (Wiley), and *Modern Infrared Spectroscopy* by Christi, Ozaki, and Gregoriou (Elsevier).

The book consists of eight chapters. In Chapter 1, Gregoriou et al. demonstrate the use of static and dynamic FT-IR spectroscopies to an interesting class of polymers: liquid crystalline polyurethanes. An in-depth analysis of the issues behind both static and dynamic linear dichroism measurements, as they are also called, is presented in this chapter with the individual responses of the different parts of the polymers clearly monitored. The following chapter by Galiotis et al. deals with the successful utilization of Raman spectroscopy in the measurement of stress or strain in both single and multiple fiber composite materials. An in-depth analysis of the use of fibre optic cables for remote measurements as well as the use of conventional Raman methods are described in this chapter. The chapter by Hasegawa and Leblanc describes the use of vibrational spectroscopy to study ultrathin materials, both biological and polymeric. Using the Langmuir-Blodgett technique, the authors were able to prepare high-quality ultrathin membranes. The principles of transmission and reflection measurements (the latter on both metallic and dielectric substrates) are very well documented, and the chapter provides a comprehensive survey of the application of these tools to a wide variety of thin films. The fourth submission by Ozaki and Šašic goes into even more depth in describing the principles of two-dimensional (2D) correlation spectroscopy, with an analysis of the recent theoretical progress in the field. In particular, the techniques of sample-sample and statistical 2D correlation spectroscopies are described, along with an array of applications to both polymeric materials (nylon 12, linear low density polyethylene, polyethylene-2,6-naphthalate, bishydroxyethyl terephthalate) and biological compounds (regenerated *Bombyx mori* silk fibroin film, -lactoglobulin).

For the chapters on biological materials in particular, we realized that no reviews of published works would likely compete with free online databases. For example, a search on the topic "FTIR" in PubMed yields over 5000 articles, including over 100 reviews, between 2000 and 2005. Therefore the authors were encouraged to focus their chapters on practical experimental considerations within their own areas of expertise. Especially in new fields, such considerations are often hard to glean from original research articles or instrumentation manuals from commercial manufacturers. In Chapter 5, Bhargava, Schaeberle, and Levin describe advances in microscopic imaging techniques that utilize molecular vibrational spectroscopy. They emphasize methods that take advantage of the recent availability of mid-IR multichannel detectors. When combined with interferometers, these allow simultaneous collection of broadband spectral data from a large number of independent pixels, and open new windows of contrast within microscopic tissue samples, without the requirement for extrinsic dyes or labels. In Chapter 6, Keiderling, Kubelka, and Hilario describe advances in vibrational circular dichroism of proteins and nucleic acids. Their work is aimed mainly towards structural analysis of large polymers that are often not amenable to high-resolution X-ray diffraction. This application depends on recent development of modeling methods, utilizing both small molecules and computations, that are described with great care. In Chapter 7, Baenzinger, Ryan and Kane describe the use of ligand-gated FT-IR difference spectroscopy and in particular its applications in neuropharmacology. This contribution is likely to appeal to a broader variety of pharmacologists who are trying to identify ligands, and modes of ligand action, for the large number of membrane receptors recently identified in the human genome. Finally, in the closing chapter, Braiman and Xiao discuss the application of time-resolved FT-IR spectroscopy to biological materials. The main focus is on the use of commercial step-scan instrumentation for examining the molecular mechanisms of chromophoric proteins that catalyze important photobiological reactions: rhodopsins, heme proteins such as the cytochromes, chlorophyll-based photosynthetic proteins, and photoactive yellow protein. However, recent adaptations of step-scan time-resolved FT-IR to samples that are driven by other triggers, such as temperature jumps and photolytic release of caged ligands, are likely to be of expanding interest, not only to biologists but also to polymer chemists.

Acknowledgments

We want to extend our sincere thanks to Anita Lekhwani, senior acquisitions editor for chemistry at Dekker, for approaching us with a book proposal; David Fausel, the production coordinator, for putting the edited chapters together; John Parthenios and Georgia Kandilioti of FORTH-ICEHT for their generous help to one of the editors (V. G.) in the final stages of the book when everything seemed like a monumental task.

Also, our greatest debt is to the authors of the individual chapters for their excellent contributions and their patience in waiting for the book to be published.

Mark S. Braiman and Vasilis G. Gregoriou
April 2005
Syracuse, New York and Patras, Greece

Editors

Mark Braiman attended Harvard University as an undergraduate; received his Ph.D. in Chemistry in 1983 from the University of California at Berkeley (Richard Mathies, advisor); was a Helen Hay Whitney Postdoctoral Fellow at MIT (1983–1986, H. Gobind Khorana, advisor); and was the recipient of a Lucille P. Markey Scholar Award (1986) that permitted both continued work as a postdoctoral fellow in the Physics Department at Boston University (Kenneth J. Rothschild, advisor) and as an independent faculty member at the University of Virginia Health Sciences Center (Assistant Professor 1988–1994; Associate Professor 1994–1998). He is also a recipient of a Whitaker Foundation Scholar Award (1990–1992). Dr. Braiman has been at Syracuse University since 1998.

Vasilis G. Gregoriou is currently a principal researcher at the Foundation of Research and Technology-Hellas (FORTH-ICEHT) in Patras, Greece. Before joining FORTH in 2000, he was a senior scientist at the Research Division of Polaroid Corporation in Boston, MA. His main research interests are focused on the discovery and physicochemical characterization of materials for use in renewable energy sources, novel optical materials and nanostructures. Dr. Gregoriou is the co-author of 48 refereed research papers, more than 100 abstracts in conference proceedings and three books. He also served as the president of the Society for Applied Spectroscopy (SAS) in 2001. He holds a Ph.D. in Physical Chemistry from Duke University (1993, Richard Palmer, advisor) and worked as an NIH fellow at Princeton University (1994, Thomas Spiro, advisor).

Contributors

John E. Baenziger
Department of Biochemistry,
Microbiology, and Immunology
University of Ottawa
Ottawa, Ontario

Rohit Bhargava
Laboratory of Chemical Physics
National Institute of Diabetes and
Digestive and Kidney Diseases
National Institutes of Health
Bethesda, Maryland

Mark S. Braiman
Chemistry Department
Center for Science and Technology
Syracuse University
Syracuse, New York

Costas Galiotis
Department of Material Science
University of Patras
Patras, Greece

Vasilis G. Gregoriou
Foundation for Research &
Technology
Hellas Institute of Chemical
Engineering and High Temperature
Chemical Processes
(FORTH/ICEHT)
Patras, Greece

Paula T. Hammond
Department of Chemical Engineering
Massachusetts Institute of Technology
Cambridge, Massachusetts

Takeshi Hasegawa
Applied Molecular Chemistry
College of Industrial Technology
Nihon University
Narashino, Japan
and
Japan Science and Technology
Agency/PRESTO
Kawaguchi, Japan

Jovencio Hilario
Department of Chemistry
University of Illinois, Chicago
Chicago, Illinois

Veronica C. Kane
Department of Biochemistry,
Microbiology, and Immunology
University of Ottawa
Ottawa, Ontario

Timothy A. Keiderling
Department of Chemistry
University of Illinois, Chicago
Chicago, Illinois

Veeranjaneyulu Konka
Department of Chemistry
University of Miami
Coral Gables, Florida

Jan Kubelka
Department of Chemistry
University of Illinois, Chicago
Chicago, Illinois

Roger M. LeBlanc
Department of Chemistry
University of Miami
Coral Gables, Florida

Ira W. Levin
Laboratory of Chemical Physics
National Institute of Diabetes and Digestive and Kidney Diseases
National Institutes of Health
Bethesda, Maryland

Bindu R. Nair
Department of Chemical Engineering
Massachusetts Institute of Technology
Cambridge, Massachusetts

Yukihiro Ozaki
Department of Chemistry
School of Science
Kwansei Gakuin University
Nishinomiya, Japan

John Parthenios
Department of Material Science
University of Patras
Patras, Greece

Sheila E. Rodman
Media Research Division
Polaroid Corporation
Waltham, Massachusetts

Stephen E. Ryan
Department of Biochemistry, Microbiology, and Immunology
University of Ottawa
Ottawa, Ontario

Slobodan Šašic
Department of Chemistry
School of Science
Kwansei Gakuin University
Nishinomiya, Japan

Michael D. Schaeberle
Laboratory of Chemical Physics
National Institute of Diabetes and Digestive and Kidney Diseases
National Institutes of Health
Bethesda, Maryland

Yao Wu Xaio
Chemistry Department
Center for Science and Technology
Syracuse University
Syracuse, New York

Table of Contents

1 Studying the Viscoelastic Behavior of Liquid Crystalline Polyurethanes Using Static and Dynamic FT-IR Spectroscopy

Vasilis G. Gregoriou, Sheila E. Rodman, Bindu R. Nair, and Paula T. Hammond

CONTENTS

1.1 INTRODUCTION

The in-depth understanding of the structure–property relationships encountered in liquid crystalline segmented copolymer systems is fundamental to establishing and exploiting the potential of these novel materials for technological applications. The underlying question of how to engineer the mechanooptic response can be answered only by understanding the response of the various parts of the macromolecular liquidcrystalline network (LC) to an applied strain. Fourier transform infrared (FT-IR) spectroscopy is a very powerful tool for the study of polymer dynamics. Infrared absorption bands arising from the functional groups are sensitive not only to the neighboring functional groups but also to physical factors such as molecular orientation and crystallinity. In recent years, FT-IR spectroscopy has been frequently used to study stress-induced molecular orientation in polymeric systems. Two variations of this experiment are used, depending on the time-dependent nature of the perturbation. In the static mode, where no time-dependence is involved, the experiment for studying the molecular response to a static or a step-wise strain is called an infrared linear dichroism experiment.[1] On the other hand, it is also possible to study the molecular response to an externally applied periodical strain. In this case, the method is referred to as a dynamic linear dichroism experiment.

Dynamic linear dichroism spectroscopy has been used successfully in the study of the molecular and submolecular (functional group) origins of macroscopic rheological properties of organic polymers in recent years. In these experiments, the responses of a variety of polymer films to modulated mechanical fields have been examined.[2] Composite films of isotactic polypropylene and poly(γ-benzyl-L-glutamate),[3] uniaxially drawn poly(ethylene terephthalate) (PET),[4] polyethylene,[5] polymethylmethacrylate,[6] polystyrene,[7,8] Kraton™,[9] and Nylon 11[10] are some of the examples that have appeared in the recent literature. Studies of conducting polymers, such as poly(p-phenylene vinylene) (PPV),[11] have also been reported. Time-resolved dynamic infrared experiments have been performed on nematic liquid crystals[12,13] to gain insight into the difference in re-orientation behavior between small molecule liquid crystals and side chain liquid crystal polymers under the influence of an electric field. In a variation of this later experiment, liquid crystal block copolymers under the influence of electric fields were studied using time-resolved techniques to determine the dynamics of electroclinic switching during the application of the field.[14] Other studies of interest included phase-modulated experiments on polyurethanes, which established that hydrogen bonds in polyurethanes are especially sensitive to the stretching of the material.[15] In addition, both static and dynamic modes were used to study a polyester/polyurethane (Estane) copolymer.[16] Finally, dynamic infrared spectroscopy was used to study the orientation and mobility of different molecular segments in a side chain ferroelectric liquid crystalline polymer.[17]

Side chain liquid crystalline polyurethanes are an exciting new class of materials developed at one of the co-authors' laboratories (P.T.H.) that offers the potential to couple the response of liquid crystals and elastomeric networks to applied mechanical strains. Advantages of these systems include the inherent film forming properties of polymers and the enhanced mechanical stability that polymers can provide. Changes in liquid crystal orientation can be induced by modifying the mechanical orientation of the thermoplastic elastomers using classic plastic processing techniques. Thermoplastic elastomers that have been designed to exhibit the mechano-optic properties of liquid crystalline crosslinked systems, have been studied by numerous researchers.[18,19] The approach taken here was to design segmented copolymers with liquid crystals pendant to low T_g siloxane soft segments, because these materials are the only reported elastomers with the responsive LC blocks above their glass transition temperature at room temperature.[20] The liquid crystalline domains in these materials could be oriented by processing, thus inducing the formation of ordered monodomains in an elastomeric matrix. The resulting films should exhibit a range of properties, including piezoelectricity, mechano-optic response, and electrorheological behavior. Since these materials are polyurethanes with the classic film-forming properties associated with this class of copolymers, they show potential as coatings for sensors or transducers.

In addition to sensors, actuators, and other devices that require a mechano-optic response, the low molecular weight series of these materials show promise as viscoelastic damping systems. Damping traditionally occurs near the glass transition temperature of a polymeric material when the ratio of energy stored to energy dissipated by the material reaches a minimum. In this viscoelastic regime, the frequency of the damped vibration coincides with the molecular motions occurring within the polymer. Interpenetrating networks (IPNs) are the state of the art in damping materials because a broad temperature or frequency range of damping can be achieved through a series of coupled molecular motions that provides additional paths to disperse energy. The segmented liquid crystal thermoplastic elastomers should exhibit a multitude of characteristic frequencies from the various segments of the macromolecule as well as from coupled interactions.

The side chain liquid crystal segmented polyurethane examined here consists of liquid crystal functionalized polysiloxane soft segments and traditional MDI/butane diol hard segments, as opposed to liquid crystalline segments in the diisocyanate or chain extender of the hard domain. The hard segments create a network of "physical crosslink" junctions in the liquid crystalline matrix that can transduce applied strains. In order to create a successful mechano-optic material, an effective conduit for the transfer of strain from the hard segments to the LC phase is desirable.

Both segmented polyurethanes and side chain liquid crystalline polymers have been extensively studied by infrared dichroism techniques. Bonart used FT-IR in conjunction with wide-angle x-ray scattering spectroscopic techniques to create a model describing the reorientation behavior of polyurethanes as a function of externally applied strain.[21] In addition, Cooper and co-workers[22a,22b] performed a series of studies using FT-IR spectroscopy to enhance this molecular model. Hsu et al. have also contributed significantly to this area.[23a–23e] In the area of molecular reorientation with strain of side chain liquid crystalline polymers, Zhao and coworkers[24a–24c] have been extremely active. The side chain liquid crystalline elastomer discussed here shows elements of both polyurethane and liquid crystalline orientation.

1.2 THEORY

1.2.1 FT-IR Static Linear Dichroism

This experiment is frequently used to measure the anisotropy in the orientation of a functional group within a sample. Anisotropy can be caused by processing conditions or by applying temperature fields or mechanical strains.

Infrared dichroism involves the orientation of the sample and the subsequent measurement of the absorbance of key vibrational bands parallel and perpendicular to the orientation direction. The dichroic ratio is defined as:

$$R = A\| / A\perp \tag{1.1}$$

where $A\|$ and $A\perp$ are the absorbances measured with radiation polarized parallel and perpendicular to the stretching direction. R can range from zero (no absorption in the parallel direction) to infinity (where there is no absorption in the perpendicular direction). When $R = 1$, the orientation is random. A useful parameter to determine the molecular orientation is the so-called order parameter S, which can be expressed as follows:

$$S = (3 < \cos^2\theta > -1)/2 \tag{1.2}$$

where θ is the angle between the draw direction and the local molecular axis. This orientation function can be related to the dichroic ratio R of the absorption band according to Equation 1.3:

$$S = (R-1)(R_o+2)/(R_o-1)(R+2) \tag{1.3}$$

where R_o, the dichroic ratio for perfect alignment, is defined as $2\cot^2\alpha$. The angle α is the angle between the dipole direction and the molecular axis of the segment under consideration. The value of α varies from zero when the dipole is parallel to the molecular axis, to $\pi/2$ when the dipole is perpendicular to the axis. Thus, the order parameter S may vary from $-1/2$ to 1, indicating perfectly perpendicular to perfectly parallel orientation of the molecule or molecular segment. S is equal to zero in the absence of any orientation. Several recently reported studies point out that this model[25,26] does not sufficiently explain the observed behavior of a number of real systems. Violations of this model have been observed in both static systems and in systems undergoing dynamic deformations.[27,28] However, this model does adequately explain the qualitative behavior of unaxial systems and was used in the experiments described below.

1.2.2 Theory of Dynamic Infrared Spectroscopy

Dynamic infrared spectroscopy is defined as the use of infrared spectroscopy to monitor a time-dependent process.[29] Because the time scale of the excitation/deexcitation process is of the order of 10^{-13} sec or less, changes in the infrared spectrum can be used to monitor the dynamics of slower processes, within the practical limits of the strength of the signal and the detector speed and electronics. Dynamic spectroscopy methods can be divided into two categories: those that use impulse-response techniques (*time-resolved* spectroscopy) and those that use synchronous modulation techniques (*phase-resolved* spectroscopy). In the first case, the dynamic response to a perturbation is monitored as an explicit function of time; in the second case, the phase and magnitude of the response with respect to that of the perturbation are measured.

A dynamic FT-IR linear dichroism experiment studies the anisotropy in a sample as a function of a sinusoidally applied perturbation. In general, infrared absorption is caused by the coincidence of the electric field vector of an infrared beam and the dipole transition moment of a molecular vibration. An infrared band that is anisotropic will have different absorbance intensities depending on the direction of the polarized infrared beam. For example, when the electric field is perfectly parallel to the transition moment of the molecule, the absorbance is maximized. The degree of anisotropy can be described by the dichroic difference $\Delta A(v)$, which is defined as the difference in absorbance of a band as probed by polarized infrared radiation.

The dynamic dichroic difference can be calculated by applying a small amplitude sinusoidal strain perturbation along one of these polarizer directions. This difference is defined as the sum of a quasi-static component and a dynamic component:

$$\Delta A(v,t) = \Delta \bar{A}(v) + \Delta \tilde{A}(v,t) \tag{1.4}$$

The quasi-static component is essentially the traditional dichroic difference described above and includes stresses that may have been introduced into the system as residual orientation from the sample preparation techniques. The dynamic component, on the other hand, is induced exclusively by the small amplitude perturbations on the sample. The small mechanical vibration of amplitude $\hat{\sigma}$ and fixed angular frequency ω_s can be described by the following equation:

$$\tilde{\sigma}(t) = \hat{\sigma} \sin \omega_s t \tag{1.5}$$

The time dependent response of the IR band absorbance to this periodic stretching can be described in terms of the dynamic component of the dichroic difference. The response is a sinusoidal function with the same periodicity as the applied stress field ω_s, an amplitude of $\hat{A}(v)$, and a phase lag $\beta(v)$, as follows:

$$\Delta \tilde{A}(v,t) = \Delta \hat{A}(v) \sin[\omega_s t + \beta(v)] \tag{1.6}$$

Equation 1.6 can be rewritten by employing trigonometric identities as the sum of two orthogonal components:

$$\Delta \tilde{A}(v,t) = \Delta A'(v) \sin \omega_s t + \Delta A''(v) \cos \omega_s t \tag{1.7}$$

where $\Delta A'(v)$ and $\Delta A''(v)$ are related to the amplitude and phase angle as follows:

$$\begin{aligned} \Delta A'(v) &= \Delta\hat{A}(v)\cos\beta(v) \\ \Delta A'(v) &= \Delta\hat{A}(v)\sin\beta(v) \end{aligned} \tag{1.8}$$

where $\Delta A'(v)$ and $\Delta A''(v)$ are the in-phase and 90° out-of-phase spectra, respectively. The in-phase spectrum is the response to the instantaneous extent of strain. The orthogonal (90°) out-of-phase or quadrature spectrum is proportional to the viscous response to the strain.

A step-scan FT-IR experiment, which permits the separation of the time of the experiment (applied perturbation) from the time of data collection, is used to collect the in-phase and out-of-phase spectra. In such an experiment, the dynamic spectra $\Delta A'(v)$ and $\Delta A''(v)$ can be considered as the difference between two static spectra collected at the peaks of the applied stress cycle.[30]

These quantities measure the viscoelastic behavior of the infrared bands in question and are analogous to the data acquired from classic viscoelasticity measurement devices such as dynamic mechanical analysis (DMA). In DMA experiments, an elastic storage (E') and a viscous loss (E'') modulus can be deconvoluted from the complex modulus (E^*). The ratio of loss to storage moduli (tan δ) is the most often reported quantity to describe degree of viscoelasticity in a system.

1.2.3 Two-Dimensional Infrared Correlation Spectroscopy

Two-dimensional infrared correlation spectroscopy (2D-IR) pioneered by Noda and coworkers has provided a powerful tool for vibrational spectroscopists studying the response of materials to an external perturbation.[31,32] 2D-IR spectra are generated as the product of a pair-wise correlation between the time-dependent fluctuations of infrared signals that occur during dynamic infrared experiments. The development of the generalized correlation spectroscopy theory eliminated the requirement that the perturbation has a sinusoidal waveform and thus extended this method to the analysis of time-resolved spectra derived from a perturbation of any arbitrary waveform.[33,34,35] An extensive discussion of 2D-IR is available in the chapter by Ozaki and Šašic in this volume, so only a few comments will be made here.

The dynamic spectrum can be defined as:

$$\bar{y}(v,t) = \begin{cases} y(v,t) \rightarrow \bar{y}(v) & \text{for} \quad T_{\min} \le t \le T_{\max} \\ 0 \quad \text{otherwise} \end{cases} \tag{1.9}$$

where $\ddot{y}(v,t)$ is the spectral intensity variation observed as a function of a spectral variable v over an interval where the external variable t ranges from $T_{\min}$ to $T_{\max}$. The external variable t can be time, or another physical perturbation such as temperature or applied strain. The reference spectrum $\ddot{y}(v,t)$ is defined as:

$$\bar{y}(v) = \frac{1}{T_{\max} - T_{\min}} \int_{T_{\min}}^{T_{\max}} y(v,t)\,dt \tag{1.10}$$

The formal definition of the generalized 2D correlation spectrum may be written as:

$$\Phi(\nu_1,\nu_2) + i\Psi(\nu_1,\nu_2) = \frac{1}{\pi(T_{\max} - T_{\min})} \int_0^{\infty} \bar{Y}_1(\omega) \bullet \bar{Y}_2^*(\omega)\, d\omega \qquad (1.11)$$

where $\Phi(\nu_1,\nu_2)$ is the synchronous correlation spectrum and $\Psi(\nu_1,\nu_2)$ is the asynchronous correlation spectrum.

The synchronous correlation spectrum offers an indication of simultaneous spectral intensity changes as a result of the perturbation. The synchronous spectrum is characterized by the presence of peaks (autopeaks) along the diagonal line defined by $\nu_1 = \nu_2$ and by the presence of off-diagonal (cross-peaks). These autopeaks indicate which transition dipoles, and thus which functional groups, have an orientational response to the perturbation. The sign of the autopeaks is always positive. Intense autopeaks are an indication of spectral bands that respond strongly to the applied perturbation. The cross-peaks illustrate the degree to which dipoles respond in phase with each other and, from their signs, the relative reorientation of these dipoles. A positive cross-peak indicates that the two corresponding dipole moments reorient parallel to each other. Mutually perpendicular reorientations are assumed when the sign of the synchronous cross-peak is negative.

The asynchronous correlation map illustrates the degree of independence between the reorientation behavior of the corresponding dipole moments. There are no diagonal peaks in an asynchronous 2D correlation spectrum. The intensity of any cross-peaks develops only to the extent that two transition dipoles reorient out of phase with each other. The signs of the asynchronous cross-peaks give the relative rates of response of the two contributing dipoles. In addition to spectral resolution enhancement due to the incorporation of the second dimension, 2D-IR spectra can provide information about the relative reorientation of transition dipole moments and the relative rates of inter- and intramolecular conformational relaxations. An advantage of the 2D-IR technique is that the deconvolution of highly overlapped absorption bands is based on physical arguments instead of mathematical data manipulation techniques (e.g., curve fitting analysis, Fourier self-deconvolution, etc.).

2D-IR spectroscopy has been proven very valuable in studies of polymer reorientation in response to applied perturbations. Some characteristic examples include the work by Noda, Palmer et al.[36,37] who studied the kinetics of reorientation of a uniaxially aligned nematic liquid crystal (4-pentyl-4′-cyanobiphenyl) under the influence of an external a.c. electrical field. The technique has also been used to study the orientation and the mobility of a ferroelectric liquid crystal dimer[38] during switching under an electrical field. Also, Siesler[39] studied the segmental mobility of a ferroelectric liquid-crystalline polymer (FLCP) in the Sc* phase under the influence of an electrical field. The reader is urged to consult the chapter by Ozaki and Šašic in this volume for more information on the technique and its applications.

1.3 EXPERIMENTAL

1.3.1 Static Linear Dichroism

The static infrared spectroscopic data was collected on a Nicolet 550 Series II FT-IR spectrometer equipped with an MCT/A detector at 2 cm^{-1} spectral resolution. A wire-grid infrared polarizer (SpectraTech Corporation, Stamford, CT) that permitted plane polarized light to reach the sample was placed in the infrared beam.

Stretching experiments were performed using a static stretcher, designed and machined for this purpose, which allows elongation of the sample up to 200% of its initial length at room temperature. A sample was clasped between a lower stationary jaw and an upper micrometer-controlled jaw. Uniaxial strains were induced in the sample by pulling the upper jaw in a controlled fashion with the micrometer.

Free-standing polyurethane films are too thick to be within the range of the Beer-Lambert Law and saturate key vibrational bands in the transmission infrared spectra. The heterogeneous nature of this segmented copolymer precluded the possibility of finding a single solvent system for the material. A solvent system which includes a nonpolar solvent for the siloxane soft segment and a polar, H-bonding solvent for the urethane hard segments, such as 50/50 mixtures of THF/DMAc or CH_2Cl_2/DMSO, was necessary for casting films. In order to examine films thin enough for FT-IR analysis, samples were prepared by casting the polyurethane on a thin poly (tetrafluoroethylene) substrate (Du Pont De Nemours & Co.) from 50/50 THF/ DMAc. The Teflon™ substrate was used for these experiments as a support for the thin polyurethane film to provide enough integrity for the film to be held by the jaws of the stretcher. The use of substrates for IR dichroism measurements has been previously reported in the literature.[3,30a,30c,40] Adhesion of the polyurethane to the Teflon™ substrate was verified by successfully passing tape peel tests and immersion in water.

There appeared to be minimal slippage between the sample and the film on stretching, and the samples viewed under the optical microscope in reflectance mode at 400x indicated no signs of cavitation. Additionally, the distance between marks placed on the sample before mechanical deformation increased with strain. The sample width decreased with increasing strain, suggesting that there is conservation of volume, and that the sample is in uniaxial tension. For this work, dichroic ratio data were collected by stretching two different samples. For the first sample, a polarizer placed parallel to the direction of stretch was positioned between the sample and the detector. A strain was applied and the FT-IR spectra were collected. The same procedure was carried out on a separate sample with the polarizer perpendicular to the stretch direction. The ratio of the parallel to perpendicular absorbance at each strain value was calculated. Since the two samples used in the experiment were of different film thicknesses due to the solvent casting technique used, the ratios were normalized by setting the 0% strain dichroic ratio to 1.0 for all the vibrational bands studied. To validate this assumption, an experiment was performed on a single sample by changing the polarizer from parallel to perpendicular at each strain point and recording two different spectra for each experimental point. By means of applying this method, the trends described for each of the bands in the results and discussion section were reproduced, and random orientation at 0% strain was confirmed. The dichroic ratio of an unstretched polyurethane cast on the Teflon™ substrate was 1.0, indicating that there was no orientation of the sample induced by solvent casting on the support. However, the data presented here is compiled from the experiments on the two different samples, because it is best to minimize absorptions from water vapor by not opening the sample chamber during the experiment.

In order to ensure that the reported trends are not a manifestation of the Teflon™ substrate, the dichroic ratio calculations were performed on the baseline absorbances for each of the polyurethane bands studied on a bare Teflon™ substrate. It was found that these ratios were close to 1.0 throughout the strain range, indicating that the changes in peak absorbance heights reported are not a consequence of the experimental technique.

Orientation function graphs were generated by using Equation 1.3 and the measured dichroic ratios. A transition moment direction (α) of 0° was assumed for the cyano band, while the carbonyl band was assumed to have an angle of 79°.[28b] The carbon–nitrogen bond of the cyano bond is sp hybridized, resulting in a linear geometry for this bond. It was assumed that the normal vibrational mode of the cyano dipole lies exactly along the molecular axis of the liquid crystalline moiety. The dichroic ratio R of the cyano band therefore maps the structure factor S of the mesogen to the strain. When studying the alignment of the hard segment, the issue is slightly more complicated. Simple ester carbonyls are often assumed to be 60° off the molecular axis in the literature because the carbon is sp^2 hybridized.[30c] The 79° angle is calculated by taking into account the geometry of the entire urethane linkage with respect to the axis of the hard segment. For this reason, the orientation function curves of the hard segment do not necessarily follow the trend of the dichroic ratio. It should be recognized that variation in the transition moment direction (α) of greater than 10° is required to significantly affect the orientation function.[22a] Qualitative comments on the reorientation of the hard segment will assume an orthogonal relationship between the carbonyl dipole and the hard segment.

1.3.2 Dynamic FT-IR Spectroscopy

Dynamic polymer rheology experiments were performed using a Nicolet Magna-IR 860 step-scan FT-IR spectrometer and a Manning Applied Technology polymer modulator, which was mounted directly on a base plate inside the sample compartment. A Molectron wire-grid polarizer was placed in front of the sample allowing only infrared radiation polarized parallel to the stress direction to reach the sample. A low pass optical filter was placed behind the sample to filter off the light above 3950 cm^{-1} or 1950 cm^{-1}. The spectra were collected at a resolution of 8 cm^{-1} wavenumber in less than one hour.

The Magna-IR 860 spectrometer employs a piezo interferometer that utilizes a set of piezoelectric transducers to actuate the phase modulation on the fixed mirror. The digital signal processors control the stepping of the moving mirror, position holding, dynamic alignment, and phase modulation. The sinusoidal signal that drives the polymer modulator is also generated internally by the same circuit that generates the signal for the phase modulation. This design has the added advantage that the phase modulation and the sample modulation by the stretcher are synchronized, reducing the sampling error and improving the signal-to-noise ratio significantly. No additional vibrational isolation is used with the polymer modulator mounted directly on a base plate inside the sample compartment.

The polymer dynamic rheology experiments described in this paper were conducted using a phase modulation frequency of 400 Hz and 3.5 λ_{HeNe} phase modulation amplitude. This modulation amplitude is chosen to cover the entire mid-IR spectral region and to enhance the modulation efficiency in the fingerprint region. The modulation of the applied strain by the dynamic stretcher was 25 Hz with a 50 μm amplitude. This small oscillation amplitude ensures operation in the viscoelastic regime. Furthermore, at these low oscillation amplitudes, the effect of sample thinning on IR band absorbances is negligible with respect to the effect of the dynamic signal. All data collection and processing, including the demodulation of the phase modulation and sample modulation frequencies, are carried out by the digital signal processors and other internal electronics of the spectrometer.

In order to check the validity of these data and the performance of the spectrometer, control experiments with isotactic polypropylene films were conducted under similar conditions. The results for isotactic polypropylene are virtually identical to those in the literature, except these data could be obtained using much shorter data collection times.[36]

The sample was clamped in the polymer modulator so that the film was taut. Although it is in the range of less than 1%, the macroscopic static strain was not determined. The material is in the linear viscoelastic regime at these low oscillation amplitudes because dynamic rheological experiments performed on a similar LC polyurethane at 1% strain indicated a sinusoidal variation of both stress and strain that satisfied the conditions for linear viscoelasticity.[41]

1.4 SYNTHESIS AND CHARACTERIZATION OF THE LIQUID CRYSTALLINE POLYURETHANES

The polyurethane under investigation is a segmented copolymer with hard segments of MDI/butane diol. The soft segments are polysiloxanes with a cyano-biphenyl mesogen attached to each repeat unit via an eight-methylene spacer. The soft polysiloxane backbone ensures that the liquid crystal is free to move in this low T_g environment; liquid crystalline mobility is further enhanced by the choice of the long spacer used to decouple the mesogen from the polymer main chain. The structure of this macromolecule as well as a predicted morphology is presented in Figure 1.1. The hard segments form hard domains (T_g = 88°C), and the soft continuous domains are in the smectic A phase over the entire liquid crystalline temperature range (T_g = −4.7°C; $T_{clearing}$ = 104.2°C).

1.4.1 OVERALL SYNTHETIC DESIGN[26]

This material was designed to provide a phase segregated morphology such that the pendant liquid crystalline groups in the soft segment would be free to respond to applied fields while anchored by the hard domains. A siloxane backbone was used in the soft segment because siloxanes are known to be very low glass transition (T_g)

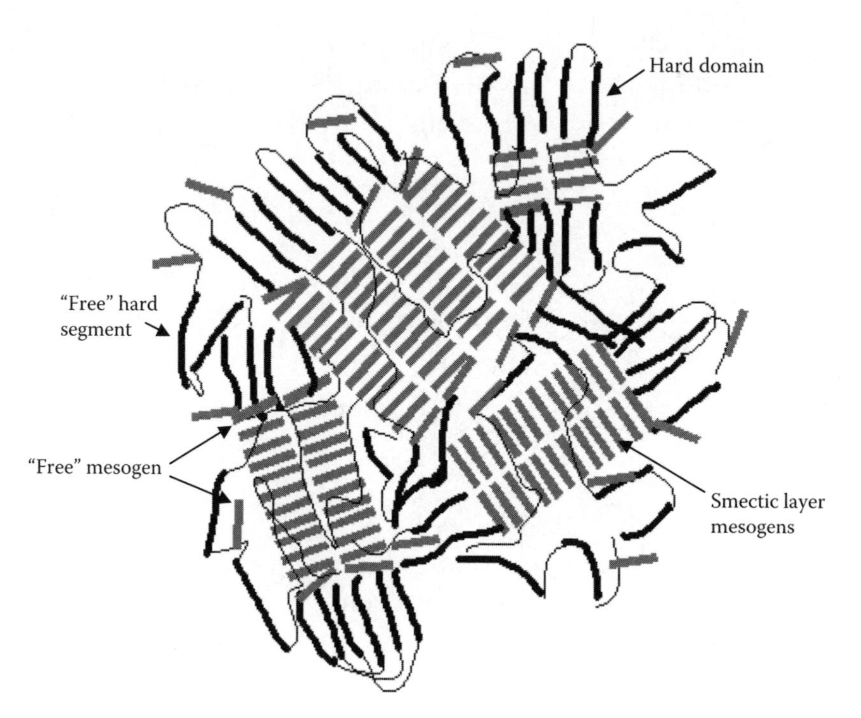

FIGURE 1.1 (a) Chemical structure of side chain liquid crystalline polyurethane. (b) Polyurethane morphology representation. (Reproduced Nair, B. R.; Gregoriou, V. G.; Hammond, P. T. *J. Phys. Chem. B.* **2000**, with permission. Copyright [2000], American Chemical Society.)

materials. For instance, nonsubstituted hydromethylsiloxanes have T_g values on the order of −110°C. Siloxane oligomers substituted with the cyano-biphenyl mesogens of the type reported in this paper have been shown to have T_g values between −16°C and 25°C, depending on the length of the spacer group.[42]

The choice of the cyano-biphenyl mesogen is governed by the well characterized nature of this mesogen, and that the cyano group is known to induce nematic liquid crystalline behavior. The relative stability of the nematic versus smectic phase is controlled by the length of the alkyl spacer, which favors smectic phases as its length increases. Another advantage of the mesogen is the clear signature that the cyano group affords in infrared spectroscopic characterizations. This identification tag allows the infrared dichroism studies to reveal the orientation of the mesogen on the application of a stress field.

The cyano-biphenyl mesogen was synthesized using well documented Williamson ether techniques,[43] and then attached to polysiloxane oligomers by the use of a traditional hydrosilylation step. The challenge in synthesizing the segmented polyurethanes was in acquiring hydromethyl siloxane oligomers that were end-functionalized with carbinol groups. The three-step approach shown in Figure 1.2 was found to be most effective in obtaining these functional siloxanes.

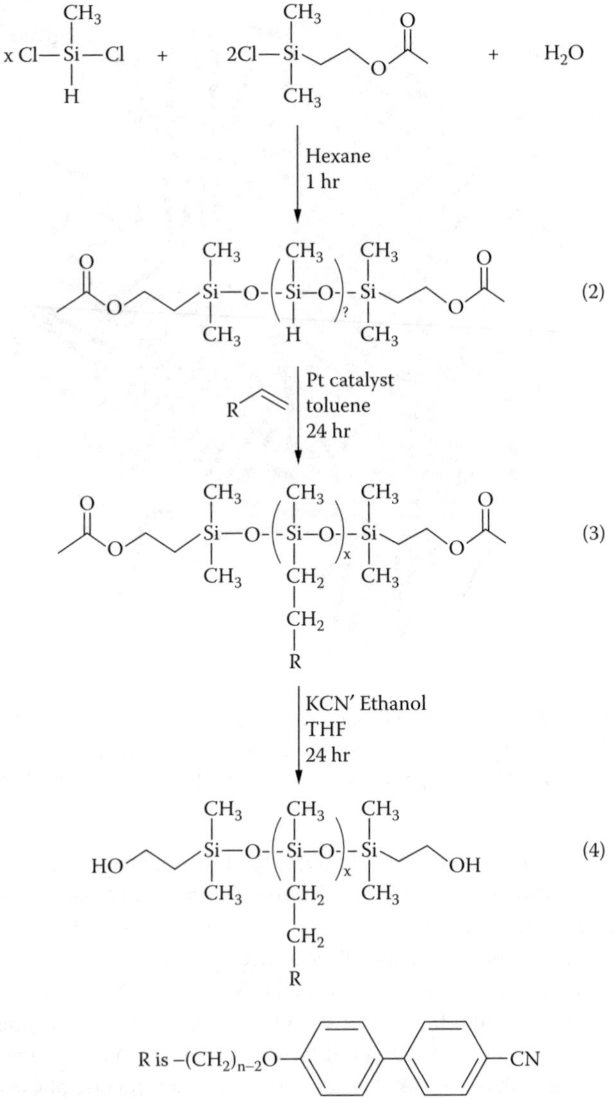

x Cl—Si(CH$_3$)(H)—Cl + 2Cl—Si(CH$_3$)$_2$–CH$_2$CH$_2$CH$_2$–O–C(=O)CH$_3$ + H_2O

↓ Hexane, 1 hr

(2)

↓ R–CH=CH$_2$; Pt catalyst, toluene, 24 hr

(3)

↓ KCN′ Ethanol, THF, 24 hr

(4)

R is $-(CH_2)_{n-2}O-C_6H_4-C_6H_4-CN$

FIGURE 1.2 Synthesis of the soft segment siloxane. (Reproduced from Nair et al. Macromolecules 1998, 31 with permission.

As illustrated in Figure 1.2, a dichlorohydromethylsilane was polymerized using step polycondensation. This approach affords some control over molecular weight and polydispersity. The step polymerization works best if sufficient water is added such that a 20% HCl solution exists at the end of the polymerization. This acid content shifts the chain/ring equilibrium such that about 30% of the monomer ends up as rings. These rings can be removed by vacuum-drying.

The endcapping agent used is an acetoxyethyldimethylchlorosilane. Because a molecule with a Si-Cl group is unstable in the presence of a carbinol group, end-capping with a functional group that could be quantitatively converted to an alcohol was necessary. The acetoxy-terminated siloxane (2) has to be deprotected using mild conditions such that the base-sensitive siloxane will not depolymerize. These criteria led to the choice of KCN as the catalyst for the ester interchange reaction that would deprotect the acetoxy group.[44] The Si-H group is heat sensitive and can cross-link at high temperatures in the presence of base; therefore, the mesogen is substituted prior to the deprotection of the alcohol. The final step of the process is the synthesis of the polyurethane using the classical two-step process shown in Figure 1.3.

The progress of the reactions was monitored primarily by use of traditional infrared spectroscopy. The hydromethylsiloxane (2) shows a strong Si-H vibration (2164 cm^{-1}) and a carbonyl ester vibration (1743 cm^{-1}) from the end-capped acetoxy groups. Upon addition of the mesogen (1), the -CN (2225 cm^{-1}) vibration appears. The acetoxy group in the substituted siloxane (3) appears at lower wavenumbers (1736 cm^{-1}) than its unsubstituted counterpart (2); this shift can be attributed to the loss of carbonyl H-bonding with Si-H groups. The deprotection shows the disappearance of the carbonyl peak associated with the end-capped acetoxy group. Finally, the appearance of the urethane carbonyl (1728 cm^{-1}) can be seen in the infrared spectra of the polyurethane (5).

1.4.2 Molecular Architecture and Structure

This segmented copolymer system is designed to utilize the thermodynamics of phase segregation to produce an environment that can combine applied strain fields to liquid crystalline mobility. Liquid crystalline mesogens are attached to a flexible siloxane chain and these LC siloxane oligomers are alternated with hydrogen bonding hard segments along the length of the polyurethane chain. Micro-phase segregation of the two types of incompatible segments results in a multiphase polymer network that is physically crosslinked by the aggregation of the hard segments in domains. These materials are low molecular weight and the hard domains are only weakly cohesive. This material behaves more like a viscoelastic gum than a thermoplastic elastomer.

Figure 1.4 is the infrared spectra of a film of the polyurethane cast from 50/50 THF/DMAc. The spectrum is tabulated with the key peaks identified in Table 1.1. This also makes note of the peaks that can be clearly assigned to the different parts of the copolymer. As previously mentioned, the cyano band at 2225 cm^{-1} provides a clear, unambiguous way of tracking the orientation of the mesogen.

The amide I carbonyl peak can be resolved into its constituent peaks using a curve fitting method.[29a,45] For urethane carbonyls, the higher the degree of hydrogen

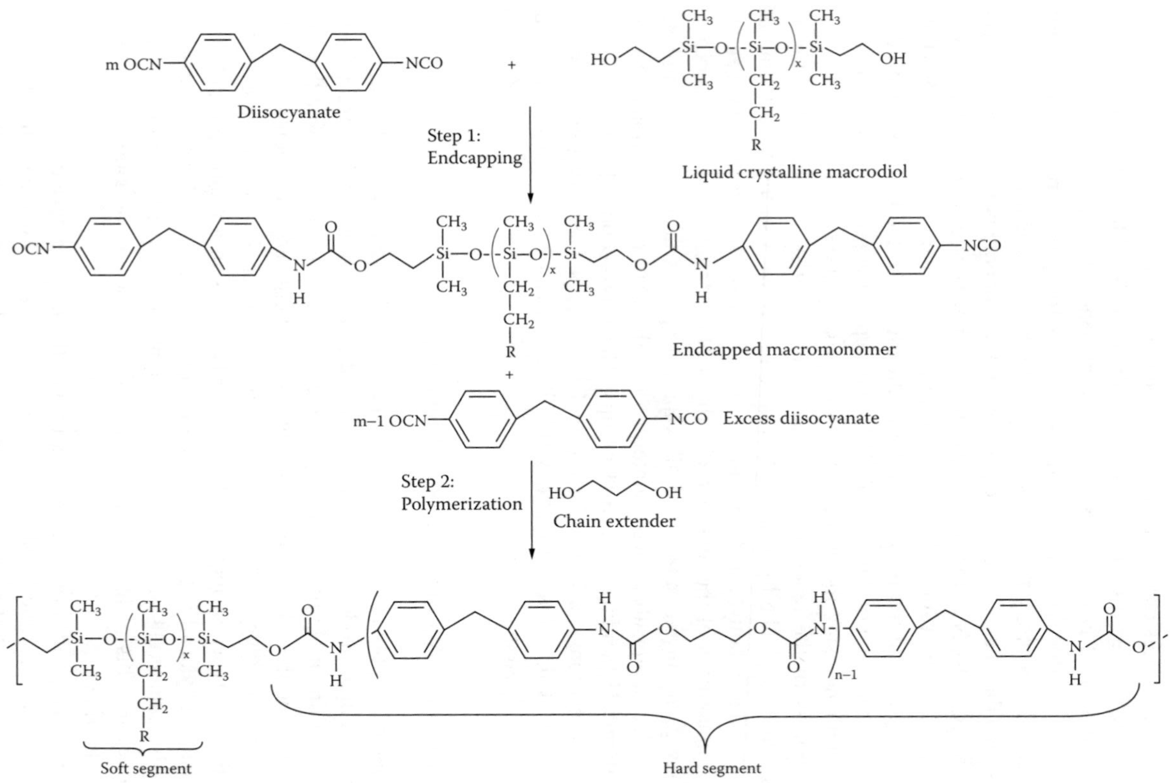

FIGURE 1.3 The two-step polyurethane synthesis. (Reproduced from Reference 20 with permission. Copyright [1998] American Chemical Society.)

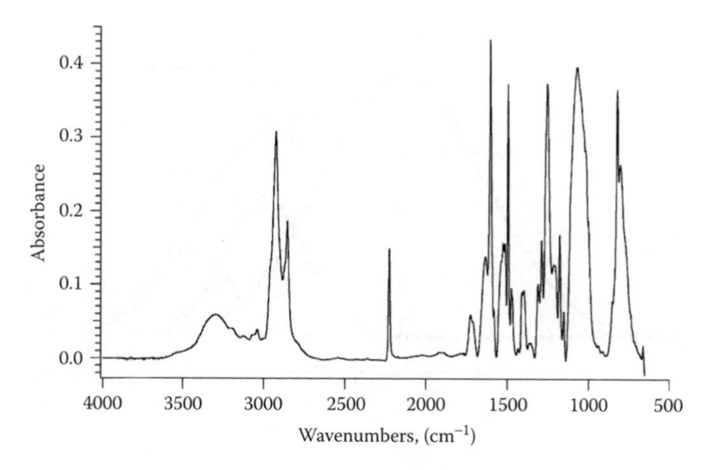

FIGURE 1.4 The FT-IR spectrum of a cast film of the liquid crystalline polymer. (Reproduced from Nair, B. R.; Gregoriou, V. G.; Hammond, P. T. *Polymer* **2000** with permission. Copyright [2000] Elsevier.)

TABLE 1.1
FT-IR Band Assignments in Polyurethane

WaveNumber [cm^{-1}]	Assignment	Uniquely Identifies
3309 (m)	NH stretching	Hard segment
3070 (wm)	CH aromatic stretch vibration	
3036 (wm)	CH aromatic stretch vibration	
2924 (s)	CH_2 asymmetric stretch vibration	
2854 (s)	CH_2 asymmetric stretch vibration	
2225 (s)	CN stretch vibration	Mesogen
1728 (s)	Non H-bonded urethane	Hard segment
1709 (s)	H-bonded urethane	Hard segment
1643 (m)		
1605 (s)	C-C aromatic stretch vibration	
1524 (s)		
1493 (s)	C-C aromatic stretch vibration	
1470 (s)	CH_3 asymmetric deformation	
1431 (m)	Amide II	Hard segment
1412 (s)	Amide	Hard segment
1362 (m)	Amide	Hard segment
1311 (s)	CH aromatic in-plane deformation	
1292 (s)	CH aromatic in-plane deformation	
1257 (vs)	Phenyl-O stretching	Mesogen
1200 (m)		
1180 (s)	CH aromatic in-plane deformation	
1068 (vs)	Si-O-Si stretching	Soft segment
941 (wm)		
822 (vs)	CH aromatic out of plane deformation	

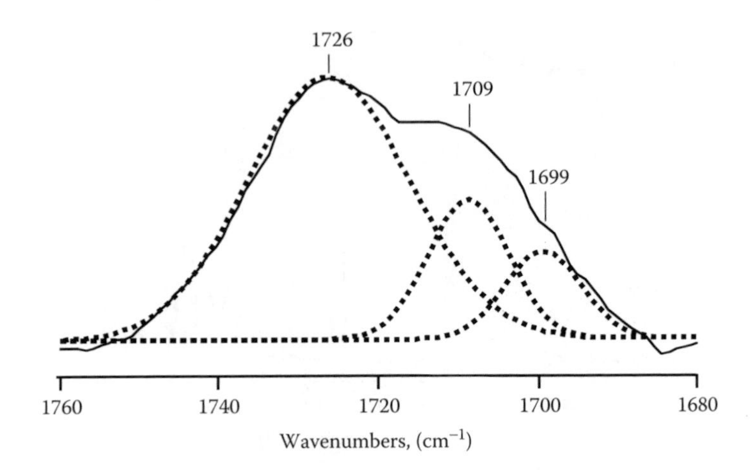

FIGURE 1.5 Amide I carbonyl band resolved into component peaks. (Reproduced from Nair, B. R.; Gregoriou, V. G.; Hammond, P. T. *Polymer* **2000** with permission. Copyright [2000] Elsevier.)

bonding and ordering of the carbonyls, the lower the frequency of the band. The three peaks identified in Figure 1.5 at 1726, 1709, and 1699 cm^{-1} correspond to free carbonyls, carbonyls at the interface between hard segments and soft segments or in less ordered regions of the hard domains, and ordered hydrogen-bonded carbonyls within the hard domains respectively. Other investigators have also ascribed these bands to these types of urethane environments in the literature.[23a,d] Similar spectra have been fitted using two Gaussian curves for the higher wavenumber bands and a Lorentzian for the ordered hydrogen bonded band to get the best fit of the data.[23a] In this case, three Gaussian bands gave the best fit, assuming for the purposes of this fit that adsorption coefficients of the hydrogen bonded bands were the same for both H-bonded bands. The ratio of H-bonded to non-H-bonded carbonyl adsorption coefficients is assumed to be 1.7.[29a]

1.4.3 Dichroic Ratio of LC Segmented Copolymer

Stretching experiments[46] were performed on polyurethane samples as described in the experimental section. Because the stretching apparatus has only one movable jaw and the infrared beam was not recentered as the sample was strained, some small error may be present due to differences in the region observed during the experiment. However, uniform strain along the sample is assumed for this analysis. The area covered by the IR beam was sufficiently removed from the jaws to minimize edge effects. Figure 1.6 shows the spectra of the 2225 cm^{-1} cyano band polarized parallel to the stretch direction at 0%, 15.7%, 39.4%, and 67.1% strains. The highest degree of orientation along the direction of strain in this sample is observed close to 40% strain. The perpendicularly polarized spectra (not pictured) show a monotonically decreasing peak absorbance with increasing strain that can be attributed to the reorientation of the mesogen along the strain direction. Sample thinning was

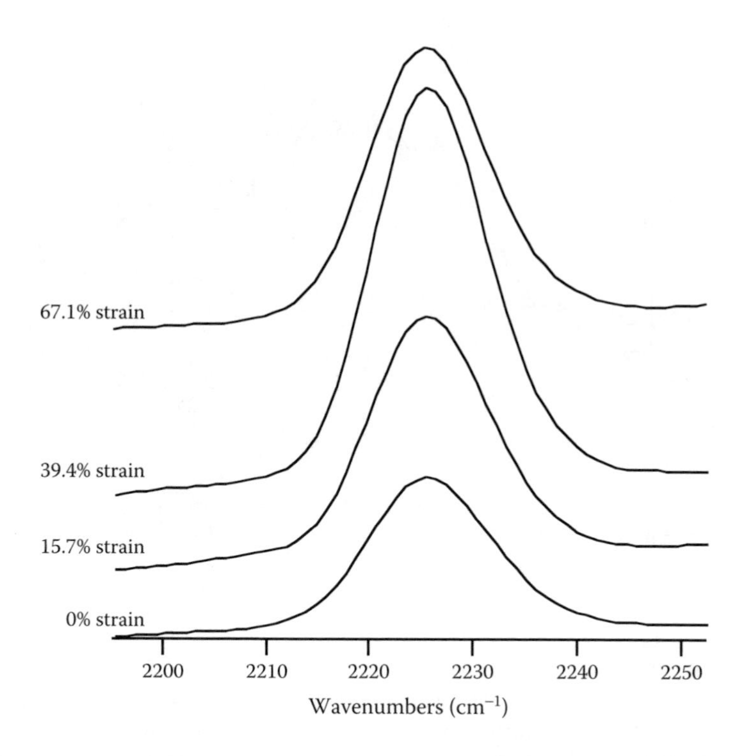

FIGURE 1.6 Parallel polarized cyano bands at 2225 cm^{-1} as a function of strain. (Reproduced from Nair, B. R.; Gregoriou, V. G.; Hammond, P. T. *Polymer* **2000** 46 with permission. Copyright [2000] Elsevier.)

observed in both parallel and perpendicular orientations. However, the fact that the dichroic ratio is the ratio of the parallel to perpendicular absorbances factors out sample thinning as a contributor to any errors in the measurements.

The dichroic ratios of the 2225 cm^{-1} cyano band as a function of strain are presented in Figure 1.7. Two different regimes of orientation behavior can be detected. At low strains, up to 40% applied strain, the cyano band orients parallel to the stretch direction, suggesting that the mesogens initially align with the mechanical field. However, at strains greater than 40%, the mesogens begin to lose their parallel orientation, and begin to align perpendicular to the mechanical field. This behavior can be observed in the peak absorbances in Figure 1.6. The parallel band absorption maximizes at 40% strain and then decreases, as can be seen in the 67.1% strain sample.

Figure 1.8a and Figure 1.8b show the urethane carbonyl band at selected strains that are polarized parallel and perpendicular to the stretch direction, respectively. As established in the previous section, three curves can be fitted to this region of the spectra. However, for the purposes of the dichroic ratio computations described in the experimental section, these bands cannot be uniquely resolved in the polarized data without the use of analytical resolution enhancement techniques. Therefore, the maximum absorption intensity at 1717 cm^{-1} was selected for study. This absorption

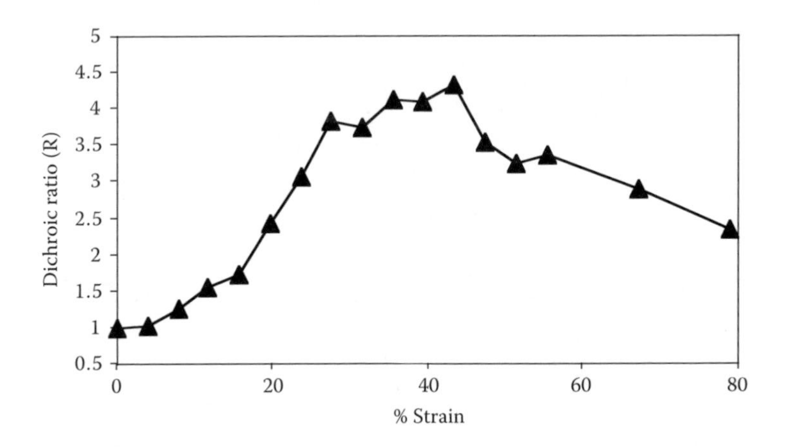

FIGURE 1.7 Dichroic ratio of cyano band (2225 cm^{-1}) as a function of the applied strain. (Reproduced from Nair, B. R.; Gregoriou, V. G.; Hammond, P. T. *Polymer* **2000** with permission. Copyright [2000] Elsevier.)

represents a convolution of the mechanical responses of both the H-bonded urethane stretch at 1709 cm^{-1} and non-H-bonded carbonyls at 1726 cm^{-1} in the segmented copolymer system. The contributions of the 1699 cm^{-1} peak are minimal.

Figure 1.9 shows the spectral response of the carbonyl as a function of applied strain. The sudden drop in the dichroic ratio at low strains (less than 8%) suggests a perpendicular orientation of the C=O band with respect to the stretch direction. At medium elongations (10%–40%), this initial perpendicular orientation is lost and the carbonyls respond parallel to the applied field. Finally, at high strains (greater than 40% strain), the dichroic ratio decreases again, suggesting that carbonyls are beginning to align perpendicular to the field once again. Examination of the spectra in Figure 1.8a and Figure 1.8b reveals that the lower frequency (lower wavenumber) shoulder at 1709 cm^{-1} appears to grow relative to the 1726 cm^{-1} peak with increasing strain, suggesting an increase in the number and degree of order of the hard segments within hard domains.

1.4.4 Orientation Function of Polyurethane Molecular Components

As previously mentioned, the dichroic ratio merely indicates the orientation of a specific functional group. The alignment of the actual molecular components of the polyurethane such as the hard segment and the mesogen can be followed by the use of the orientation function S. Table 1.2 is a tabulation of the order parameter, dichroic ratio, and the average angle between the stress direction and the molecular axis for this sample. Figure 1.10 is a depiction of the orientation parameter S of both the mesogen and the hard segments. The cyano band is representative of the orientation of the liquid crystalline mesogens, while the carbonyls are indicative of the orientation of the hard segments of the polyurethane.

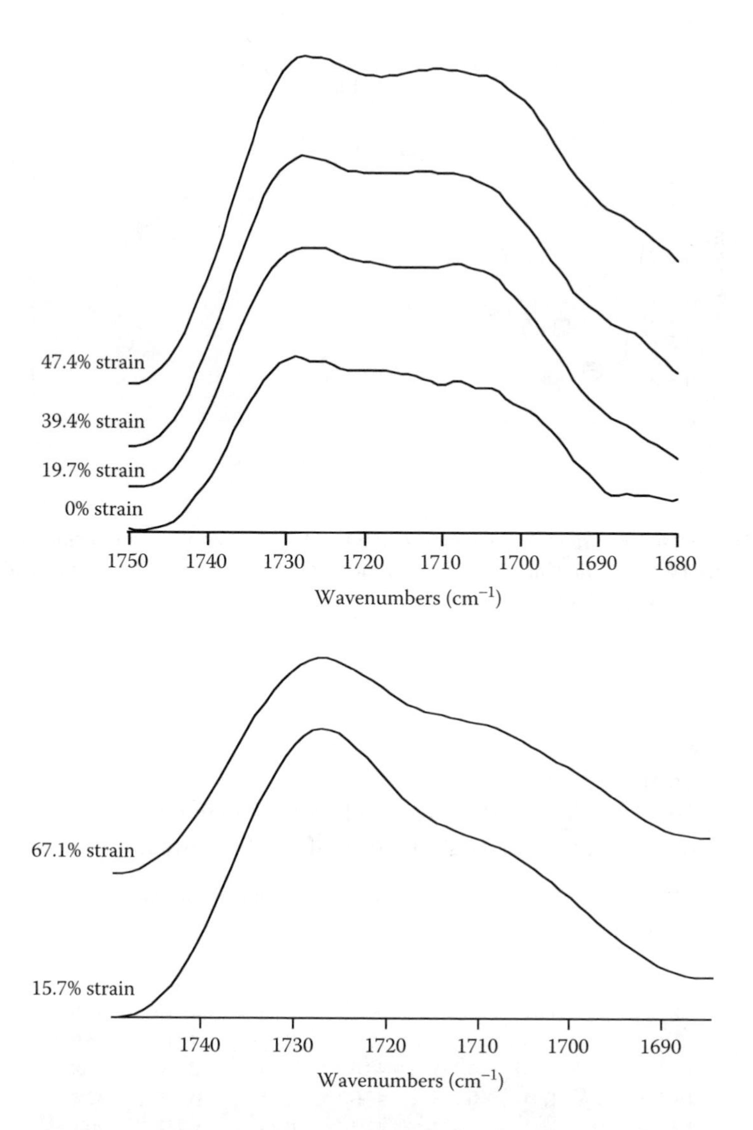

FIGURE 1.8 (a) Carbonyl band with polarizer placed parallel. (b) Carbonyl band with polarizer perpendicular to direction of strain. (Reproduced from Nair, B. R.; Gregoriou, V. G.; Hammond, P. T. *Polymer* **2000** with permission. Copyright [2000] Elsevier.)

From the data in Figure 1.10, it can be observed that the mesogens align first parallel and then perpendicular to the field. Previous small-angle x-ray scattering (SAXS) and optical microscopy studies of this material[26] have shown that it is in the smectic A state over its entire LC temperature range. Mesogens initially align with their molecular axis along the direction of strain, resulting in the smectic layers stacked with their long axis perpendicular to the direction of strain. However, at 40% strain, some of the smectic layers begin to undergo shear and re-align parallel to the field, inducing a perpendicular alignment in the mesogens.

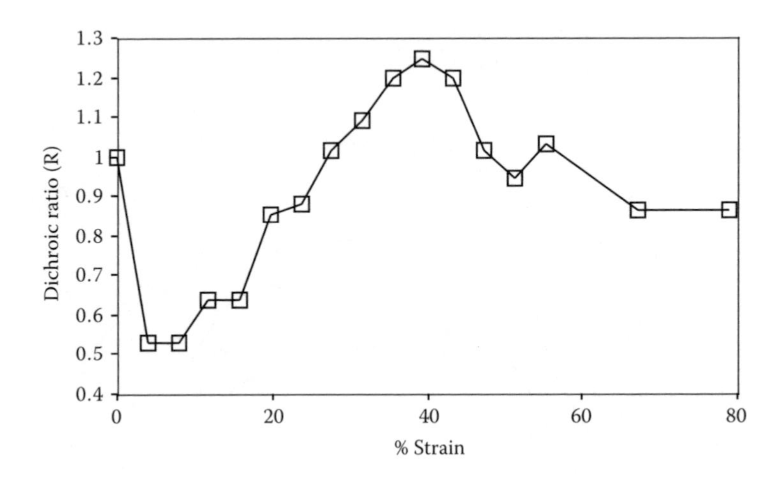

FIGURE 1.9 Dichroic ratio of urethane carbonyl band (1717 cm^{-1}) as a function of applied strain. (Reproduced from Nair, B. R.; Gregoriou, V. G.; Hammond, P. T. *Polymer* **2000** with permission. Copyright [2000] Elsevier.)

TABLE 1.2
Dichroic Ratios, Structure Factors, and Average Angles of Disorientation as a Function of Applied Strain

	Cyano Band (2225 cm^{-1})			Carbonyl Band (1717 cm^{-1})		
Strain (%)	**R**	**f**	**q**	**R**	**f**	**q**
0.0	1.00	0.000	54.7	1.00	0.000	54.7
3.9	1.03	0.009	54.4	0.53	0.416	38.6
7.9	1.25	0.076	51.7	0.53	0.416	38.6
11.8	1.55	0.155	48.6	0.64	0.306	42.8
15.7	1.74	0.198	47.0	0.64	0.306	42.8
19.7	2.44	0.324	42.2	0.85	0.118	50.1
23.7	3.07	0.408	38.9	0.88	0.094	51.0
27.6	3.83	0.485	35.9	1.01	–0.009	55.1
31.6	3.73	0.477	36.2	1.09	–0.065	57.4
35.5	4.13	0.510	34.8	1.20	–0.139	60.6
39.4	4.10	0.508	34.9	1.25	–0.171	62.1
43.4	4.33	0.526	34.2	1.20	–0.139	60.6
47.4	3.55	0.459	36.9	1.01	–0.009	55.1
51.3	3.23	0.427	38.2	0.94	0.042	53.0
55.3	3.37	0.441	37.6	1.03	–0.022	55.7
67.1	2.89	0.387	39.7	0.86	0.108	50.5
79.0	2.33	0.307	42.8	0.86	0.108	50.5

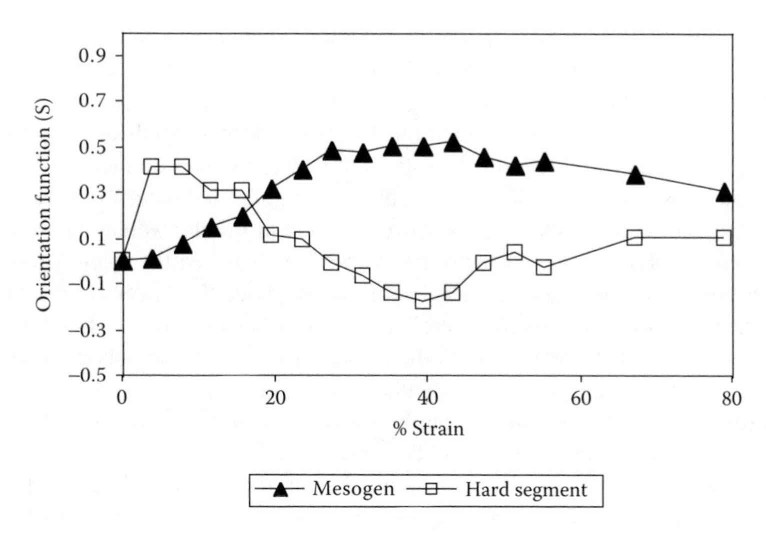

FIGURE 1.10 Structure factor as a function of applied strains demonstrating the orientation behavior of the mesogenic groups and the hard segments. (Reproduced from Nair, B. R.; Gregoriou, V. G.; Hammond, P. T. *Polymer* **2000** with permission. Copyright [2000] Elsevier.)

The plot in Figure 1.10 tracks the reorientation of the hard segments as a function of the externally applied mechanical field. Two different hard segment populations can be visualized. The first population consists of the "lone" hard segments that are embedded in the soft siloxane matrix. The second population includes the hard segments that are intermolecularly hydrogen-bonded in hard domains.

Figure 1.10 suggests three different regimes of behavior for the hard segment. Both populations of hard segments are randomly oriented at 0% strain, giving a net orientation factor of zero. Between 0% and 8% strain, the population of lone hard segments initially undergoes orientation, while the hard segments in the mechanically cohesive hard domains remain randomly arranged. Thus, the lone segments dominate the orientation factor at low strains. It has been established that the lone segments, which are characterized by non-H-bonding bands (1728 cm^{-1}), are representative of the movement of the soft, rubbery domains.[29d] Because the peak heights of saturated Si-O bands cannot be directly measured to describe soft segment orientation, it must be assumed that the lone hard segment reorientation is representative of the soft segment movement. These hard segments are fairly independent of the arrangement of the LC layers at low strains. At this strain range, the "lone" segments align elastically with the applied strain resulting in a sharp increase in the orientation function.

The alignment of the hydrogen bonded hard domains dominates the convoluted carbonyl response at higher strains, as these components also begin to orient in the strain field. From 8% up to 40% strain, the orientation function of this band decreases with strain, indicating a net perpendicular alignment of the hard segments. At higher strains, the orientation function increases again, suggesting that the hard segments are now aligning with the field. Many researchers have seen this type of negative to positive orientation shift in polyurethanes at low strains.[27,22b] Bonart and coworkers[21] proposed the widely accepted model that the perpendicular alignment of the hard segments results from an alignment of the long axis of the aggregated hard domains with the strain field. At extremely high strains, the domains break up, the hard segments align along the strain direction, and the hydrogen bonds are reformed in this new "domain" configuration. The strain at which the hard domains re-orient in past studies has been near 300%, which is much higher than the 40% reorientation strain observed with this material. However, the low molecular weight nature of this segmented copolymer, coupled with the interactions between the liquid crystalline smectic layers and the hard segment, could account for the lower strain value at which reorientation is observed. In earlier studies, it has been found that this reorientation strain is highly dependent on the crystalline nature of the polyurethane hard domains, and in some cases, perpendicular orientation is not observed at all.[22a,b,23a]

Further support for this model can be found in samples where we fit Gaussian curves to the amide I carbonyl band. The spectra in Figure 1.8b were successfully fitted for the carbonyl bands polarized perpendicular to the stretch direction at 15.7% and 67.1%. These two points were chosen because they are above and below the transition strain of 40% where the hard segments go from a parallel to a perpendicular arrangement. For the 15.7% strain sample, it was found that the free carbonyls at 1726 cm^{-1} contribute 58.3% to the total area, while the hydrogen bonded carbonyls at 1709 and 1699 cm^{-1} contribute 26.7% and 15.0%, respectively. In the 67.1% sample, the relative areas are 48.7%, 34.0%, and 17.3% for the free and the two H-bonded carbonyls. As mentioned earlier, the shoulder at lower wavenumber is more pronounced in the more highly strained sample. The increase in the H-bonded fraction suggests that as strain is applied, the original hard segments break up to arrange in more highly ordered domains, aligned such that the segments are along the direction of stress. The population of lone hard segments appears to decrease upon incorporation into ordered domains. This sort of breakup and reordering has also been previously observed by other researchers in common segmented polyurethanes.[5,28b]

1.4.5 Interdependence of the Orientation Behavior of LC Layers and Hard Segment Domains

An interesting feature of Figure 1.10 is the observation that the reorientation of the smectic layers and the hard domains occur at the same strain. It has been shown in a previous paper that the polyurethane hard segment thermally stabilizes the smectic phase.[26] In these studies, polysiloxane oligomers with only nematic phases (those having three methylene unit spacers between the siloxane and the LC mesogen) formed smectic phases below the dissociation temperature of the hard segment when placed in the polyurethane. For the polysiloxane oligomer studied in this work, which has both smectic and nematic phases (eight methylene spacers), the nematic phase did not occur in the polyurethane. This can be clearly shown by optical microscopy. A film of this polymer was heated at 10°C/min. At 125°C, all birefringence disappeared. Following annealing in the isotropic melt for several hours, the film was cooled. Needlelike batonnets began to form at 105ºC. As the film continued to cool, the needles coalesced and focal conical fans appeared. This batonnet to fan transition is classic smectic A behavior.[47] For a polyurethane that has both a smectic and a nematic phase, the onset of the nematic phase is the same as the temperature at which the hard segments dissociate. Thus, the behavior of the hard domains and the LC smectic layers seem to be correlated.

1.4.6 In-Phase and Out-of-Phase Dynamic Spectra[48]

As discussed in the theory section, the in-phase response of a material to a cyclic stress field is proportional to the extent of strain. Thus, this component is the elastic component of the total material response. On the other hand, the strain rate dependent quadrature spectrum is related to the viscous component of the total response. The spectra were collected with the polarizer parallel to the direction of strain. Both in-phase (0°) and out-of-phase (90°) spectra were recorded at each step of the step scan. The difference spectra, which measures the difference between the stretched and relaxed sample, will show a positive increase in intensity when molecules align with the direction of stress and a decrease in intensity when molecules align perpendicular to the applied stress.[36]

The in-phase and the out-of-phase responses of the side chain liquid crystalline polyurethane to the applied cyclic strains are shown in Figure 1.11a. Figure 1.11b and Figure 1.11c are expansions of the cyano and the carbonyl regions respectively of the spectra in Figure 1.11a. It is clear that the in-phase response is much stronger than the quad response. Furthermore, it can be seen that the ratio of elastic to viscous components is much higher in the cyano band than in the urethane band. The cyano bands are indicative of the mesogen that are found in the soft segment. Because the soft segment is above its T_g of –4°C, the mesogens are expected to respond rapidly to the imposed stress. However, the urethane bands should have a more viscous response because of their location in the hard domains, which are below their T_g of 88°C.

A strong positive in-phase response is observed for the cyano band, as shown in Figure 1.11b. Thus, the data indicates that the mesogens are aligning parallel to the direction of the strain. This is in good agreement with the static experiments shown earlier, which demonstrated that the mesogens align along the direction of the stress at low strains.

The cyano band shows a bisignate signature in the quad spectrum. There are two possible explanations for this type of band shape in the dynamic spectra. The first is that the band shifts to a new position during the deformation process. The second is that the original band is representative of more than one population that responds to the perturbation in different ways. For instance, if one response is parallel to the deformation and the other perpendicular, the resulting dynamic spectrum will show a bisignate.

The bisignate in the quadrature spectrum of the cyano band can be attributed to the existence of two distinct populations. One population consists of the mesogens existing in smectic layers. The lower wavenumber (lower energy) band of 2224 cm^{-1} can be assigned to cyano groups within the smectic layer. Infrared band frequencies of a bond are proportional to the square root of the Hookian force constant of the bond. When the –CN bond is placed in an environment that diffuses its strong dipole, such as the smectic layer with interlocking antiparallel mesogens (as previously

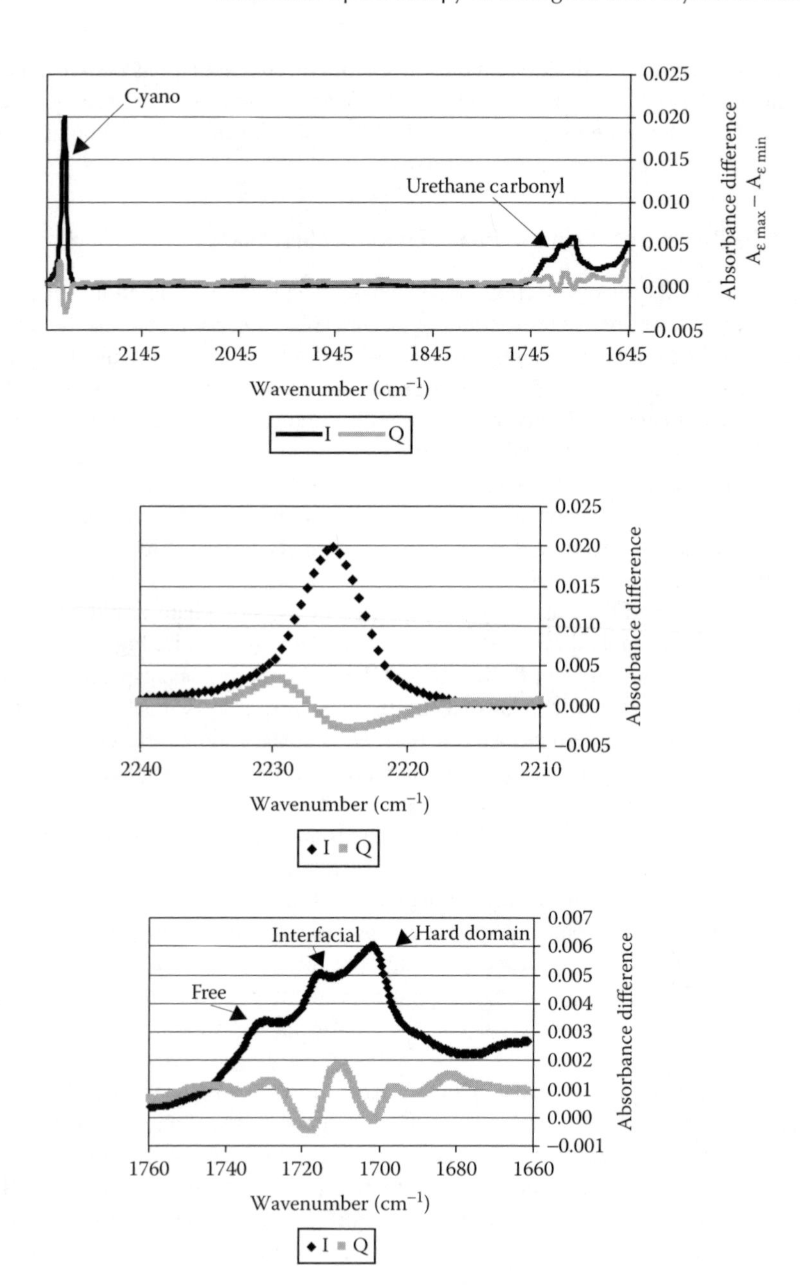

FIGURE 1.11 (a) In-phase (I) and quadrature (Q) spectra. Note the strength of the elastic response as compared to the viscous. (b) Expanded cyano band region. (c) Expanded urethane carbonyl region. (Reproduced from Nair, B. R.; Gregoriou, V. G.; Hammond, P. T. *J. Phys. Chem. B.* **2000**. Copyright [2000] American Chemical Society.)

demonstrated for this material[15]), the force constant for each dipole is reduced and the frequency of vibration is lowered.[22a] The higher wavenumber population (2229 cm^{-1}) is assigned to the -CN groups that are not involved in the formation of smectic layers and therefore are less ordered.

As can be seen in Figure 1.11b, the quadrature spectrum has a positive component in the higher wavenumber band (free mesogens) and a negative component in the lower wavenumber band (smectic layers). The latter suggests that the viscous response of the ordered smectic layers is to align the mesogen perpendicular to the direction of stress. This reorientation of the mesogen leads to the smectic layers aligning parallel to the direction of the stress. Previous work by Zhao[24a,b] suggests that the smectic layer can be considered as a mechanical entity that likes to align along a mechanical field. As shown in the static section, at an elongation of 40% strain, the average mesogen orientation switches from parallel to the strain field to perpendicular to the field.

The higher wavenumber population on the other hand shows a positive quadrature peak. This population consists of "free" mesogens that are not involved in smectic layers. The quadrature response of this small population suggests that since these mesogens do not exist in smectic layers, the viscous component of the stress continues to align the free mesogens along the direction of applied strain. It is of interest to note that these free mesogens can only be observed in the viscous response, since this population cannot be resolved in the in-phase spectrum. It is possible that these mesogens are found near the viscous hard domains where the smectic layers are least perfect, and that the hard domain interfaces slow down the response time of these liquid crystals.

2D-IR correlation spectroscopy can be used to further visualize these results. For the cyano bands, the 2D-IR asynchronous correlation is in excellent agreement with the results of the dynamic spectra. The presence of the cross-peak in the -CN region (Figure 1.12) confirms the two populations of mesogens implied by the bisignate in the dynamic spectrum. The signs of the two cross-peaks show that the free mesogens (2229 cm^{-1}) respond to the strain before the less ordered groups at 2224 cm^{-1}.

Figure 1.11c shows the carbonyl region of the dynamic spectra. Here a positive in-phase response is observed for all three bands. The positive response correlates with the perpendicular alignment of the hard segments because the carbonyl dipole is orthogonal to the hard segment axis, as discussed above. Despite the expected elastic behavior of the free carbonyls within the soft matrix, these free carbonyls show only a small positive dichroic difference. Previous work suggests that the free carbonyls respond elastically parallel to the applied stress field.[26,22a,b] It is therefore possible that there is a parallel (or negative) alignment of the free hard segments that is masked by the convolution of the free urethane bands with the hydrogen-bonded urethane bands.

The carbonyl region of the quad spectrum shows the strongest viscous responses in the entire material for reasons discussed above. The "free" carbonyls at 1730 cm^{-1}, which are not found in the sluggish hard domains, are the only carbonyls without much activity in the quadrature spectrum. These carbonyls primarily show an elastic response, suggesting that the hard segments embedded in the soft matrix, and the denied hydrogen

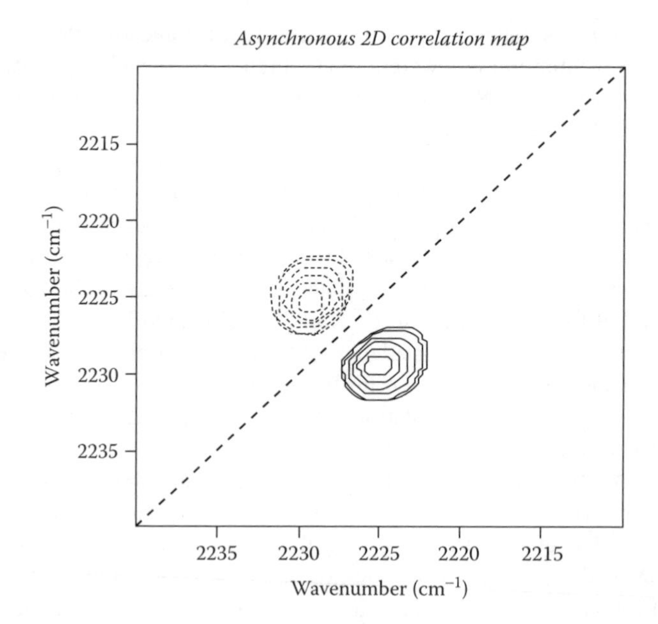

FIGURE 1.12 Asynchronous correlation map of the -CN region.

bonding interactions with its neighbors, orient with the flexible soft segment. This observation is consistent with the widely accepted models of polyurethane reorientation behavior presented by Bonart et al.[27] and expanded by Cooper et al.[22a,b]

In the case of the carbonyl bands, the power of 2D-IR correlation spectroscopy to resolve overlapping bands is evident. The asynchronous correlation map of the carbonyl region is shown in Figure 1.13. By drawing asynchronous correlation squares between the three sets of cross peaks, the presence of four bands (1702, 1710, 1718, and 1728 cm^{-1}) can be distinguished. In the previous section it was shown that three distinct bands may be fit to the broad carbonyl band. These were ascribed to the free carbonyls (1728 cm^{-1}), the H-bonded carbonyls in the hard segments (1701 cm^{-1}), and a set of more loosely H-bonded carbonyls at 1718 cm^{-1}. The 2D-IR correlation provides some evidence that there may be two populations (1710 and 1718 cm^{-1}) of slightly less ordered carbonyls at the interface between the hard and soft segments, in addition to the free carbonyls at 1728 and the H-bonded carbonyls at 1701cm^{-1}. However, for the purpose of this work, it is sufficient to treat the less ordered carbonyls as a single population. The signs of the cross-peaks reveal that the band at 1701 cm^{-1}, the H-bonded carbonyls in the hard segments, responds to the applied strain before the bands at 1710 cm^{-1} and 1718 cm^{-1}.

The "free" or non-H-bonded carbonyls do not show much viscous response. On the other hand, the H-bonded carbonyls do show a strong negative band. Thus, the data indicates that the viscous component of the stress induces the hard segments to reorient parallel to the direction of stress. Bonart[27] proposed a model for MDI/butane diol polyurethanes wherein hard segments align perpendicular to the stretch direction at low strains. This orthogonal orientation was attributed to the long

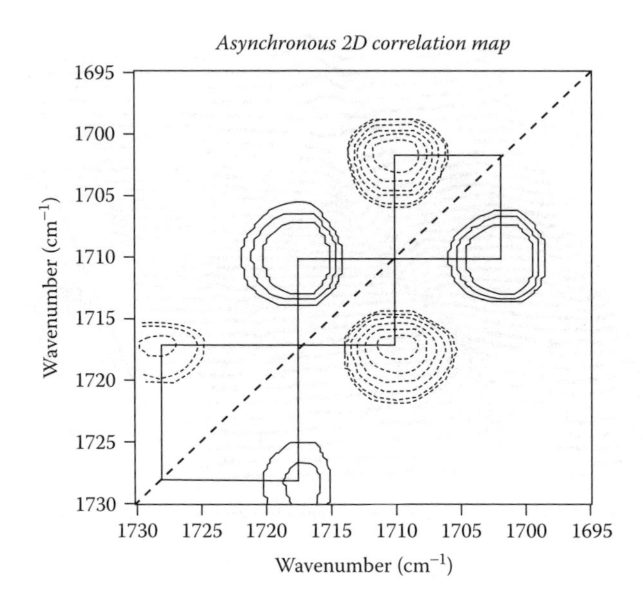

FIGURE 1.13 Asynchronous correlation map of the -CO region.

direction of the hard domain acting as a mechanical unit to orient within the stress field. At high strains of greater than 300%, the polyurethane hard domains were assumed to slip and deform, with hydrogen bonding now occurring between hard segments aligned parallel to the direction of strain. At this reorientation strain, the chains have undergone permanent deformation, and the material has lost its complete reversibility of strain.

1.4.7 Viscoelastic Response of the SCLCP

Using the identities developed in the theory section (Equation 1.4), in-phase and quadrature spectra can be used to create plots of the dynamic response $\Delta\tilde{A}(v,t)$ as a function of both wavenumber and phase angle. Figure 1.14 and Figure 1.15 show the cyano and the carbonyl regions respectively. These figures allow us to visualize how the bands are affected during each cycle of applied strain. Since a 25 Hz modulation frequency is applied, the period of each cycle is 40 ms.

Figure 1.16a is a plot of a perfectly elastic and a perfectly viscous response to an externally applied oscillatory strain. The elastic component is exactly in-phase with the sinusoidal variation, whereas the viscous component has a phase lag of exactly 90°. Figure 1.16b depicts the actual variations of the different bands of interest with the applied strain. Thus, this figure can be considered as a series of cross sections of Figure 1.14 and Figure 1.15 at the infrared frequencies of interest. All the curves have been normalized to an amplitude of 1.0 in order to represent them on the same scale. The data show that all the bands have some phase lag when compared to the perfectly elastic response. The two categories of responses based on the phase lags suggest that there are two rates of time-dependent behavior that can be found in the material. The

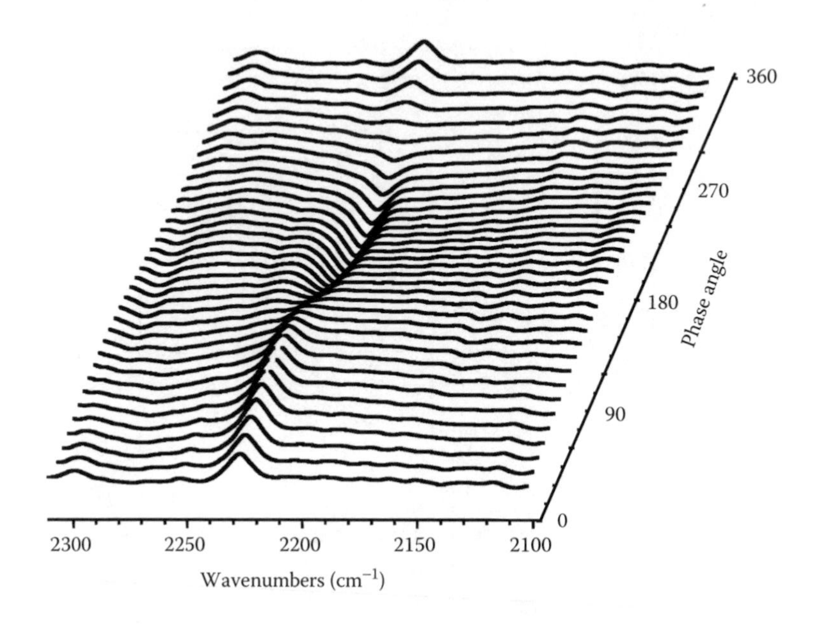

FIGURE 1.14 CN band as a function of wavenumber and phase angle. This graph was compiled using linear combination methods described in the theory section from $\Delta A(\nu)$ and $\Delta A(\nu)$. (Reproduced from Nair, B. R.; Gregoriou, V. G.; Hammond, P. T. *J. Phys. Chem. B.* **2000** with permission. Copyright [2000] American Chemical Society.)

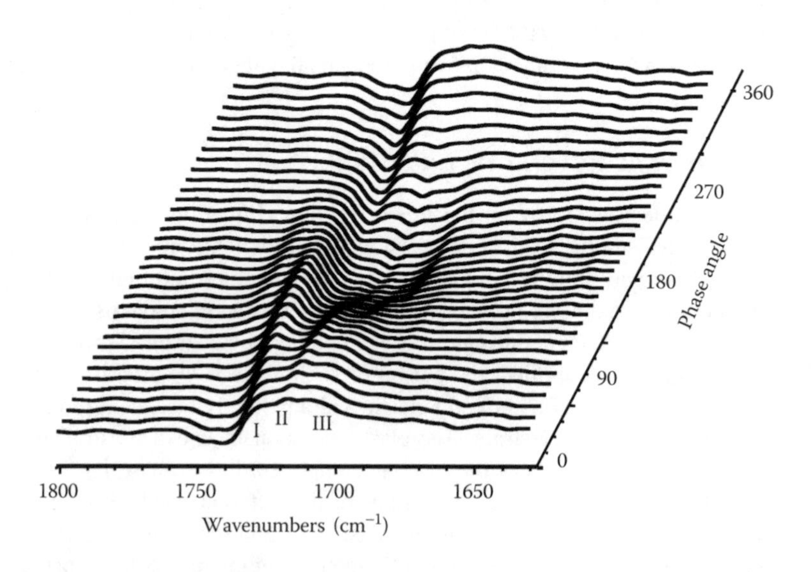

FIGURE 1.15 Carbonyl band as a function of wavenumber and frequency. Graph compiled in the same manner as Figure 1.14. (Reproduced from Nair, B. R.; Gregoriou, V. G.; Hammond, P. T. *J. Phys. Chem. B.* **2000** with permission. Copyright [2000] American Chemical Society.)

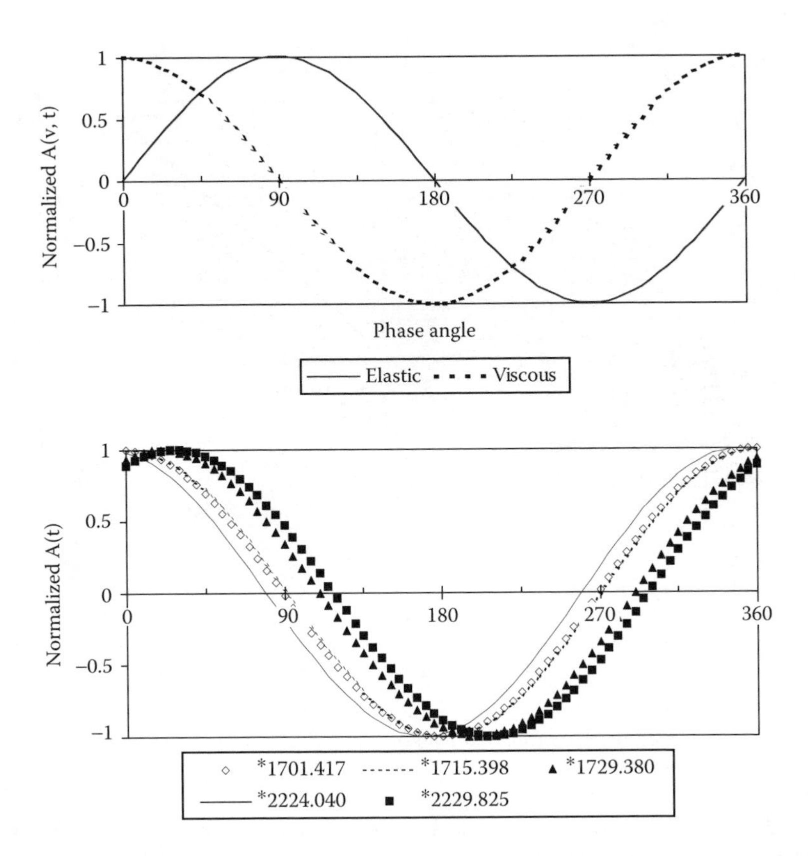

FIGURE 1.16 (a) Perfectly elastic and perfectly viscous response curves. (b) Dynamic dichroic response of polyurethane created by compiling cross sections of Figures 1.14 and 1.15. (Reproduced from Nair, B. R.; Gregoriou, V. G.; Hammond, P. T. *J. Phys. Chem. B.* **2000** with permission. Copyright [2000] American Chemical Society.)

"free" hard segments and the "free" mesogens appear to respond at the same rate. These curves are clearly viscoelastic within the time frame of this experiment (40 ms). The second group of responses can be found in the H-bonded hard segments that are indicative of the hard domains, and the mesogens in smectic layers. These responses are 90° out of phase, indicating that the hard domains exhibit a perfectly viscous response within the time frame of this experiment.

Thus, two different response patterns in the complex macromolecular system under investigation can be elucidated. The smectic layers and the hard domains appear to respond sluggishly at similar rates, while the "free" mesogens and the "lone" hard segments both re-orient in a more elastic manner.

The results of the infrared stretching experiments confirm that the smectic layer and the hard segment orientations must be cooperative. The synchronous correlation map for the spectral region from 2300 cm^{-1} to 1690 cm^{-1} is shown in Figure 1.17. The presence of the strong autopeaks at 2225 cm^{-1} and at 1705 cm^{-1} indicates that both the mesogens (CN band) and the hard segments (CO band) respond to the

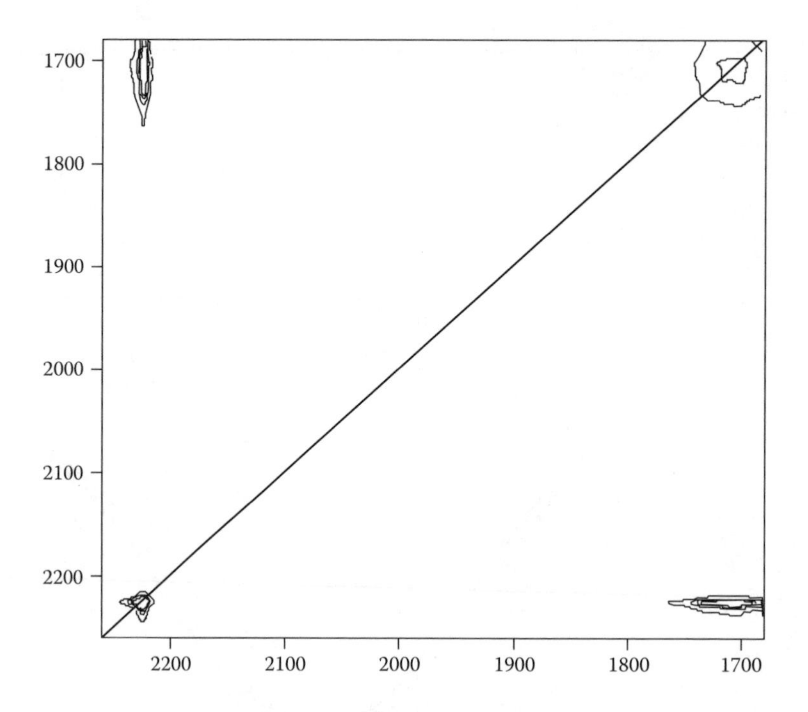

FIGURE 1.17 Synchronous correlation map of the -CN and -CO spectral regions.

applied strain. The positive cross-peaks confirm that the mesogens and the hard segments respond to the applied strain at similar rates.

1.5 CONCLUSIONS

FT-IR dichroism studies have been used to determine the relative orientations as well as the relative rates of deformation for the hard segments and liquid crystalline mesogens during deformation. We find that the elastic component of the strain aligns smectic layers parallel and hard domains perpendicular to the direction of strain. The viscous strain component, on the other hand, induces a perpendicular smectic layer and parallel hard domain orientation behavior. In addition, we find through the use of dynamic FT-IR and 2D-IR techniques that the smectic layers and the hard domains appear to respond sluggishly at similar rates, while the "free" mesogens and the "lone" hard segments both reorient in a more elastic manner.

It has been determined that upon application of tensile strain, the lone hard segments within the soft segment continuous phase are the first elements to orient within the applied mechanical field. The hard segments incorporated into H-bonded domains and the liquid crystalline smectic layers undergo reorientation and alignment above approximately 8% strain, initially aligning perpendicular to the stress direction. At strains above 40%, both the liquid crystalline smectic layers and hard segments reorient parallel to the stress direction. The elastic component of the strain aligns smectic layers parallel and hard domains perpendicular to the direction of strain. The viscous strain component, on the other hand, induces a perpendicular smectic layer and a parallel hard domain orientation behavior.

Figure 1.18 shows a schematic of a proposed model that is consistent with the results. The picture at 0% strain suggests that the hard domain interface acts as an anchoring surface for the smectic layers, with the smectic layers perpendicular to the long axis of the hard domains. The configuration of the soft siloxane backbone is shown to be relatively decoupled from the arrangement of the liquid crystalline layers. The eight methylene spacer groups connecting the mesogens to the siloxane

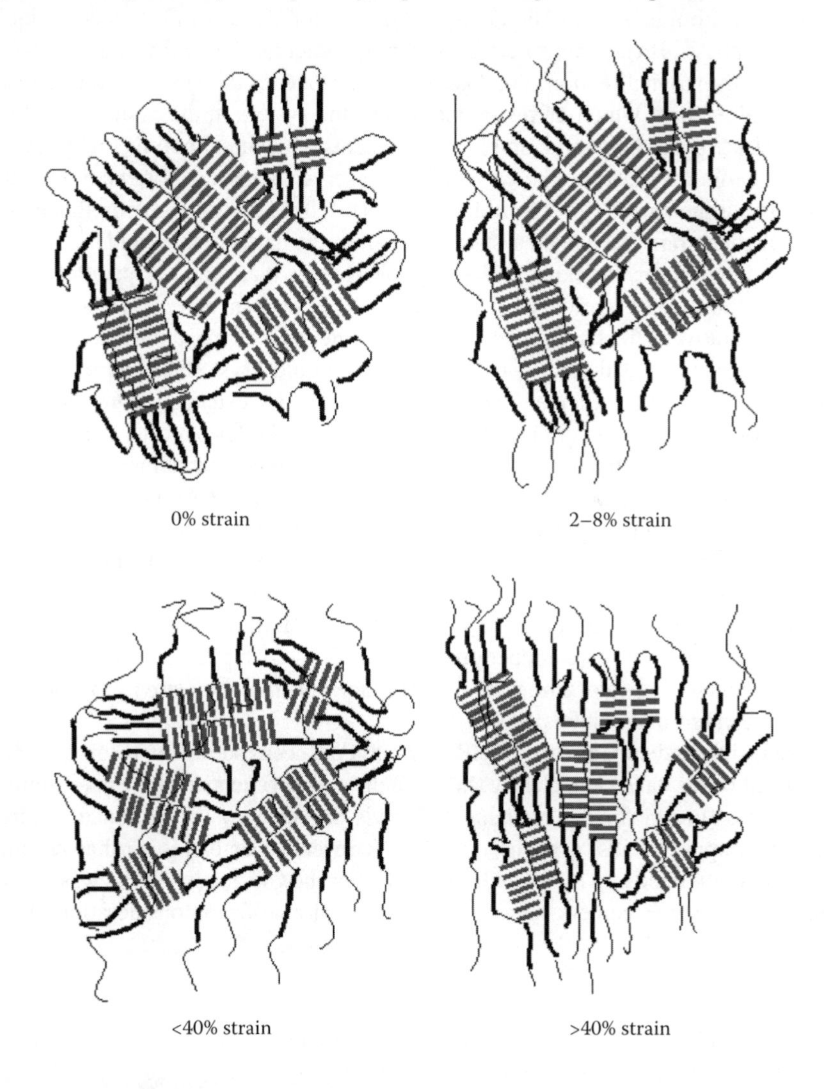

FIGURE 1.18 Proposed model of cooperative deformation of hard segments and smectic layers as a function of strain. (Reproduced from Nair, B. R.; Gregoriou, V. G.; Hammond, P. T. *Polymer* **2000** with permission. Copyright [2000] Elsevier.)

main chain are expected to allow independent arrangements of the main chain and side group at the lower strain levels.

As strain is applied to the randomly oriented polyurethane, the lone hard segments align with the stress field almost instantaneously (see Figure 1.18, upper right). The hard domains containing H-bonded hard segments align and rotate due to torsion at intermediate strains from approximately 8% to 40%, with their long axis perpendicular to the direction of strain (Figure 1.18, lower left). At this point, the smectic layers rearrange so that the individual mesogens are parallel to the stress field, and the smectic layers are perpendicular to the strain direction. It has been observed that fibers drawn from main chain liquid crystalline homopolymers in the smectic liquid crystalline phase also contain smectic layers perpendicular (i.e., individual mesogens parallel) to the strain direction;[49] in this case the mesogen is thought to be the primary mechanical element which drives orientation of the system in the liquid crystalline phase, despite potentially unfavorable arrangements of the homopolymer main chain. Similar factors may also be the cause of the observed transverse orientation of the liquid crystalline smectic layers in the liquid crystalline siloxane polyurethane system. At higher strains (above 40% strain), however, the hard domains and the smectic layers undergo shear, breakup, and realignment such that they are parallel to the stress direction (see Figure 1.18, lower right). This rearrangement, which appears to be cooperative, effectively involves two interdependent mesophase systems — the polyurethane hard domains and the smectic liquid crystalline phase. At 40% strain and higher, the polymer main chain exists in its extended state along the direction of stress; thus at very high strain levels, molecular level reorientation results in extension of the main chain of both segments. It is interesting to note that throughout the reorientation process, the relative arrangements of the liquid crystalline and hard segments within domains remained the same. The transverse anchoring of the liquid crystalline layers with respect to the hard domains is maintained, despite the shearing and reordering processes of both the liquid crystalline and polyurethane morphologies. This interdependence suggests that the deformation behavior of these novel materials may involve cooperative motions of the two phases, driven by liquid crystalline anchoring interactions at the interface and the connectivity of the hard and soft segment main chains.

Finally, the advantages of 2D-IR correlation spectroscopy to resolve highly overlapped bands have been clearly shown in these experiments. Where curve-fitting the dynamic spectra for the urethane carbonyls can identify the presence of three highly overlapped bands, asynchronous 2D-IR reveals four bands, further refining the concept of the loosely H-bonded carbonyls. For the CN bands, 2D-IR was shown to be in excellent agreement with the conclusions drawn from consideration of the dynamic spectra only.

REFERENCES

1. Siesler, H. W.; Holland-Moritz, K. *Infrared and Raman Spectroscopy of Polymers,* Marcel Dekker, New York, 1980, p. 266.
2. Palmer, R. A.; Gregoriou, V. G.; Fuji, A.; Jiang, E. Y.; Plunkett, S. E.; Connors, L.; Boccara, S.; Chao, J. L. *ACS Symp. Ser.* **1995**, *598*, 99.

3. Palmer, R. A.; Manning, C. J.; Chao, J. L.; Noda, I.; Dowry, A. E.; Marcott, C. *Appl. Spectrosc.* **1991**, *45*, 12.
4. Sonoyama, M.; Shoda, K.; Katagiri, G.; Ishida, H. *Appl. Spectrosc.* **1996**, *50*, 377.
5. Gregoriou, V.; Noda, I.; Dowrey, A. E.; Marcott, C.; Chao, J. L.; Palmer, R. A. *J. Polym. Sci.: Part B: Polym. Phys.* **1995**, *31*, 1769–1777.
6. Noda, I.; Dowrey, A. E.; Marcott, C. *Polym. Preprint, ACS — Poly Sec.* **1990**, *31*, 576.
7. Noda, I.; Dowrey, A. E.; Marcott, C. *Appl. Spectrosc.* **1988**, *42*, 203.
8. Chase, D. B.; Ikeda, R. M. *Appl. Spectrosc.* **1999**, *53*, 17.
9. Noda, I.; Dowrey, A. E.; Marcott, C. *Fourier Transform Infrared Characterization of Polymers*, Plenium Press, New York, 1987, Vol. 36, pp. 33–59.
10. Singhal, A.; Fina, L. J. *Vib. Spectrosc.* **1996**, *13*, 75.
11. Shah, H. V.; Manning, C. J.; Arbuckle, G. A. *Appl. Spectrosc.* **1999**, *53*, 1542.
12. Gregoriou, V. G.; Chao, J. L.; Toriumi, H.; Palmer, R. A. *Chem. Phys. Lett.* **1991**, *179*, 491.
13. Czarnecki, M. A.; Okretic, S.; Siesler, H. W. *J. Phys. Chem. B.* 101, 1997, 374.
14. Merenga, A.; Shilov, S. V.; Kremer, F.; Mao, G.; Ober, C. K.; Brehmer, M. *Macromolecules* **1998**, *31*, 9008.
15. Kischel, M.; Kisters, D.; Strohe, G.; Veeman, W. S. *Eur. Polym. J.* **1998**, *34*, 1571.
16. Wang, H.; Graff, D. K.; Schoonover, J. R.; Palmer, R. A. *Appl. Spectrosc.* **1999**, *53*, 687.
17. Shilov, S. V.; Skupin, H.; Kremer, F.; Gebhard, E.; Zentel, R. *Liq. Cryst.* **1997**, *22*, 203..
18. Davis, F. J. *J. Mater. Chem.* **1993**, *3*, 551.
19. Finkelmann, H.; Kock, H.-J.; Rehage, G. *Makromol. Chem., Rapid Commun.* **1981**, *2*, 317.
20. Nair, B. R.; Osbourne, M. A. R.; Hammond, P. T. *Macromolecules* **1998**, *31*, 8749.
21. Bonart, R. *J. Macromol. Sci. B Phys.* **1968**, *B2*, 115.
22. (a) Estes, G. M.; Seymour, R. W.; Cooper, S. L. *Macromolecules* **1971**, *4*, 452. (b) Seymour, R. W.; Allegrezza, A. E.; Cooper, S. L. *Macromolecules* **1973**, *6*, 896.
23. (a) Tang, W.; MacKnight, W. J.; Hsu, S. L. *Macromolecules* **1995**, *28*, 4284. (b) Nitzsche, S. A.; Hsu, S. L.; Hammond, P. T.; Rubner, M. F. *Macromolecules* **1992**, *25*, 2391. (c) Tao, H. J.; Meuse, C. W.; Yang, X. Z.; Macknight, W. J.; Hsu, S. L. *Macromolecules* **1994**, *27*, 7146. (d) Lee, H. S.; Hsu, S. L. *J. Polym. Sci. Part B — Polym. Phys.* **1994**, *32*, 2085. (e) Tao, D. J.; Rice, D. M.; Macknight, W. J.; Hsu, S. L. *Macromolecules* **1995**, *28*, 4036.
24. (a) Zhao, Y.; Lei, H. *Polymer* **1994**, *35*, 1419. (b) Zhao, Y. *Polymer* **1995**, *36*, 2717. (c) Zhao, Y.; Roche, P.; Yuan, G. *Macromolecules* **1996**, *29*, 4619.
25. Frasier, R. D. B. *J. Chem. Phys.* **1953**, *21*, 1511.
26. Stein, R. S. *J. Appl. Phys.* **1961**, *32*, 1280.
27. Cael, J. J.; Dalzell, W.; Trapani, G.; Gregoriou, V. G. *Pittsburgh Conference in Analytical Chemistry and Applied Spectroscopy, Abst. #1204*, March, Atlanta, GA, 1997.
28. Noda, I.; Dowrey, A. E.; Marcott, C. *Polym. Prepr.* **1984**, *5*, 167.
29. Gregoriou VG, Fuji A, Jiang E.Y., Plunkett S.E., Connors L.M., Boccara S., and Chao J.L., American Chemical Society Symposium Series M.W. Urban and T. Provder Eds., No. 598, Ch. 6, p. 99 (1995).
30. Budevska, B. O.; Manning, C. J.; Griffiths, P. R.; Roginski, R. T. *Appl. Spectrosc.* **1993**, *47*, 1843.
31. Noda, I. *J. Am. Chem. Soc.* **1989**, *111,* 8116.
32. Noda, I. *Appl. Spectrosc.* **1990**, *44*, 550.
33. Noda, I. *Applied Spectrosc.* **1993**, *47*, 1329.

34. Noda, I.; Dowrey, A. E.; Marcott, C.; Story, G. M.; Ozaki, Y. *Appl. Spectrosc.* **2000**, *54*, 236A.
35. Noda, I. *Appl. Spectrosc.* **2000**, *54*, 994.
36. Gregoriou, V. G.; Chao, J. L.; Toriumi, H.; Palmer, R. A. *Chem. Phys. Lett.* **1991**, *179*, 491.
37. Gregoriou, V. G.; Chao, J. L.; Toriumi, H.; Marcott, C.; Noda, I.; Palmer, R. A. *Proc. SPIE—Int. Soc. Opt. Eng.* **1575**, (Int. Conf. Fourier Transform Spectrosc., 8th, 1991), 209 (1992).
38. Shilov, S. V.; Skupin, H.; Kremer, F.; Wittig, T.; Zentel, R. *Pittsburgh Conference in Analytical Chemistry and Applied Spectroscopy,* March, Atlanta, GA, 1997 abstract *119.*
39. Czarnecki, M. A.; Okretic, S.; Siesler, H. *Vib. Spectrosc.* **1998**, *18*, 17.
40. Marcott, C.; Dowrey, A. E.; Noda, I. *Anal. Chem.* **1994**, *66*, 1065A.
41. Ward, I. M.; Hadley, D. W. *An Introduction to the Mechanical Properties of Solid Polymers*, John Wiley and Sons, New York, 1993.
42. Mano, J. F.; Correia, N. T.; Moura-Ramos, J. *J. Liq. Cryst.* **1996**, *20*, 201.
43. Heroguez, V.; Schappacher, M.; Papon, E.; Def-fieux, A. *Polym. Bull.* **1991**, *25*, 307.
44. Mori, K.; Tominga, T.; Matsui, M. *Synthesis* **1973**, 790.
45. Papadimitrakopolos, F.; Sawa, E.; MacKnight, W. J. *Macromolecules* **1992**, *25*, 4682.
46. Nair, B. R.; Gregoriou, V. G.; Hammond, P. T. *Polymer* **2000**, *41*, 2961.
47. Noel, C.; Blumstein, A. *Polymeric Liquid Crystals*, Noel, C. (Ed.), Plenum Press, New York, 1985.
48. Nair, B. R.; Gregoriou, V. G.; Hammond, P. T. *J. Phys. Chem. B.* **2000**, *104*, 7874.
49. Tokita, M.; Osada K.; Kawauchi, S.; Watanabe, J. *Polym. J.* **1998**, *30*, 687.

2 Stress/Strain Measurements in Fibers and Composites Using Raman Spectroscopy

Costas Galiotis and John Parthenios

CONTENTS

2.1 BACKGROUND

The determination of the state of stress in structural materials during service has always been one of the key issues that the structural engineer or designer has to address. The lack of experimental techniques in this area often forces the engineer to resort to analytical or numerical methods to assess the overall stress distribution within a structural component. As a result, stringent design rules have to be applied to ensure safety in a structural assembly. This is particularly well demonstrated in the case of advanced fiber composite materials; the lack of knowledge of the complex state of stress generated by the anisotropy and quite often inhomogeneity in these materials leads to overdesign and hence high component costs. Thus, in polymer based composites, the savings gained in moving parts as a result of their light weight and correspondingly high specific properties, can quite often be offset by the volume of material required to address the safety design limits.

Another very important issue is the detection of the propagation of damage in service. In metals, visual inspection combined occasionally with nondestructive diagnostic techniques, can provide information about the integrity of the material at various stage of its life cycle. In composites that incorporate brittle polymer matrices reinforced with brittle fibers, toughness is normally attained by the complexity of the propagation of damage at the microscopic scale [1]. Careful control of the strength of the fiber/matrix interface can also enhance tensile toughness and strength by increasing the overall crack propagation and by diluting the effect of stress concentration emanating from isolated fiber breaks [2,3]. In all cases, the exact knowledge of the local stress field generated by an applied external load is of paramount importance for determining the efficiency of stress transfer and for monitoring the propagation of damage at various load increments and time intervals.

A new technique for stress and strain measurements in composites has been developed over the past few years. The stress/strain dependent property is the frequency of the atomic vibrations in a crystalline material, which can be probed with the spectroscopic technique of laser Raman spectroscopy. The principle of this technique is based on the fact that when a crystalline material is stressed, the equilibrium separation between its constituent atoms is altered in a reversible manner. As a result, the interatomic force constants, which determine the atomic vibrational frequencies, also change since they are related to the interatomic separation. In general, as the bond lengths increase with tensile load the force constants and, hence, the vibrational frequencies decrease [4], while the reverse effect is present when the

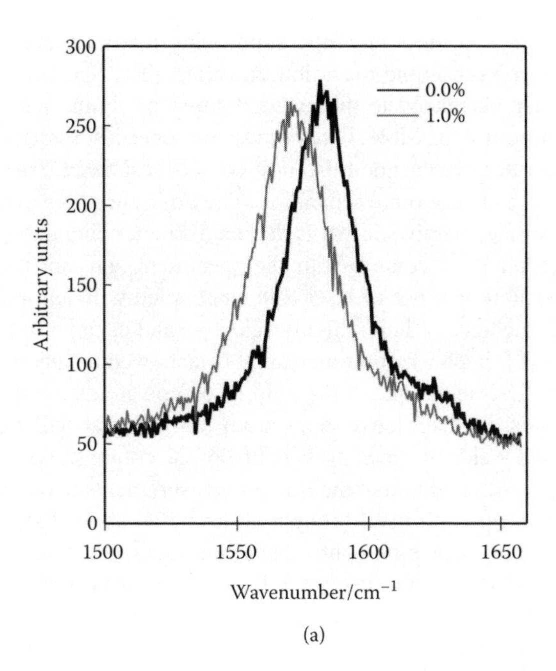

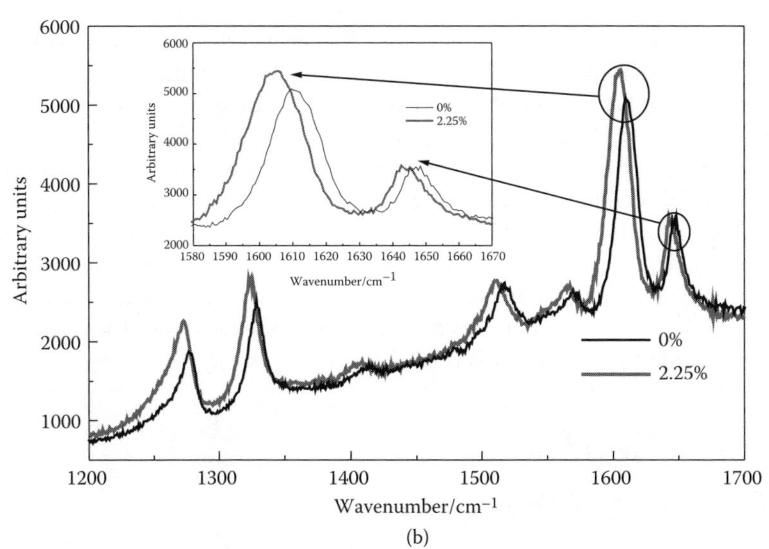

FIGURE 2.1 Characteristic Raman spectrum of (a) E_{2g} (1580 cm^{-1}) band of carbon fibers and (b) Kevlar 29 fibers. The shifted spectrum corresponds to tensile deformation of (a) 1.00% and (b) 2.25%.

material is subjected to hydrostatic [5] or mechanical compression [6]. A Raman spectrum that presents the strain dependence of certain vibrational modes of Kevlar® 29 fiber is shown in Figure 2.1. This effect has universal applicability in materials as diverse as commercial polymers, such as polyethylene [6], and even inorganics such as silicon [7]. In composites, most reinforcing fibers are crystalline materials and, therefore, have been found to exhibit this effect. The magnitude of this Raman frequency shift can be related to the external stress or strain, hence making stress and strain measurements possible. Other physical properties, such as electrical resistivity, which has been put in good use in electrical resistance "strain gauges," are also a function of the interatomic separation. The difference between Raman (or IR) and electrical resistivity strain sensors is that the former, being an optical technique, does not require physical contact with the specimen, and can therefore be used remotely. Furthermore, the use of laser Raman coupled with an optical microscope has the added advantages of being highly selective and capable of probing areas as small as 1–2 μm [8]. Such a resolution cannot be achieved by any other nondestructive evaluation (NDE) technique in the field of composite materials.

For composites, its usefulness stems from the fact that it is the only existing method to date that yields the fiber stress in one of the composite principal directions (fiber axis). It is worth adding that the Raman measurements correspond to physical changes of the measure and itself brought about by an applied stress field. On the contrary, electrical resistance or light modulation measurements are performed on attached or embedded external devices, which are assumed to be strained equally with the surrounding composite material, and to have no effect on the mechanical properties of the host structure.

By loading individual fibers either in uniaxial tension or compression, the researcher may measure the magnitude of the Raman wavenumber shift from the value of the free-standing fiber. In a composite material the inverse methodology is applied; the magnitude of the wavenumber shift is measured along the reinforcing fibers and converted to axial stress or strain via a fiber-specific calibration curve. All fibers situated near the surface of a polymer composite can be interrogated remotely (and nondestructively) provided that the matrix is reasonably transparent. Fibers that are located in the bulk of a composite can be interrogated by means of a wave guide, such as a fiber optic cable [9] but in this case a certain amount of stress perturbation is expected to occur. This is of extreme importance to composites manufacturers and users, and paves the way for the developments of laser Raman Spectroscopy (LRS) as a nondestructive method for routine inspection of composite panels.

2.2 INSTRUMENTATION

2.2.1 Conventional Measurements

The experimental requirements of a conventional laser Raman setup are shown in Figure 2.2. The monochromatic light normally in the visible range produced by a gas laser (Ar^+ or He-Ne) is directed to a Raman modified microscope via a system of mirrors and focused down to the specimen, which is housed on the microscope platform. The 180° back-scattered radiation is sent to the spectrometer by a beam splitter through a variable size pinhole to attain confocality. A charge-coupled device (CCD) converts the optical signal into an electrical output, which is subsequently stored in a PC. This experimental arrangement is ideal for small specimens, which can be translated in the laser beam with a suitable micromanipulator.

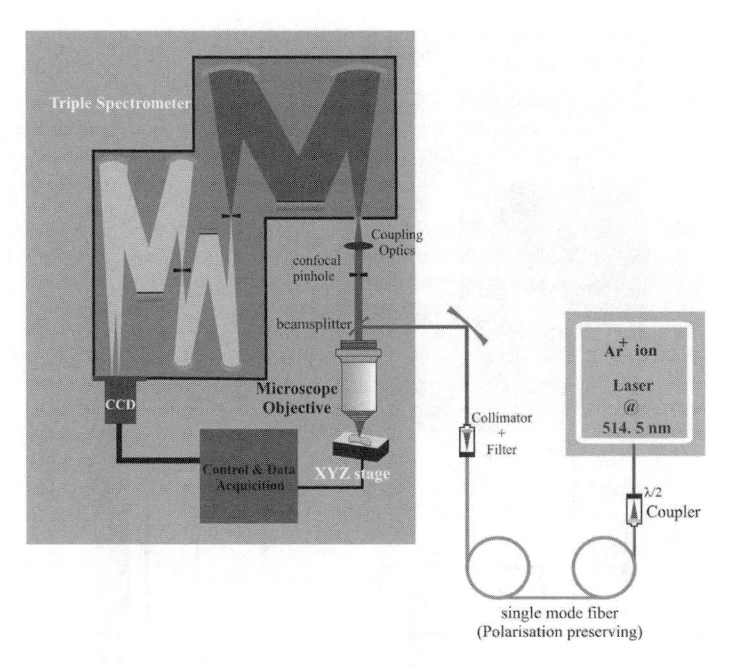

FIGURE 2.2 Schematic of the conventional laser Raman experimental setup, comprising a laser source, a microscope, a spectrometer, a CCD device and a data acquisition system.

2.2.2 Remote Laser Raman Measurements

To avoid the space restrictions imposed by the conventional setup and enhance its experimental capabilities, the spectroscopic assembly (laser plus spectrometer) can be decoupled from the testing area by employing fiber optic cables.

2.2.2.1 Fiber Optic Microprobes

The application of fiber optic cables in a flexible Raman microprobe is shown in Figure 2.3 [10]. The main difference between the two setups lies in the design of the Raman microprobe; the incorporation of flexible fiber optic cables for laser light delivery and collection allows the positioning of the microprobe at any angle with respect to a system of reference such as the work bench. Furthermore, the length of the fiber optic can be as long as 300 m, and therefore specimens at large distances from the laser source or monochomator can be interrogated (Figure 2.3). The details of the design of the remote Raman microprobe (ReRaM) itself are shown in Figure 2.4. Tailor-made optics at both input and output positions of each fiber optic ensure laser collimation, maximum efficiency, and enhancement of the Raman scattering radiation [10]. Finally, the use of an incorporated CCTV camera permits optical observations of the specimen during Raman spectrum acquisition. This remote setup is particularly useful for composite materials, which are subjected to mechanical loading, but it can also be used in a whole variety of other technological applications, such as conducting remote Raman measurements in chemically hostile environments or elevated temperatures; monitoring the curing of polymer resins and the crystallisation of polymers during solidification;

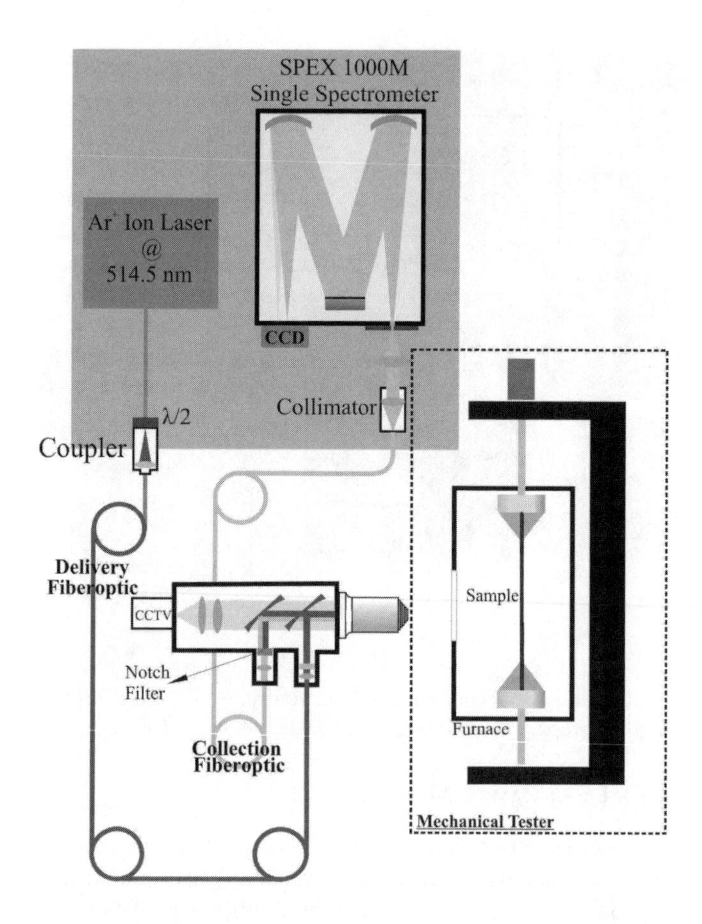

FIGURE 2.3 Schematic of a remote Raman experimental setup comprising one laser source, two fiber optic cables for light delivery and collection, a modified microscope, a CCD for spectrum recording, and associated optical components.

providing nondestructive "health" monitoring of sections of large structures (e.g., aircrafts, ships, and bridges); and assessing the quality of oil supplies or other chemical media.

2.2.2.2 Microprobes Incorporating Solid-State Lasers

A more recent development is the introduction of a miniature diode pumped solid state (DPSS) laser into the main body of the microprobe to reduce the main components of the system and related peripherals from three (laser–microprobe–spectrometer) to two (microprobe–spectrometer). This gives tremendous advantages for remote external measurements as it enhances the flexibility and the portability of the setup assembly. Such a Raman microprobe, which incorporates a miniature DPSS laser system emitting at 532 nm, was first designed at FORTH-ICHT and was developed by DILOR–Jobin Yvon [11,12]. A schematic representation of the system is shown in Figure 2.5. The use of

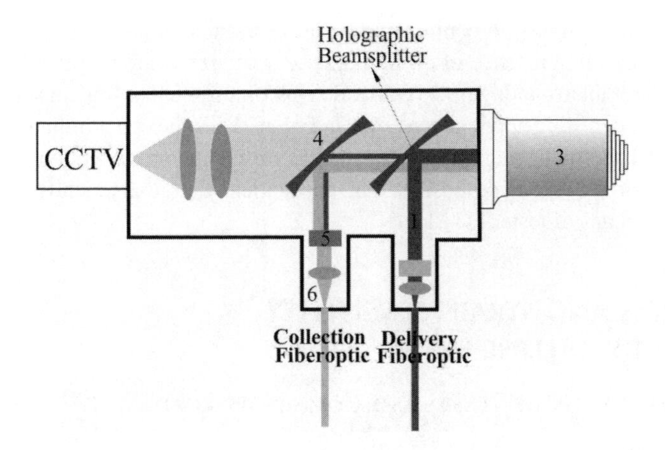

FIGURE 2.4 Schematic of the remote Raman microscope (ReRaM) comprising tailor made optics at both input and output positions, a CCTV video camera, and associated optical components to ensure laser beam collimation, maximum efficiency, confocality, and enhancement of Raman scattering radiation. The various components shown are (1) laser beam collimation unit (input), (2) bandpass filter, (3) microscope objective, (4) mirror, (5) notch filter, and (6) laser collimation unit (output).

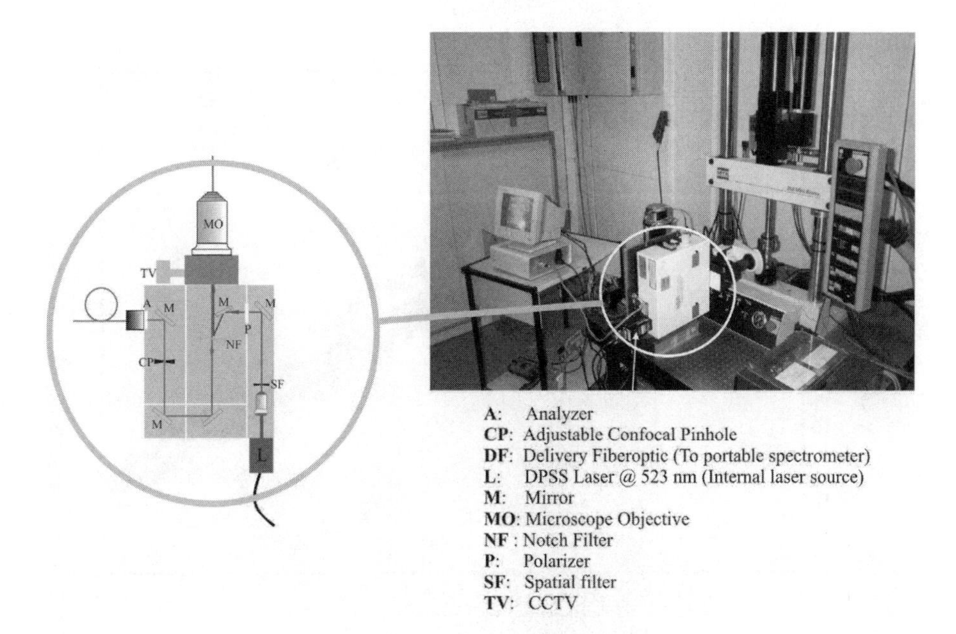

FIGURE 2.5 Experimental setup of the concurrent use of a ReRaM probe and a hydraulic mechanical tester for the simultaneous inspection of stress of polymer composites at micro and macromechanical level. The schematic of ReRaM with various optical components is also shown.

higher laser wavelengths, in conjunction with the an internal attenuator for controlling laser power and the presence of an internal variable size confocal pinhole, make this system ideal for interrogating fibers embedded into fluorescent or Raman active polymer matrices. An interesting application shown in Figure 2.5 is the concurrent use of ReRaM IV and an IR thermal imaging camera for the simultaneous inspection of stress and temperature of polymer composites that are statically or dynamically loaded on a hydraulic mechanical tester [11].

2.3 STRESS AND STRAIN SENSITIVITY OF CRYSTALLINE FIBERS

2.3.1 Application of Tensile and Compressive Loading

The relationship between Raman wavenumber and tensile stress or strain is obtained by stressing a single fiber in air on a suitable mechanical tester operating at a low strain rate, while Raman spectra are taken at any position along the fiber length. The difference between the Raman wavenumber values obtained at each level of strain and that of the free-standing fiber represents the Raman wavenumber shift. In Figure 2.6 the Raman wavenumber shift is plotted as a function of applied stress for the 1611 cm^{-1} Raman band of Kevlar® 29 and 49 fibers and, as can be seen, a linear relationship is obtained:

$$\Delta \nu = k_{1611}\sigma \tag{2.1}$$

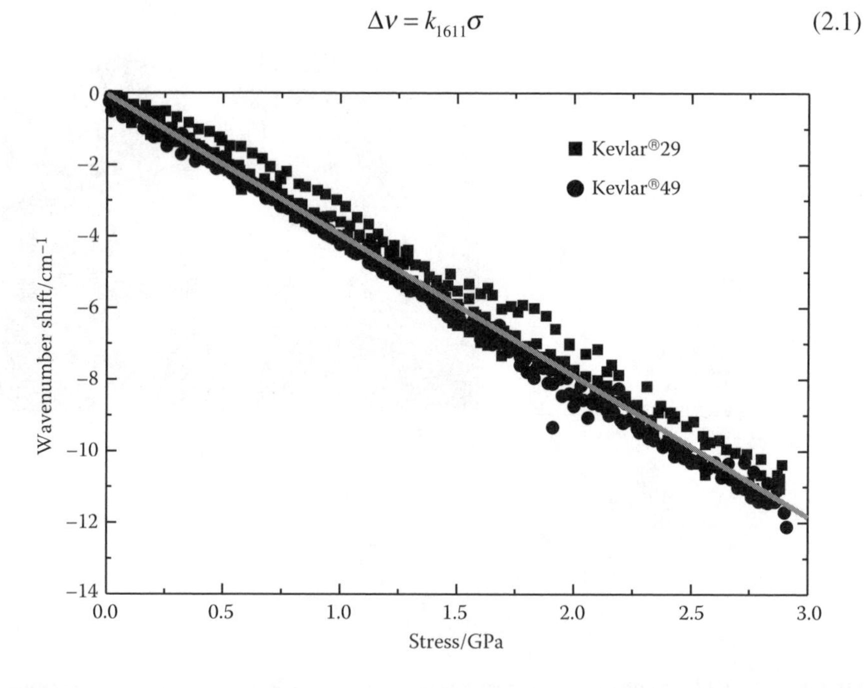

FIGURE 2.6 Raman wavenumber shift as a function of tensile stress for the Kevlar® 29 and 49 fibers.

where k_{1611} represents the sensitivity of the Raman frequency shift to an applied stress. The negative slope of −4.1 + 0.4 cm^{-1}/GPa of the least-squares-fitted straight line represents the sensitivity of the stress sensor. This value allows the conversion of the Raman frequencies obtained from fibers embedded in polymer-based composites into values of stress.

High performance fibers are highly anisotropic and their behavior in tension and compression may be different particularly under deformation-controlled experiments. Normally, in order to calibrate the Raman wavenumber as a function of tensile or compressive strain, single filaments are bonded to the surface of an elastic polymer cantilever beam which can be flexed up or down to subject the fibers to tension or compression [13]. Provided that no slippage takes place between the fiber and the bar, the strain (tensile or compressive) varies linearly along the length of the beam and is only a function of the position on the bar as determined by the elastic beam theory. Relevant experiments on Kevlar® fibers and carbon surface-treated high-modulus (HMS) fibers have been performed and the corresponding typical graphs of Raman frequency versus strain are presented in Figure 2.7 and Figure 2.8a and b, respectively. It is interesting to note that the Raman wavenumber scales nonlinearly with strain, whereas for aramid, as well as carbon fibers, the corresponding relationship with stress is always linear [13]. Thus, since most polymer or carbon fibers can be considered as equal-stress solid bodies, it can be said that these fibers act, within the context of the Raman technique, as stress sensors.

For strain controlled experiments, spline polynomial functions are used for converting the Raman wavenumbers into strain. Important mechanical parameters such as the critical compressive strain to failure and the molecular deformation of

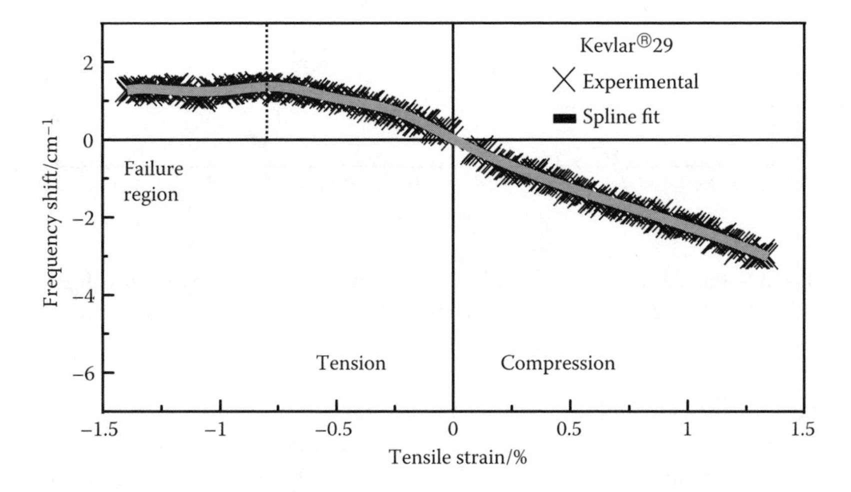

FIGURE 2.7 Raman wavenumber shift as a function of strain for the Kevlar® 49 fiber. The solid line represents a cubic spline fit to the experimental data.

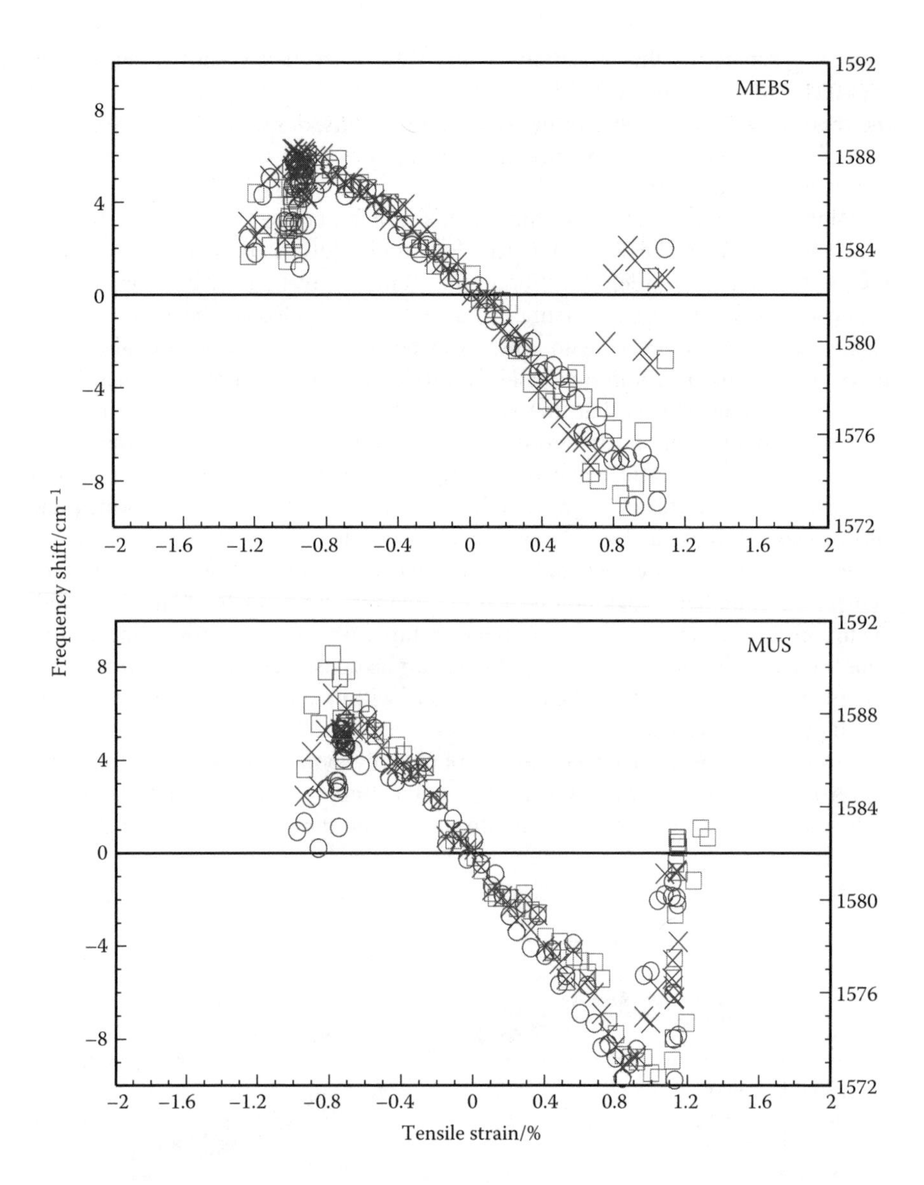

FIGURE 2.8 Raman wavenumber shift as a function of strain for the sized (MEBS) and unsized (MUS) M40 carbon fibers supplied by Courtaulds Grafil plc (U.K.).

the fiber in the post-failure (post-buckling) region can also be determined from Figure 2.6 and Figure 2.7. The nominal values of stress or strain sensitivity of the Raman wavenumber for a number of commercial fibers are given in Table 2.1. Finally, it must be added that in all cases for which the degree of nonlinearity for strain measurements is small, an approximate linear relationship between Raman wavenumber still holds, particularly for small strains.

TABLE 2.1
Stress/Strain Sensitivities for the Most Important Modes of Kevlar® 29 Fibers

Mode	1611 cm^{-1}		1648 cm^{-1}	
	Sensitivities			
Fiber Type	Strain (cm^{-1}/%)	Stress(cm^{-1}/GPa)	Strain (cm^{-1}/%)	Stress(cm^{-1}/GPa)
Kevlar® 29	−2.8 ± 0.07	−4.1 ± 0.4	−1.32 ± 0	−2.0 ± 0.1
Kevlar® 49	−4.21	−4.0		

2.3.2 Effect of Temperature

Raman band positions are dependent on temperature, since temperature can induce lattice changes in crystalline materials. It has been demonstrated that certain peaks of the Raman spectrum of crystalline fibers (carbon or aramid) are dependent on temperature variations. Generally speaking, the effect of temperature upon certain vibrations is similar to that of stress; i.e., the Raman wavenumber shift is a decreasing linear function of temperature. Laser-induced heating of crystalline fibers can cause additional band shifts in addition to those induced by sample stress or strain. The need for careful control of laser beam power at the fiber surface is of paramount importance when calibrating certain Raman peaks as functions of the applied stress and strain. Furthermore, in stress and strain measurements under nonisothermal conditions, the bandshift due to temperature has to be taken into account.

2.3.2.1 Carbon Fibers

Experiments for the dependence of the E_{2g} band position of M40 fibers on temperature has been performed as a function of temperature as well as incident laser power. The effect of laser-induced heating on this band is shown in Figure 2.9a. The fibers were tested over a power range from 0.5 to 10 mW and the power dependence is found to be −0.35 cm^{-1}/mW, as indicated by the least-square-fitted line. Laser powers lower than 1 mW do not cause significant band shifts.

Hence, when the laser power was kept lower than 1 mW, the temperature sensitivity of the M40 carbon fibers using the E_{2g} band was found to be 0.026 cm^{-1}/°C (Figure 2.9b). Similar results for other types of carbon fibers have been presented elsewhere [14].

TABLE 2.2
Stress/Strain Sensitivities for Carbon Fiber $E2_g$ Raman Band of Carbon Fibers

Mode	1580 cm^{-1}	
	Sensitivities in	
Fiber Type	Strain (cm^{-1}/%)	Stress(cm^{-1}/GPa)
Carbon M40-40B	−11.4 ± 0.06	−3.0 ± 0.1

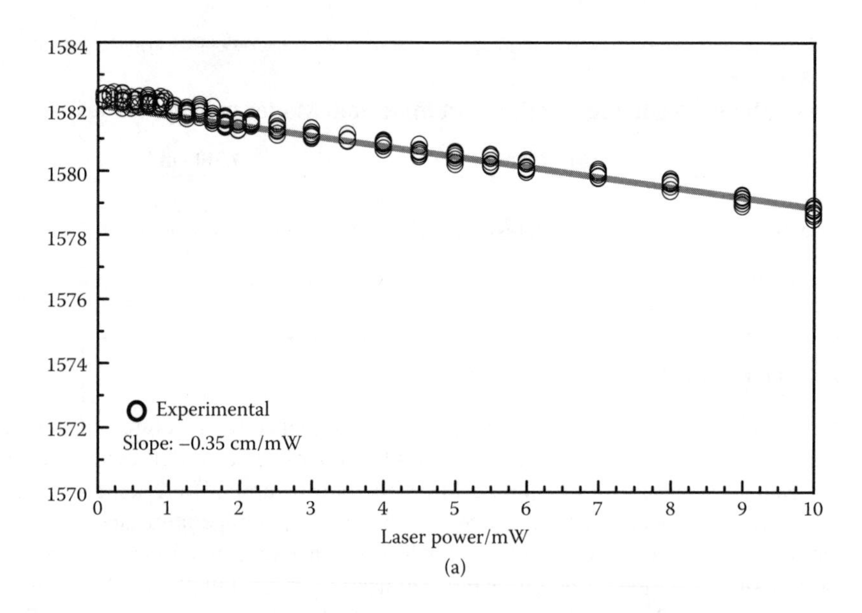

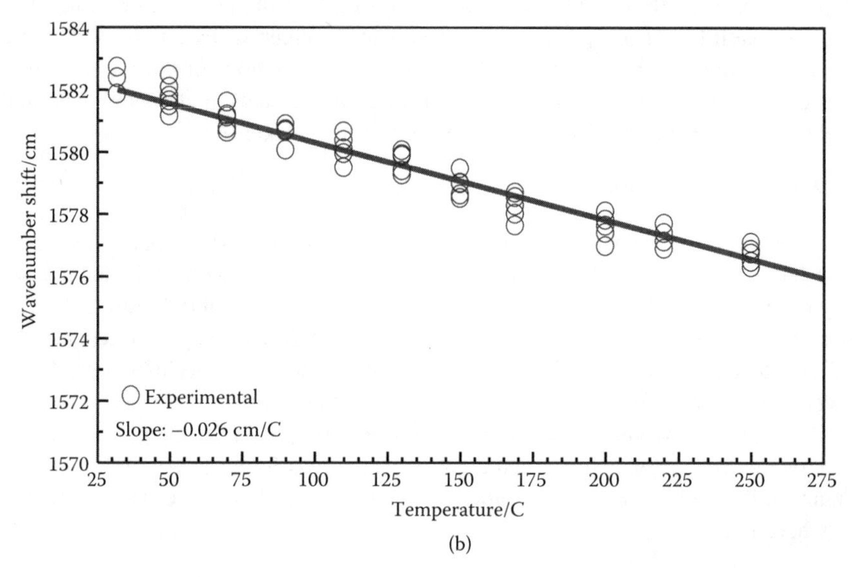

FIGURE 2.9 (a) The effect of laser-induced heating of the E_{2g} band in M40 carbon fibers. (b) Temperature calibration curve of E_{2g} band. (Laser power was kept lower than 1 mW.)

2.3.2.2 Aramid Fibers

As expected, the effect of temperature upon the skeletal vibrations of poly (p-phenylene tere phthalamide) (PPTA) repeat unit of aramid fibers is similar to that of stress. The expected shift to lower wavenumbers has been confirmed for most peaks within the 1200–1800 cm^{-1} spectral region. As an example, the effect of

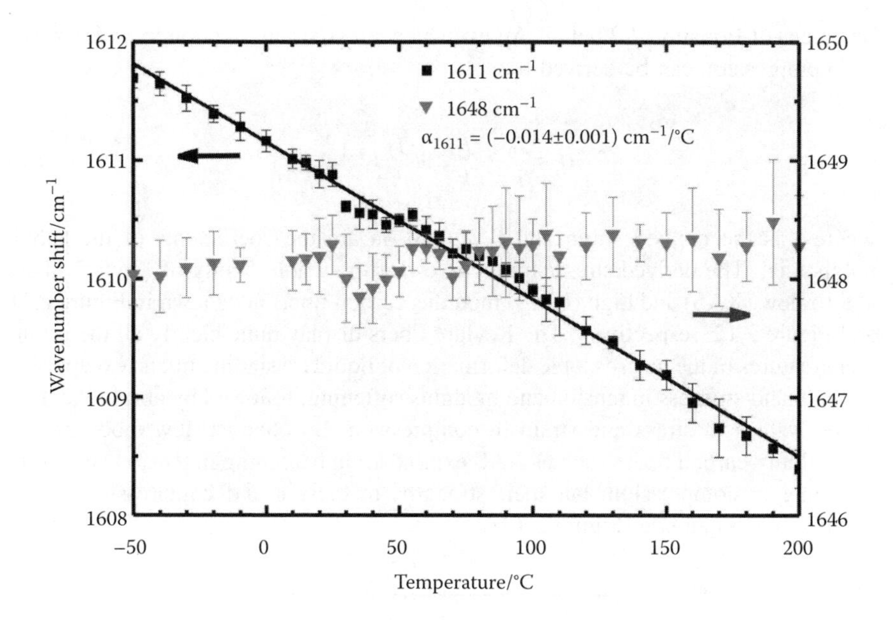

FIGURE 2.10 Raman band peak position as a function of the temperature, for the 1611 and 1648 cm^{-1} Raman vibrational modes within the –50 to 200°C range.

temperature upon the wavenumbers of two specific stress/strain sensitive vibrations of the PPTA polymer backbone is presented here. These vibrations recorded at 1611 and 1648 cm^{-1} (Figure 2.1) correspond mainly to ring/C–C stretching and to amide/C=O stretching, respectively [15]. The influence of temperature is determined by monitoring the induced changes in the vibration peak wavenumber by the applied thermal field in individual Kevlar® 29 fibers. The results of this investigation are depicted in Figure 2.10, where it is clear that the position of the 1611 cm^{-1} depends on the fiber temperature, while the 1648 cm^{-1} band can be considered temperature insensitive. The Raman wavenumber shift of the 1611 cm^{-1} band is least-squares-fitted with a straight line of slope of -0.015 ± 0.001 cm^{-1}/°C (Figure 2.10). If the temperature sensitivity of these two peaks is considered constant for any level of stress or strain, then the recorded shift of the 1611 cm^{-1} band can be used for fiber temperature measurements [16–18].

2.4 CONVERTING SPECTROSCOPIC DATA INTO FIBER STRESS/STRAIN CURVES IN TENSION AND COMPRESSION

Raman spectroscopic data can be converted into mechanical stress/strain curves, either in tension or in compression, using the prevailing approach that aramid or carbon fibers can be considered as aggregates of crystallites connected in series [13,19,20]. The conversion methodology combines the stress-controlled [13] (see Equation 2.1 and Figure 2.6) and the cantilever strain-controlled experiments [13] by means of Equation 2.2 below. An estimated stress/strain function in both tension and compression can be derived by:

$$\sigma = \frac{f_0}{k} + \frac{f_1}{k}\varepsilon + \frac{f_2}{k}\varepsilon^2 + \ldots \tag{2.2}$$

where ε is the applied strain and f_0 and so on are the coefficients of the spline polynomial. The derived stress/strain curves for the whole family of Kevlar® fibers and for low (XAS) and high (HMS) modulus carbon fibers are shown in Figure 2.11 and Figure 2.12, respectively. The Kevlar® fibers display quite clearly all the prominent features of the macroscopic deformation of liquid crystalline fibers: exceptional strength and stiffness in tension and modulus softening, followed by abrupt yielding at low values of stress and strain in compression. By contrast, low modulus and crystallinity carbon fibers such as XAS exhibit strain hardening in tension and strain softening in compression, but their strengths in tension and compression are of comparable magnitude (Figure 2.12).

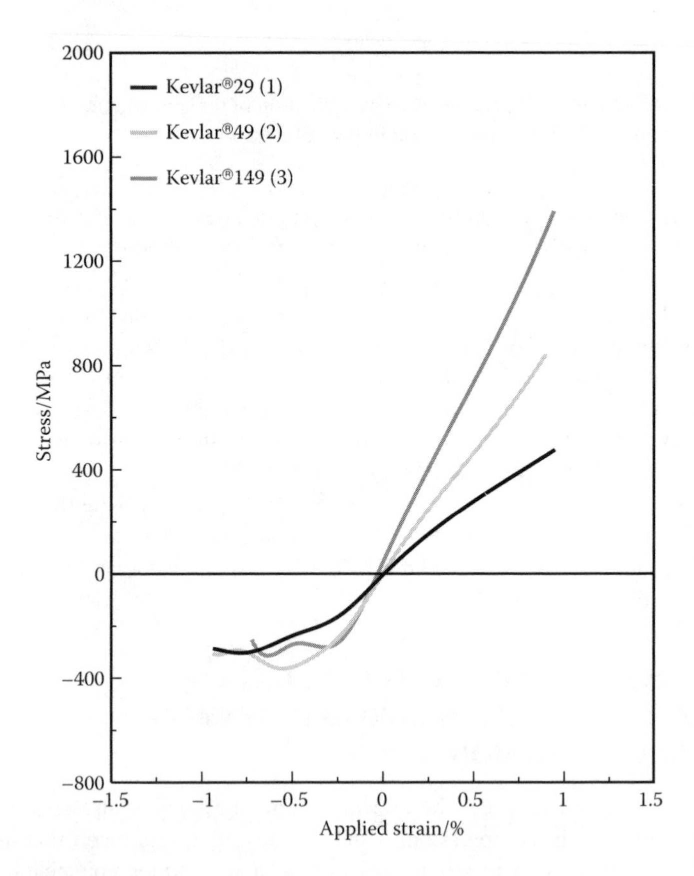

FIGURE 2.11 Spectroscopically derived stress-strain curves in tension and compression for commercial Kevlar® fibers.

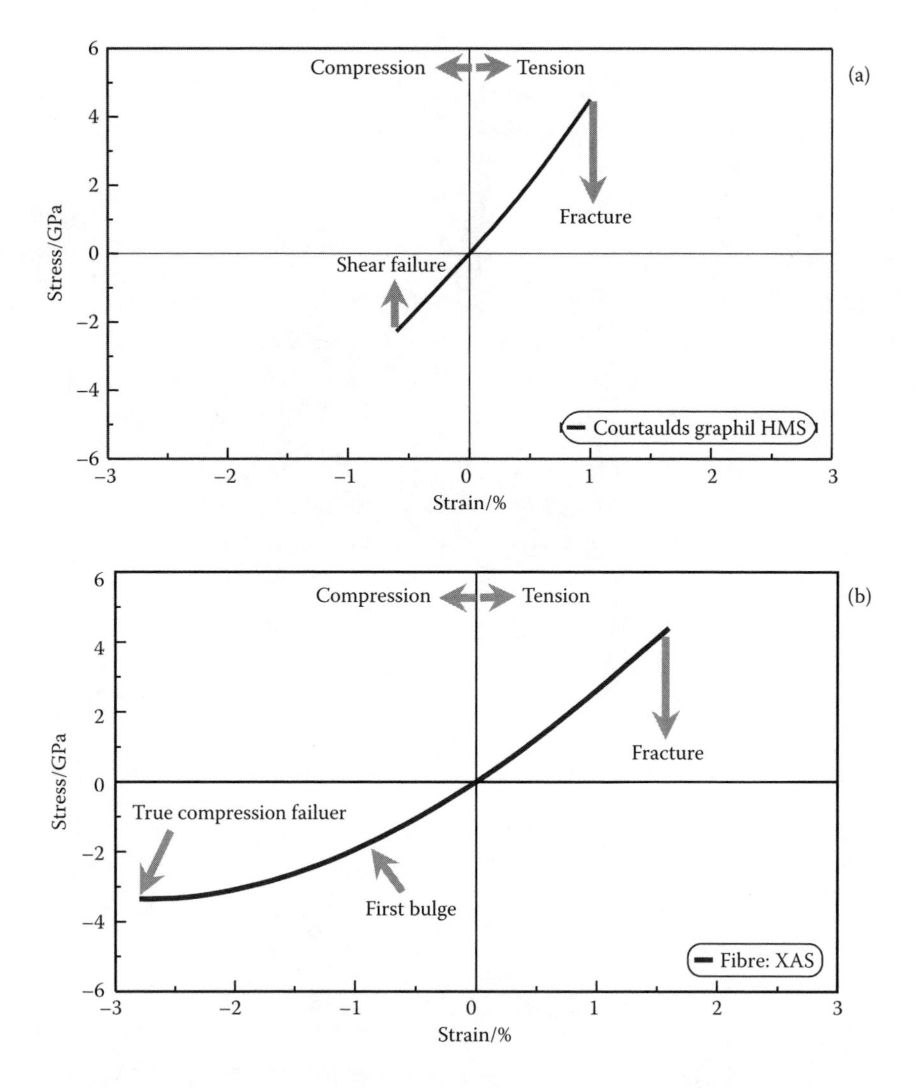

FIGURE 2.12 Spectroscopically derived stress-strain curves in tension and compression for the (a) HMS and (b) XAS Courtaulds Graphil carbon fibers.

2.5 ASSESSING THE EFFICIENCY OF THE FIBER/MATRIX BOND IN COMPOSITES

The architecture of the fiber/matrix interface in composites is presented in Figure 2.13. At the macroscopic level, the bond in the representative volume element (RVE) is considered for design purposes to be perfect [21]. However, at the microscopic level the picture is extremely complex, due to the existence of an interphase of variable thickness comprising fiber surface chemistry and topography; sizing, wetting, and other coating agents; and diffused matrix materia [22]. The macroscopic,

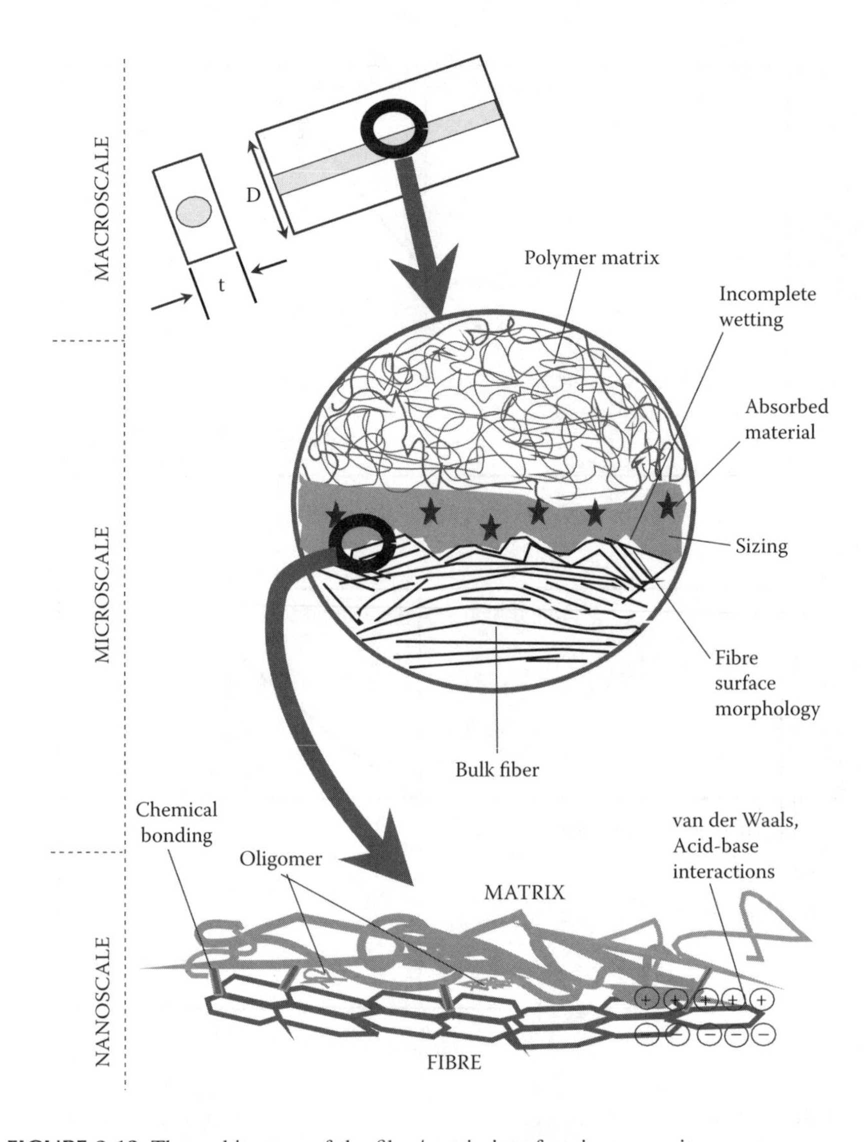

FIGURE 2.13 The architecture of the fiber/matrix interface in composites.

microscopic, and molecular phenomena are strongly interrelated [23] and therefore the overall physico-mechanical performance can be successfully tailored by means of suitable manipulation of the various critical variables at the molecular or microscopic levels.

One of the most important functions of the interface is the efficient transfer of stress from the matrix to the reinforcing fibers. The degree of efficiency of stress transfer will depend primarily upon the fiber and matrix chemistry and the existence of residual thermal stresses at the interface. The latter is related to the geometry of the test coupon, the associated curing process, and the fiber volume fraction.

2.5.1 Methodology

The stress-transfer mechanism in fiber reinforced composites is activated in the vicinity of discontinuities such as fiber ends and fiber breaks. Fiber ends are normally found in discontinuous fiber composites, while fiber breaks are formed as a result of fiber fracture during fabrication or the influence of an externally applied stress field. Until the advent of the laser Raman technique [24,25], the stress transfer profiles activated at a fiber discontinuity could only be derived by means of analytical [26] or even photo-elastic modeling [27]. Now the only requirement for polymer-based composites is the presence of a fiber discontinuity (end or break). Raman data are normally collected by scanning the fiber point by point along the broken fiber on either side of the discontinuity. The strain-transfer profiles obtained can be converted to interfacial shear stress (ISS) profiles along the length of the fiber by means of a straightforward balance of shear to axial forces argument. This leads to a simple analytical expression between the ISS $\tau_{(rx)}$ and the gradient $d\sigma_f/dx$ of the stress transfer profiles:

$$\tau_{(rx)} = -\frac{r}{2}\left(\frac{d\sigma_f}{dx}\right) \tag{2.3}$$

where σ_f is the axial stress in the fiber, r the fiber radius, and x the distance along the fiber length. The ISS profiles $\tau_{(rx)}$ are derived by (a) fitting a cubic spline to the raw data; (b) calculating the derivatives $d\sigma_f/dx$ from the spline equations; and, finally, (c) employing Equation 2.3. For fibers exhibiting elastic stress/strain characteristics the $\tau_{(rx)}$ distribution can also be obtained from the axial strain distribution by substituting σ in Equation 2.3 with the corresponding product $E\cdot\varepsilon$, where E is the Young's modulus and ε the axial fiber strain.

2.5.2 Model Composite Specimens

In Figure 2.14 the most important test geometries for the determination of interfacial parameters in composites are shown. In general, single fiber test coupons can be quite useful in the detection of true interfacial phenomena without the influence of fiber–fiber interactions, but their applicability is limited, as they cannot be considered truly composite tests. Tests on full composites can only provide indirect evidence on the strength of the fiber–matrix interface through measurements of the off-axis properties, such as the in-plane shear strength. As shown schematically in Figure 2.14, laser Raman microscopy yields the axial stress distribution in the embedded fibers, through which the stress transfer efficiency is easily determined. Hence, the desirable link between single-fiber and multiple-fiber test geometries is established. The results obtained to date on single- and multiple-fiber test methods are briefly reviewed below.

2.5.2.1 Effect of Fiber Treatment

Laser Raman spectroscopy has been used to monitor fragmentation processes in carbon fiber and epoxy resin systems [28,29] and to review the suitability of the

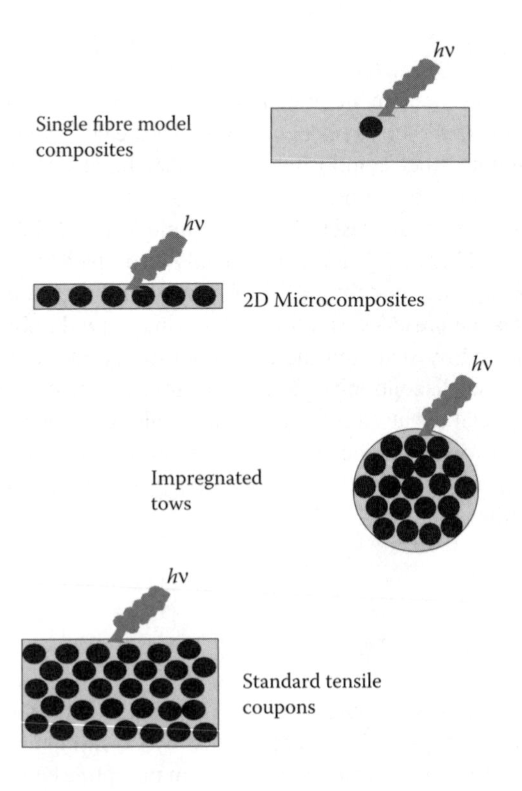

FIGURE 2.14 The most important test geometries for the determination of interfacial parameters in composites.

so-called fragmentation test as a valid test for measuring the interfacial shear strength (IFSS) of single fiber model composites. The fiber strain distributions of representative fragments of similar lengths for each fiber/resin system are shown in Figure 2.15. In all cases the maximum strain supported by the fiber is 1%. As can be seen, the strain profile for a high-modulus carbon fiber with an untreated surface (HMU) is virtually linear, indicating a frictional type of reinforcement. For surface-treated high-modulus (HMS) and intermediate-modulus (IMD) fiber systems, the strain take up is more or less in accordance with the elastic stress transfer models [26]. These fiber strain distributions are converted to ISS distributions via Equation 2.3, and the resulting curves are shown in Figure 2.15. The following observations can be made at this point: (a) the interfacial shear stress is nearly constant along the HMU/epoxy fragment; (b) the surface treated fiber and resin systems, HMS/epoxy and IMD/epoxy, exhibit distributions that reach a maximum value near the fiber end and decay to zero towards the middle of each fragment; and, finally, (c) the higher the maximum ISS value for the HMS and IMD/MY-750 systems, the shorter the distance from the fiber end where this maximum appears. The interfacial or interphasial shear strength (IFSS) of a fiber and resin system is normally defined as the maximum value of ISS developed throughout the fragmentation test. It should be stressed, however, that there is a statistical distribution of the ISS maxima at each level of

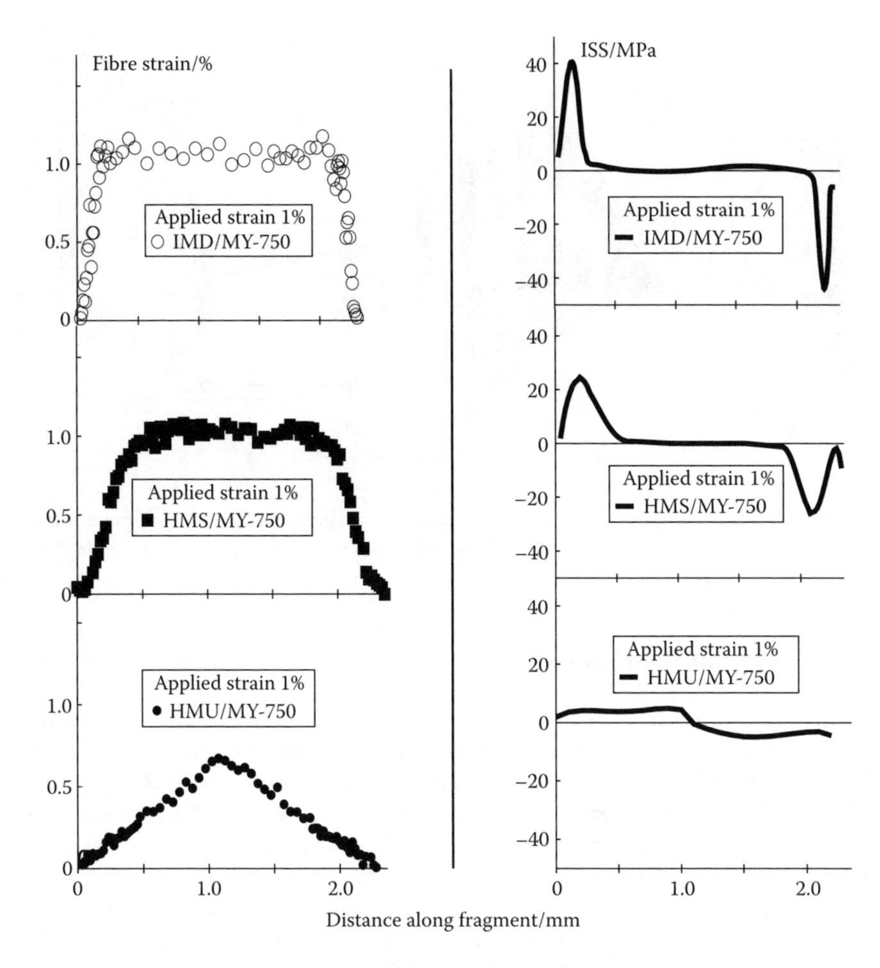

FIGURE 2.15 Fiber strain (left) and interfacial shear stress (right) distributions of representative fragments of similar length for three different carbon fiber/epoxy systems. All fibers were supplied by Courtaulds Grafil plc (U.K.), and MY-750 is a trademark of Ciba-Geigy plc (U.K.).

applied strain, as large numbers of fragments are sampled along the gauge length. It is, therefore, more appropriate to derive an average maximum ISS value at each level of applied strain, along with the standard deviation of the mean. A plot of average maximum ISS as a function of applied strain for all three systems examined in this project is shown in Figure 2.16. As can be seen, the average maximum ISS increases with applied strain for both systems and reaches an upper limit of 36 ± 6 MPa and 66 ± 15 MPa for the HMS/epoxy and IMD/epoxy systems, respectively. These values are good estimates of the IFSS of the two systems. The average maximum ISS for the untreated HMU/MY-750 system appears to be insensitive to applied strain and is approximately six times lower than that of the treated HMS/MY-750 system.

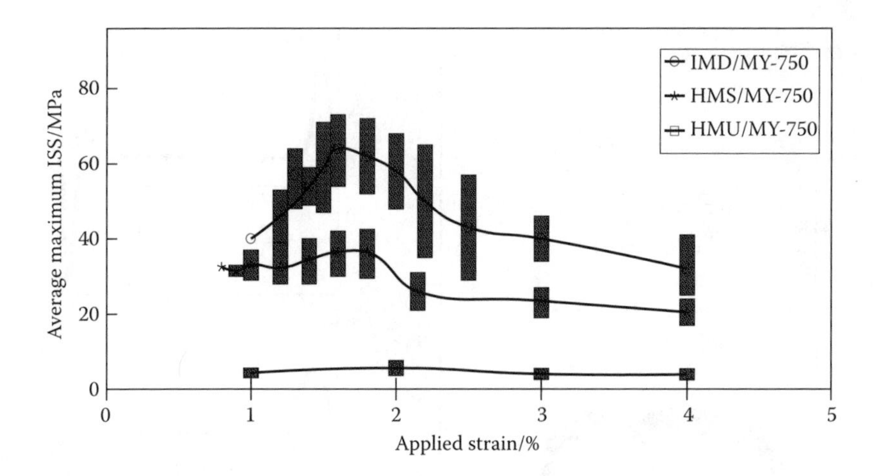

FIGURE 2.16 Average maximum interfacial shear stress (ISS) as a function of applied strain for three different carbon fiber/epoxy systems. All fibers were supplied by Courtaulds Grafil plc (U.K.), and MY-750 is a trademark of Ciba-Geigy plc (U.K.).

2.5.2.2 Effect of Fiber Sizing

To demonstrate the effect of fiber sizing upon the stress transfer efficiency, single fiber model systems incorporating sized and unsized carbon fibers have been studied. The model composites consist of the sized and unsized M40 carbon fiber supplied by Soficar embedded in the Ciba–Geigy MY-750 epoxy. In Figure 2.17 and Figure 2.18, the axial fiber stress profiles of representative fragments at various increments of applied composite strain for the sized and unsized fiber system, respectively, are presented. In the case of the sized fiber system (Figure 2.17), the propagation of interfacial damage with applied strain leads to the progressive reduction of the effective length for stress transfer and to the growth of zones of zero stress transfer on either side of the fragment [30]. These zones emanate from the fiber breaks and grow towards the middle of the fragment, resulting in the characteristic S-shaped profiles of Figure 2.17b and Figure 2.17c. On the contrary, in the case of the unsized system (Figure 2.19), there are no areas in the fiber where the fiber axial stress is zero. This indicates that the stress is transferred efficiently along the whole fragment and, therefore, the interfacial damage zone is adequately bridged. As a result of this, the effective length required for stress transfer is reduced only marginally and therefore a new fiber fracture event occurs at higher strains (Figure 2.18c).

The average maximum IFSS of the two systems is plotted as a function of applied strain in Figure 2.19. The values of IFSS for each system do not seem to vary considerably with applied strain. Overall, the maximum IFSS values for the sized fiber system are higher than those of the unsized system, at a confidence level of 95%, as determined by the Student t-test [30]. At the point of saturation, the IFSS of the sized and unsized systems are approximately 42 MPa and 35 MPa, respectively. The average value of approximately 42 MPa measured for the sized M40/MY-750 system compares well with the shear yield strength of the resin, which is

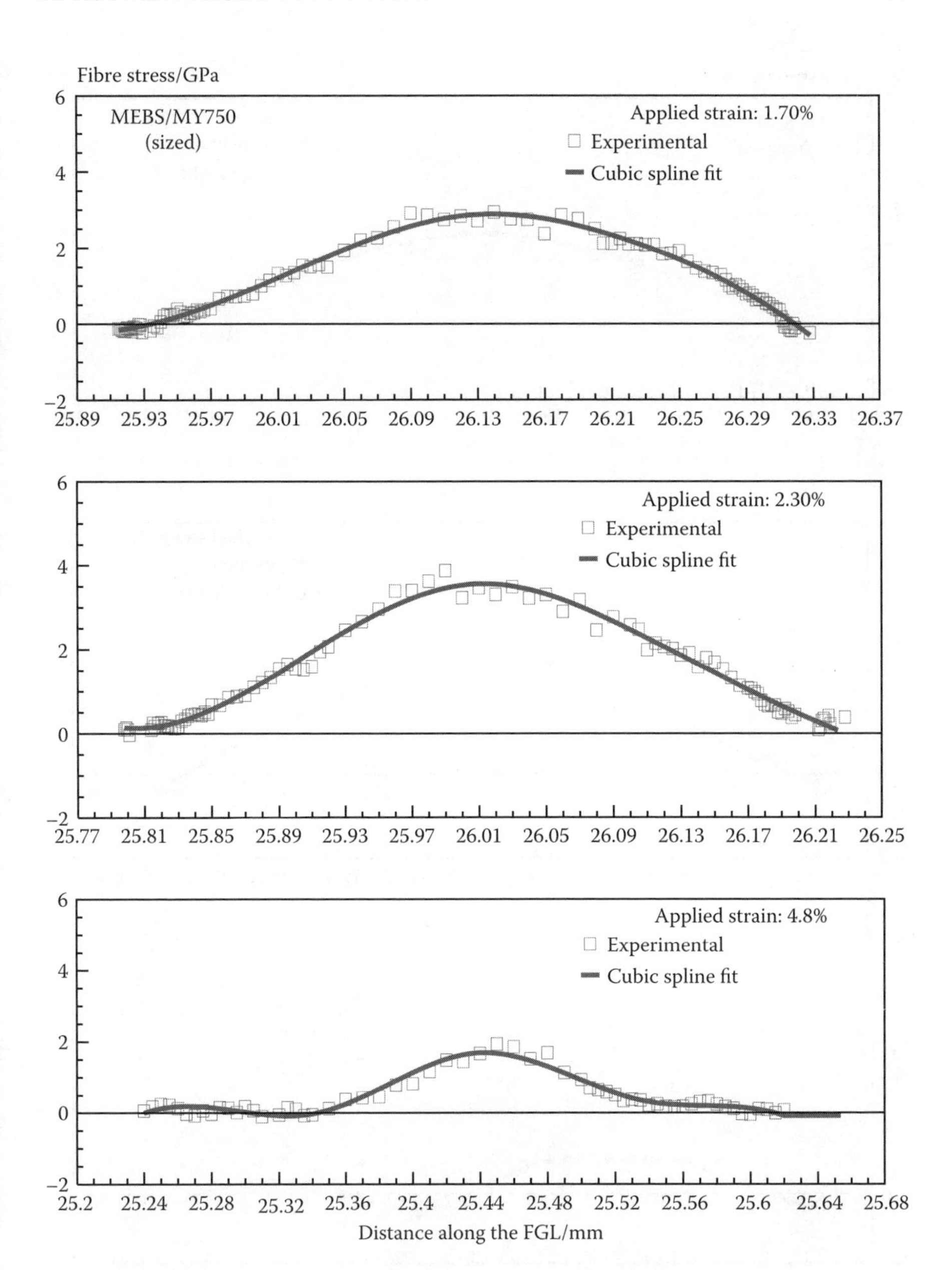

FIGURE 2.17 Fiber stress as a function of position for a representative fragment of a sized M40-3k-40B fiber at (a) 1.7%, (b) 2.3%, and (c) 3.8% applied composite strain.

estimated to be on the order of 40 MPa [31]. SEM evidence points to two distinct modes for interfacial failure for the two systems; whereas in the unsized fiber system clear debonding can be seen at high strains, in the case of the sized system, the plane of interfacial damage appears shifted towards the matrix material in a mixed-mode fashion.

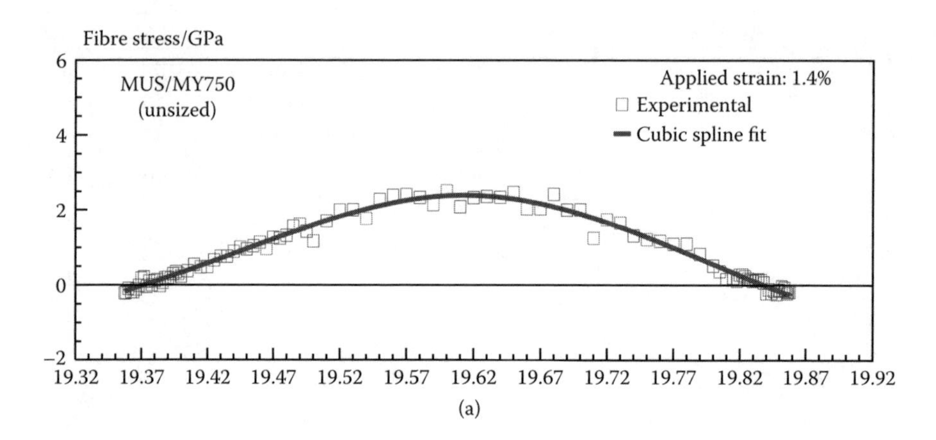

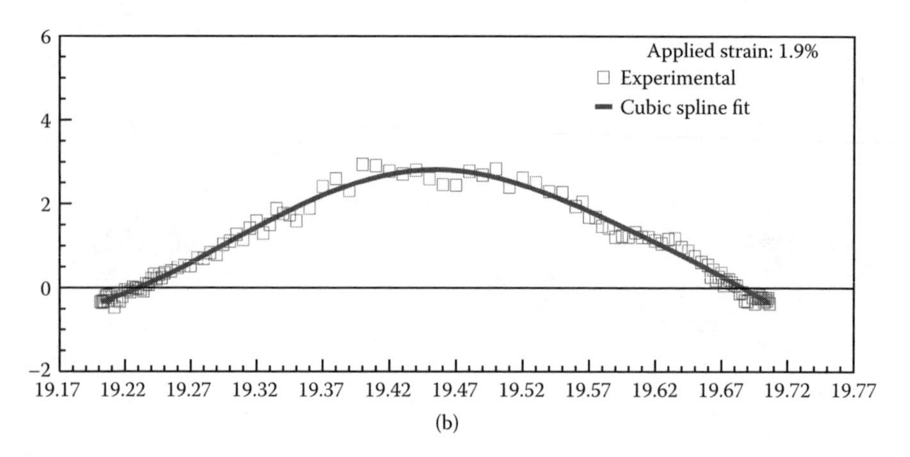

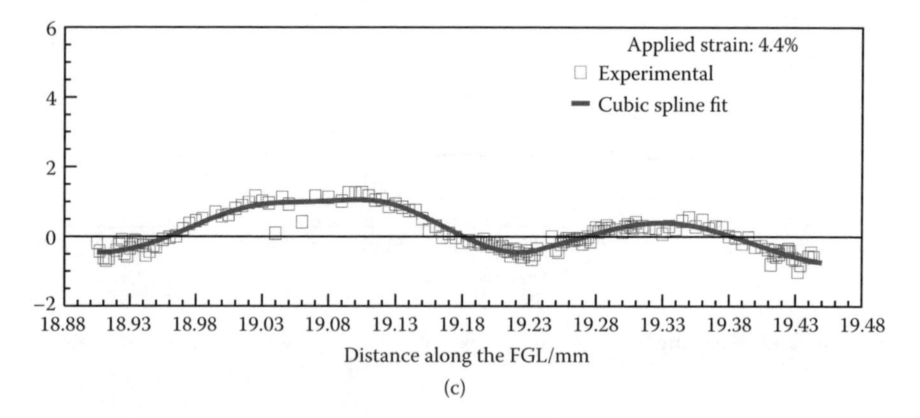

FIGURE 2.18 Fiber stress as a function of position for a representative fragment of an unsized M40-3k-40B fiber at (a) 1.6%, (b) 1.9%, and (c) 3.4% applied composite strain.

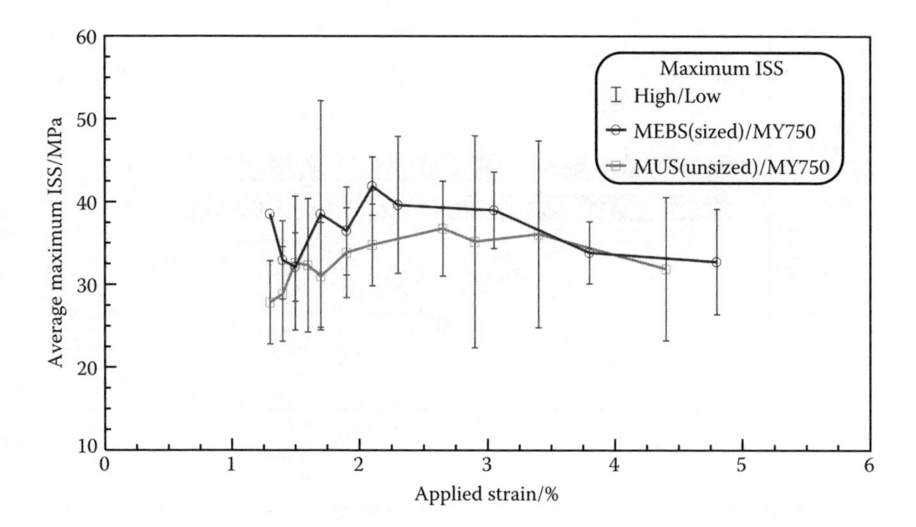

FIGURE 2.19 Average maximum ISS as a function of applied strain for the sized (MEBS) M40/MY-750 and the unsized (MUS) M40/MY-750 systems.

In conclusion, it can be said that the sized fiber system can sustain higher shear stresses at the interface (Figure 2.19), which trigger a mixed mode type of failure. This is clearly undesirable, at least at very high levels of applied strain, since there is no adequate bridging between matrix and fiber, and, therefore, the effective length of the fiber (Figure 2.17) is severely reduced. On the other hand, the relatively weaker interface observed in the unsized system fails by means of fiber/matrix debonding. This does not impair the ability of the interface to transfer stress since the two surfaces are adequately bridged, due to the presence of a compressive radial stress field.

2.5.3 Full Composites

2.5.3.1 Stress Transfer in Composites

When an individual fiber breaks the shear field perturbation, which is generated at the point of fiber failure will cause a stress redistribution in the neighbouring fibers. In the fully elastic case, the shear forces are maximum at the plane of fiber fracture ($\sigma = 0$) and decay to zero at some distance away from the fiber break. As shown schematically in Figure 2.20, the distance over which the axial stress reached its maximum value (positively affected length [PAL]) is identical in magnitude to the transfer or ineffective length of the fractured fiber.

The stress concentration factor K_q for a fiber adjacent to q fractured fibers is defined as:

$$K_q = \frac{\sigma_n^{\xi=0}}{\sigma_{applied}} \quad (2.4)$$

where $\sigma_n^{\xi=0}$ is the stress of the intact fiber at the plane of first fracture ($\sigma = 0$) and σ_{applied} is the far field stress in the fiber. Similar to the procedure for the derivation

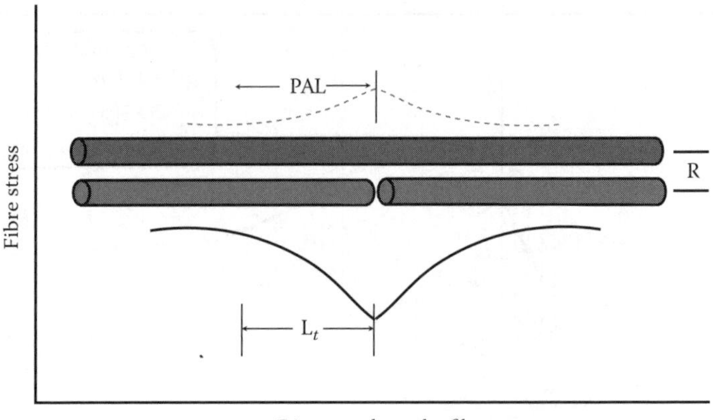

FIGURE 2.20 Schematic representation of the stress distributions in a fractured fiber and its nearest neighbor. The fracture "well" and the zone of positively affected length are clearly shown.

of the ISS mentioned earlier, the fiber stress profiles are fitted with cubic spline polynomials and the maximum stress is defined at the point for which the derivative $d\sigma_f/dx$ becomes zero. In almost all cases presented here, the maximum stress concentration is obtained at a displacement ξ from the plane of fracture, but it is interesting to note that the a kind of "hill" of stress concentration is observed rather than the "spike" predicted analytically. The existence of a "hill" of stress concentration is indicative of the presence of interfacial failure in tandem with the fiber fracture. Finally, it is worth mentioning that the balance-of-forces argument of Equation 2.3 is of general validity and can be also applied to the stress field acting on fibers adjacent to a fiber break. Thus, the interfacial shear stress, and its decay in fibers located at a radial distance R from a given fiber break, can also be derived [32].

2.5.3.2 As-Received Composite Plates and Tows

2.5.3.2.1 Two-Dimensional Microcomposites

The 2D microcomposite tapes typically consist of regular arrays of three to a maximum of seven individual carbon fibers lying on the same plane of uniform interfiber distances. The advantages for investigating these model geometries are twofold: Firstly, for a given fiber/matrix bond strength, the fracture behavior can be monitored as a function of interfiber distance, which can be controlled and varied using special devices [33]. Secondly, the stress concentration in a fiber adjacent to a fiber break can be measured without the presence of fibers lying on planes underneath the plane of observation. In contrast, the interfiber distance in full commercial composites cannot be adequately controlled and local variations near fiber breaks can be significant [34]. Also the presence of fiber breaks lying on planes underneath the plane of observation can affect the overall stress field and, hence, the exact value of the stress concentration factor [34].

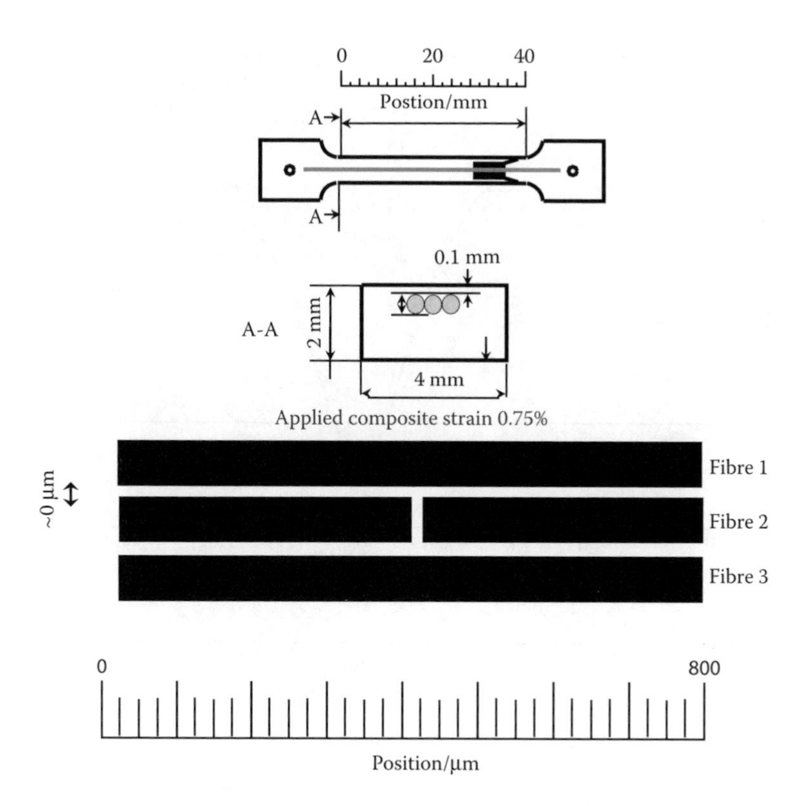

FIGURE 2.21 Schematic illustration of a three-fiber microcomposite specimen and of the corresponding area of stress measurements at 0.75% of applied strain.

In Figure 2.21, a representative three-fiber microcomposite at effectively zero interfiber distance is schematically shown. The fiber stress distribution as a function of position along the length of a fractured fiber (fiber 2, Figure 2.21) and the corresponding interfacial shear stress profile at an applied composite strain of 0.75%, are shown in Figure 2.22. As can be seen, the fiber stress at the fracture point builds from a compressive stress of approximately -0.70 GPa to the far field value of 2.5 GPa at a "transfer length" distance of about 200 µm from the fractured point (Figure 2.22a). The maximum ISS that this particular combination of fiber/matrix can sustain at 0.75% of applied strain was approximately 40 MPa (Figure 2.22b). It is worth noting, however, that the ISS maxima obtained on either end of the fiber fracture were not located on the plane of fracture but at a finite distance of about 40 µm away from it. As reported elsewhere [35], this is indicative of the onset of interfacial failure at the vicinity of a discontinuity such as a fiber break.

The corresponding axial stress and ISS profiles for the two fibers (fibers 1 and 3, Figure 2.21) adjacent to the fractured fiber, are given in Figure 2.23 and Figure 2.24. In both cases, the axial stress builds from a far field value of 2.5 GPa to a maximum value of about 3.3 GPa at the plane of fracture ($\sigma = 0$). For both adjacent fibers, the PAL is identical in magnitude with the transfer or ineffective length of the fractured fiber. A value of stress concentration factor, K_1 ($q = 1$), of 1.36 is estimated for the

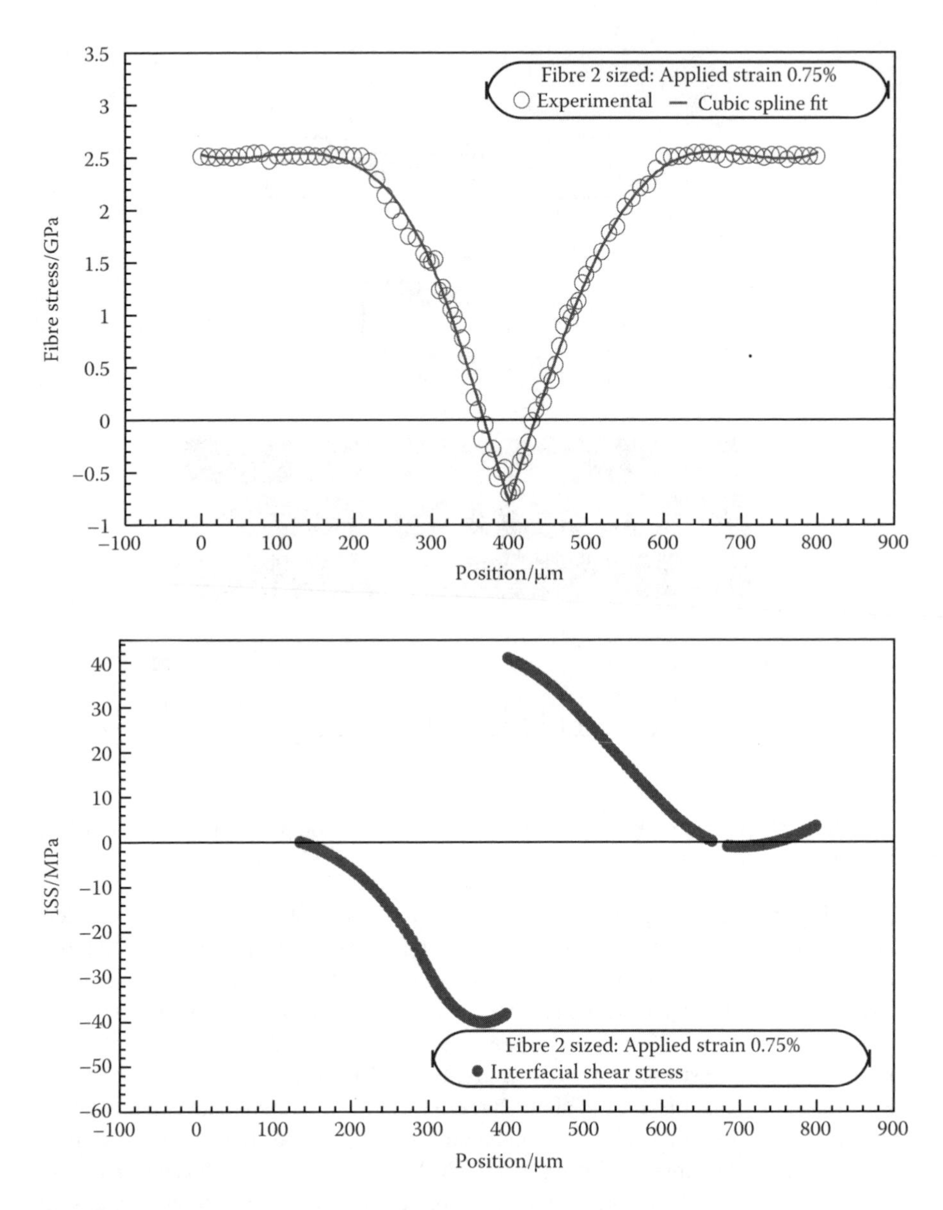

FIGURE 2.22 (a) Stress profile and corresponding cubic spline fit for fiber 2 at 0.75% applied composite strain. (b) Interfacial shear stress (ISS) profile for fiber 2 at 0.75% applied composite strain.

two nearest neighbors. The maximum value of ISS for the two nearest neighbors decreases dramatically to a value of about 8 MPa in spite of the close proximity of the three fibers (Figure 2.21). Finally, the area of interfacial damage of approximately 80 μm observed in the fractured fiber results in the smooth decrease of the ISS distribution in the two nearest neighbors (Figure 2.23b and Figure 2.24b) on either side of the fracture plane ($\sigma = 0$).

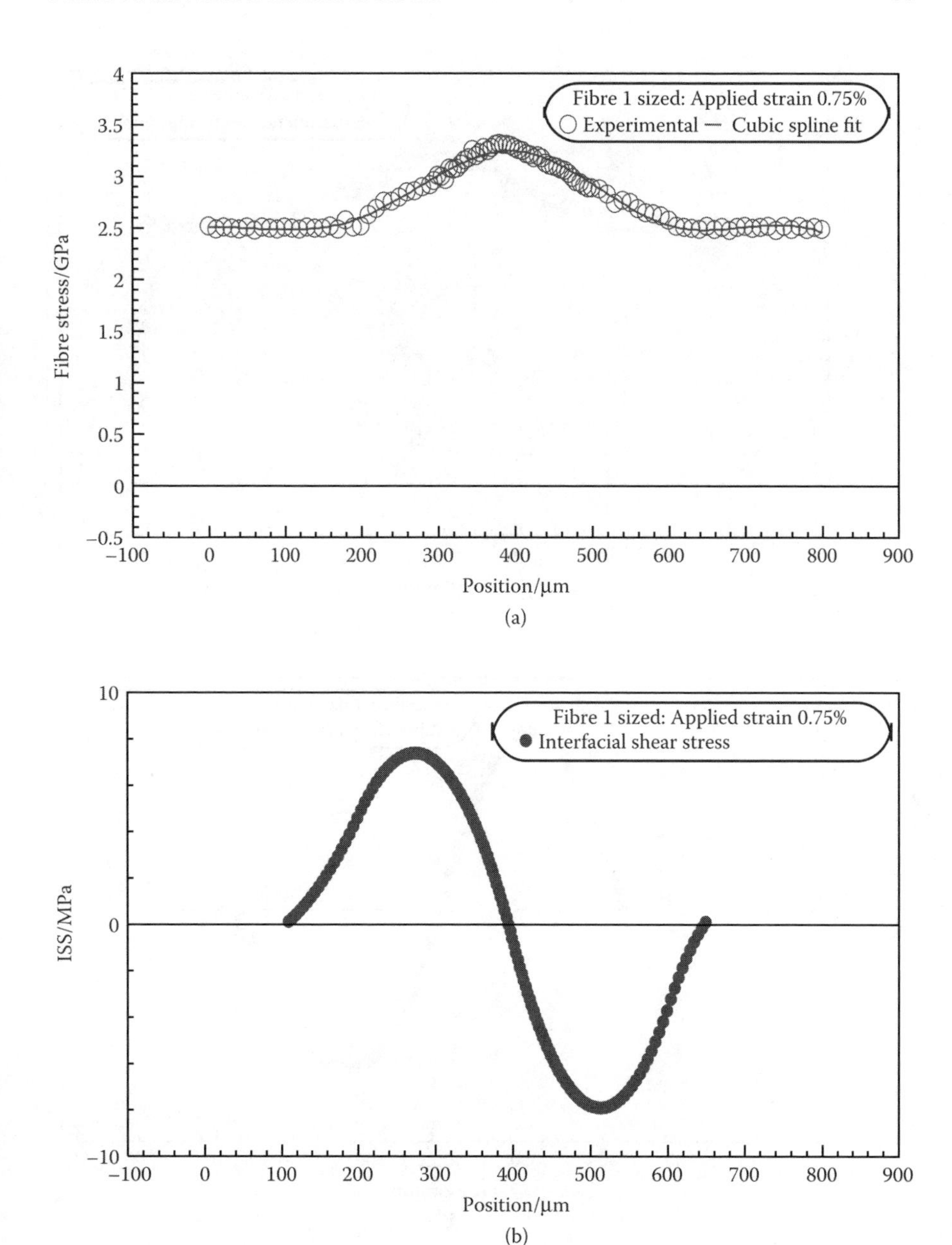

FIGURE 2.23 (a) Stress profile and corresponding cubic spline fit for fiber 1 at 0.75% applied composite strain. (b) Interfacial shear stress (ISS) profile for fiber 1 at 0.75 % applied composite strain.

2.5.3.2.2 Full Unidirectional Coupons

The work on full composites is performed on ASTM-standard unidirectional tensile coupons (Figure 2.25) [34] and fiber tows [36], which are loaded incrementally in

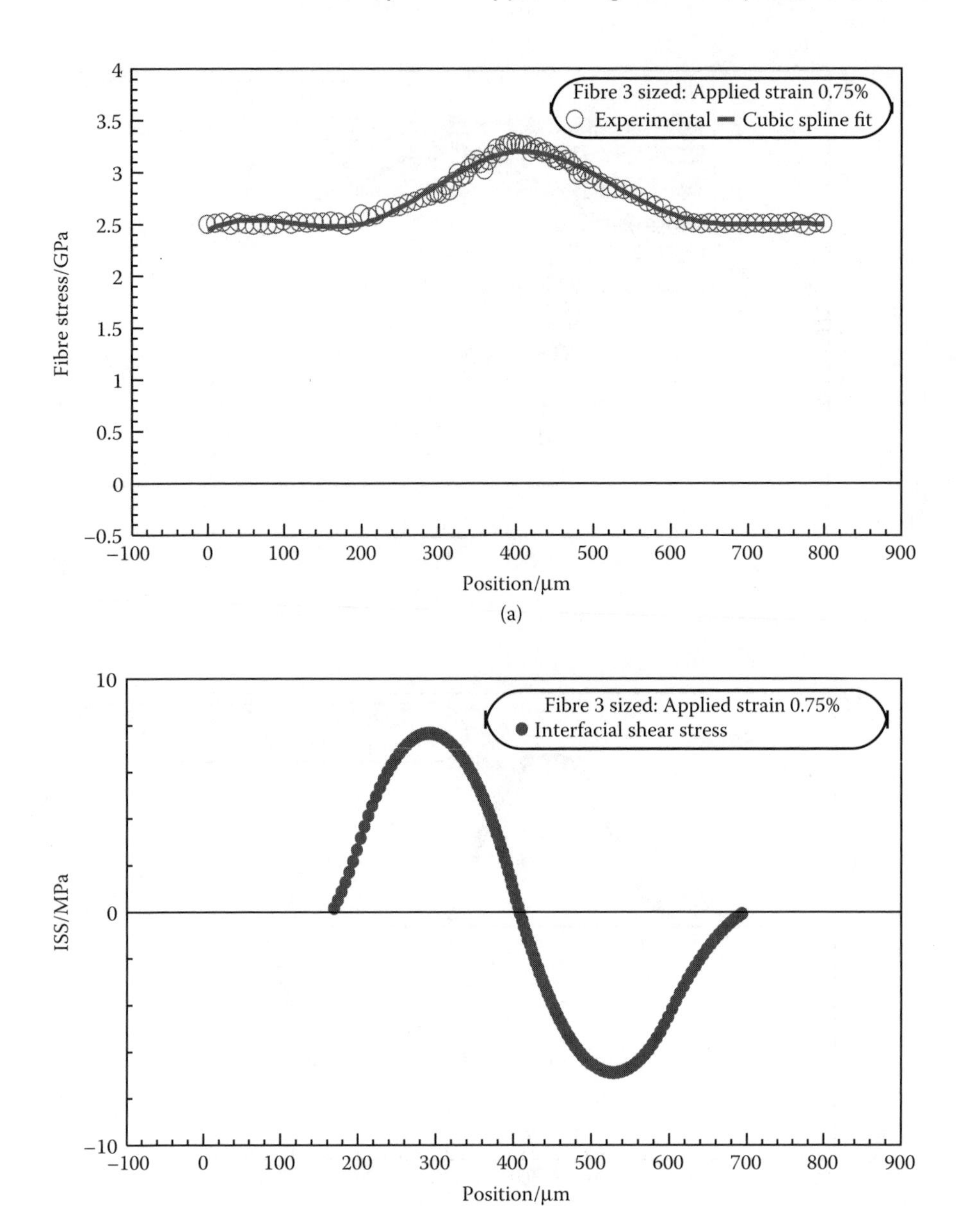

FIGURE 2.24 (a) Stress profile and corresponding cubic spline fit for fiber 3 at 0.75% applied composite strain. (b) Interfacial shear stress (ISS) profile for fiber 3 at 0.75 % applied composite strain.

tension up to fracture. The procedure involves the identification of a single fiber fracture at a certain increment of applied load and the point-by-point measurement of the fiber stress within a "window" of seven fibers (Figure 2.25) around the locus of first fiber failure. A representative axial stress profile along the fractured fiber at

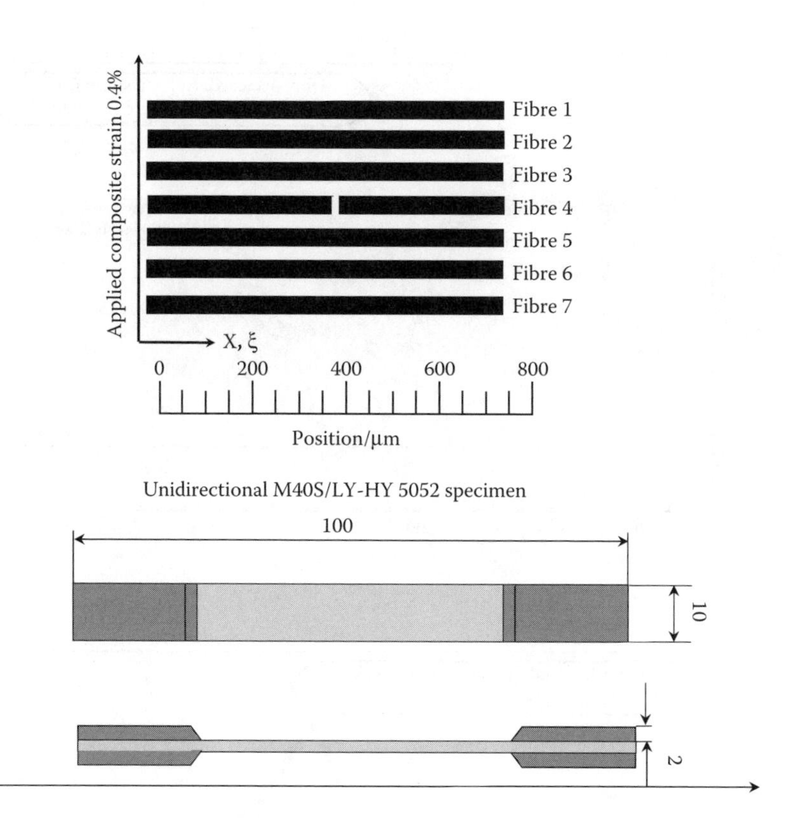

FIGURE 2.25 Schematic illustration of the four-ply unidirectional composite and of the corresponding area of stress measurements. The dimensions of the ASTM standard coupon are given in mm. The interfiber distance of the near-surface fibers was variable.

an applied composite strain of 0.4% is shown in Figure 26. As expected, the fiber stress drops to zero at the fiber fracture and then reaches a maximum on either side of the fiber break. As in the case of the 2D microcomposites, the transfer length on either side of the fracture point is of the order of 200 μm. The maximum ISS developed in the case of the full composite on either side of fiber fracture at 0.4% applied strain, was of the order of 30 MPa. Finally, the observed significant "knee" of the ISS distribution on either side of the fiber fracture indicates that quite considerable interfacial damage was initiated in tandem with the fiber fracture process. The size of this zone as defined by the separation of the two ISS maxima, was of the order of 150 μm (Figure 26).

In Figure 2.27 and Figure 2.28 the fiber stress distributions in the remaining six fibers of the window of measurements are given. The considerable scatter of the data points is not surprising, since in a full composite, the axial stress values along any individual fiber are affected by shear field pertubations present not only in the plane of Raman measurements but also in planes immediately beneath it [34]. As can be seen in Figure 2.27 and Figure 2.28, the stress magnification was particularly evident in fibers 3 and 5, located at interfiber distances of +10.8 and −11.8 μm,

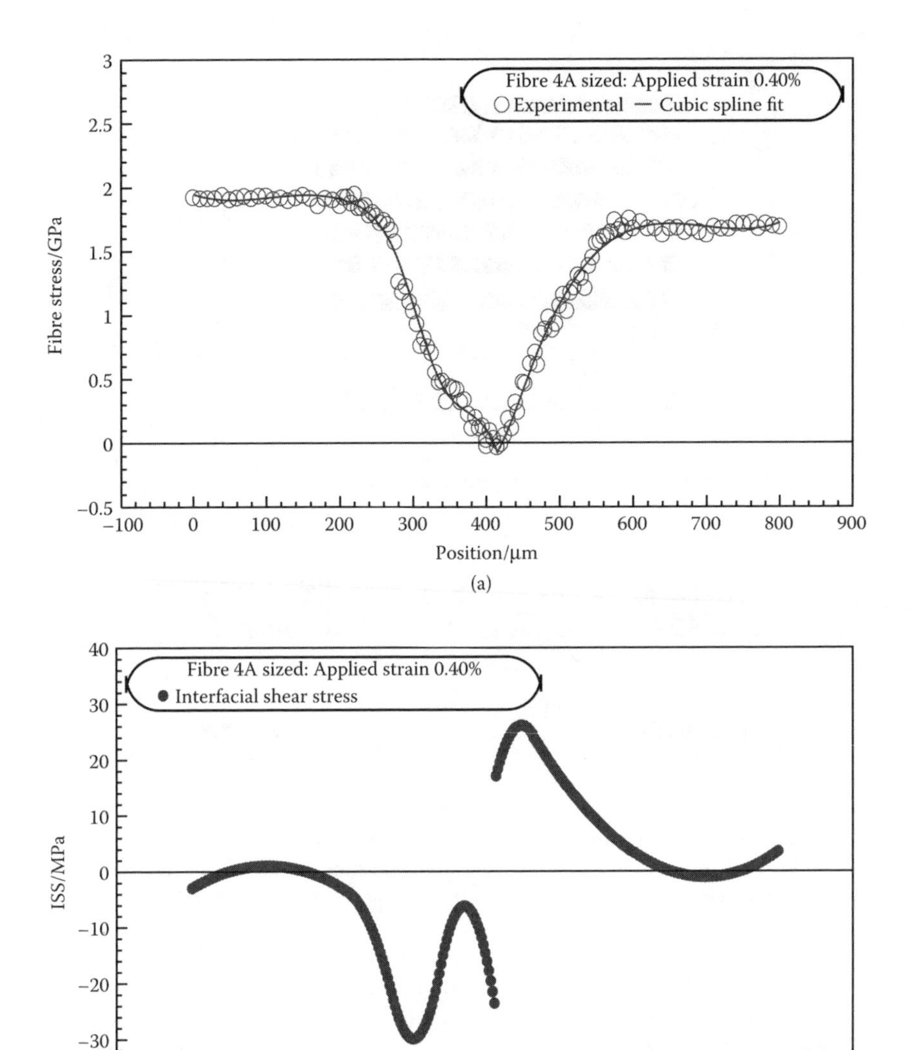

FIGURE 2.26 (a) Stress profile and corresponding cubic spline fit for fiber 4 at 0.40% applied composite strain. (b) Interfacial shear stress (ISS) profile for fiber 4 at 0.40 % applied composite strain for the full composite coupon.

respectively. As expected, the axial stress profile of fiber 3, which was closest to the fractured fiber 4, showed a more intense stress magnification effect. Farther away from the locus of fiber failure the stress concentration clearly diminished (Figure 2.27 and Figure 2.28, fibers 1, 2, 6, 7). Values for stress concentration factor K_q of 1.18 and 1.11 were measured for fibers 3 and 5, respectively.

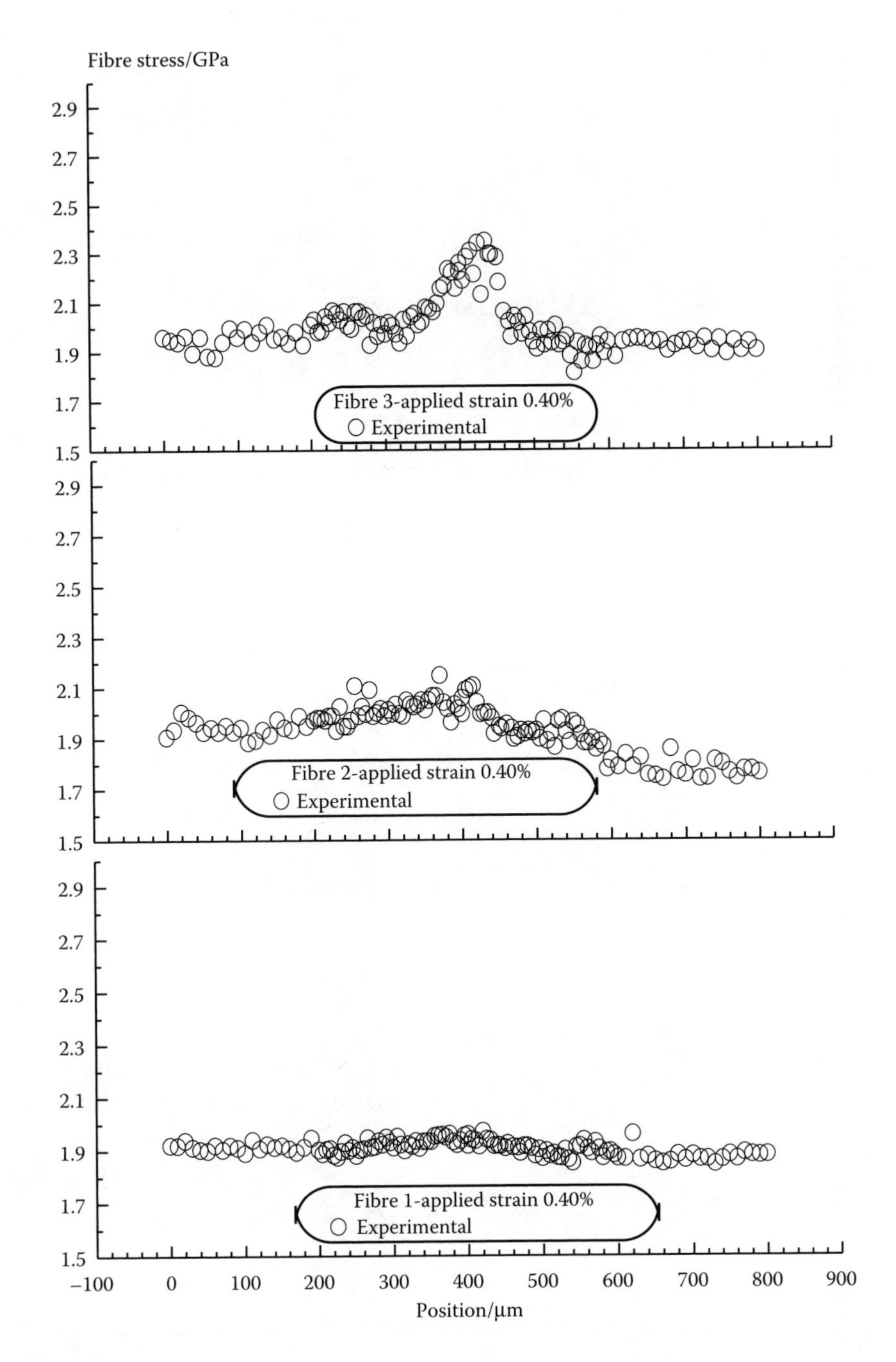

FIGURE 2.27 Stress profiles for fibers 1, 2 and 3 at 0.40% applied composite strain.

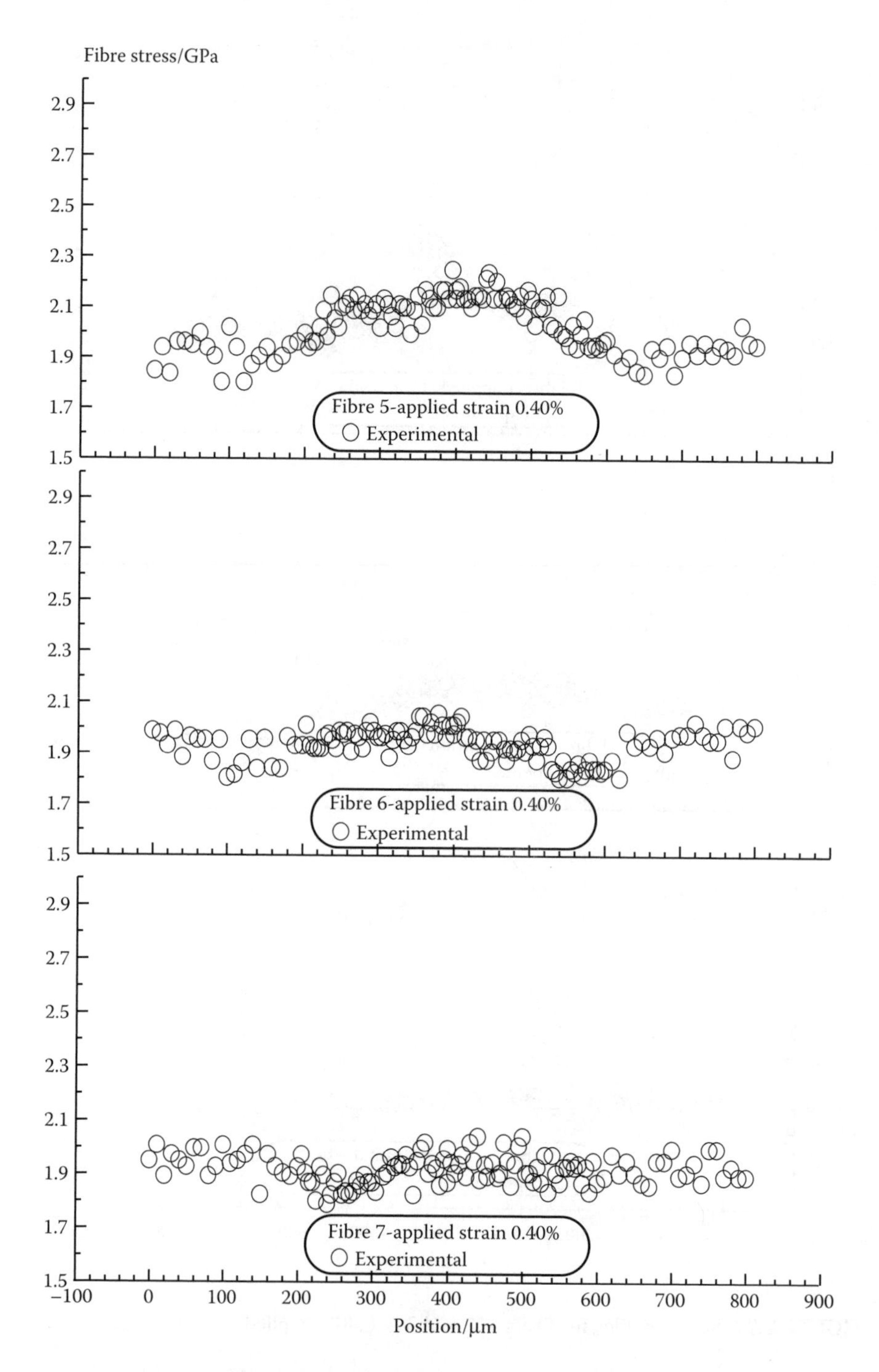

FIGURE 2.28 Stress profiles for fibers 5, 6 and 7 at 0.40% applied composite strain.

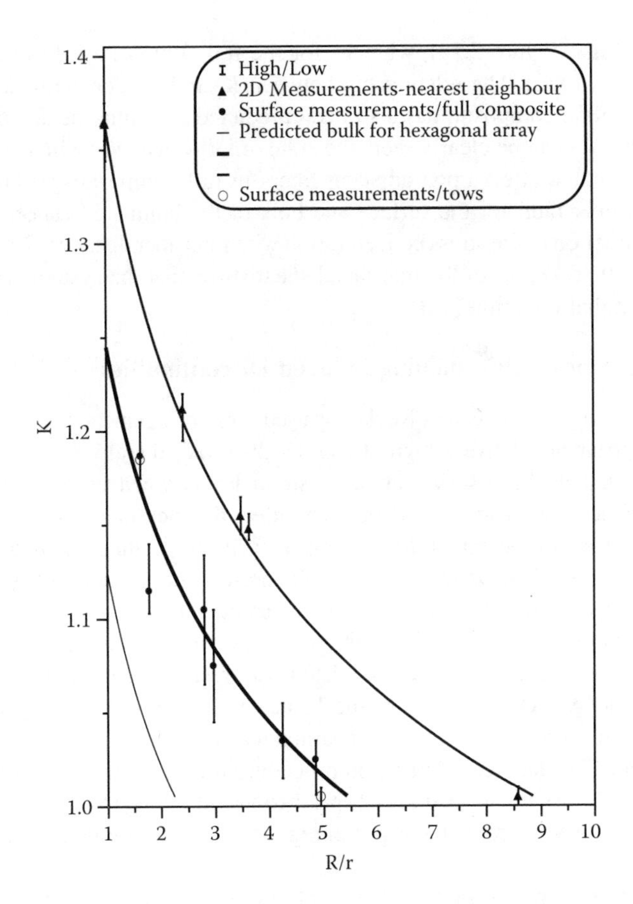

FIGURE 2.29 Graph of stress concentration for a single fracture ($q = 1$) for all geometries as a function of normalized interfiber distance R/r, where R is half the center-to-center distance and r is the radius of the fiber.

The two sets of data for K_1 (single fiber fracture, $q = 1$) as a function of the normalized interfiber distance R/r, where R is the center-to-center distance and r is the radius of the fiber, are presented in Figure 2.29. As can be seen, two distinct families of data are obtained for each respective specimen geometry. The data derived from planar fiber arrays lie higher than those obtained from the surface of full composites. For either set of values the stress concentration value K_1 seems to decay exponentially with R/r in agreement with previous predictions [37]. Regression analysis performed on this set of data [32] has shown that the stress concentration value K_1 relates to R/r via the equation:

$$K_1 = K_1^{r=R} \left[\frac{R}{r} \right]^{-0.14} \tag{2.5}$$

where $K_1^{r=R}$ is the maximum stress concentration at close proximity to the broken fiber. For the case of the 2D array, the value of $K_1^{r=R}$ has been measured experimentally

(Figure 2.23 and Figure 2.24), whereas for the full composite it has been obtained by extrapolation [32]. The relationship between K_1 and R/r for the bulk of the sized M40/LY-HY5052 composite has also been predicted assuming hexagonal geometry (Figure 2.29). As can be clearly seen, the zone of influence over which a broken fiber has no measurable effect upon adjacent fibers in full composites, is approximately 5.5 and 2.5 fiber radii for the surface and bulk of the laminate, respectively. Such a dramatic reduction of the stress concentration with interfiber distance is a consequence of the dramatic decrease of the interfacial shear stress that the system can accommodate in the radial direction [32].

2.5.3.3 Composites Containing Induced Discontinuities

If a relatively tough fiber such as Kevlar® is used for reinforcement, the first fiber breaks normally appear at relatively high strains (> 2%) and, therefore, the stress transfer efficiency can only be assessed at those strain levels when employing the Raman technique. Since nonlinear phenomena may affect the measured properties, there is a clear need for measurements at low strains as well. To do this, a new technique was introduced recently [38,39] whereby a small surgical cut is made by a sharp knife in the upper prepreg layer of either Kevlar® or carbon/epoxy laminates. The resulting discontinuity only affects a small number of fibers and after curing and postcuring in the autoclave the created gap is consolidated with resin material. Such a minute fiber severance is not expected to affect the mechanical properties of the composite laminate.

The technique was demonstrated in laminates incorporating Kevlar® 29 (DuPont) aramid fibers. The detailed preparation procedure for lamination and curing is given in Reference 39. As mentioned above, prior to vacuum bagging a small surgical cut was made on the surface of the top prepreg layer using a surgical scalpe as shown in Figure 2.30.

Initially, the stress transfer profiles at certain levels of applied tensile stress were obtained. Indicative results are presented in Figure 2.31a and Figure 2.32a; the position of the fiber break in these figures is considered to be the origin of

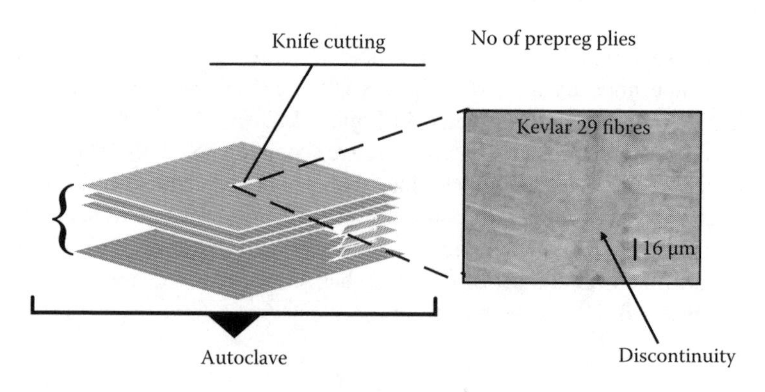

FIGURE 2.30 Schematic representation of the technique used in order to induce a fiber discontinuity into the composite coupon. Indicative microphotograph (after surface polishing) from aramid fiber epoxy resin laminate is also shown.

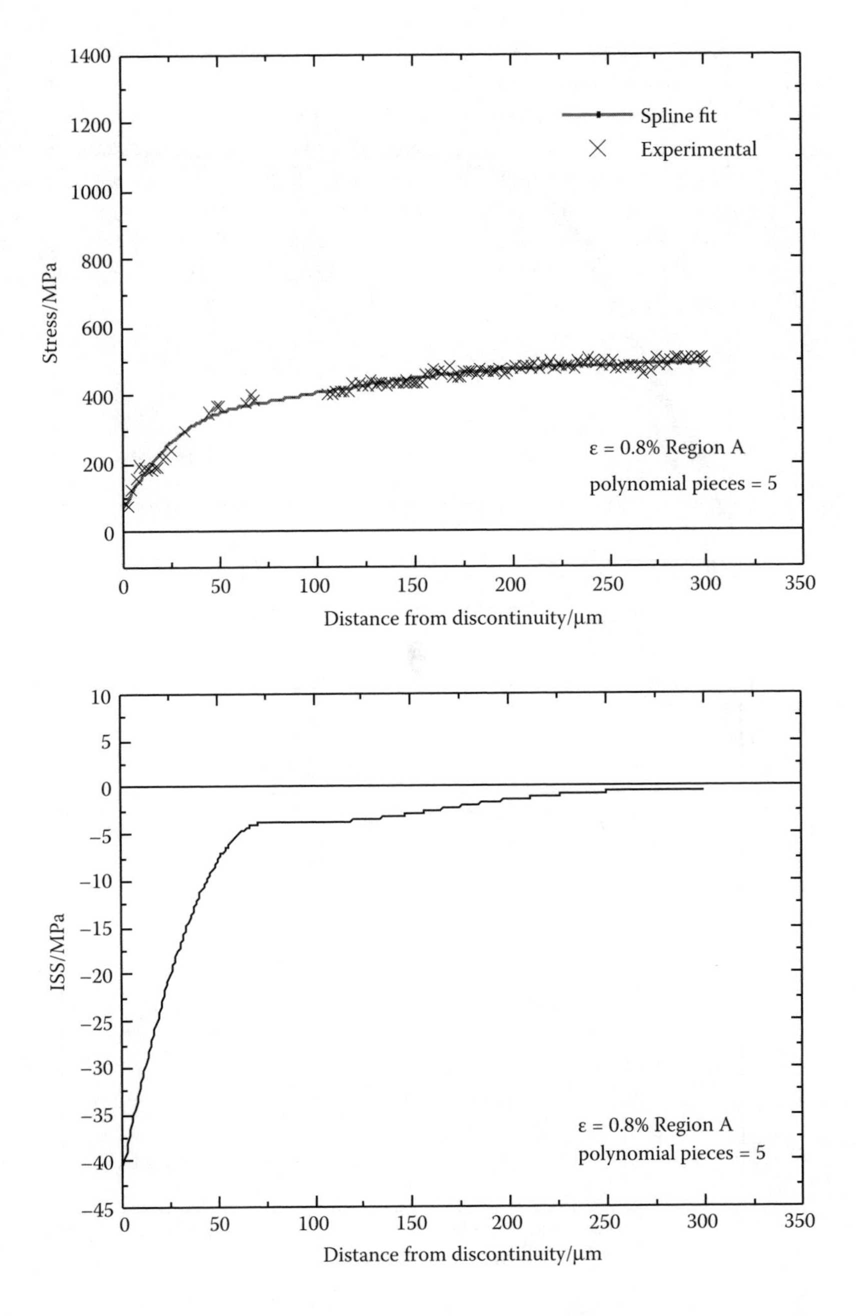

FIGURE 2.31 (a) Stress profile and spline fit of the examined Kevlar® fiber at 0.8% applied composite strain. (b) The corresponding interfacial shear stress profile.

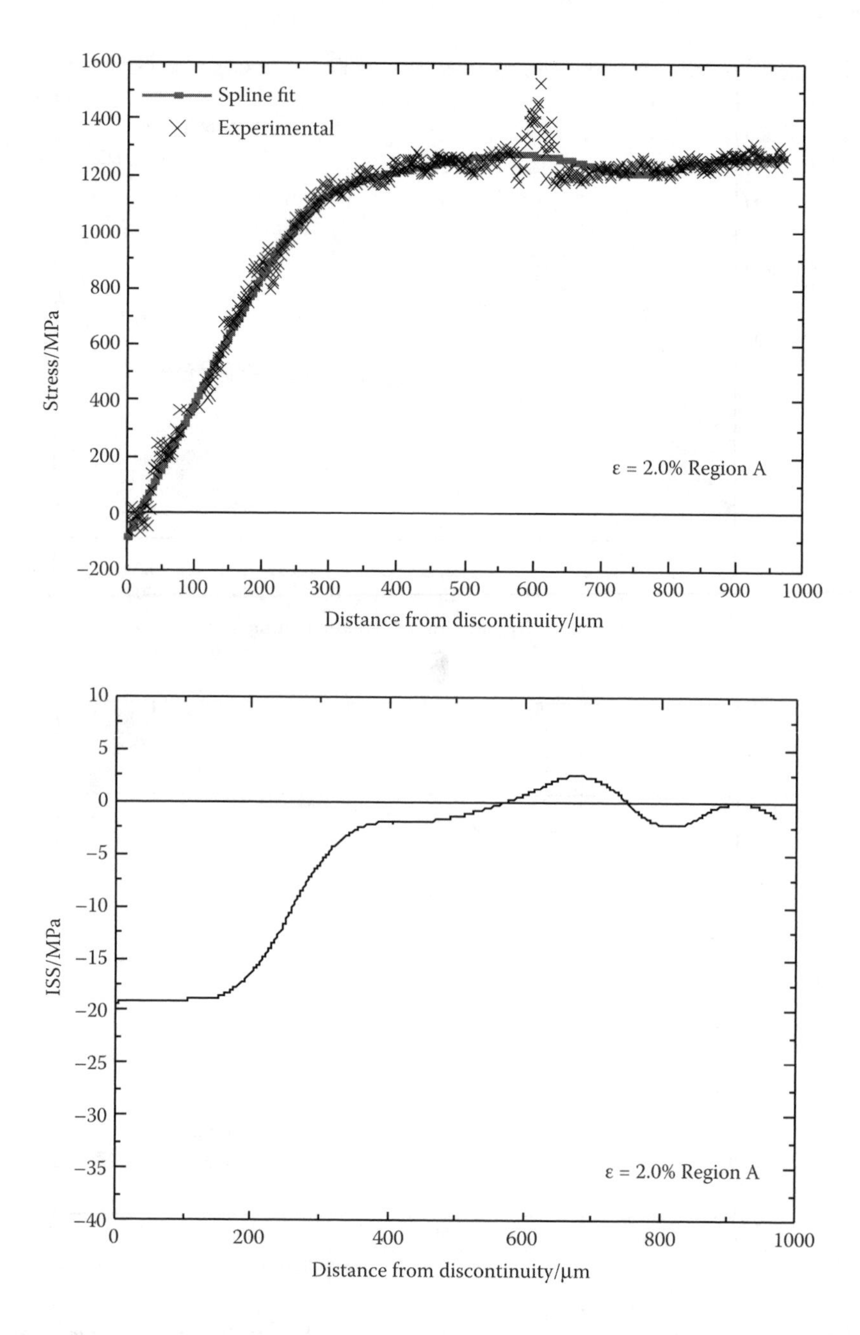

FIGURE 2.32 (a) Stress profile and spline fit of the examined Kevlar® fiber at 2.0% applied composite strain. (b) The corresponding interfacial shear stress profile.

the x-axis. As expected, the stresses build up from the discontinuity and reach a maximum value (far field stress) at a certain distance away from it. At 0.8% of applied strain (Figure 2.31a), the fiber stress builds to a maximum plateau value of 500 MPa at a distance of approximately 250 μm. When the applied strain is increased to 2.0% (Figure 2.32a), the axial stress builds to a maximum plateau value of about 1250 MPa at a distance of approximately 400 μm from the discontinuity.

The resulting ISS profiles at 0.8% and 2.0% of applied strain are shown in Figure 2.31b and Figure 2.32b, respectively. At 0.8% of applied strain, the ISS maximum (ISSmax) of about 40 MPa appears at the fiber discontinuity and decays to zero at a distance of about 250 from it (Figure 2.31b). The corresponding ISS profile at 2.0% of applied strain is markedly different (Figure 2.32b); the ISSmax is about 18 MPa, remains almost constant for 200 μm, and then decays to zero (Figure 2.32b). In Figure 2.33a the resulting ISSmax from both specimens tested are summarized as a function of the applied tensile strain. It is clearly seen that ISSmax increases with strain, reaching a maximum of almost 40 MPa. At a strain level of 1.5%, ISSmax starts to decrease, reaching a plateau of approximately 18 MPa for high values of applied strain.

A parameter that characterizes the integrity of the interface is the transfer length L_t, defined as the distance from the discontinuity where the ISS reaches 1 MPa. In Figure 2.33b the dependence of transfer length on the applied strain is presented. As can be seen, L_t increases with applied strain and reaches a maximum at a strain level of 1.5% beyond which the ISSmax starts decreasing. For higher strain levels the transfer length remains constant at about 600 μm.

The ISS_{max} results presented in Figure 2.33a show that for Kevlar® 29/epoxy systems the stress transfer characteristics exhibit elastic behavior up to an applied strain of 1.4%. The maximum ceiling of ISS_{max} at that level of strain ranges between 40–45 MPa, which is of the same order of magnitude as the shear yield strength of the resin [40]. At higher levels of strain the stress transfer data and associated ISS distributions show that ISS_{max} is not obtained at the tip of the induced break, but a characteristic plateau of ISS_{max} is formed over a certain distance from the induced break. For 2% of applied strain shown in Figure 2.32b, the fiber length over which ISS_{max} is almost constant is of the order of 200 μm. This type of interface failure is typical for aramid-based composites and has also been seen previously in short single-fiber specimens [41]. It has, in fact, been modelled successfully by means of finite element analysis [42], and has been assigned to local matrix yielding emanating from the discontinuity. The presence of matrix plasticity near the fiber end in tandem with the viscoelastic properties of both resin and fibers leads to a drop in the ISS_{max} that the system can sustain at higher levels of strain (Figure 2.33a). Similarly, the values of L_t (Figure 2.33b) increase approximately linearly with applied strain in the elastic region but then considerable fluctuations are observed at higher strains. Since L_t is defined as the length over which the ISS reaches 1 MPa, these fluctuations are attributed to corresponding ISS fluctuations as a result of the presence of interface damage beyond 1.4%.

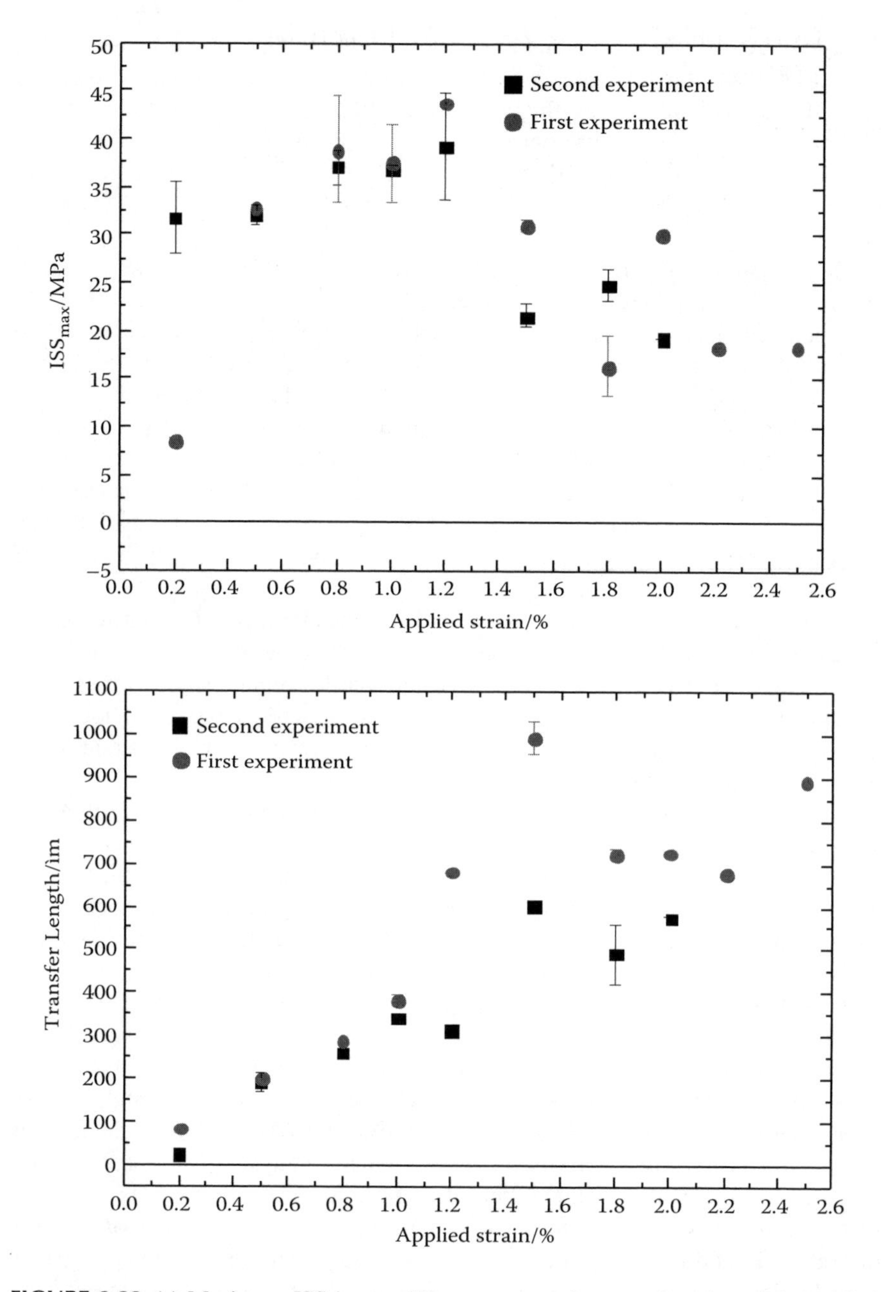

FIGURE 2.33 (a) Maximum ISS in aramid/epoxy composites as a function of the applied tensile strain, (b) Transfer length in aramid/ epoxy composites as a function of the applied tensile strain.

2.6 DETERMINATION OF THERMAL RESIDUAL STRESSES IN COMPOSITES

Inherent to the manufacturing of composites with polymeric matrices are high processing temperatures, which give rise to the development of residual thermal stresses upon cooling and solidification [43,44]. These stresses result from the thermal mismatch between fiber and matrix and at room temperature the fibers are subjected to residual compression, whereas the matrix is in tension [43]. The level of these stresses affects the composite mechanical response and thus has to be determined by an appropriate methodology. The distribution of residual stresses and strains in composites can be derived directly at the embedded fiber level by employing Raman spectroscopy. The advantages of this technique over other methods [45–47] are twofold: (1) the stress or strain can be measured directly in the embedded fibers at steps as small as 2 μm and, therefore, a detailed point-to-point distribution can be obtained; and (2) there is no need to resort to analytical models (i.e., laminate theory) in order to predict the macroscopic residual stresses of the laminate [48–50].

2.6.1 METHODOLOGY

Measurement of the residual thermal stresses with laser Raman microscopy involves the detailed mapping of fiber stresses within an area of variable size (e.g., 100 mm^2) of the composite specimen along and across the fiber direction. The fiber stress mapping was performed at some distance from the coupon ends in order to avoid "end effects." Raman spectra were taken from 200 points within the predefined scanned region at steps of approximately one fiber diameter. The Raman wavenumber distribution of the free-standing fibers in air is obtained by collecting 100 Raman spectra from random positions within a batch of fibers. Since the errors encountered at this stage are nonbiased, the results are normally expressed with a Gaussian function. The same procedure is repeated for the embedded fibers and the results are also fitted at a first approximation with Gaussian functions.

The Gaussian (normal) frequency distribution function *y(x)* can be expressed by [51]:

$$y(x) = \frac{Ne^{-\frac{(x-\mu)^2}{2s^2}}}{s\sqrt{2\pi}} \tag{2.6}$$

where N is the number of measurements, μ is the arithmetic mean value, and s is the standard deviation. Equation 2.6 can be transformed to a normalized Gaussian distribution function, where the probability (density) function *P(x)* is written as:

$$P(x) = \frac{e^{-\frac{(x-\mu)^2}{2s^2}}}{s\sqrt{2\pi}} \tag{2.7}$$

It is evident that the sum of all probabilities is equal to unity and hence the area enclosed by the curve of Equation 2.7 should also be equal to one. In order to derive the net Raman wavenumber shift due to curing and post-curing processes, the resulting distributions of fibers in air and of fibers embedded in the composite must be statistically subtracted [22]. The probability density function of the difference of two sets of measurements described by Gausssian distributions is given by [51]:

$$P(z) = \frac{e^{-\frac{(z-\mu)^2}{2\left(s_1^2+s_2^2\right)}}}{\sqrt{2\pi\left(s_1^2+s_2^2\right)^{1/2}}} \tag{2.8}$$

where $z = x_2 - x_1$ is the difference of the two variables; $\mu = \bar{x}_2 - \bar{x}_1$ is the difference of the corresponding arithmetic means, with $\bar{x}_1$ being the smaller value; and s_1, and s_2 are the standard deviations of the initial distributions. In the final step the resulting Raman shift distribution is converted into stress through the predetermined Raman stress sensitivity of the corresponding stress sensitive Raman peak. In the case of carbon fibers the stress sensitive peak is the 1580 cm^{-1} (G line), while for aramid fibers the peak is 1611 cm^{-1}. The whole procedure for the calculation of the residual stress distribution is given schematically in Figure 2.34.

2.6.2 Residual Thermal Stress in Thermosetting and Thermoplastic Composites

2.6.2.1 Thermoplastic Composites

The previously described methodology was applied for the determination of the residual strain distributions in thermoplastic composites. As an example, carbon fiber (P75)/poly(ether ether ketone) (PEEK) single- and eight-ply composites were employed. As explained above, the first step required in order to obtain the residual strain distributions is the subtraction of the Raman wavenumber distribution of the free-standing fibers in air from the corresponding distributions of the embedded fibers. The resulting distributions are also Gaussian, with mean values of 0.09 cm^{-1} and 0.16 cm^{-1} for the prepreg and the eight-ply composite, respectively. Their corresponding standard deviations were 0.977 cm^{-1} for the prepreg and 0.997 cm^{-1} for the eight-ply composite. The resulting Raman wavenumber shift distributions were multiplied by the gauge factor of –22.40 cm^{-1}/%, which corresponds to the strain sensitivity of a second-order Raman band at 2705 cm^{-1} for carbon fibers. Thus, they are converted to the residual strain distributions, shown in Figure 2.35a. These distributions are also Gaussian, with mean values of –0.004% and –0.007% and standard deviations of 0.033% and 0.034% for the single and the eight-ply composites, respectively (Figure 2.35b). The residual thermal stresses in the P75/PEEK composite are expected to be compressive for the fibers and tensile for the matrix. The experimentally derived mean values of longitudinal residual thermal strains in the embedded fibers of –0.004 % (prepreg) and –0.007 % (eight-ply) compare well [48] with analytically computed values [52] based on the mismatch of the thermal expansion coefficients of fibers and matrix. Previous work [48] has verified that the observed broadening of the Raman wavenumber distributions of the embedded fibers

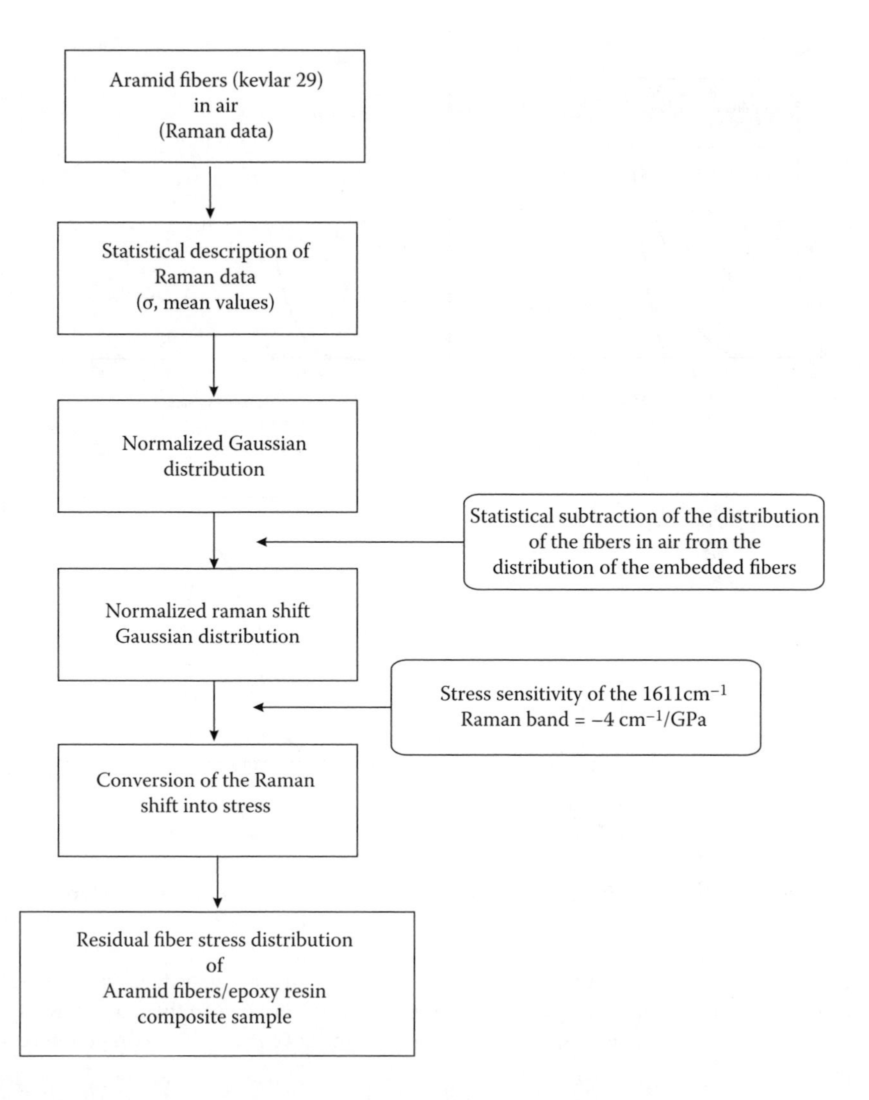

FIGURE 2.34 Flowchart presenting schematically the procedure used to calculate residual thermal stresses and strains in aramid or carbon fiber composites.

and, hence, of the resulting residual strain Gaussian curves (Figure 2.35b), do not stem from errors in defining the fiber Raman band position due to the presence of an interfering resin layer of variable thickness, but it represents an inherent property of the composite.

2.6.2.2 Thermosetting Composites

Similar methodology [50] is used for the determination of the residual stresses in thermosetting composites. For example, the results from the four-ply Kevlar®

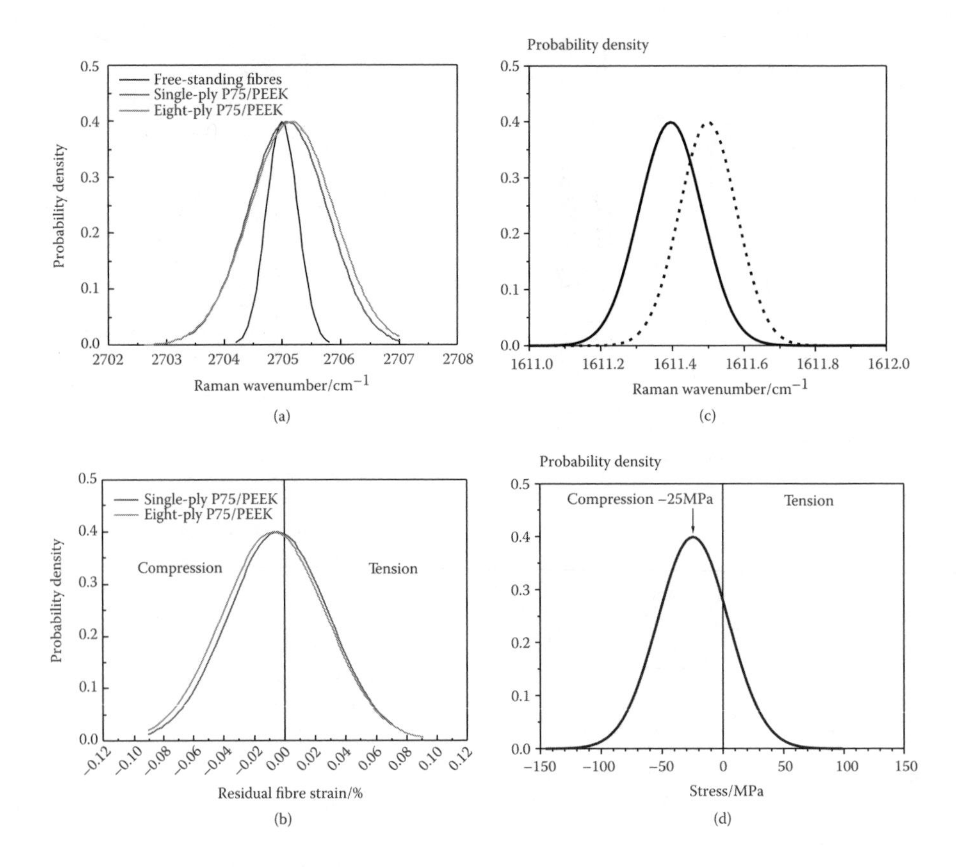

FIGURE 2.35 Gaussian frequency distributions of (a) Raman wavenumber shift of P75 carbon fibers in air and embedded in the single and eight-ply composite coupons; (b) residual fiber strain in the P75 carbon fibers of single and eight-ply composite coupons; (c) Raman wavenumber shift of Kevlar® 29 fibers in air and embedded in a four-ply composite coupon; and (d) residual fiber stress in Kevlar® 29 fibers in the four-ply Kevlar® 29/epoxy resin composite coupon. In all cases the quantity $P(x) - s\sqrt{2\pi}$ is represented in the graphs.

29/epoxy resin (LTM217, ACG) composite specimen at applied strains of 0.4%, 0.7%, 0.9%, 1.3%, and 1.7% are presented in Figure 2.35c and Figure 2.35d. Comparing the normalized Gaussian distribution of the Raman response of free-standing aramid fibers with that obtained from the embedded Kevlar® fibers (Figure 2.35c and Figure 2.35d), it becomes apparent that there is a relative shift of the 1611 cm^{-1} wavenumber band to higher values. This behavior clearly implies that the aramid fibers in the composites are subjected to compression as a result of the curing and postcuring procedures. The derived mean values of the thermal stresses for an aramid fiber/epoxy resin composite specimen is of the order of –25 MPa (Figure 2.35d).

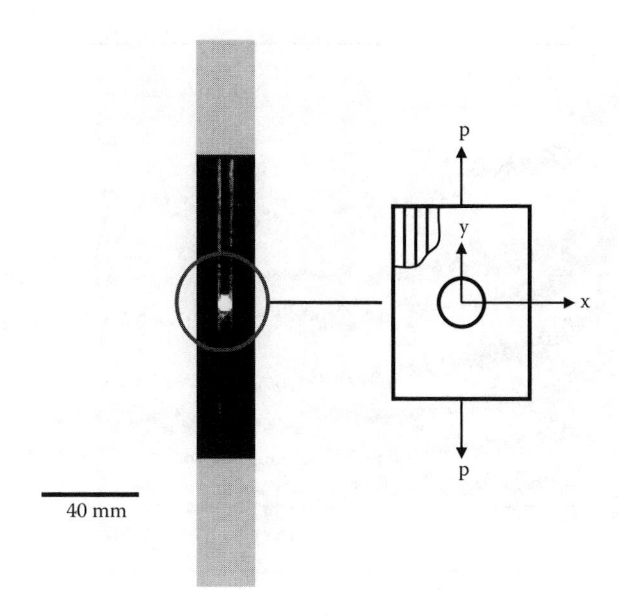

FIGURE 2.36 Photograph showing a typical fracture of a notched standard composite Kevlar® 49/epoxy tensile coupon and a sketch of the imposed loading.

2.7 STRESS-CONCENTRATION MEASUREMENTS IN NOTCHED LAMINATES

Discontinuities such as joints are required for assembling a structure and imperfections such as voids, cracks, and cutouts are unavoidable when working with composites. Knowledge of the magnitude and extent of the stress concentration in these structures and materials is crucial as it assists in the prediction of the location of first failure and the subsequent damage propagation. The major difficulty in calculating the state of stress near a circular notch or a crack tip is the presence of a singularity of $1/\sqrt{r}$, where r is the distance from the notch boundary [53]. Laser Raman microscopy has been employed to determine the exact fiber strain distribution at the edge of a circular notch in a Kevlar® 49/epoxy unidirectional (0°) composite laminate (Figure 2.36). The elastic properties of the employed composite laminate are summarized in Table 2.3.

TABLE 2.3
Elastic Properties of the Kevlar® 49/914k-49-54.8% Composite Laminate

E_{xx}/GPa	E_{yy}/GPa	G_{xy}/GPa	xy
80.3	5.5	2.3	0.34

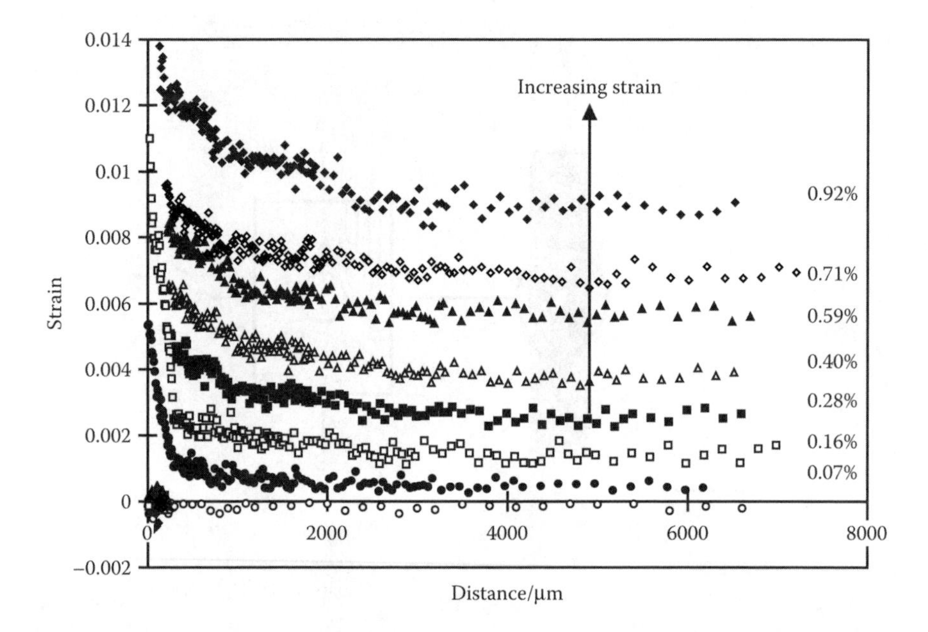

FIGURE 2.37 Strain distribution as a function of distance from the notch boundary, at the directions 90° and 270°, to the fiber axis. Each graph denotes a different applied strain level.

The strain distribution profiles for different applied strain levels are shown in Figure 2.37. Measurements at the stress-free state suggested a residual thermal strain in the fibers of about –0.02% ± 0.01%. The strain magnifications at each level of applied strain were calculated as follows:

$$\text{Strain magnification} = \frac{\text{Maximum strain at the crack plane} + \text{Residual strain}}{\text{Average far} - \text{field strain measurement} + \text{Residual strain}}$$

The average far-field strain measurement is the arithmetic mean of the strain measurements obtained using ReRaM in the Kevlar® 49 fiber, away from the hole boundary. To obtain the values of strain magnification, a number of Raman spectra are taken along the direction of maximum stress concentration at 90° or 270° to the loading (fiber) axis, first at the stress-free state and then at discrete levels of applied strain (inclusive of the initial residual strain) of 0.07%, 0.16%, 0.27%, 0.40%, 0.59%, 0.71%, and 0.91%. The fiber strains are mapped at steps as small as 10 μm starting from the vicinity of the hole and terminating at the edge of the coupon. The first fiber fracture at the notch boundary is observed at 0.27% applied strain. A typical fiber fracture at the notch's boundary is depicted in Figure 2.38. At 0.59% applied strain, matrix cracks began to appear along the fiber direction on either side of the hole. At 0.71% applied strain, the cracks were fully developed but the specimen could still bear the applied load. Finally, at 0.92% applied strain the specimen fractured.

The plot of strain magnification at the hole boundary versus far-field strain (Figure 2.39) indicates that the strain magnification is at its highest at 0.07% applied

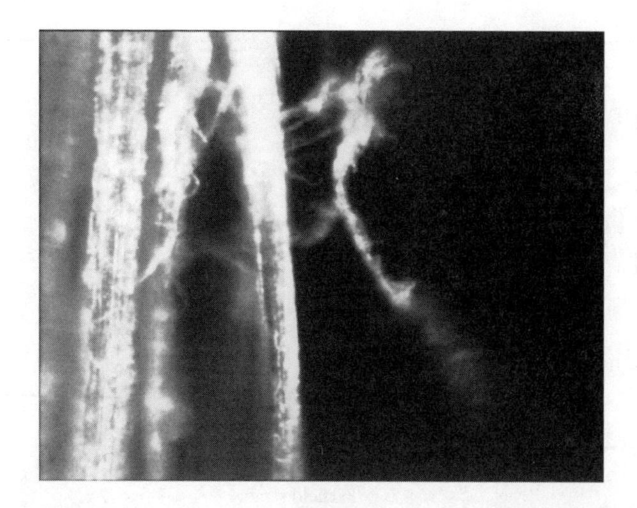

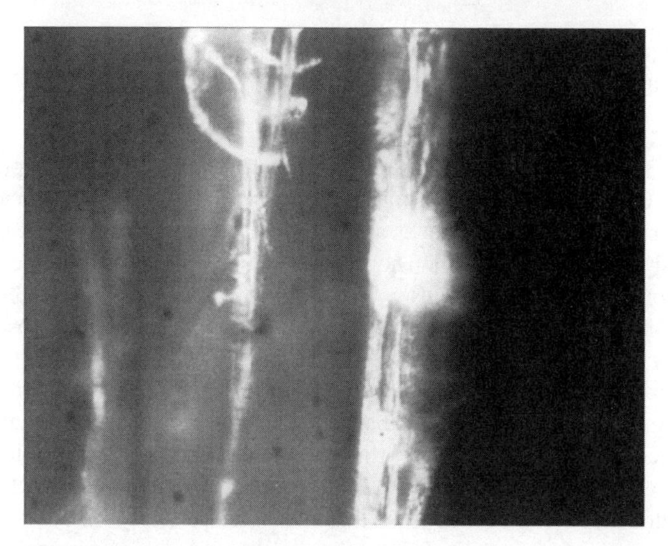

FIGURE 2.38 Micrographs showing fiber breakage at the notch boundary at 0.16% applied strain.

strain, where a magnification factor of 8.06 was obtained. The strain magnification factor decayed to 7.08 at 0.16% applied strain, and 1.69 at 0.40% applied strain. For applied strains higher than 0.40%, the strain magnification factor at the notch boundary decayed to approximately 1.4–1.5. The error bars of Figure 2.39 correspond to the sum of the standard deviations of the experimental values that have been employed to yield the strain magnification (see above) at each level of far-field strain; i.e., the strain at the notch boundary and the average value of the far field strain as measured by laser Raman probe at distances greater than 2000 μm. The magnitude of the error in estimating the strain magnification is governed by the variability of the far-field fiber strain which, in percentage terms, is at its highest at low values of applied strain

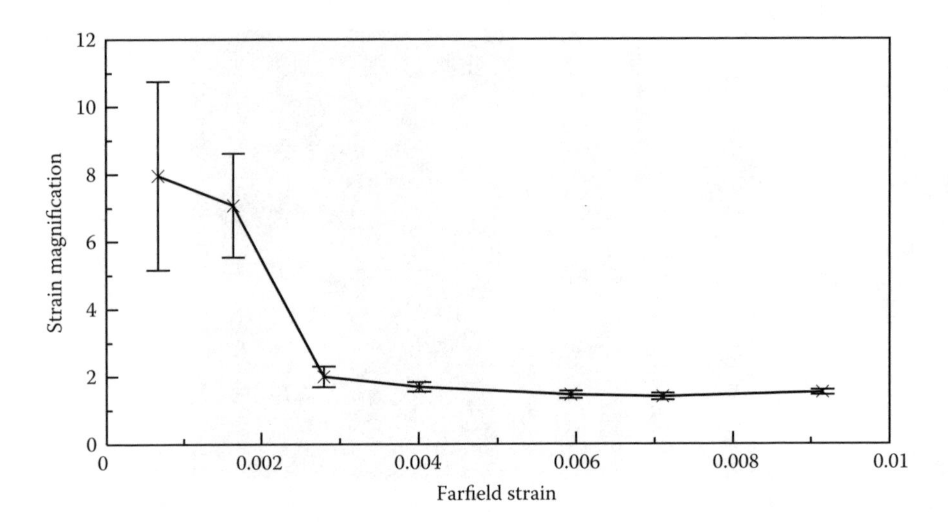

FIGURE 2.39 Strain magnification factor as a function of the far-field strain value at the notch boundary.

(Figure 2.37 and Figure 2.39). The thermal strain on the fibers of approximately –0.02% ± 0.01% is due to the mismatch of the thermal expansion coefficients of the fiber and matrix and comes about as a result of curing and postcuring the composites at elevated temperatures [48,49]. The fracture of fibers located at the vicinity of the opening for applied tensile strains ≥ 0.16% (Figure 2.37 and Figure 2.39) confirms that there should be a significant reduction of the apparent value of composite strength induced by the stress concentration even at relatively low applied strains.

The experimental data are compared with an exact analytical stress solution employing a complex variable method and fundamental concepts of elasticity in anisotropic media [54–57] for an orthotropic plate with a circular hole subjected to a remotely applied load. Finally, a finite element based on the MSC.NASTRAN (a computer aided engineering tool that manufacturers rely on for their critical engineering computing needs) code is also employed for the numerical calculation of the stress concentration factor. Figure 2.40 shows a comparison of the experimental stress distribution near the hole vis-à-vis the analytical stress solution and the best fit of the finite elements (FE) results for an applied strain of 0.07%. The applied strain is low enough to ensure that the strain magnification is equal to the stress concentration factor [58]. The experimental Raman measurements yield a maximum K_c^{exp} of 8.06 (Figure 2.37 and Figure 2.39) whereas a value of $K_c^{an} = 7.46$ is obtained with the analytical solution (after multiplying it by the FWC factor). The corresponding value for parabolic triangle elements of the FE approximation is $K_c^{num} = 7.06$. Both the experimental measurements and the analytical results suggest that at distances greater than about two times the radius of the hole the stress concentration diminishes (Figure 2.40) whereas the numerical treatment presents an upper bound to the data with the stress concentration diminishing at a normalized distance greater than 3. The experimental data at low strains, however, have a large

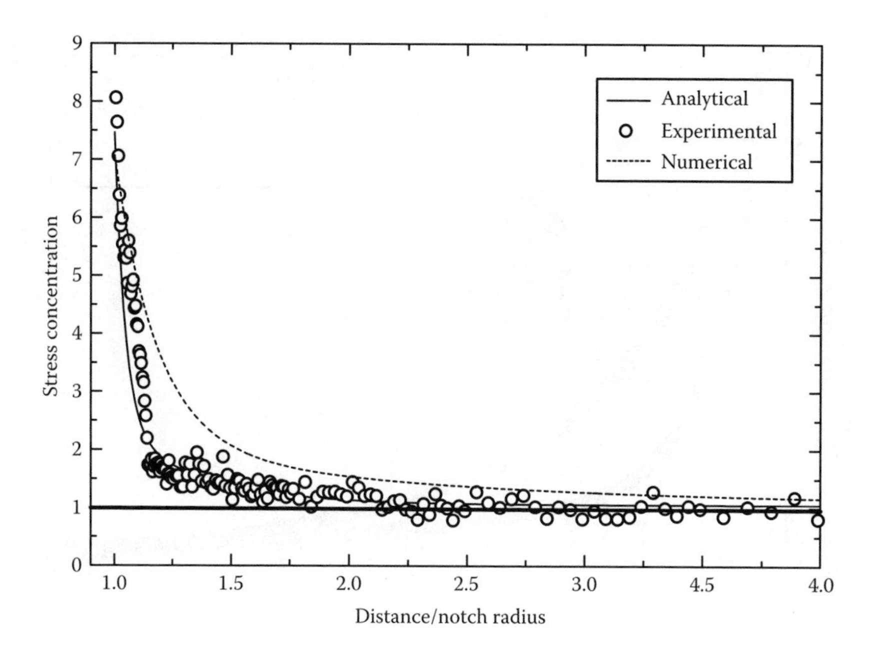

FIGURE 2.40 Analytical and numerical calculations plotted against the Raman experimental data.

error bar associated with them (Figure 2.39), which is related to the scatter of the far field strain data and the resolution of the technique itself. Hence, it is reasonable to assume that both the analytical and the numerical methods predict quite well the maximum value of stress concentration at the edge of the circular notch. However, the power of the experimental technique is its ability to provide values of fiber strain (or stress) and, therefore, stress concentration values at any level of applied strain. The analytical or numerical treatments represent elastic approximations and are therefore unable to predict the stress concentration in a damaged specimen. The strain magnification diminishes as the applied strain increases. This is due to damage propagation, mainly fiber fracture and matrix microcracking, as a result of the very high values of stress concentration that the material is subjected to near the notch. The induced damage brings about a redistribution of stresses and strains and a corresponding reduction of the peak stresses observed (Figure 2.39). Evidently, this can be classified as a toughening mechanism and it leads to the arrest of the damage propagation as shown in Figure 2.37. Hence, the coupon does not fail in Mode I but it finally fails in shear as indicated in Figure 2.34.

As mentioned in a previous section, the finite element results appear to be sensitive to the geometry of the element employed for the FE mesh. The comparative results shown in Figure 2.41 indicate clearly that the parabolic triangle elements provide a much better fit to the experimental and analytical data in the elastic regime. This is explained by the fact that in these elements the parabolic equations, which are used as shape functions, exhibit smaller error than the linear ones that are used for the three-node triangles. Also the use of triangles instead of quadrants leads to greater accuracy, because the points where the Gauss integration is taking place are

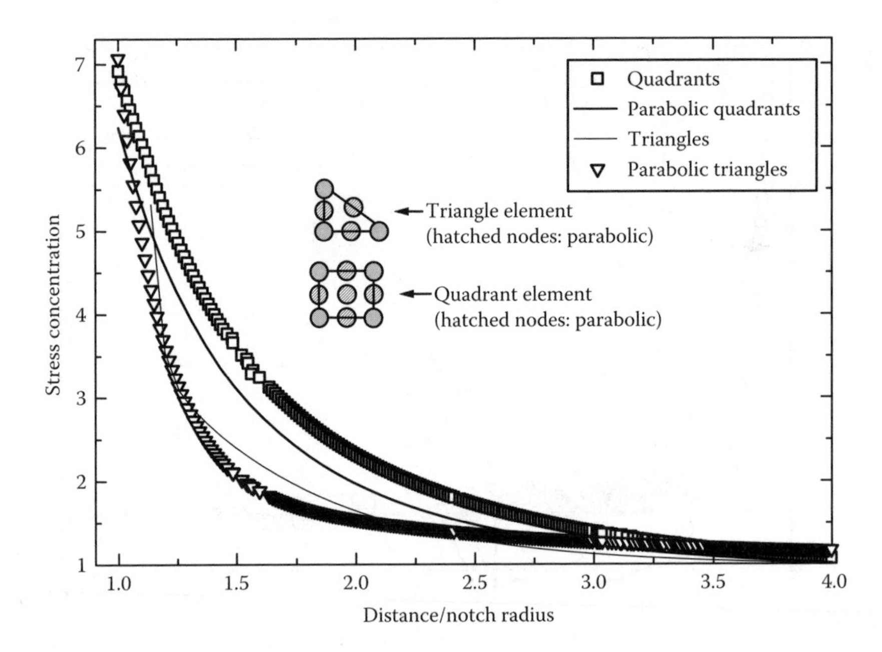

FIGURE 2.41 A comparative presentation of the results obtained by different types of elements used in the FE analysis calculation.

closer to the notch boundary than those for the quadrants. The observed differences between the numerical on the one hand and the analytical or experimental results on the other hand, are due to the precision of the FEM compared with the other methods. Numerical procedures are in general inferior, as they employ approximate solutions and are not as rigorous as the analytical method presented here.

The Raman method of stress/strain measurements appears to be a powerful tool for the measurement of stress concentration around and near a notch. In the past, Moiré interferometry and strain gauge techniques have been applied successfully to the same problem albeit with different spatial resolutions. In particular, the Moiré mesh gives a 25 μm resolution whereas the best resolution of a commercially available strain gauge is 200 μm. In addition, the strain gauge technique is destructive as it involves the adhesion of tapes to the host material. The Raman technique is not destructive and its best resolution is approximately one wavelength of light within the visible range (~0.5 μm). Finally, both Moiré and strain gauge methods relate to surface measurements, whereas Raman provides values of stress or strain in the reinforcing fibers which are the load-bearing elements.

2.8 STRESS-CONCENTRATION MEASUREMENTS IN CROSS-PLY COMPOSITES

Matrix cracking is the predominant mechanism at the initial stage of degradation of cross-ply composite laminates under mechanical loading. The matrix crack density increases as a consequence of increasing applied strain or increasing number of

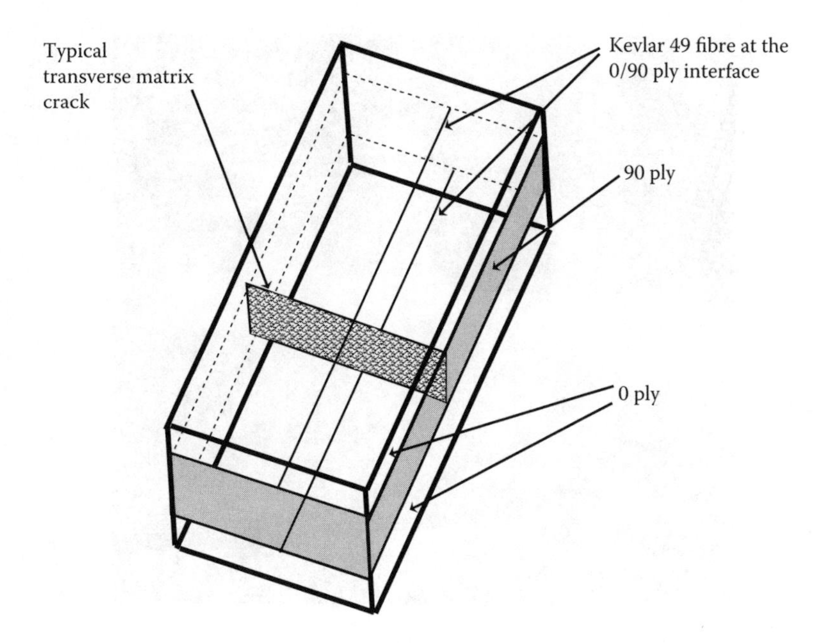

FIGURE 2.42 Sketch of a typical (0°/90°) cross-ply composite incorporating a Kevlar® 49 filament at one 0°/90° interface as an embedded strain sensor. The single transverse crack in the 90° ply is shown.

fatigue cycles, with a consequent change in such laminate properties as Young's modulus, Poisson's ratio, and residual strain after unloading. Recent work [59] has shown that if the glass fiber composite is made transparent by matching the refractive indices of glass fibers and the epoxy matrix, then individual Kevlar® fibers can be placed at the ply interfaces to act as strain sensors. This way the Raman strain sensor technique is capable of measuring the strain magnification in the 0° ply as a consequence of cracking in the 90° ply. The placement of the aramid fiber strain sensor at the 0°/90° interface of the cross-ply composite is shown schematically in Figure 2.42 [60]. A closer survey of the position of the strain sensor is shown in Figure 2.43, which represents an actual micrograph of the embedded aramid fiber (12 μm in diameter) in the composite with a matrix crack in the background. As can be seen, the size of the fiber strain sensor is compatible with the opening of the transverse crack and, therefore, the exact location of the transverse crack can be found by Raman measurements on the illuminated section of the fiber, which is immediately adjacent to the crack.

The transparent cross-ply laminates containing the Kevlar® 49 fibers were manufactured using a modified frame-winding technique [59]. The details of the fabrication procedure and the placement of the single Kevlar® 49 fibers at the 0°/90° interface are given elsewhere [59,60]. The panel was 360 mm square; from the panel coupons 230 mm long by 20 mm wide were cut, each specimen containing a centrally positioned Kevlar® 49 fiber (Figure 2.42). The final specimens had a transverse ply thickness of 1.22 mm and a 0° ply thickness of 0.64 mm.

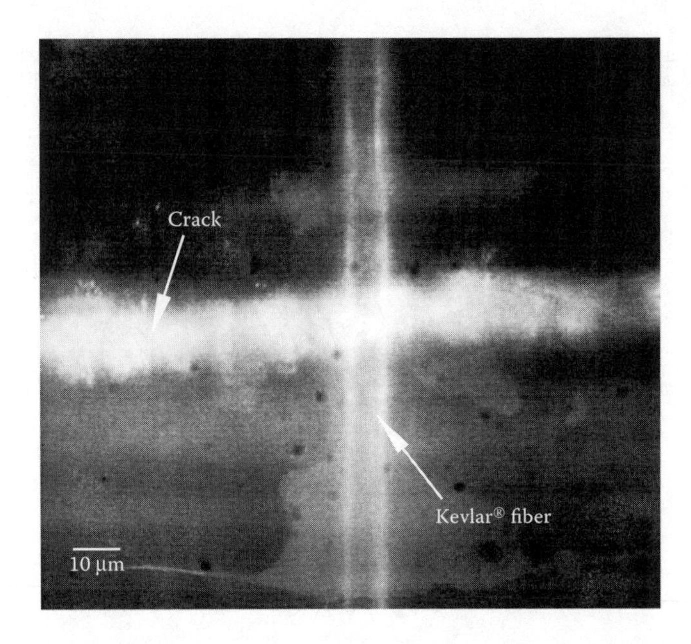

FIGURE 2.43 The view through microprobe showing the Kevlar® 49 fiber adjacent to the crack plane. The diameter of the fiber is 12 mm.

In all cases measurements were made within the Raman window (distance of 20.95 mm separated by two cracks) at applied strains required for transverse crack formation. To avoid unstable crack growth during Raman data acquisition all measurements were made by unloading the specimens to (a) 0.2% applied strain (which is below the crack initiation threshold value) and then (b) 0% applied strain.

The peaks in the Raman strain distribution measurements accurately pinpoint the positions of the crack planes even at 0% applied strain [59]. Although the position of the cracks can also be observed with an optical microscope, the exact location of the center of the crack is difficult to define accurately as the width of a typical crack at even 0.2% strain is of the order of 10–20 μm (Figure 2.43). Therefore any measurement of strain with a travelling microscope based on the relative displacement of adjacent cracks is prone to large margins of error. On the contrary, with the Raman technique one can determine the exact location of the crack center with a typical accuracy of 1 μm. Hence, strain measurements can be performed by monitoring the relative movement of the peaks of the fiber strain distributions.

The procedure for residual strain measurements required strain mapping along the length of the aramid sensor at various increments of applied strain. As explained earlier, one crack at a time is generated within the Raman window. Each time a new crack is generated, the displacement in the length of the Raman window due to the crack is measured. The residual strain for the section of the laminate within the Raman window is obtained by dividing the change in the displacement by the original Raman window gauge length. The process is repeated each time a new crack is generated.

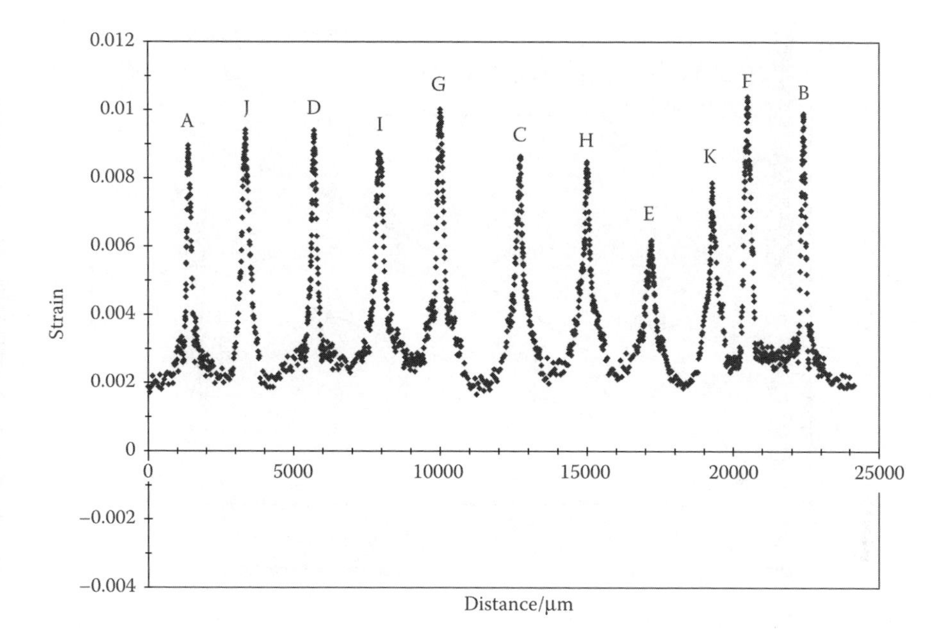

FIGURE 2.44 Strain distribution in the embedded Kevlar® 49 fiber sensor within the 20 mm gauge length at the point of transverse crack saturation. The Raman measurements have been performed at 0.2% applied strain.

The reduction in stiffness of the specimen is also calculated using the displacement of the peaks in the strain profile [59]. The initial modulus of the laminate is calculated by dividing the applied stress (for 0.2% laminate strain) by the initial laminate strain. The initial laminate strain is calculated by dividing the change in displacement of the original length of the Raman window at 0.2% laminate strain by the original length of the Raman window. This way, the modulus for the section of the laminate within the window was measured for the increasing crack density. The process was repeated and the modulus was measured each time a new crack appeared till the window was saturated with cracks. The reduction in stiffness is obtained by dividing the modulus obtained for increasing crack densities by the initial modulus.

Figure 2.44 shows the results of increasing the strain on the laminate containing the initial cracks A and B to the point where an additional nine cracks (labeled C to K) have been generated between these two initial cracks. The strain profile shown in Figure 2.44 has been taken at an applied strain of 0.2%. When the applied load is removed, the peaks in the Raman measurements pinpoint accurately the position of the cracks [59]. The change in the displacement of the peaks in Figure 2.44 can therefore be used to measure the residual strain as a function of the increasing number of cracks between cracks A and B. Previous work has shown that these residual strains, generated as a consequence of the local release of the balanced thermal stresses in the 0° and 90° plies generated during manufacture, are relatively easy to measure in glass fiber reinforced polymer (GFRP) laminates using a long

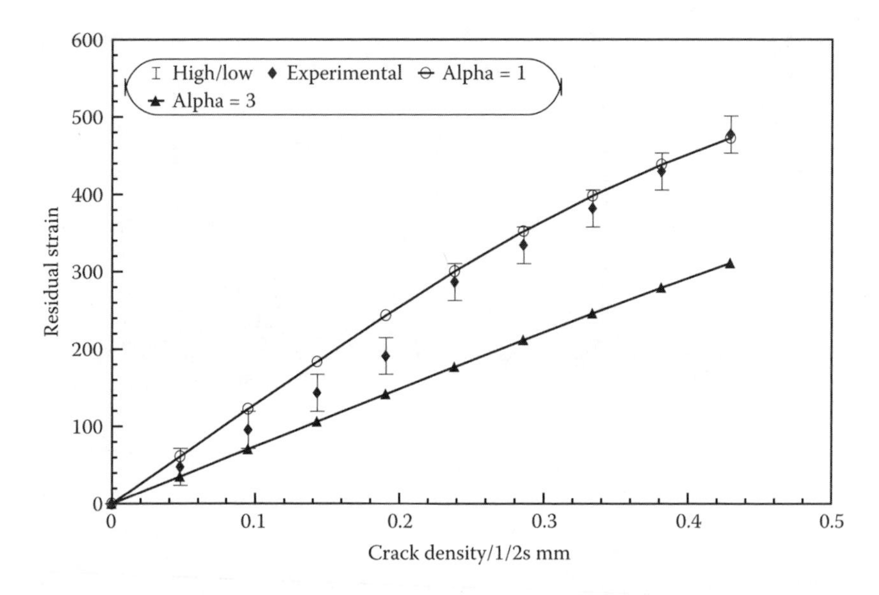

FIGURE 2.45 Residual strain as a function of crack density. The solid lines represent predictions of the residual strain as a function of crack density based on shear lag analysis (for $a = 1$, $a = 3$).

gauge length extensometer [60]. The Raman technique enables similar measurements to be made over small distances and few cracks. The progressive increase in the separation of cracks A and B for the unloaded laminate is measured as function of the increasing number of cracks (labelled C to K). The results are shown in Figure 2.45 as residual strain (using the initial spacing of cracks A and B as the gauge length) as a function of average crack spacing. The residual strain increases approximately linearly with increasing crack density, which is in agreement with previous work [60] (the crack density is given by $^1/_2$s, where 2s is the average crack spacing).

2.9 BULK STRAIN MEASUREMENTS IN COMPOSITES USING FIBER OPTICS

Over the last two decades, a lot of work has been put into developing structural materials that can incorporate functions of biological systems such as sensing, actuation, and control. Advanced fibrous composites already possess many desirable characteristics for a whole range of structural applications. By adding the biological features of sensing, actuation, and control, they can be taken a step further and can be given certain "true life" features and "intelligence" [61]. Needless to say, such a development would substantially lessen concerns over the introduction of composite materials into aerospace, automotive, and civil industries, as well as significantly reduce the cost [62]. Already, there is a great deal of activity in developing optical fibers as strain and temperature sensors, shape memory alloys (SMA) as actuators [63], and piezoelectric materials as both strain sensors and actuators [64]. Neural networks are looked at as possible self-learning control systems for composite "smart" structures [65].

For a structural composite, monitoring of strain represents one of the most important sensing endeavors. The standard method to measure strain in laminates still involves using intrusive electrical resistance strain gauges attached to the surface of the structure, and then determining the ply-by-ply strains using laminate theory [66]. The technique of incorporating fiber optics in composite materials for nondestructive evaluation or strain measuring purposes has progressed rapidly since the late 1970s. Fiber optics sensors are being used for "smart" structures as they have numerous advantages, the main one being their ability to serve the dual role of sensor and pathway for the signal. Since the optical fiber is a dielectric fiber, it is quite compatible with the composite material and avoids creating electrical pathways within the structure. They are also of comparable size to a single ply of unidirectional laminate and have a relatively high modulus. These features enable the fiber optic to integrate into the structure without affecting the structural properties, particularly when it is placed parallel to the ply direction, whereas there is a slight decrease of strength in the transverse direction [67].

An optical sensor may be defined as a device in which an optical signal is modulated in response to a measurand field. An optical signal is characterized by its wavelength, phase, intensity, and polarization. In an optical sensor any one of these parameters, or a combination of them, may be modulated in response to external influence; e.g., strain, temperature, or pressure [62,68,69]. However, independent concurrent measurements of all the external parameters are not easy to perform [70]. In composites, intensiometric (microbend) [67], interferometric (Mach-Zehnder, Michelson, Fabry-Perot, Bragg grating) [64,69,71,73], and polarimetric (high and low birefringence) [64], fiber optic sensors have been evaluated primarily for strain measurements and detection of damage generated by impacts, manufacturing flaws, excessive loading, or fatigue. The main fundamental difference between the Raman sensor examined here and all other existing sensors and techniques is that measurements are conducted at the reinforcing fiber itself, at a resolution which is determined by the wavelength of the exciting laser light (typically 0.5–1 μm). The Raman sensor is advantageous over conventional electrical resistance strain gauges and fiber optics as it provides higher resolution tailored to the testing requirements, it is not intrusive, and it is not limited to a specific sampling area (see sections 2.1, 2.2 of this chapter). It is worth adding that the Raman measurements correspond to physical changes of the measurand itself brought about by an applied stress field. On the contrary, electrical resistance or light modulation measurements are indirectly related to the measurand under the assumption of equal strain with the host material.

An attempt has been made to obtain high quality Raman data directly from the reinforcing fibers in the bulk of the composite by channelling the laser light via fiber optic cables [74]. Fiber cables are embedded into laminate in unidirectional and multidirectional Kevlar®/epoxy-laminated composite coupons. The work on multidirectional composites opens up the possibility of using this technique to measure ply-by-ply point stress measurements in composites in service. Finally, the effect that an embedded fiber optic cable has upon the mechanical properties of the composite is assessed. Eight-ply laminates were produced from unidirectional Kevlar® 49 /epoxy resin preimpregnated tapes supplied by Hexcel Composites (type 914k-49-54.8%, Table 2.4).

TABLE 2.4
Elastic Parameters for the Fiber, the Matrix, and the Hexcel Unidirectional 914k-49 Composite

				E_f/GPa				
E_{11}/GPa	E_{22}/GPa	G_{12}^*/GPa	ν_{12}	≤0.5%	>0.5%	E_m/GPa	$\alpha_f/10^{-6}\ K^{-1}$	$\alpha_m/10^{-6}\ K^{-1}$
80.3	8.0	2.1	0.35	110	125	3.5	6.5	60

*Supplied by manufacturer.

To produce the bulk-perpendicular configuration, the prepreg tapes were cut to strips of 203 mm in length and 12.7 mm in width and were then laid on top of each other with a cleaved fiber optic embedded in the center (Figure 2.46). The multimode fiber-optic cables employed in this work were supplied by Newport Co. (model F-MLD)

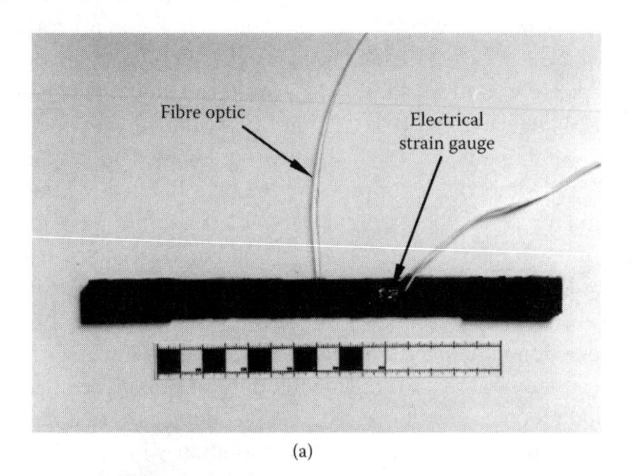

(a)

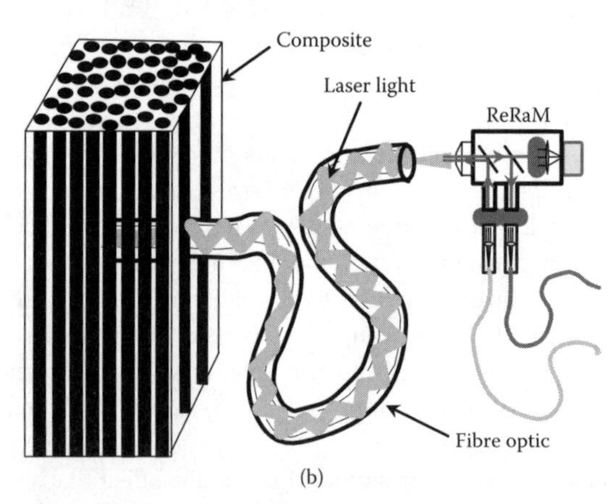

(b)

FIGURE 2.46 (a) Photograph of bulk-perpendicular standard Kevlar® 49/epoxy composite tensile coupon. (b) Schematic of bulk-perpendicular configuration. The items are not shown to scale.

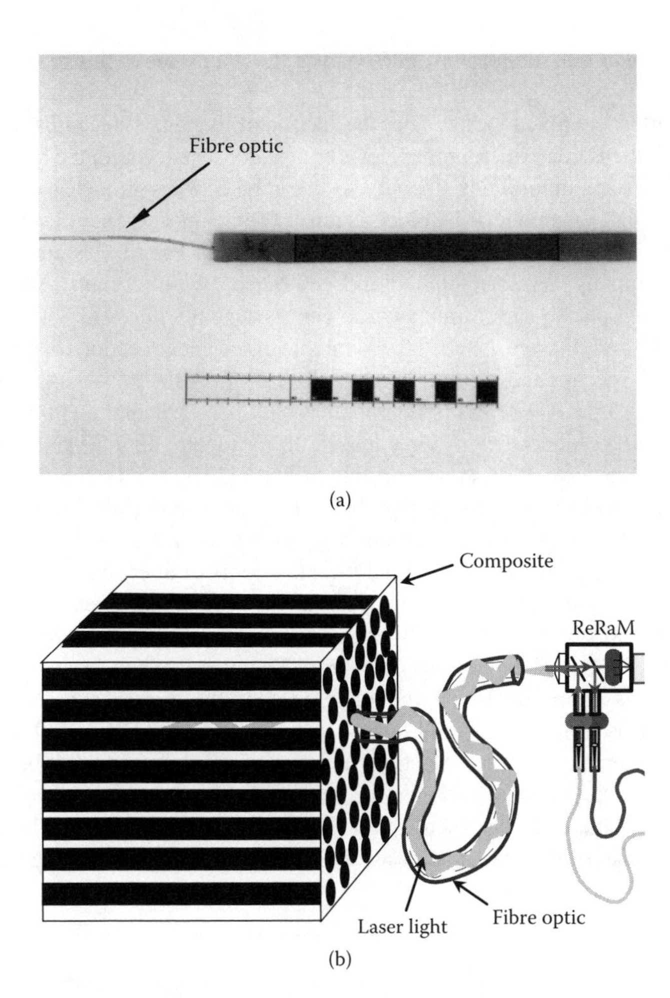

FIGURE 2.47 (a) Photograph of bulk-parallel standard composite Kevlar® 49/epoxy tensile coupon. (b) Schematic of bulk-parallel configuration. The items are not shown to scale.

and had a nominal core diameter of 100 μm and a numerical aperture of 0.29. To produce the bulk-parallel configuration the cleaved fiber optic cable was wrapped in a strand of Kevlar® fibers and placed in the center of laminates in a direction parallel to the axis of the reinforcing fibers (Figure 2.47). To produce the multidirectional $[0_2, -45, +45]_s$ configuration the laminates were cut to dimension and cleaved fiber optic cables were placed in between the two +45° plies at the very center of the laminate and at a direction perpendicular to the axis of the reinforcing fibers. Simultaneous measurements on the 0° ply were conducted by either interrogating the fibers at the surface of the laminate or by embedding a second fiber optic between the 0° plies at a direction perpendicular to the reinforcing fibers. In all cases the preimpregnated tapes were laid on the top of each other and cured in an autoclave for one hour at 175° C under a pressure of 420 kNm^{-2}, according to the manufacturer's instructions. To avoid fracture of the cleaved fiber optic during lay-up in the autoclave by resin bleeding onto the fiber, care was taken to seal the nonembedded part of the fiber optic within sleeves made of peel-ply.

The reinforcing fibers located near the surface of the laminate can be interrogated directly with the Raman microprobe. However, for consistency with the bulk measurements as above, the reinforcing fibers were interrogated through a cleaved fiber optic bonded to a small squared PMMA block (8 mm). The composites were tested in tension following the ASTM D3039-76 standard procedure [75]. Prior to testing, the ends of the tensile coupons were sandblasted and end-tabbed with standard, 2.4 mm thick, glass-reinforced plastic tabs. Strain gauges with a gauge resistance of 350 ± 1.0 Ω and a gauge factor of 2.03 were attached to the middle of the gauge section for each coupon. A total of 15 specimens were tested in tension; 5 of those specimens contained no embedded fiber optics, whereas 5 incorporated bulk-perpendicular and 5 bulk-parallel fiber optic cables. All specimens were loaded up to fracture on a 20 kN screw-driven Hounsfield mechanical tester at a strain rate of approximately 0.002 min^{-1}.

The fiber strain measured by the ReRaM is compared with composite strain, measured by means of an attached strain gauge, in Figure 2.48, Figure 2.49, and Figure 2.50. On average, 100 individual measurements of fiber strain vis-à-vis strain gauge were taken during tensile testing of the specimen. In all three cases (Figure 2.48, Figure 2.49, and Figure 2.50) the fiber strain measured using ReRaM increases in tandem with the strain gauge measurements. The slope of the least-squares-fitted straight lines is approximately 1 for all three configurations (Table 2.5). Another important observation made from Figure 2.7, Figure 2.8, and Figure 2.9 is that prior to loading the tensile coupons, the strain in the reinforcing Kevlar® 49 fibers is found to be compressive. Residual compressive strain values were measured by taking the average of 10 Raman measurements prior to stressing the laminate. For bulk-perpendicular and bulk-parallel configurations, fiber compressive strains of magnitude 400

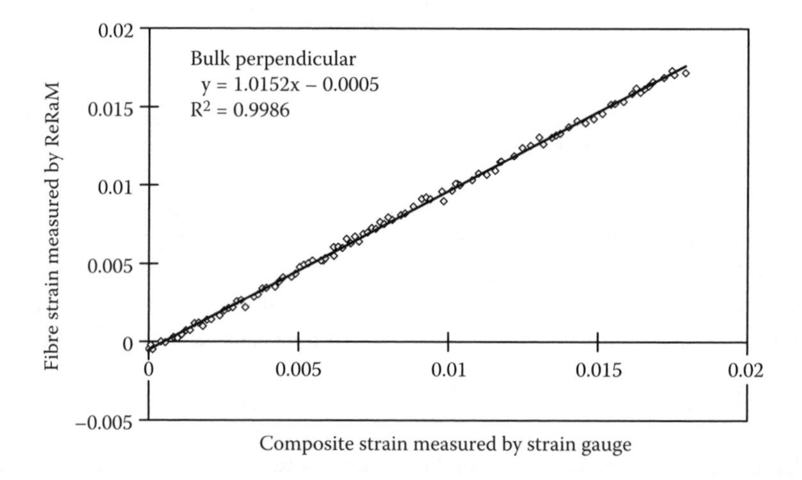

FIGURE 2.48 Fiber strain measured by the remote Raman microprobe using an embedded fiber optic in a direction perpendicular to the reinforcing fibers of the composite, versus the strain in the composite measured by means of an attached strain gauge. The solid line is a least-squares-fit to the experimental data.

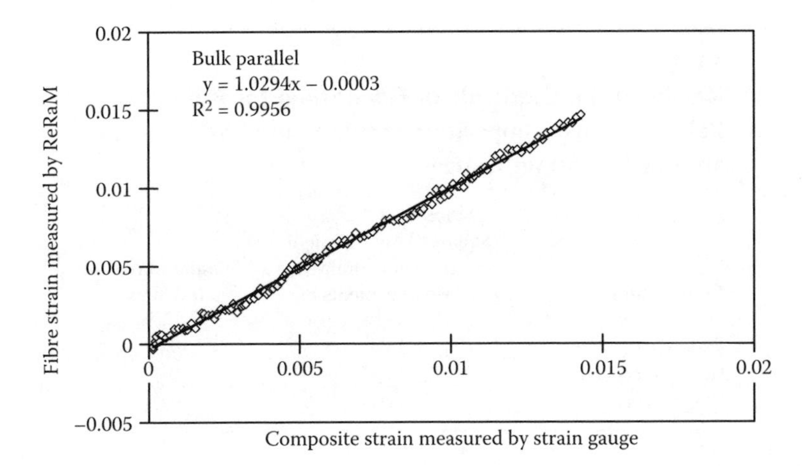

FIGURE 2.49 Fiber strain measured by the remote Raman microprobe using an embedded fiber optic in a direction parallel to the reinforcing fibers of the composite, versus the strain in the composite measured by means of an attached strain gauge. The solid line is a least-squares-fit to the experimental data.

and 300 $\mu\varepsilon$, respectively, have been obtained. The corresponding compressive strain for the surface measurements is 400 $\mu\varepsilon$.

In Table 2.6, the effect of the presence of an embedded fiber optic upon the integrity of the composite is investigated. As can be seen, the presence of a fiber optic cable perpendicular to the reinforcing fibers reduces the tensile strength by approximately 10%, whereas the presence of a fiber parallel to the reinforcing fibers has no practical effect upon the tensile strength of the composite. Careful monitoring of the fractured process for the perpendicular configuration revealed that the failure

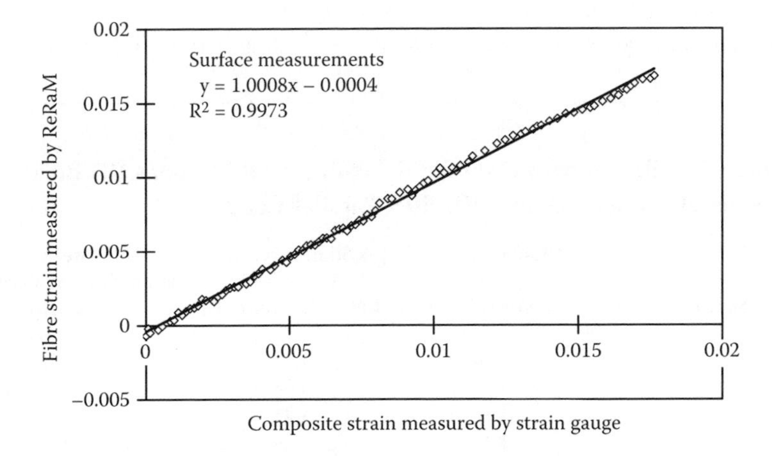

FIGURE 2.50 Fiber strain measured by the remote Raman microprobe near the surface of the laminate versus the strain in the composite, measured by means of an attached strain gauge. The solid line is a least-squares-fit to the experimental data.

TABLE 2.5
Results of the Gradients of Fiber Strain Measured using ReRaM versus Composite Strain Measured by Means of an Attached Strain Gauge

Configuration	Slope of the Least-Squares-Fitted Straight Lines for Strain Measurements	Standard Error of the Slope
Surface measurements	1.00	0.01
Bulk-perpendicular	1.02	0.02
Bulk-parallel	1.03	0.01

in two out of five specimens initiated at the point where the fiber optic was embedded (Figure 2.51). However, the post-mortem examination in all five specimens showed a typical "broom-like" fracture. In the case of the *bulk-parallel* configuration (Figure 2.52), failure initiation was random as in the case of dummy coupons that contained no fiber optics. Figure 2.53 shows the measured stress by ReRaM on fibers embedded perpendicular to 0° and to +45° plies as a function of the applied stress over the whole [02, –45, +45]s composite coupon. As expected, the axial fiber stress in the 0° plies increases steeply with the applied stress, whereas the principal fiber stress in the +45° ply is only a fraction of the applied composite stress (Figure 2.53).

The results obtained from unidirectional composites (Figure 2.48, Figure 2.49 and Figure 2.50) show clearly that the fiber strain values obtained with the technique of Raman spectroscopy are in agreement with theoretical predictions, as well as conventional strain gauge measurements. The superiority of the Raman sensor over other existing sensors is its ability to provide independently values of fiber strain, and stress if required, from composite sample volumes as small as 1 mm^3. In addition, this is

TABLE 2.6
Ultimate Tensile Strength Data for (I) As-Received Coupons (II) Bulk Perpendicular Coupons and (III) Bulk Parallel Coupons

No. of Specimens	Ultimate Tensile Strength of Bulk Á Coupons/GPa	Ultimate Tensile Strength of Bulk° Coupons/GPa	Ultimate Tensile Strength of As-Received Coupons
1	1.18	1.33	1.43
2	1.25	1.38	1.39
3	1.32	1.41	1.35
4	1.23	1.35	1.41
5	1.19	1.42	1.27
Mean/GPa	1.23	1.38	1.37
Standard deviation/GPa	0.06	0.04	0.06

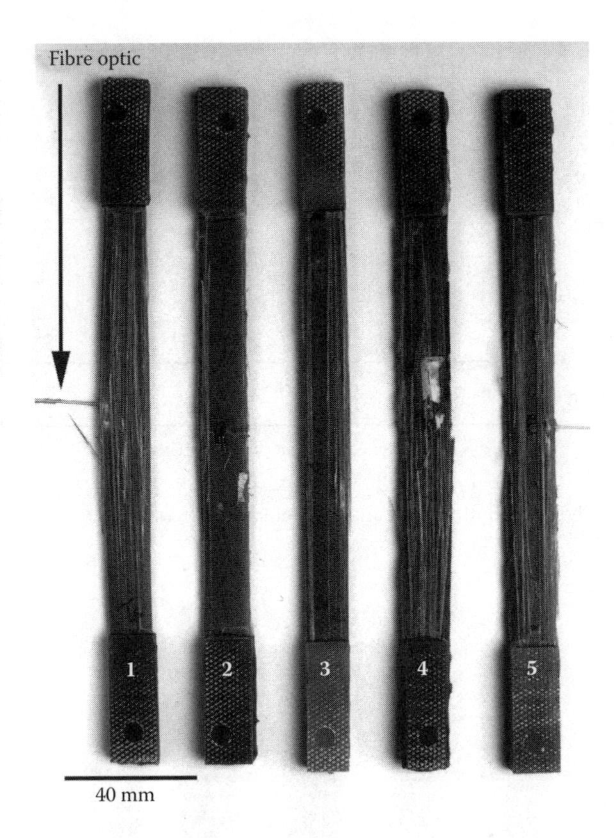

FIGURE 2.51 Photograph of failed bulk-perpendicular standard composite Kevlar® 49/epoxy tensile coupons.

the only technique that can directly measure stress in composites as most of the currently available nondestructive methods can only provide strain measurements [61].

As has been demonstrated here, the Raman stress/strain sensor can be employed for unidirectional as well as multidirectional composites. The stress measurements using ReRaM on the fibers in the 0° and +45° plies have been compared against the calculated fiber stress using the laminate analysis program [76], which employs standard laminate theory to calculate stress on each individual ply. The program requires knowledge of E_{11}, E_{22}, G_{12}, v_{12}, ply thickness, ply orientation, and curing temperature to calculate the theoretical stress along the fiber direction in any given ply. Fiber stress measurements using ReRaM are compared with the calculated fiber stress using LAP for the +45° ply and the 0° ply respectively in Figure 2.54a and Figure 2.54b. The slopes of the least-squares-fitted straight lines for measurements in the 0° and +45° plies are 1.05 ± 0.01 and 0.97 ± 0.02, respectively. The ReRaM fiber stress measurements in the 0° ply and the +45° ply, prior to loading the tensile coupons, show compressive stresses in the reinforcing Kevlar® 49 fibers of magnitude 34 MPa and 126 MPa, respectively. The high compressive stresses in the principal (+45°) direction in an angle ply are generated during cooling of the laminate from the curing temperature (175° C) down to room temperature [23]. A closer inspection

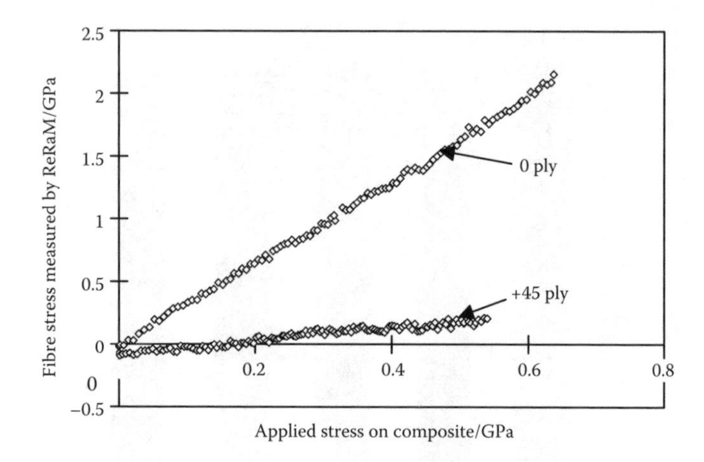

FIGURE 2.52 Photograph of failed bulk-parallel standard composite Kevlar® 49/epoxy tensile coupons.

FIGURE 2.53 Fiber stress in the 0° and +45° plies, measured by the remote Raman microprobe using embedded fiber optics in a direction perpendicular to the fibers. The geometry of the coupon is [02, −45, +45].

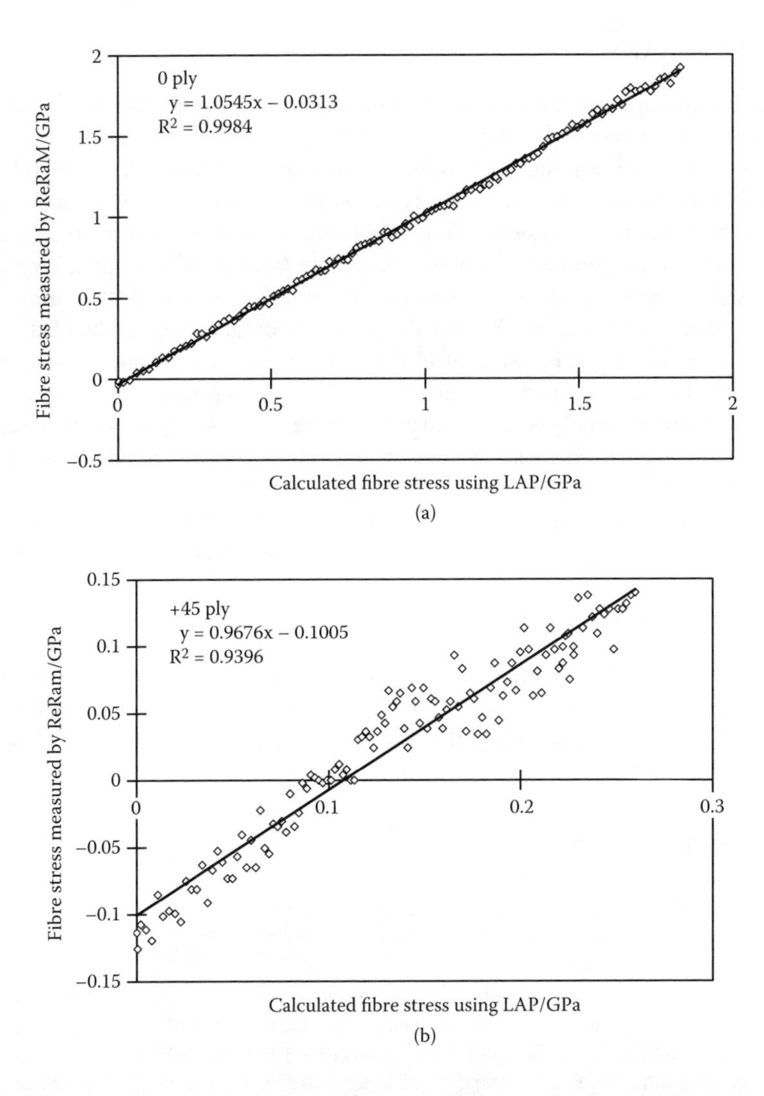

FIGURE 2.54 (a) Fiber stress measured by the remote Raman microprobe in the 0° ply versus the calculated fiber stress obtained by LAP. The geometry of the coupon is [02, –45, +45]s. The solid line is a least-squares-fit to the experimental data. (b) Fiber stress measured by the remote Raman microprobe on the +45° ply versus the calculated fiber stress using LAP. The geometry of the coupon is [02, –45, +45]s. The solid line is a least-squares-fit to the experimental data.

of the data points revealed that the relationship is not quite linear. Useful information upon the deformation mechanisms in these geometries can be extracted from the relationship between the stress in the principal fiber direction within an angle ply to that of the applied stress in the 0° direction (Figure 2.54). Conventional laminate analysis does not account for nonlinear deformations in the principal direction of an angle ply, possibly due to matrix yielding or other nonlinear effects, as observed in +45° ply examined here.

2.10 SUMMARY

Raman microscopy can be employed to determine the stress state in single fiber composites, in planar fiber arrays, and in full composites. The measurements are conducted at the reinforcing fiber itself, at a resolution which is determined by the wavelength of the exciting laser light (typically 0.5–1 μm). Thus, mechanical quantities in the microscopic scale, such as interfacial shear stress and interfacial shear strength, can be determined. Furthermore, remote Raman microscopy allows us to make measurements of stresses and strains in composite structures that are located at larger distances from the Raman detector. An optical fiber embedded in the composite laminate can be coupled to the remote microscope, and thus adequate measurements can be extracted from the bulk of the composite.

Carbon and aramid fibers have been proven to be excellent stress and strain sensors through their Raman response. Moreover, exploiting certain vibrational modes of aramid fibers can extend their self-sensing capabilities to temperature measurements inside the composite. Thus, a novel spectroscopic sensor could be developed, capable of being used for a large number of applications in the field of "smart materials."

ACKNOWLEDGMENTS

The authors would like to thank the following researchers for conducting most of the experiments presented here: G. Anagnostopoulos, B. Arjyal, D. Bollas, V. Chohan, G.D. Filliou, D.G. Katerelos, C. Koimtzoglou, C. Marston, N. Melanitis, A. Paipetis, G.C. Psarras, and C. Vlattas.

REFERENCES

1. P. C. Powel, Engineering with Fibre-Polymer Laminates, Chapman and Hall, 1994.
2. A. M. Sasrty and S. L. Phoenix, Shielding and magnification of loads in elastic, unidirectional composites, *Soc. Adv. Mater. Proc. Eng. (SAMPE) J.* 1994 **30**(4), 61–67.
3. C. Sweben, Advanced composites for aerospace applications: a review of current status and future prospects, *Composites* 1981 **12**(4), 235–240.
4. S. R. Bretzlaf and P. R. Wool, *Macromolecules* 1983 **16**, 1907–1917.
5. A. C. Cottle, W. F. Lewis, and D. N. Batchelder, *J. Phys. C-Solid State Phys.* 1978 **11**, 605–610.
6. N. Melanitis and C. Galiotis, *J. Mater. Sci.* 1990 **25**, 5081–5090.
7. E. Anastassakis, A. Pinczuk, E. Burstein, F. H. Pollak, and M. Cardona, *Solid State Comm.* 1970 **8**, 133–138.
8. C. Galiotis, *Comp. Sci. Tech.* 1991 **42**, 125–150.
9. B. Arjyal and C. Galiotis, *Adv. Comp. Lett.* 1995 **4**, 47–52.
10. A. Paipetis, C. Vlattas, and C. Galiotis, *J. Raman Spectros.* 1996 **27**, 519–526.
11. Parthenios, J. et al., in preparation.
12. G. A. Voyiatzis and K. S. Andrikopoulos *Appl. Spectr.* 2002 **56**(4), 528–535.
13. C. Vlattas and C. Galiotis, *Polymer* 1994 **35**, 2335–2347.
14. N. Everall and J. Lumsdon, *J. Mat. Sci.* 1991 **26**, 5269–5274.
15. K. Kim, C. Chang, and S . L. Hsu, *Polymer* 1986 **27**, 34.
16. G. C. Psarras, J. Parthenios, D. Bollas, and C. Galiotis, *Chem. Phys. Lett.* 367/ 3–4, 270–277, 2003.

17. J. Parthenios, G. C. Psarras, and C. Galiotis, *Comp. Part A: Appl. Sci. Manuf.* 2001 **32**(12), 1735–1747.
18. G. C. Psarras, J. Parthenios, and C. Galiotis, *J. Mat. Sci.* 2001 **36,** 535–546.
19. S. R. Allen and E. J. Roche, *Polymer* 1989 **30**, 996.
20. M. G. Northolt, *Polymer* 1980 **21**, 1199.
21. C. C. Chamis, Mechanics of load transfer at the interface, *Composite Materials*, Vol. 6, Academic Press, New York, 1974.
22. L. T. Drzal, M. J. Rich, and F. F. Loyd, *J. Adhes.* 1982 **16**, 1–30.
23. D. Hull and T. W. Clyne, *An Introduction to Composite Materials*, Cambridge University Press, Cambridge, U.K., 1996.
24. L. S. Schadler and C. Galiotis, *Internat. Mater. Rev.* 1995 **40**(3), 116–134.
25. C. Galiotis, Micromechanics of reinforcement using laser Raman spectroscopy, in *Microstructural Characterisation of Fibre-Reinforced Composites*, ed. J. Summerscales, Woodhead Publishing, Cambridge, England, 1998, 224–253.
26. H. L. Cox, *Br. J. Appl. Phys.* 1952 **3**, 72–79.
27. T. F. McLaughlin, *J. Comp. Mater.* 1968 **2**(1), 44–45.
28. N. Melanitis and C. Galiotis, *Proc. Roy. Soc. London* 1993 **440**, 379–398.
29. N. Melanitis, C. Galiotis, P. L. Tetlow, and C. K. L. Davis *J. Composite Mater.* 1992 **26**(4), 574–610.
30. A. Paipetis and C. Galiotis *Composites* 1996 **27A**(9), 755–767.
31. N. Melanitis, C. Galiotis, P. L. Tetlow, and C. K. L. Davis *J. Mater. Sci.* 1992 **28**, 1648–1654.
32. V. Chohan and C. Galiotis, *Compos. Sci. Technol.* 1997 **57**(8), 1089–1101.
33. H. D. Wagner and L. W. Steenbakkers, *J. Mater. Sci.* 1989 **24**, 3956–3975.
34. C. Galiotis, V. Chohan, A. Paipetis, and C. Vlattas, Interfacial measurements in single and multi-fiber composites using the technique of laser Raman spectroscopy, Eds. *Fiber* J. C. Spragg and L.T. Drzal, ASTM-STP 1290, American Society for Testing and Materials, 1996, 19–33.
35. V. Chohan and C. Galiotis, *Composites* 1996 **27A**(9), 881–888.
36. C. Marston, B. Gabbitas, J. Adams, S. Nutt, P. Marshall, and C. Galiotis, *Composites* 1996 **27A**(12), 1183–1194.
37. H. D. Wagner and A. Eitan, *Composite Sci. Technol.* 1993 **46**(4), 353–362.
38. J. Parthenios, D. G. Katerelos, G. C. Psarras, and C. Galiotis, *Eng. Fract. Mech.* 2002 **69**(9), 1067–1087.
39. D. Bollas, C. Koimtzoglou, G. Anagnostopoulos, G. C. Psarras, J. Parthenios, and C. Galiotis, ECCM10, Brugge Belgium 2002, June 3–7. Proceedings of the conference on CD-ROM.
40. J. Schrooten, V. Michaud, J. Parthenios, G. C. Psarras, C. Galiotis, R. Gotthardt, J. Anders Månson, and J. Van Humbeeck, *Mater. Trans. JIM* 2002 **43**(5), 961–973.
41. V. Vlattas, A study of the mechanical properties of liquid crystal polymer fibres and their adhesion to epoxy resin using Laser Raman Spectroscopy, Ph.D. Thesis, U. London, 1995.
42. F. J. Guild, V. Vlattas, and C. Galiotis, C., Modelling of stress transfer in fibre composites, *Compos. Sci. Technol.* 1994 **50**, 319–332.
43. F. N. Cogswell, *Thermoplastic Aromatic Polymer Composites*, Butterworth Heinemann, Oxford, U.K., 1992.
44. G. Jeronimidis and A. T. Parkyn, *J. Compos. Mater.* 1988 **22** 401.
45. J A. Nairn and P. Zoller, in *Toughened Composites*, ed. N. Johnston, ASTM-STP 937, American Society for Testing and Materials, 1987, 328.
46. J. E. Bailey, P. T. Curtis, and A. Parvisi, *Proc. Royal Soc. Ser. A (London)* 1979 **366**, 599.
47. F. R. Jones, M. Mulheron, and J. E. Bailey *J. Mater. Sci.* 1983 **18**, 1533.

48. C. D. Filiou, C. Galiotis, and D. N. Batchelder, *Composites* 1992 **28**(1), 28–37.
49. C. Filiou and C. Galiotis, *Composites Sci. Technol.* 1999 **59**(14), 2149–2161.
50. G. C. Psarras, J. Parthenios, and C. Galiotis, *J. Mat. Sci.* 2001 **36,** 535–546.
51. C. F. Dietrich, Uncertainty, calibration, probability, in *The Adam Hilger Series on Measurements Science and Technology*, 2nd ed., A. Higler, Bristol, U.K., 1991.
52. J. A. Nairn and P. Zoller, *J. Mat. Sci.* 1985 **20**, 355–367.
53. T. L. Anderson, *Fracture Mechanics: Fundamentals and Applications*, CRC Press, Boca Raton, FL, 1995.
54. B. P. Arjyal, D. G. Katerelos, C. Filiou, and C. Galiotis, *J. Exp. Mech.* 2000 **40**(3), 248–256.
55. S. G. Lekhnitskii, *Anisotropic Plates*, 2nd ed., trans. S. W. Tsai and T. Cheron, Gordon and Breach Science, New York, 1968.
56. G. N. Savin, *Stress Concentration around Holes*, trans. E. Gros, Ed. W. Johnson, Pergamon Press, Oxford, U.K., 1961.
57. S. C. Tan, *J. Compos. Mater.* 1987 **21**, 750–780.
58. S. C. Tan, *Stress Concentrations in Laminated Composites*, Technomic, New York, 1994.
59. B. P. Arjyal, C. Galiotis, S. L. Ogin, and R. D. Whattingham, *J. Mater. Sci.* 1998 **33**(11), 2745–2750.
60. B.P. Arjyal, C. Galiotis, S. L. Ogin, and R. D. Whattingham, *Composites A* 1998 **29**(11), 1363–1369.
61. F. Bassam, L. Boniface, K. Jones, and S. L. Ogin, *Composites A* 1998 **29A**, 1425–1432.
62. C. A. Rogers, *Smart Composites*, ECCM-6, ed. P. Gardiner, A. Kelly, and A. R. Bunsell, Woodhead Publishers, Cambridge, U.K., 1993, 3–9.
63. R. M. Measures, *Prog. Aerosp. Sci.* 1989, **26**, 289.
64. J. Parthenios, G. C. Psarras, and C. Galiotis, *Compos. Part A: Appl. Sci. Manuf.* 2001 **32**(12), 1735–1747.
65. M. C. Friend, *Interdiscipl. Sci. Rev.* 1996 **21**(3), 195.
66. C. Boller, *Proceedings, Third International Conference on Intelligent Materials / Third European Conference on Smart Structures and Materials*, Ed. P. F. Gobin and J. Tatibouet, SPIE Publishers, Philadelphia, 1996, 16.
67. S. W. Tsai and H. T. Hahn, *Introduction to Composite Materials*, Technomic, Lancaster, PA, 1980.
68. R. O. Claus, K. Bennett, and B. Jackson, *Rev. Prog. Quant. NDE* 1985 **5**, 1149.
69. R. S. Medlock, *J. Opt. Sens.* 1986 **1**, 43.
70. B. Culshaw and J. Dakin, *Optical Fiber Sensors,* Artech House, Norwood, MA, Vol. 1, 1988, Vol. 2, 1989.
71. K. T. V. Grattan and B. T. Meggitt, *Optical Fiber Sensor Technology,* Chapman & Hall, London, 1995.
72. D. A. Jackson and J. D. C. Jones, *Fibre Opt. Sens. Opt. Acta* 1986 **33**, 1469.
73. P. De Paula and E. L. Moore, Fibre optic sensors overview, *SPIE* 1985 **566**, Fibre Optic and Laser Sensors III, 2.
74. B. P. Arjyal, P.A. Tarantili, A.G. Andreopoulos, and C. Galiotis, *Composites* 1999 **30**(10), 1187–1195.
75. American Society for Testing and Materials, *Standard Test Method for Tensile Properties of Fiber-Resin Composites*, Designation D3039-76, ASTM, Philadelphia, 118 (Re-approved 1989).
76. Laminate Analysis Program (LAP), DOS version, Centre of Composite Materials, Imperial College, London.

3 FT-IR Spectroscopy of Ultrathin Materials

Takeshi Hasegawa, Veeranjaneyulu Konka, and Roger M. Leblanc

CONTENTS

3.1 INTRODUCTION

Two-dimensional molecular assemblies that involve monolayers (monomer aggregates) and polymer membranes are expected to be promising nano soft-materials. Monolayers formed as monomer assemblies are of particular importance because of the following characteristics: (1) thickness can be controlled by monolayer level in nanometer scale; (2) functionalized monolayers can be stacked, and designing stratified layers is relatively easy; (3) no covalent bond formation is necessary to assemble the monomer molecules to form a monolayer, and the monolayer construction employs a "force balance" between some physical parameters, such as in-plane hydrophobic interactive force and interlayer interactive force. Since the monolayer formation requires no covalent bonds among the monomers, monolayer fabrication has two major benefits: production energy can be reduced, and resource compounds can be recycled after use with low energy. Very recently, therefore, these two benefits have been recognized in terms of new materials science and energetic interests. As for polymer layers, the interlayer interaction depends on noncovalent bonds, while the monomer molecules in the polymer layer are covalently bonded.

The concept of the strong inter-electron correlation used in physics has recently been introduced into soft-material chemistry, so that the force balance of chemical correlation factors could be discussed effectively, and a new concept of strongly correlated soft materials (SCS) has been created.[1] As a chemical correlation factor, hydrogen (H) bonding has particularly been focused on, since H-bonding has relatively high energy (~5 kcal mol^{-1}) in comparison to that of ionic bonds (4~7 kcal mol^{-1}).[2] In fact, in biomaterials like proteins, H-bonding plays a crucial role, correlating with hydrophobic interactive forces to form three-dimensional structures and functions.[3–5] In the case of nanomaterial science, studies of SCS would be even more important.

Physicochemical studies of soft materials have been considered more difficult than those of hard materials, since theories and experimental approaches of solid-state physics are difficult to apply to the soft materials.[6] It is true that soft materials, including polymer materials, have relatively poor periodicity of molecular arrangement in comparison with inorganic crystals, and the characteristics of phonons become complicated. Nevertheless, the feasibility of assembly of organic compounds based on the concept of SCS would be more important for new material formation. To physically evaluate new or unknown materials, infrared spectrometry is a very powerful technique, since it extracts anisotropic dielectric properties through vibrational normal modes.[6] Anisotropic phonon analysis via the dielectric properties is quite useful for discussions of the molecular structure in the material of interest, which is strongly correlated to inner molecular arrangement and molecular orientation. It is worth noting that infrared spectroscopy is a useful technique for anisotropic phonon analysis in soft materials, a fact that is not well recognized at present. In this chapter, the theoretical basics of various infrared spectrometries for surface chemistry and studies of SCS are described by introducing application studies.

3.2 MONOLAYERS ON WATER SURFACE

3.2.1 Langmuir Technique

The Langmuir technique provides a promising approach for understanding *in vivo* membrane properties because it allows for fine control of the molecular constituents and their relative orientations that are similar to those found in natural membranes. In addition, this technique permits us to study the effects of environmental parameters such as temperature, pH, and ionic strength, and to use those parameters to control the desirable formation of films. Nevertheless, to utilize this technique with confidence, the optimization of the Langmuir technique for different samples and the development of systematic methods for film formation are required.

As examples of complex biological molecules, photosystem II (PS II) of higher plants and acetylcholinesterase (AChE) were selected as model systems, and the following information explains the significance and applications of Langmuir technique. By following the monolayer properties of PS II and AChE, we can understand clearly the governing principles of monolayers at the interface.

3.2.1.1 Langmuir Films of Photosystem II

Using the Langmuir technique, we investigated the surface pressure and surface potential properties of monolayers of PS II membranes and PS II core complexes, as well as the mixed monolayers of PS II core complexes and supporting lipids monogalactosyl diacylglycerol (MGDG) formed at the air–water interface. The optimization of utilizing Langmuir technique for PS II systems is essential for further investigations using surface spectroscopy and atomic force microscopy.[7]

3.2.1.2 Surface Pressure Studies

The surface pressure (π) as a function of molecular area (A), or the π-A curve, as the sample at the air–water interface being compressed, is an important indicator of the quality of the monolayer. A sudden change in surface pressure usually signals the formation or collapse of a stable layered structure. Although the PS II membrane proteins possess a pair of hydrophobic and hydrophilic parts, which are appropriate for forming a monolayer at the air–water interface, their ability to form a stable monolayer structure is hindered by their massive size and density. The complex structure of the PS II membrane and the interactions among its subunits further complicate the interpretation of its π-A curve. The π-A curve of PS II membrane proteins, although quite typical of other proteins (Figure 3.1), lacks some sharp features that are representative for those more rigid and well-structured materials such as the pure lipids. Nevertheless, the fact that stable π-A curves can be produced with reliability and repeatability supports the idea that the stable monolayer of PS II membranes can be formed reliably at the air–water interface.

It is also interesting to note that only a fraction of PS II membrane proteins formed a stable monolayer structure and rest of them fell into the water subphase. This can be seen directly by the naked eye during compression. Furthermore, if we

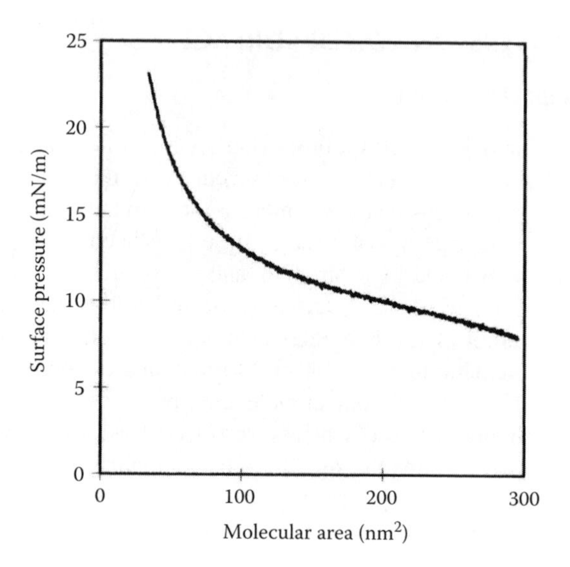

FIGURE 3.1 Surface pressure-area isotherm of PS II membranes.

use the total amount of PS II membrane proteins to calculate the average particle size from the π-A curve, we obtain an area of about 200 nm^2. This value is very small when compared with that of the PS II core complex (320 nm^2, as discussed in a subsequent section), which is a smaller subunit of the PS II membrane. A PS II membrane fragment contains PS II core complex and several LHC II proteins, and is much larger than a PS II core complex particle. Using the actual particle size along with the π-A curve, we can estimate that only a fraction of PS II membrane proteins stay on the surface and form a stable monolayer structure.

Compression of the PS II membrane monolayer shows that the monolayer collapses at a relatively low surface pressure, around 20 mN m^{-1}. This can be attributed to the formation of a multilayered structure,[7] and to the fact that some of the PS II membrane fragments diffuse into the subphase. This observation further indicates that PS II membranes can only marginally stay at the air–water interface and one must be very careful in choosing the experimental parameters.

In the case of PS II membrane proteins, as discussed above, the hydrophobic and hydrophilic pairs of attached lipids can partially support the protein complex at the air–water interface, despite their large size and density. However, in the case of the PS II core complex, the detergent strips the attached lipids and some extrinsic proteins. The remaining protein complex is water-soluble. It is very difficult to prepare a stable monolayer of water-soluble proteins with the Langmuir method. Indeed, it is difficult to prepare a stable monolayer of PS II core complex directly, because of its water solubility as well as its density. One possible solution is to change the density and ionic strength of the subphase.[8]

Leblanc et al. studied the π-A isotherms of the PS II core complex at different concentrations of NaCl in the subphase (Figure 3.2). The addition of NaCl solution greatly enhanced the stability of monolayers of PS II core complex particles at the

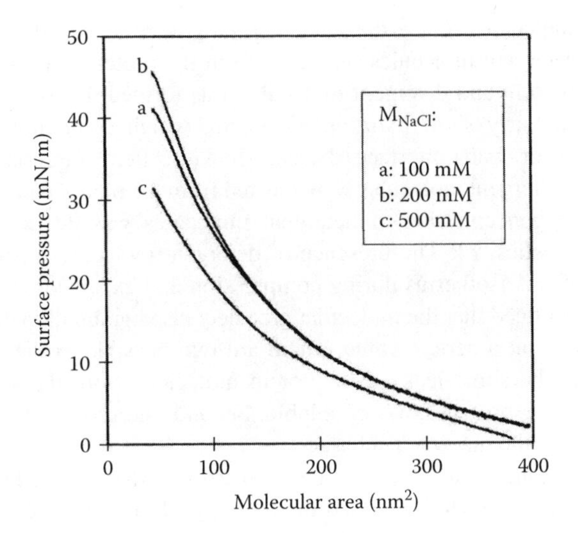

FIGURE 3.2 The surface pressure-area isotherms of PS II core complex with different concentrations of salt in the subphase.

air–water interface. The π-A curves at subphases of 100 mM and 200 mM NaCl solutions clearly demonstrated that PS II core complexes could be compressed to a relatively high surface pressure (40 mN m^{-1}) before the monolayer collapsed, under our experimental conditions. Moreover, the average particle size calculated from π-A curves using the total amount of protein complex is about 320 nm^2. This observation agrees well with the particle size directly observed using atomic force microscopy,[7] and indicates that nearly all the protein complexes stay at the water surface and form a well-structured monolayer.

However, one must be careful when using a high concentration of salt to increase the density of the water subphase. There are two important disadvantages in using a high-concentration salt solution as subphase: (1) A high concentration of salt in the subphase can change the protein conformation at the air–water interface; (2) The salt particles can be deposited together with the protein complex monolayer onto the substrate when Langmuir-Blodgett (L-B) films are prepared.[9] An optimal condition must be attained in order to use this technique effectively. It was noticed that there were few differences in the π-A curves of 100 mM and 200 mM of NaCl in the subphase. However, for 500 mM of NaCl in the subphase, the average particle size calculated from the π-A curve decreased when compared with that of 100 mM and 200 mM of NaCl in the subphase. This indicates that the conformation of the PS II core complex is altered in a high-concentration salt solution. Therefore, 100 mM NaCl concentration in the subphase was chosen for studies.

It was also noticed that the initial surface pressure is relatively high for all subphase compositions. This is because a rather small Langmuir trough was used in the presence of residual detergent in the sample.

The π-A isotherm was not reproducible during compression and expansion. In the sample solution, the hydrophobic regions of protein subunits were covered by the detergent molecules. Once the PS II samples were spread onto the air–water interface, the detergent molecules separated from the protein particles and a mixed monolayer of protein and detergent molecules was formed. However, the detergent molecule had a fairly short hydrophobic chain, which could not form a stable monolayer at the air–water interface (data not shown). After compression to relatively high pressures, detergent molecules were ejected from the monolayer to the subphase and the surface concentration of detergent molecules was decreased due to the formation of micelles.[10–12] The presence of detergent molecules resulted in nonreproducibility of π-A isotherms during compression and expansion.

It was also noticed that the molecular area decreased gradually when the surface pressure was held at a certain value. There are two possible explanations for this. First, there may be some leakage of protein molecules from the surface into the subphase, since the protein is water-soluble. Second, the protein denaturation may be taking place at the air–water interface.

The studies on the surface pressure-area isotherms of MGDG and the mixture of PS II core complex and MGDG suggested the presence of both PS II core complex and MGDG in the monolayer. MGDG molecules diluted the PS II core complex concentration in the monolayer. MGDG lipid functioned as a support for the protein complex and the resulting mixture formed higher quality films than PS II core complex alone.

3.2.2 Surface Potential Studies

The surface potential as a function of particle area (ΔV-A isotherm) is another indicator of the quality of the monolayer structure. The surface potential at an air–water interface changes as the film-forming molecules reorient themselves during the compression process. For a closely packed monolayer, the surface potential is directly proportional to the surface dipole moment ($\mu_\perp$), as shown by:[13]

$$\Delta V = 12\pi\mu_\perp / A \tag{3.1}$$

where ΔV is the surface potential change in millivolts when a monolayer is spread at the air–water interface, A is the molecular area in Å^2 molecule^{-1}, and $\mu_\perp$ is the surface dipole in milliDebye (mD) units.

As the surface area is compressed from 400 to 175 nm^2, the surface potential remains at a constant value, whereas the surface pressure increases slightly from 6.5 to 10 mN m^{-1} (Figure 3.3). Further compression results in a decrease in the surface potential and a sharp increase in the surface pressure. Both the decrease in surface potential and the sharp increase in surface pressure after 175 nm^2 clearly indicate the collapse of monolayer structure and the formation of the multilayer at a surface pressure larger than 10 mN m^{-1}. The formation of a multilayered structure partially cancels out the molecular dipole moment of the PS II membrane protein and results in the decrease in surface potential. Independent atomic force microscope (AFM) studies confirmed that the monolayer structure of the PS II membrane protein collapsed under the high surface pressure.[7] From the ΔV-A curve of PS II membrane protein, we can calculate the surface dipole moment at about 1.58×10^5 mD.

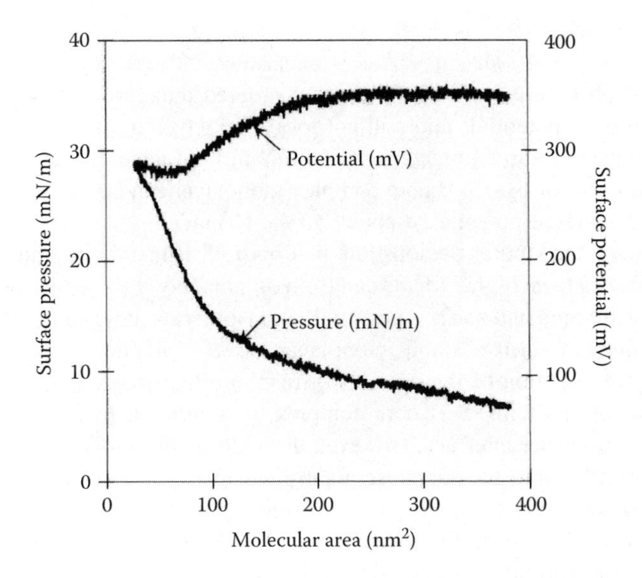

FIGURE 3.3 The surface pressure-area (π-A) and surface potential-area (ΔV-A) isotherms of PS II membranes.

The ΔV-A isotherm of PS II core complex is rather different from that of PS II membrane (Figure 3.4). The surface potential of a monolayer of PS II core complex increases slightly as the molecular area is compressed from 600 to about 150 nm^2, while surface pressure changes from 5 to 35 mN m^{-1}. Further compression results in a sharper increase in surface potential. The surface potential starts to decrease

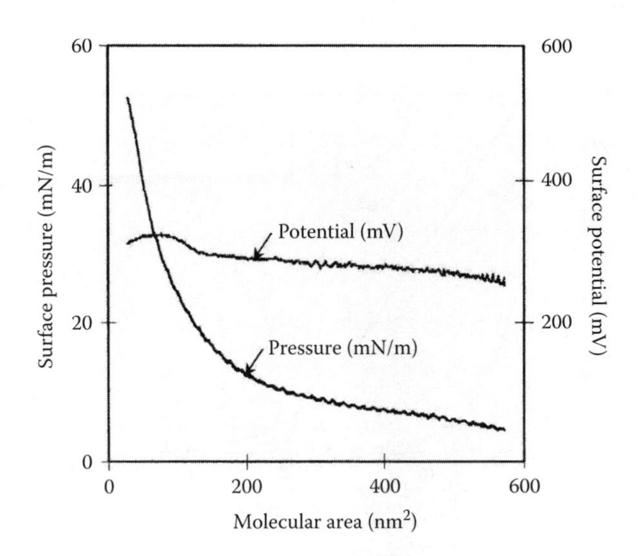

FIGURE 3.4 The surface pressure-area (π-A) and surface potential-area (ΔV-A) isotherms of PS II core complex particles.

only after the surface area is compressed to about 80 nm^2 or the surface pressure becomes larger than 40 mN m^{-1}. This is consistent with previous discussion indicating that PS II core complexes form more ordered monolayer structures at relatively high surface potentials and will not form multilayered (collapsed) aggregates until the surface pressure is greater than 40 mN m^{-1}. At a surface area of 320 nm^2, the dipole moment of the PS II core complex monolayer can be calculated from the corresponding surface potential at about 2.38×10^5 mD.

In summary, the studies demonstrate that both PS II membranes and PS II core complex particles from higher plants can form monolayers at the air–water interface. Surface pressure-area and surface potential-area isotherms indicate that PS II membrane proteins can form a stable monolayer directly at the air–water interface, although a large fraction of the proteins diffuse into the aqueous subphase. As for the water-soluble protein PS II core complex, it is difficult to form a monolayer directly at the air–water interface. However, the addition of a moderate concentration of NaCl (100 mM) into the subphase greatly enhances the film-forming ability of PS II core complex and the quality of the resulting monolayer. It was also demonstrated that the addition of lipid MGDG to the PS II core complex also greatly enhances the film-forming ability, and the mixture monolayer shows good miscibility at all surface pressures. These results confirm that lipid MGDG can function as a supporting material in the mixture monolayer. Thus, studies with PS II monolayers evidently enlighten us as to the requirements for developing stable monolayers at the interface.

3.2.3 Langmuir Films of AChE

The surface pressure-area isotherms (Figure 3.5) of AChE monolayers under different subphase conditions (subphase composition and pH) were studied at different surface

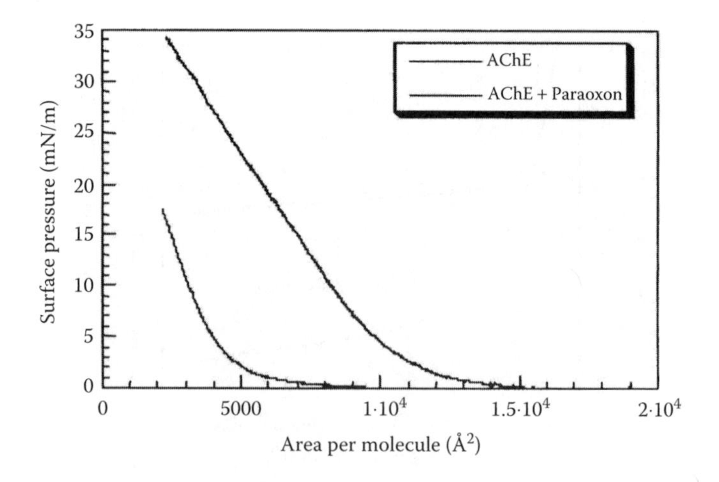

FIGURE 3.5 The surface pressure-area isotherm of AChE in the presence or absence of paraoxon.

pressures. The UV absorption spectra of the enzyme monolayer were recorded at different surface pressures and the spectra showed a peak around 200 ± 2 nm, which was assigned to the absorption band of peptide bonds.

The data clearly showed that AChE forms a highly stable monolayer at the air–aqueous interface. Compressions of this enzyme on the surface of a subphase containing salt led to a decrease in its solubility in the bulk medium, resulting in the formation of more closely packed and organized monolayers, similar to the observations made with photosystem particles. This notable stability of the enzyme was attributed to its polar and nonpolar nature, and also to its high molecular weight. Studies on the pH-dependent process of enzyme stability showed that the organized AChE monolayers at the interface require a pH value of 6.5 in the bulk medium.

3.3 LANGMUIR-BLODGETT FILMS

The L-B technique is a powerful tool for investigating the models of biological membranes and for understanding the molecular structure and function of the proteins.[9,14–16] This method allows for the creation of complicated molecular systems; i.e., to realize the ideas of molecular architecture. In recent years, a new wave of interest arose from the possibility of using protein L-B films as biosensors in molecular electronics.[9,14,17–19] Several studies demonstrate that the stability and activity of complex biological molecules arranged in closely packed two-dimensional arrays in L-B films remain unchanged from the *in vivo* state.[19–23] L-B film technology allows for convenient creation and deposition of oriented protein films with a high density of active sites of the enzyme molecules. It offers the advantage of tailoring sensitivity of sensors via control of the number of deposited films. It also offers a reproducible casting of the receptor layer. Hence, this technique is well suited to understanding the molecular interactions between two compounds. Some of the details of L-B films of chlorophyll-*a* (Chl-*a*) and AChE molecules will be introduced in the next section.

3.3.1 LANGMUIR-BLODGETT FILMS OF CHLOROPHYLL-A

In photosynthetic pigment protein complexes, Chl-*a* molecules are organized with a well-ordered spacing and orientation for the gathering of light energy and for the efficient transfer of absorbed energy to the site of charge separation.[24] This charge separation occurs in the reaction centers of photosystems. In the PS II reaction center, charge separation occurs at a specialized Chl-*a* molecule called P680. The oligomeric nature of P680 *in vivo* is a matter of considerable interest. As mentioned, L-B films have the potential to assemble individual molecules into a highly ordered architecture.[25] This exciting aspect has motivated researchers to simulate the models of natural membranes. With the development of scanning probe microscopy, attempts have been made to understand the organization of L-B films of the molecules at a nano-scale level. In order to understand the oligomeric nature of the Chl-*a* molecules in PS II reaction centers, Leblanc et al. have developed mono- and multilayers of Chl-*a* molecules and observed their topography using scanning probe microscopy.

The images of Chl-*a* molecules in multilayers were recorded using scanning tunneling microscopy (STM) and atomic force microscopy (AFM). The images

FIGURE 3.6 AFM image of acetylcholinesterase L-B film.

show that films undergo a reconstruction process following their fabrication and transfer onto the substrate in order to reach a more stable state. Although the reconstruction differs from one film to another, the repeated unit in the reorganization of the films is always found to be a dimer of Chl-*a*. In summary, the studies with mono- and multilayer L-B films of Chl-*a* presented a strong evidence for the dimeric nature of Chl-*a in vivo*. The model presented explains the high-energy transfer efficiency in the pigment protein complexes of photosystems of photosynthetic apparatuses.

3.3.2 Langmuir-Blodgett Films of AChE

The tapping mode AFM (TMAFM) images of AChE L-B films (Figure 3.6) show that AChE might exist in the form of monomer, dimer, or tetramer. Considering this observation, the size of the small particles observed by the TMAFM is within the range of the dimensions estimated by X-ray data for an AChE monomer (175,000 $Å^3$). On the other hand, the dimensions obtained with the AFM (720,000 $Å^3$) for the large and medium size of AChE particles are within the range of the estimated dimensions of an AChE tetramer (700,000 $Å^3$). Therefore, we conclude that the most abundant population of the L-B film is the globular AChE monomer. Moreover, the larger and medium-sized particles of AChE represent the tetramer form of this enzyme. We made a comparative study of AChE interaction with substrate acetylthiocholine (ATChI).

For a better understanding of the complex formation mechanism between AChE and ATChI, AChE L-B film was prepared and examined with TMAFM in two steps. The monolayer was first compressed on the substrate-free buffered subphase. After reaching a surface pressure of 25 mN m^{-1}, the acetylthiocholine was injected into the subphase. The TMAFM images (Figure 3.7a) of a transferred monolayer, 6 minutes after the injection, show the presence of acetylcholinesterase-acetylthiocholine complex and a homogeneous monolayer composition. However, the images of a second transferred monolayer at the same surface pressure, but 15 minutes after the injection, show the formation of a mixed monolayer, due to the presence of both the enzyme-substrate complex and free enzyme (Figure 3.7b). This was the first time that such a visual interaction between AChE and the substrate was noticed.

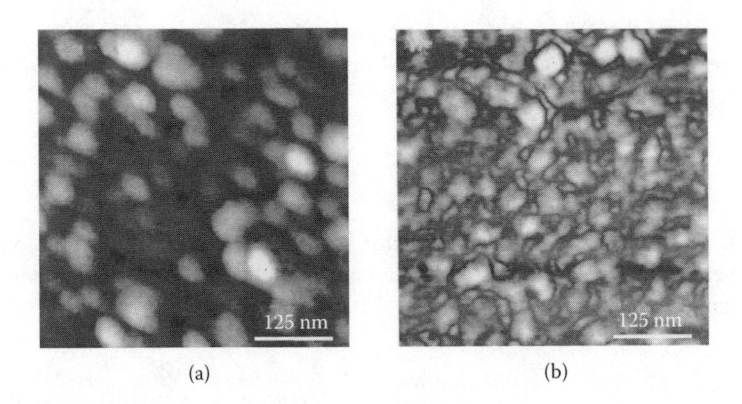

FIGURE 3.7 AFM image of L-B film of acetylcholinesterase (a) and L-B film of acetylcholinesterase-acetylthiocholine complex (b).

3.4 TRANSMISSION AND RAS TECHNIQUES

As condensed materials, Langmuir and L-B films are of much interest because of the high anisotropy in molecular film structure. The assembled molecules are highly ordered in the film plane, and molecular conformation is often fairly ordered, especially when the surface pressure is high. In this situation, molecular groups are also well oriented, at least in uniaxial molecular topography.[26] In regard to the dielectric properties of the films, the highly condensed L-B films provide materials with physically unique characteristics.[6,26] Since there is no electric transition in a wavenumber region of infrared absorption, the dielectric dispersions are very simple, and they are easily monitored by FT-IR spectroscopy.

The uniaxial dielectric properties are represented by the two transition-moment dependent spectra: in-plane (IP) and out-of-plane (OP) spectra. When a spectrum specifically responds to IP vibrational modes only, the spectrum is observed by transmission spectrometry. On the other hand, when a spectrum specifically responds to OP vibrational modes, the spectrum is observed on a metallic substrate by reflection spectrometry.[27,28] The latter measurement technique performed on metallic substrates is called reflection–absorption spectrometry (RAS).[26] In this section, the basic theory of the transmission and RAS will be discussed.

3.4.1 Surface-Normal Incidence Transmission Spectroscopy

For quantitative analysis of surface monolayers by means of FT-IR spectroscopy, the core theory is based on Beer's law.[29] For example, when the number of monolayers becomes doubled, the absorbance of every band is doubled as well. This linear relationship is intuitively easy to follow. It is especially used for the evaluation of concentration of a chemical constituent in liquid. Nevertheless, the law, though it seems very easy to understand, offers deeper insights, particularly for anisotropic condensed matter, since the absorption coefficient is not a scalar, but a tensor. To explore the characteristics of the absorption coefficient, a simplified Maxwell's equation is useful.[30] When the magnetic permeability is unity and electric current is not present in the isotropic matter, Maxwell's equation yields the following equation:

$$\nabla^2 \mathbf{E} - \frac{\varepsilon}{c} \frac{\partial^2 \mathbf{E}}{\partial t^2} = 0 \tag{3.2}$$

where, **E**, ε, and c are electric-field vector, dielectric function, and light constant, respectively. Simply stated, this equation can be represented in a one-dimensional vector manner:

$$\left(\frac{\partial^2}{\partial x^2} - \frac{\varepsilon(\omega)}{c^2} \frac{\partial^2}{\partial t^2} \right) E(x,t) = 0 \tag{3.3}$$

The solution of this equation is easily deduced as the following:

$$E(x,t) = E_0 \exp(\mathrm{Im}\, \tilde{\nu} x - \mathrm{Im}\, \omega t) \tag{3.4}$$

Here, $\tilde{\nu}$ and ω are wavenumber and angular frequency, respectively. The tilde (~) indicates the variable under this symbol has a complex value. The wavenumber $\tilde{\nu}$ is formulated by the following equation.

$$|\tilde{\nu}|^2 = \frac{\varepsilon(\omega)\omega^2}{c^2} \tag{3.5}$$

Now, the next relationship is introduced in Equation (3.5).

$$\sqrt{\varepsilon} = \tilde{n} = n + ik \tag{3.6}$$

The real and imaginary parts of complex refractive index ($\tilde{n}$) are represented by n and k, respectively. With this relationship, Equation 3.4 changes to the following:

$$E(x,t) = E_0 \exp(-2\pi k \tilde{\nu} x) \exp(\mathrm{Im}(2\pi n \tilde{\nu} x - \omega t)) \tag{3.7}$$

The intensity of electric field is calculated as the squared value of the amplitude of the electric field; the intensity I is deduced as:

$$I = |E|^2 = I_0 \exp(-4\pi k \tilde{\nu} x) \tag{3.8}$$

With this equation, we can derive Lambert's equation by applying the natural logarithm:

$$A = \alpha x \tag{3.9}$$

Here, the absorbance A is defied as $A = \ln(I_0/I)$. Note that Lambert's law is slightly different from Beer's law ($A = \varepsilon cx$), in which the "common" logarithm was applied to Equation 3.9. The physical meaning of Lambert's law is, however, almost identical to that of Beer's law. From Equation 3.8 and Equation 3.9, the following important relationship is yielded:

$$\alpha = 4\pi k\tilde{\nu} = 4\pi k\lambda^{-1} \tag{3.10}$$

This equation implies that the absorptivity (α) is not an identical value for the matter, and it depends on the direction of a transition moment (or the absorption index [k] vector). In this manner, the infrared absorption by matter is confirmed by the interaction angle between the electric field and the direction of the transition moment. This makes infrared spectroscopy a powerful tool for evaluating the anisotropic structure in ultrathin materials quantitatively.

Figure 3.8a presents a schematic view of infrared transparent material with L-B films of 11-monolayer cadmium stearate deposited on the surfaces. As presented in this figure, the infrared ray is ordinarily irradiated perpendicularly to the film surface when transmission spectra are measured. Since most of the infrared spectrometers provide an uncollimated infrared beam that has a cone shape of approximately 5°, the sample is often placed at the focus point of the infrared beam. At the focal point, the diameter of the infrared spot is approximately 0.1 mm. Of note is that the detection area on the detector surface is very small, approximately 1.0 mm^2 especially for an MCT detector. Therefore, the infrared spot on the detector surface sometimes overfills[31] the detector area, when the optical alignment is largely changed by the sample thickness and its refractive index. This overfilling causes loss of a portion of infrared power, which results in dark measurements. To avoid this overfilling, the appropriate positioning of the sample in the light path is very important.

Newer FT-IR spectrometers are equipped with a high-output light source, and the detector is often saturated when the transmitted beam is introduced directly to

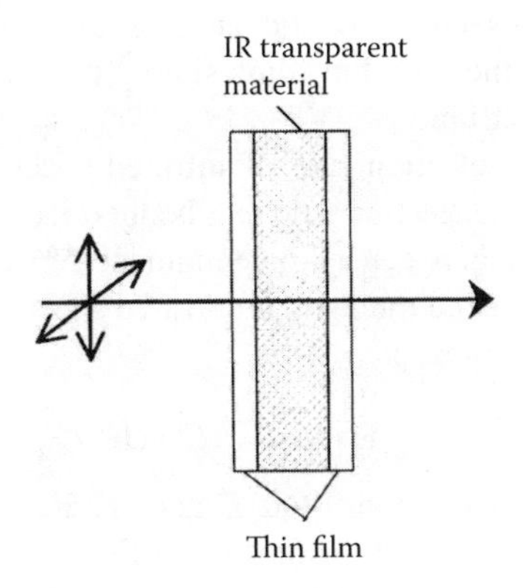

FIGURE 3.8 Schematic of IR transparent measurements of L-B films deposited on an IR transparent material.

the detector. By watching the maximum voltage of the interferogram during the setup, one can manually avoid saturation. A wire grid is simply placed on the detector-side window to reduce the intensity. A few metallic wire grids are included with newer FT-IR spectrometers for the purpose. Another technique to reduce the beam intensity is to change the aperture size. The diameter of the aperture in the optics can be changed by software, which is ordinarily controlled automatically. When the detector is saturated even after the automatic control, one can decrease the size manually. This is particularly important when the modulation frequency[32] is set to a low frequency to earn throughput. For example, when we want to measure monolayer level spectra with a high signal-to-noise ratio, the velocity of the moving mirror of the interferometer is decreased as much as possible, since the low-velocity movement of the mirror earns time for one scan, which enables us to collect a bright signal.

It is of interest to compare a transmission spectrum of the L-B film to a KBr pellet transmission spectrum (data not shown). In the L-B film, the film molecules are highly oriented to yield anisotropic dielectric properties. In the potassium bromide (KBr) pellet, on the other hand, microcrystallines are randomly dispersed in the pellet, which results in a nonoriented status, even if each microcrystalline has highly oriented molecules.

The most significant difference in the spectra is found for the antisymmetric and symmetric COO^- stretching vibration bands that are located at 1540 cm^{-1} and 1430 cm^{-1}, respectively. In the KBr spectrum, both bands are observed as comparative bands in intensity. On the other hand, in the normal-incident transmission spectrum of the L-B film, only the antisymmetric COO^- stretching vibration band appears strongly, while the symmetric band is largely suppressed. This change in relative band intensity suggests that the transition moments of the bands are highly oriented in specific directions: surface parallel and surface normal. When an infrared ray is perpendicularly irradiated on a film, the electric field vector is directed parallel to the film surface, no matter what polarization is used. Therefore, in the normal-incident transmission spectra of L-B film, the surface parallel component of transition moments is selectively observed, while the surface parallel component is largely suppressed.[33] This characteristic enables us to qualitatively discuss the molecular orientation in the ultrathin films simply by comparing the transmission spectrum to the KBr spectrum.

This surface-specific selection rule of infrared spectroscopy is called surface selection rule. The surface selection rule can be theorized in a very simple way by considering the interaction between a dipole moment (**M**) and an electric-field vector (**E**). When the observable band intensity in infrared spectra is noted as I, the surface selection rule can be expressed as:

$$I = C(\mathbf{E} \bullet \mathbf{M})^2 = (EM\cos\theta)^2 \tag{3.11}$$

Here, C is a proportional constant, and E and M are amplitudes of **E** and **M**, respectively. The angle of **M** with respective to **E** is expressed by θ. This simple relationship implies that the molecular orientation angle would be evaluated quantitatively by theoretically connecting I to absorbance A. Details of this evaluation technique will be discussed below.

In addition, it is of importance to note that the surface selection rule of the transmission spectroscopy enhances transverse optic (TO) modes only. Since the infrared ray of the surface-perpendicular incidence generates the so-called transverse waves of electric fields in the film media, the infrared ray induces alternately directed dipole zones that are perpendicular to the wave-propagation vector.[34] Such a dipole zone can interact with TO modes only, and the longitudinal optic (LO) modes are not observable. When we want to observe the LO modes in the film, the most convenient way is to employ reflection–absorption (RA) spectrometry, as discussed in the next section.

It is also of interest to see the dependence of absorbance on the infrared-transparent materials. If the film of interest is a freestanding film, the absorbance is obtained as a unique value, which is dominated by the optical constants of the film only. When the film is deposited on another material, however, the optical situation changes. For example, when an infrared ray is irradiated on a flat dielectric material surface, the reflected ray would have a different phase from π, which would reduce the electric field intensity near the surface. This tendency becomes stronger when the dielectric constant of the material is high. The evaluation of the reduction is available in a study by Umemura et al.[35] A table after his paper is presented as Table 3.1. It is a surprise to see the result that the absorbance becomes 60% less than the freestanding film when a Ge substrate is used for the L-B film, since the refractive index of Ge is very high (ca. 4.0). This indicates that a high dielectric-constant material makes the transmission measurements of thin L-B films insensitive. In this sense, CaF_2 would be a good material to sensitively measure the L-B films. Note that this theory holds for very thin films only, and if a relatively thick film like polymer cast films were measured, the influence of this effect would be minor.

Since the ordinary transmission technique does not involve optical enhancement, this technique may be difficult to apply for analysis of monolayer-level thin films. In recent years, FT-IR has significantly progressed in both sensitivity and signal-to-noise ratio, and monolayer work can be done with the transmission technique. When we used an old FT-IR, however, it was generally difficult to measure monolayers on an infrared transparent substrate. In such a case, monolayers are accumulated to form a multilayer L-B film in order to achieve signal intensity.

Dhanabalan et al.[36] studied conductive polymer (polyaniline) monolayers mixed with cadmium arachidate by using π-A isotherm, X-ray diffraction, UV-visible, and FT-IR transmission spectrometric techniques. The X-ray analysis of an 11-monolayer mixed L-B film (Figure 3.9a) presented clear diffraction peaks when the L-B film was not chemically treated, while all the peaks were lost when the film was treated with HCl vapor (Figure 3.9b). X-ray diffraction peaks appeared when a periodical structure in the film had a strong electron-diffractive property. Metal ions with high electron density caused the diffraction, and even the thin film exhibited strong diffraction peaks.[37] Therefore, the disappearance of the peaks suggests two possibilities: (1) the periodic structure has been lost by the chemical treatment, or (2) the metallic ion has been removed from the film. A good reason for using an IR technique for the combination analysis with X-ray technique is that both techniques can be directly applied to common multilayer L-B films. In other words, an identical L-B film can be analyzed by the different techniques, which provides us with multiple

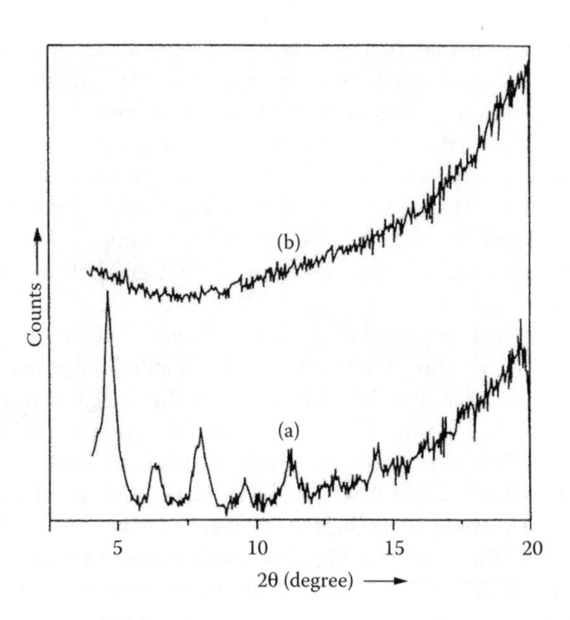

FIGURE 3.9 X-ray diffraction patterns of polyaniline and cadmium arachidate mixed L-B film on quartz (11-layer). (a) As deposited; (b) after HCl vapor treatment. [Reprinted with permission from *Thin Sol. Films* 295: 255–259, 1997. Copyright (1990) Elsevier Science.]

pieces of chemical information. They observed FT-IR transmission spectra of multilayer L-B films, as presented in Figure 3.10. The spectra were very clear in the entire mid-infrared region, and it was clearly found that the antisymmetric COO^- stretching vibration band strongly appears at 1545 cm^{-1} in both cadmium arachidate and the mixed L-B films. After the HCl treatment, however, this band is largely diminished, and a new band appears at 1702 cm^{-1} instead. Since this new band is attributed to the C=O stretching vibration band, the spectra strongly suggest a highly effective conversion from arachidate salt to arachidic acid by means of HCl vaporing. This discussion is supported by another point: both O-H and N-H stretching vibration bands at about 3440 cm^{-1} and 3240 cm^{-1} are significantly increased in intensity.

Zhao et al.[38] studied the anisotropy of the molecular arrangement of an azobenzene derivative in an L-B film. They prepared a six-monolayer L-B film of 4-didodecylamino-4′-(3-carboxypropyl)azobenzene on a CaF_2 plate by the ordinary vertical dipping technique. Since this molecule has a large planer moiety, it was expected to have anisotropic molecular arrangement, which depends on the dipping direction of the substrate during the preparation of the L-B film. They measured FT-IR normal-incident polarized transmission spectra at different polarization angles to the film, so that dichroic analysis would be performed (Figure 3.11). It is of interest to note that the C=O stretching vibration band dominantly appears at 1694 cm^{-1} in the spectrum at 0° to the dipping direction, whereas the same band strongly appears at 1708 cm^{-1} at 90°. They evaluated the orientation angle of the two C=O bonds as −2° and 87° with respect to the dipping direction. This suggests that the two C=O

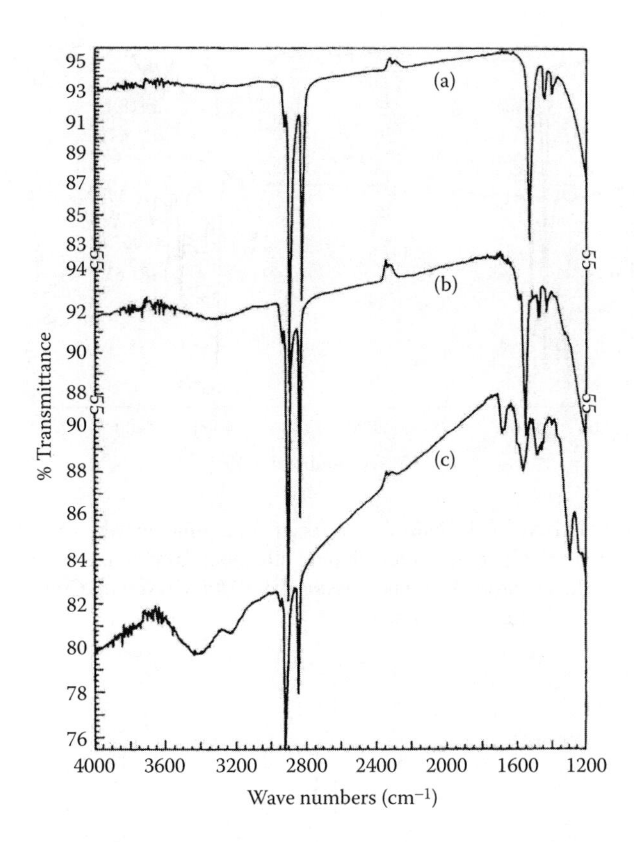

FIGURE 3.10 FT-IR transmission spectra of (a) as-deposited cadmium arachidate L-B film (21 layers); (b) as-deposited mixed L-B films of cadmium arachidate and polyaniline (15 layers); and (c) HCl vapor treated mixed L-B film of cadmium arachidate and polyaniline (15 layers). [Reprinted with permission from *Thin Sol. Films* 295: 255–259, 1997. Copyright (1990) Elsevier Science.]

bonds are oriented nearly perpendicularly (Figure 3.12) to each other. In this manner, infrared dichroism is a powerful means of investigating molecular anisotropy in thin-layer materials.

Recently, the number of papers that report transmission spectra of single monolayers is rapidly increasing, and it has become a relatively easy task to measure transmission spectra of ultrathin layers. For example, Johal et al.[39] reported a very beautiful FT-IR transmission spectrum of a single-layer L-B film of 4-eicosyloxo-(E) stilbazolium iodide. The spectrum is presented in Figure 3.13. Although the spectrum was derived from a single monolayer, the quality of the spectrum is high. Another surprise is that this spectrum is obtained by use of a DTGS (not MCT!) detector with a moderate modulation frequency of 10 kHz. In this manner, particularly when no polarizer is used, bright measurements can be performed, and a DTGS detector is adequate for the measurements. It should be noted, however, that perfect air purge inside the spectrometer is an inevitable key to obtaining such a good

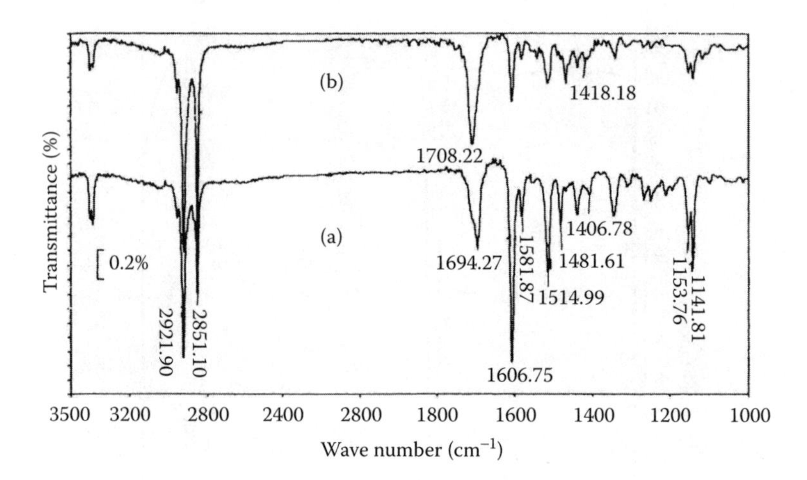

FIGURE 3.11 Polarized FT-IR transmission spectra of a 6-monolayer azo L-B film on CaF_2 substrate. (a) Polarization parallel to the dipping direction; (b) polarization perpendicular to the dipping direction. [Reprinted with permission from *Mol. Cryst. Liq. Cryst.* 294: 185–188, 1997. Copyright (1990) Taylor and Francis.]

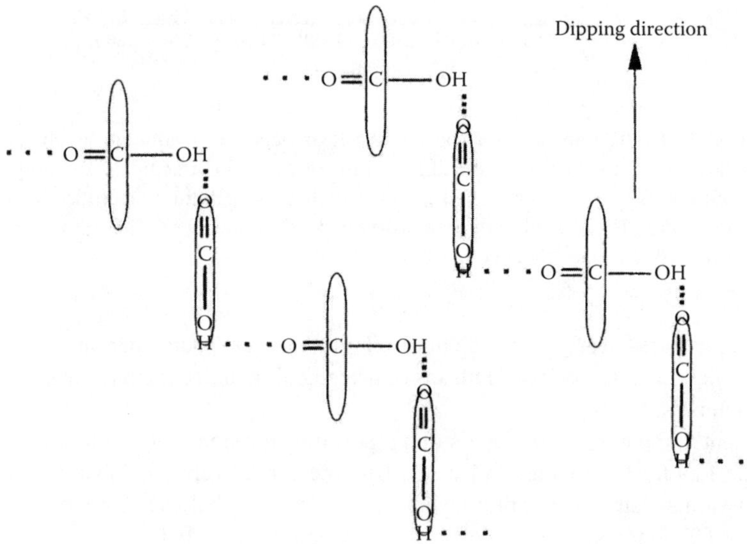

FIGURE 3.12 Projection of the two-dimensional hydrogen-bonded structure on the substrate surface, where the narrow rectangles represent the azobenzene group. [Reprinted with permission from *Mol. Cryst. Liq. Cryst.* 294: 185–188, 1997. Copyright (1990) Taylor and Francis.]

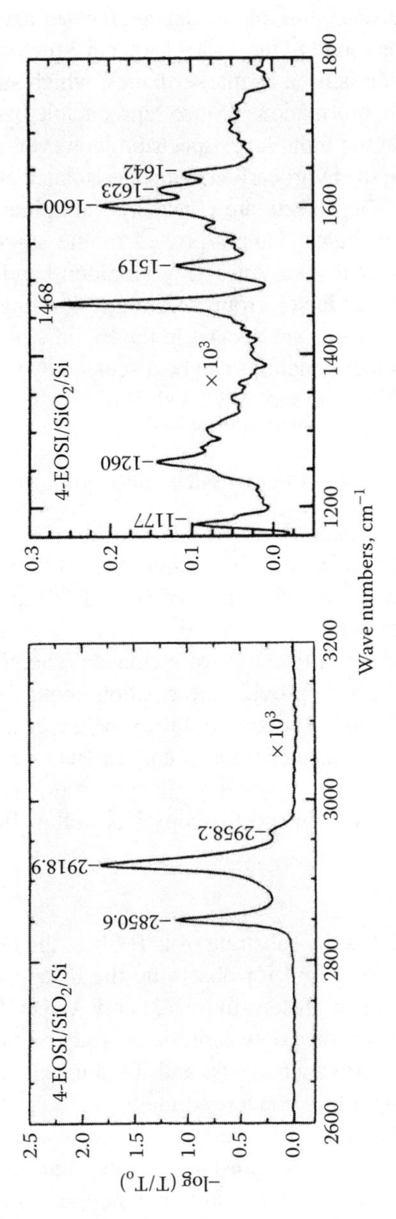

FIGURE 3.13 Transmission FT-IR spectrum of the single-layer L-B film of 4-EOSI. [Reprinted with permission from Langmuir 15: 1275–1282, 1999. Copyright (1999) American Chemical Society.]

spectrum. Water absorption bands at about 1600 cm^{-1} can be arithmetically subtracted from the observed spectra, which does not work well when the band intensity is very weak, because strong water absorption often influences the interferogram collection and disturbs the monolayer measurements. Once a high quality spectrum is readily obtained, we can discuss fine molecular aggregates and packing. In the paper, the series of weak bands found in the 1150–1300 cm^{-1} region were discussed. In general, the band progression is used as marker bands, which suggest the hydrocarbon chain has an all-*trans* conformation.[21] Since reproducible bands at 1303 cm^{-1} and 1341 cm^{-1} are observed in the monolayer spectrum, however, it was concluded that this spectrum suggests that the hydrocarbon chain has isolated chain-kink (*tgtg't*) and end-*gauche* (*tg*) defects. The defects are considered marginally to perturb the lateral packing of the film molecules. This is proved by the singlet band at 1467 cm^{-1}, which is assigned to the methylene scissoring vibration band. Since this band is known to be a good indicator of factor group splitting,[40] this singlet band directly indicates that the hydrocarbon chains are packed in the triclinic or hexagonal form. In this manner, very fine molecular structures can be discussed by use of high-quality transmission spectra even in single-monolayer level films.

3.4.2 Oblique-Angle Incidence Transmission Spectroscopy

Normal-incidence transmission spectroscopy is a powerful means of retrieving the chemical information of IP modes selectively. Tilting the L-B film in relation to the incidental beam, however, causes mixed signals of IP and OP modes as observed by the transmission optical geometry. The mixed spectra reflect dichroic ratios of the IP and OP modes as a function of the angle of incidence. Therefore, the oblique-angle incidence transmission spectra provide information about the molecular orientation, although "pure" OP mode spectra could not be experimentally resolved from the observed spectra. Since this technique is convenient for discussing molecular orientation, which requires no complex experimental setup and analytical theories, it is widely used with UV-visible spectroscopy[41] as well as IR spectroscopy.[42]

3.4.3 RA Spectrometry

When a metallic surface is used as the substrate of L-B films, the reflection–absorption (RA) spectrometry[26] is quite useful for observing the IR spectra of thin films, because the RA technique has (1) high sensitivity, (2) an RA-specific surface selection rule, and (3) an LO modes-sensitive optical property. The first two characteristics are particularly important for surface chemists, and the third characteristic is very useful for physicists dealing with thin condensed matter.

Figure 3.14 presents a schematic of specular reflection measurements of a thin film deposited on a metallic surface. As presented in the figure, p-polarization is usually used for the reflection measurements, although nonpolarization light can be used for the measurements when the film is thin enough.[26] Although the schematic presents a rough model, it is useful to consider the state of polarization near the surface. Strictly speaking, the phase-angle change depending on the angle of incidence can be calculated as presented in Figure 3.15. Four curves calculated for a

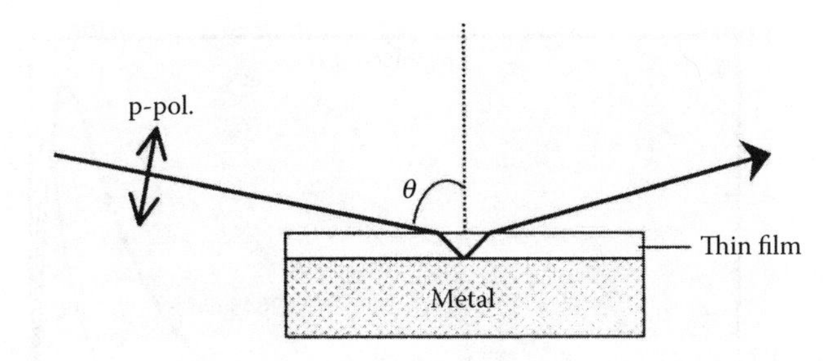

FIGURE 3.14 Schematic of IR reflection-absorption (RA) measurements of L-B films deposited on a metallic substrate.

gold surface are presented for two wavenumbers and two polarizations. It is found that the phase shift near the metal surface strongly responds to polarizations. The phase shift drastically changes for the p-polarization, especially near the grazing angle, while it slightly changes for the s-polarization. Of note is that the phase shift for s-polarization is always near 180°. irrespective of the angle of incidence. This means that the s-polarization ray is diminished near the metal surface, and almost no chemical information could be observed by the s-polarization. On the other hand, the phase shift of p-polarization varies from 180° to nearly zero. For example, when a p-polarized ray travels parallel to a metal surface that corresponds to the angle of incidence of 180°, the electric field would vanish due to a mirror image effect. When the same polarization is irradiated on the surface normally, however, no mirror image effect would influence the reflection, and the surface-parallel electric field would interact with the surface species like transmission measurements. Nevertheless, when the angle of incidence is near 85°, the electric fields of incidence and reflected rays

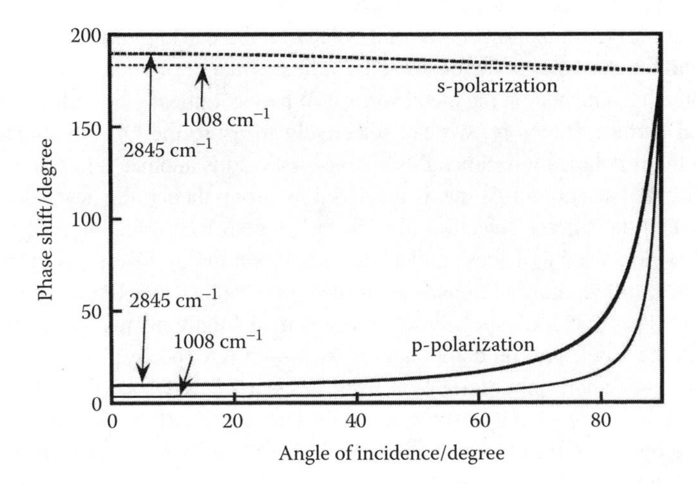

FIGURE 3.15 Diagram of phase shift of at a reflection on a gold surface calculated for s- and p-polarizations and two wavenumbers (2845 cm^{-1} and 1008 cm^{-1}).

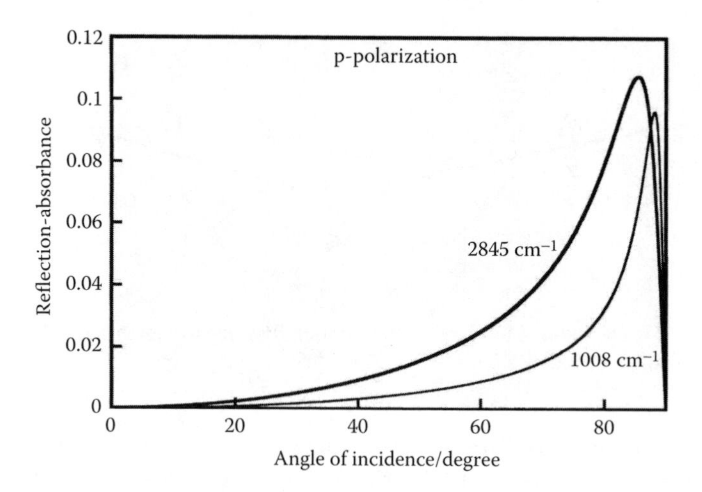

FIGURE 3.16 Reflection-absorption plot against the angle of incidence calculated for two wavenumbers (2845 cm^{-1} and 1008 cm^{-1}).

are strongly correlated to show strong intensity. Figure 3.16 presents angle-of-incidence-dependent reflection–absorbance change for p-polarization, which was calculated for two wavenumbers. It was found that absorption by the thin film (22.5 nm in thickness) was most enhanced when the angle of incidence was selected to be near 85°. The angle of incidence that gives the maximum reflection–absorbance corresponds to the angle that gives the phase shift of 90° in Figure 3.15. Although it is difficult to intuitively understand the correlation between the enhancement and the phase shift, it is important to notice that the sensitivity of the RA technique depends on the phase shift on the metal surface. This concept will be of more importance in ER spectrometry, as described later.

In Figure 3.16, simulation curves of s-polarization are also plotted, but they are too weak to find for all angles of incidence. This indicates that the s-polarization component is automatically vanished on the metal surface. When an optically thin film is deposited on a metal surface, therefore, we can selectively measure the OP vibrational modes only with the p-polarization. Since this characteristic adds another selection rule to the intrinsic infrared selection rule that is theorized by group theory, the new selection rule is often called the surface selection rule. In fact, the surface selection rule of the RA method is widely used to discuss molecular orientation in thin layers. As mentioned in Section 3.4.1, surface-normal transmission spectrometry provides IP vibrational modes in the films. Therefore, RA spectrometry complements the transmission technique. Of note is that the s-polarization component is useless in RA spectrometry.

Regardless, when nonpolarization light is used for RA spectrometry, we have to note that linearity of the absorbance is lost to some extent. Simply stated, the absorbance by use of p-polarization is correlated to that by nonpolarization with the following equation:[26]

$$A_p = (1+\rho)A_{\text{non}} \tag{3.12}$$

The coefficient ρ is a function defined as $\rho = \chi\Pi$, in which χ and Π are the intrinsic reflectivity ratio of s- and p-polarizations and the spectrometer-specific polarization response, respectively. When the sample is fixed at 45° from the horizontal position, the problem of the polarization response is eliminated. In this case, only the reflectivity ratio would remain, which is calculated as 1.14 (2845 cm^{-1}) at the optimized angle of incidence, 86.0. Consequently, A_p is found to be near twice A_{non}. For quantitative study, however, the use of p-polarization is recommended.

Another important characteristic of the RA technique is that LO modes in thin films are selectively observed. In general, light provides transverse-wave electric field in matter. This transverse wave generates polarized zones with alternative directions with respect to the light direction, which are associated with TO modes only. When p-polarization light is irradiated on a metallic material at a grazing angle, on the other hand, a strongly surface-perpendicular oriented electric field is generated in the vicinity of the surface. If the sample film is adequately thin in terms of wavelength, the film is irradiated in the electric field that has a conserve field, which excites LO modes in the film.[6] Since TO–LO splitting is useful for molecular orientation analysis, infrared RA spectrometry is a very important technique for examining structures in thin films.

Meucci et al. performed infrared RA measurements for a comparative study of self-assembled monolayers (SAM) and L-B films on metal surfaces.[43] Although both preparation techniques of monolayers on metal surfaces have traditionally been used for forming supramolecular assemblies, it has been pointed out that the different techniques would yield different film structures and properties. They used a single-chain thiol (octadecylmercaptan [ODM]) and a double-chain compound (dioctadecyldithiocarbammate [DODTC]), both of which form a single monolayer bound to a gold surface by direct covalent bonding through the sulphur atoms in the head groups.

The infrared RA spectra of the SAM and L-B films of the two compounds clearly suggested similarity and difference between the SAM and L-B films. It is unfortunate that the quality of the RA spectra was not so high, probably because air purge in the sample room was not adequate, and strong infrared absorption by remaining water vapor influenced the interferogram collection, which often occurs in measurements of ultrathin films. Regardless, the C-H stretching vibration region presented apparent peak shifts for DODTC films. Table 3.2 presents major bands of RA spectra of L-B and SAM films for DODTC and ODM, accompanied by bands of liquid and crystal. It is found that there is no significant change between L-B and SAM for ODM, while a large difference is found for DODTC especially for the symmetric CH_2 stretching vibration band. The band positions suggest that the SAM film of DODTC is similar to the crystalline state, whereas the L-B film is close to the liquid state. As for the ODM film, the molecular configuration of both SAM and L-B films is similar to that of the liquid state. The L-B film was prepared at a surface pressure of 25.0 mN m^{-1}, which corresponded to the solid state of the film. In this situation, double-chain molecules generally form a highly stable and fully packed L-B monolayer. Therefore, it is of interest to find that the L-B film has a disordered film structure in comparison to the SAM film. The peak shifts are consistent with the change of band intensity. The L-B film exhibited significantly larger band intensities for both symmetric and antisymmetric CH_2 stretching vibration bands than SAM, which suggests that the hydrocarbon chains are disordered to give a large average value as the tilt angle. This difference between L-B and SAM is not a general tendency, but it is important to note that the film structure that depends on preparation methods can be easily checked by infrared RA spectrometry.

Kawai[44] studied the control of liquid-crystal molecular alignment on a photochromic polyion complex L-B film by infrared RA spectroscopy. The polyion-complex monolayer was comprised of 4-octyl-4′-(5-carboxypentamethyleneoxy) azobenzene (8A5H) and 1,5-dimethyl-1,5′-diazaundecamethylene polymethobromide (PB), which was used as a command layer.[45] On the command layer, a liquid-crystal layer of 4′-dodecyl-4-cyanobiphenyl (12CB) mixed with stearic acid-d_{35} (DSt) layers were deposited. The molecular alignment of this L-B film can be controlled by irradiating UV/visible light and by heating, since the command layer has a photochromic moiety (8A5H) and thermal responsible layer (upper layer). Kawai observed very high-quality RA spectra of the stratified L-B film, and found that the alignment change of the film molecules did not simply obey the Friedel-Creagh-Kmetz (FCK) rule. Since RA spectroscopy has the useful surface selection rule, the absorbance plot revealed that the molecular alignment change exhibited a unique profile, as presented in Figure 3.17. By means of a deuterated compound for the upper layer, the configuration change can be followed by measurements of infrared RA spectra.

Itadera et al.[46] employed RA spectrometry to study charge transfer between a monolayer and a surface adsorbed species. The gold surface for the RA measurements can also be used as an electrode for electrochemical analyses. One limitation of RA technique is that IR measurements cannot be performed *in situ* during electrochemical analyses, since water subphase disturbs the measurements. Therefore, L-B technique becomes a compensatory method for measuring the surface adsorbates on the gold surface after the electrochemical reactions. In addition, surface adsorbates are often

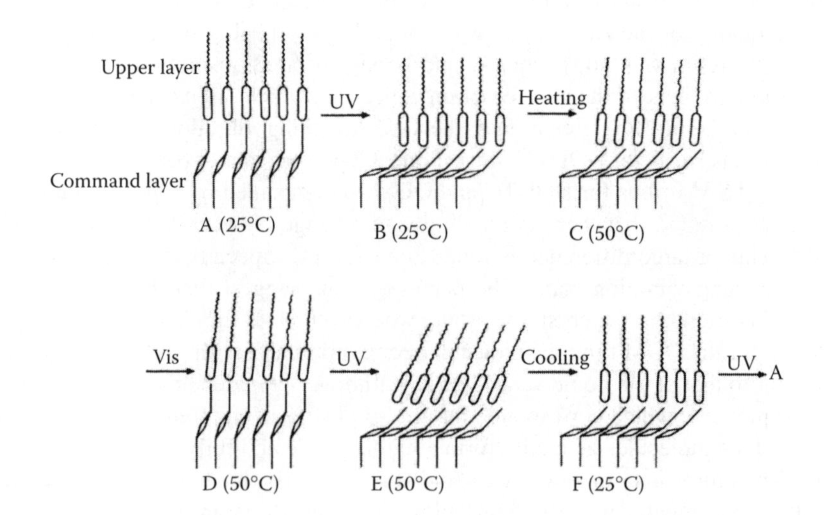

FIGURE 3.17 Schematic illustration of molecular orientation of the upper layer and command layer for each set of conditions. [Reprinted with permission from *Thin Sol. Films* 352: 228–233, 1999. Copyright (1999) Elsevier Science.]

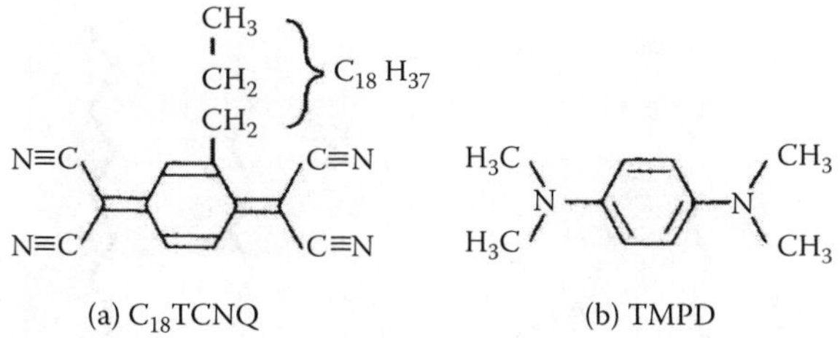

(a) $C_{18}TCNQ$ (b) TMPD

FIGURE 3.18 Chemical structures of the materials used: (a) $C_{18}TCNQ$ and (b) TMPD. [Reprinted with permission from *Synth. Metals* 86: 2261–2262, 1997. Copyright (1997) Elsevier Science.]

one monolayer or less, which requires a highly sensitive measurement technique. In this sense also, RA technique is quite suitable.

They prepared $C_{18}TCNQ$ (Figure 3.18) monolayer L-B film on the gold substrate in a tetramethyl-p-phenylenediamine (TMPD) (Figure 3.18) aqueous solution at different electric potentials. Representative RA spectra are presented in Figure 3.19.

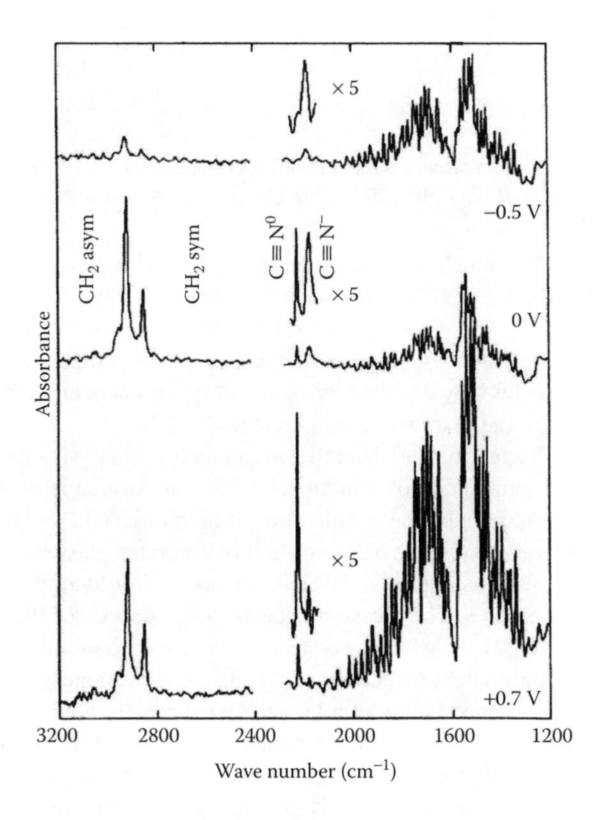

FIGURE 3.19 FT-IR RA spectra of TMPD adsorbed TCNQ L-B films. [Reprinted with permission from *Synth. Metals* 86: 2261–2262, 1997. Copyright (1997) Elsevier Science.]

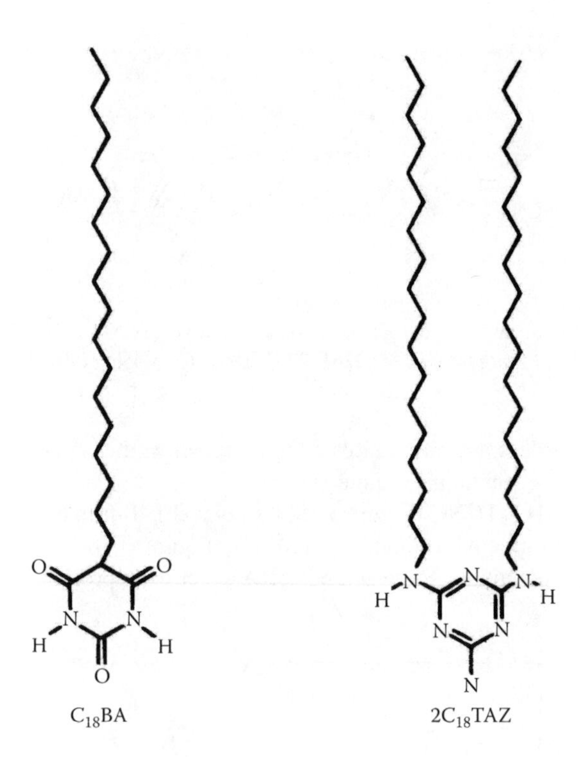

FIGURE 3.20 Schematic molecular structures of $C_{18}BA$ and $2C_{18}TAZ$. [Reprinted with permission from *J. Phys. Chem. B* 103: 7505–7513, 1999. Copyright (1999) American Chemical Society.]

The TCNQ moiety has an electron accepter group, CN. This group gives an infrared absorption band, ν(CN), at 2222 cm^{-1} when it is electronically neutral, while it goes down to 2180 cm^{-1} when it is negatively charged to form CN^-. The RA spectra clearly show relative intensity changes of the two bands, which strongly suggests that an electron transfer between the two chemical species happens, and this transfer can be controlled by the external electric potential.

In addition, RA spectrometry is used for quantitative analysis of molecular orientation, if optical parameters of vibrational modes are available.[48] Hasegawa and Leblanc et al. prepared double-layer L-B films of barbituric acid (BA) and triaminotriazien (TAZ) derivatives (Figure 3.20) on a gold-evaporated glass slide covered with a deuterated cadmium stearate monolayer that worked as a hydrophobic surface at different surface pressures. The pressure-dependent RA spectra of the L-B films are presented in Figure 3.21, in which two striking changes are found. One of them is that the C=O stretching vibration bands greatly change with an increase of the surface pressure. When the surface pressure is low, the band mainly appears at 1690 cm^{-1}, while it strongly appears at 1745 cm^{-1} for the high-pressure film. This corresponds to the change of the C=O group from a hydrogen-bonded state to a hydrogen-bonding-free state. The other significant change is that the hydrocarbon-chain related regions (C-H and C-D stretching) exhibit large intensity changes. This implies that the film molecules change the tilt angle to the surface normal with the increase of the surface

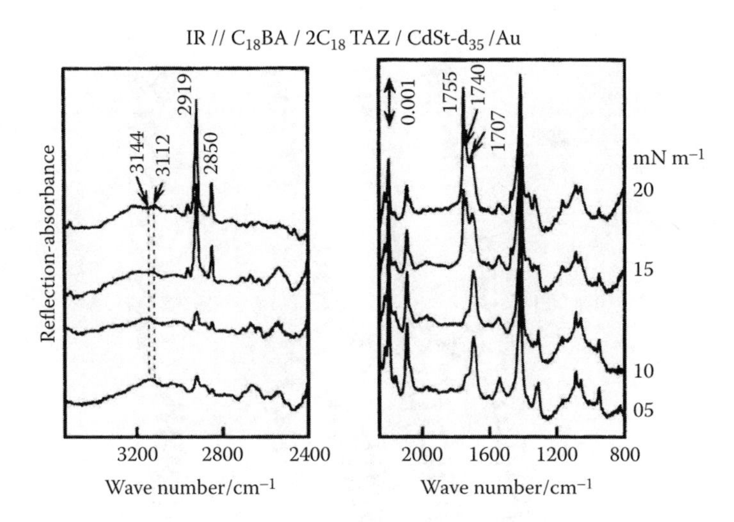

FIGURE 3.21 IRRA spectra of accumulated $C_{18}BA/2C_{18}TAZ$ L-B films deposited on Cd stearate-d_{35} monolayer on a gold-evaporated glass slide. The $C_{18}BA$ monolayers were prepared at various surface pressures on $2C_{18}TAZ$ monolayer that was prepared at the fixed surface pressure of 20 mN m^{-1}. [Reprinted with permission from *J. Phys. Chem. B* 103: 7505–7513, 1999. Copyright (1999) American Chemical Society.]

pressure. Of note is that the wavenumber positions are kept unchanged during the intensity changes, which suggests that molecular configuration is not changed, while only the tile angles change. After the calculation of the orientation angles, a schematic picture has been obtained as presented in Figure 3.22. This picture clearly suggests that one of the C=O group in the BA moiety has lost hydrogen bonding due to the change of the tilt angle. This is a good example of how the evaluation of molecular orientation through infrared RA spectra is a powerful means of analyzing L-B films.

3.5 ER TECHNIQUE

When thin films are deposited on dielectric (nonmetallic) materials that are not transparent to infrared rays, reflection measurements should be performed. The reflection measurement technique is distinguished from that performed on metallic substrates (RA), and is called external reflection (ER) technique. ER technique is quite important for practical application studies — for example, on semiconductors and liquid surfaces. Nevertheless, the analytical technique is much more complicated than RA technique.

Figure 3.23 shows a phase-shift diagram for the two polarizations when an infrared ray of 2845 cm^{-1} is irradiated on a GaAs wafer. The characteristic of the s-polarization is similar to that of RA, while the p-polarization exhibits a largely different character. The significantly changing point is not related to the grazing angle; instead, it appears at Brewster's angle of the wafer.

Figure 3.24 presents the reflection-absorbance changes when a thin isotropic film (22.5 nm) is deposited on GaAs. This corresponds to Figure 3.16 for RA, but the results for ER are found to be highly complicated. It is of interest that the absorbance

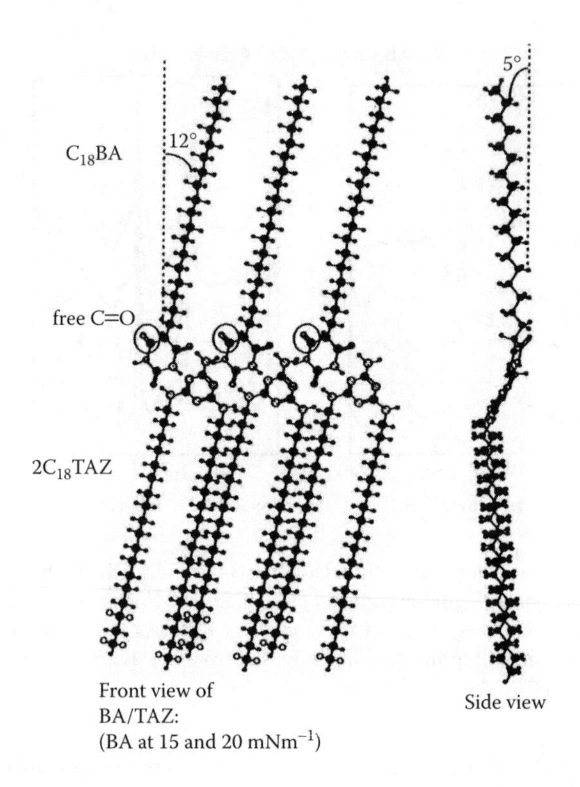

FIGURE 3.22 Schematic views of the accumulated layers. [Reprinted with permission from *J. Phys. Chem. B* 103: 7505–7513, 1999. Copyright (1999) American Chemical Society.]

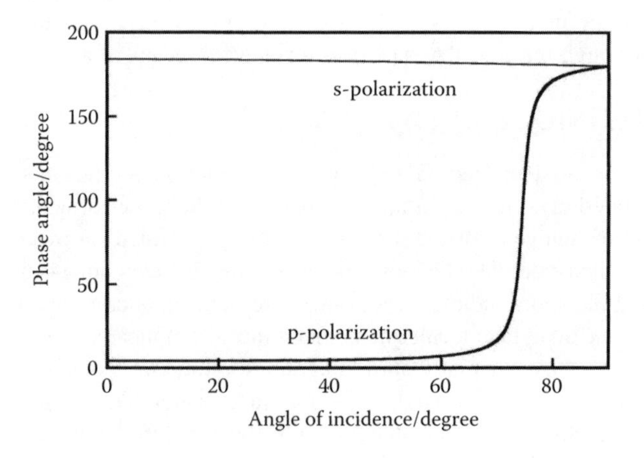

FIGURE 3.23 A phase-shift diagram against the angle of incidence of s- and p-polarizations on a GaAs wafer (2845 cm^{-1}).

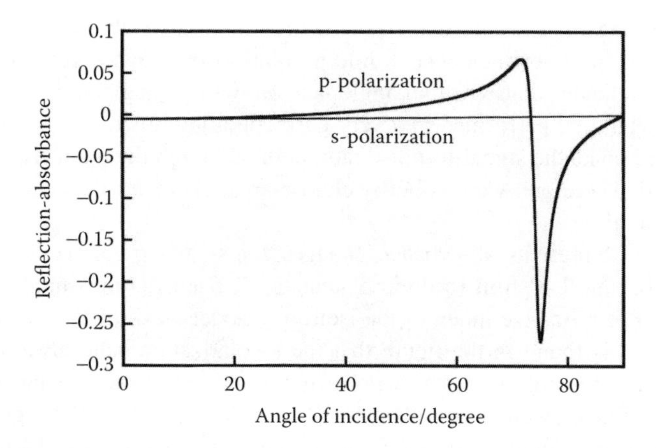

FIGURE 3.24 Reflection-absorption plot as a function of the angle of incidence calculated for s- and p-polarizations on a GaAs wafer.

of s-polarization is always negative for all the angles of incidence, which is definitely different from RA and transmission spectra. The intensity of the s-polarization spectra is very small, and it monotonously decreases to zero with an increase of the angle of incidence. On the other hand, p-polarization has a large dispersion curve, which has a transition point. The transition point corresponds to Brewster's angle, which also appeared in the phase-shift diagram. This result means that p-polarization spectra change the sign when the angle of incidence changes. As described later, this sign and intensity also depend on optical anisotropy that reflects molecular orientation.

The ER technique has another experimentally unique characteristic. As expected from Figure 3.24, the sensitivity of p-polarization measurements would be enhanced when an angle near Brewster's angle would be used. It is theoretically true, but it simultaneously has a disadvantage. Figure 3.25 presents reflectivity changes for the

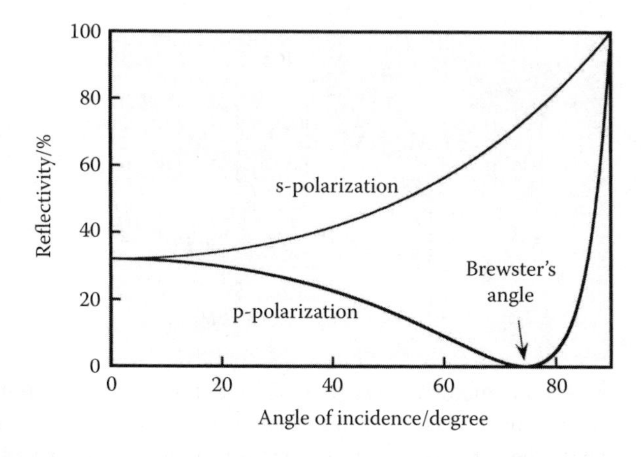

FIGURE 3.25 Reflectivity changes against the angle of incidence on GaAs.

two polarizations. It is found that s-polarization monotonously increases its reflectivity with the angle of incidence, while p-polarization reaches zero at Brewster's angle. This indicates that when an angle near Brewster's angle is chosen, the measurements would be performed in a very dark condition due to the low reflectivity, which would make the signal-to-noise ratio poor, although the intensity itself would be enhanced. Therefore, we had better choose an angle of incidence different from Brewster's angle.

Figure 3.24 presents absorbance changes for s- and p-polarizations when an optically isotropic L-B film (cadmium stearate, 22.5 nm) is measured on GaAs at 2850 cm^{-1}. The refractive index of the isotropic model used for the calculation was 1.50 + 0.3i. It is found in the figure that the s-polarization band always has small negative values regardless of the angle of incidence, while the p-polarization band changes its sign as well as its intensity. The critical point of the sign of the p-polarization band is found at Brewster's angle. This figure indicates that s-polarization spectra have minor absorbance in comparison to p-polarization spectra. In this sense, p-polarization measurements seem to be better in terms of sensitivity. Nevertheless, as found in Figure 3.25, the reflectivity of s-polarization is much higher than that of p-polarization, which gives the s-polarization spectra a good signal-to-noise ratio. In this way, in general, s-polarization measurements are much easier to perform.

Regardless, p-polarization spectra are still practically quite useful, since they respond well to optical anisotropy in the film, which directly reflects molecular orientation. Reflection absorbance of p-polarized ER spectra that take optical anisotropy into account is available by employing the Hansen-Hasegawa equations.[48]

$$\tilde{\xi}_j = \left(\tilde{n}_{je}^2 - \tilde{n}_{1o}^2 \sin^2\theta_1\right)^{1/2} \tag{3.13}$$

$$\tilde{q}_j = \frac{\tilde{\xi}_j}{\tilde{n}_{jo}\tilde{n}_{je}} \tag{3.14}$$

$$\tilde{\beta}_j = \frac{2\pi}{\lambda} h_j \frac{\tilde{n}_{jo}}{\tilde{n}_{je}} \tilde{\xi}_j \tag{3.15}$$

$$\begin{bmatrix} U_1 \\ V_1 \end{bmatrix} = \tilde{\mathbf{M}}_2\tilde{\mathbf{M}}_3\cdots\tilde{\mathbf{M}}_{N-1}\begin{bmatrix} U_{N-1} \\ V_{N-1} \end{bmatrix} = \tilde{\mathbf{M}}\begin{bmatrix} U_{N-1} \\ V_{N-1} \end{bmatrix} \tag{3.16}$$

$$\tilde{\mathbf{M}}_j = \begin{bmatrix} \cos\tilde{\beta}_j & -\dfrac{\mathrm{i}}{\tilde{q}_j}\sin\tilde{\beta}_j \\ -\mathrm{i}\tilde{q}_j\sin\tilde{\beta}_j & \cos\tilde{\beta}_j \end{bmatrix} \tag{3.17}$$

These equations are constructed on a schematic model, in which the stratified layers are numbered from 1 to N. The absorbing media correspond to the 2nd through N - 1th layers, while the 1st and the Nth layers correspond to the air and supporting material of the films, in most cases. The anisotropic refractive indices of each (jth) layer are represented by $\tilde{n}_{jo}$ and $\tilde{n}_{je}$, which correspond to the surface parallel and surface normal components, respectively. The parameters $\tilde{\xi}_j$, $\tilde{q}_j$, and $\tilde{\beta}_j$ are necessary to calculate a matrix $\tilde{\mathbf{M}}_j$, but it is not necessary to consider their physical meanings here. The most important equations are Equation 3.16 and Equation 3.17, which relate tangential components of field intensities to an interface: U_1 and V_1 to U_{N-1} and V_{N-1}. For s-polarization, U and V correspond to electric and magnetic fields, respectively, while they correspond to magnetic and electric fields for p-polarization. With the use of matrix components in $\tilde{\mathbf{M}}_j$ $(m_{11}, m_{12}, m_{21}, m_{22})$, the reflection coefficients for s- and p-polarizations are calculated as follows where

$$\tilde{r}_s = \frac{(m_{11} + m_{12}\tilde{p}_N)\tilde{p}_1 - (m_{21} + m_{22}\tilde{p}_N)}{(m_{11} + m_{12}\tilde{p}_N)\tilde{p}_1 + (m_{21} + m_{22}\tilde{p}_N)} \tag{3.18}$$

$$\tilde{r}_p = \frac{(m_{11} + m_{12}\tilde{q}_N)\tilde{q}_1 - (m_{21} + m_{22}\tilde{q}_N)}{(m_{11} + m_{12}\tilde{q}_N)\tilde{q}_1 + (m_{21} + m_{22}\tilde{q}_N)} \tag{3.19}$$

With these coefficients, reflectances are finally available in the following way.

$$R_s = |\tilde{r}_s|^2 \tag{3.20}$$

$$R_p = |\tilde{r}_p|^2 \tag{3.21}$$

The equations 3.18 through 3.21 are the conventional equations, while the equations 3.13 through 3.17 are key equations for introducing the optical anisotropy. Application studies that employ the equations above will be presented in a later section.

3.6 TRANSMISSION AND RA SPECTRAL ANALYSES OF L-B FILMS

As stated earlier, transmission and RA spectrometries are complementary techniques for capturing vibrational modes in thin films. For a better understanding of the physical properties of anisotropic films, measurements of both spectra are ideal. When an identical vibrational mode is observed by the two techniques, we would be able to calculate the orientation angle by making a ratio of the band intensities, particularly for a uniaxial system. Nevertheless, the absolute value of band intensity depends on optical properties of the substrate for the L-B film, and we have to know the precise wavenumber dispersion of optical parameters for the two different kinds of substrate used for the transmission and RA measurements. Regardless, even if we had no knowledge of the optical property, the combination technique of transmission and RA spectrometries provides us with an excellent way to qualitatively study molecular orientations, which is much clearer than using one of the methods only.

Umemura et al. developed a powerful technique that can directly use the band intensities of infrared transmission and RA spectra, so that the orientation angle of a vibrational mode can be evaluated quantitatively.[35] This technique is based on the

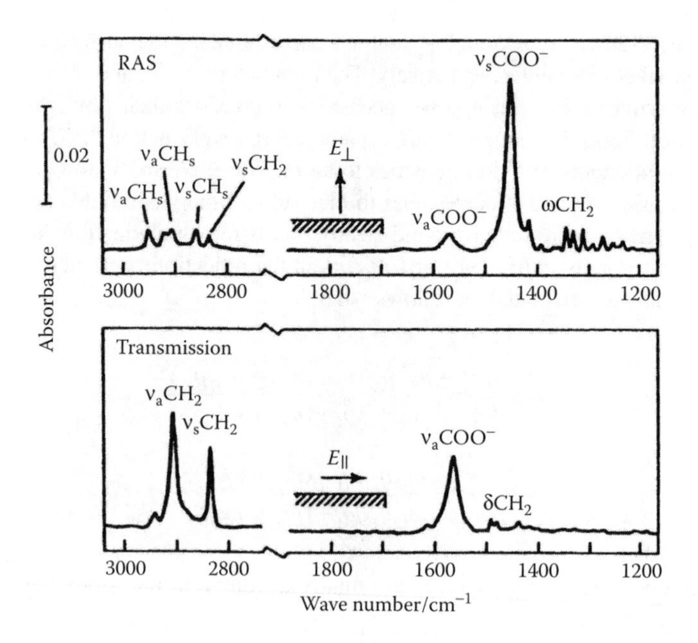

FIGURE 3.26 Infrared RA and transmission spectra of a 7-layer LB film of Cd stearate. [Reprinted with permission from *J. Phys. Chem.* 94: 62, 1990. Copyright (1990) American Chemical Society.]

transmittance and reflectance calculations for stratified layers, which was developed by Hansen. The theory is a basis for developing the anisotropic theory described in the previous section, and the reader is referred to the literature.[26,49] A representative set of transmission and RA spectra is presented in Figure 3.26. These spectra have arisen from 7-monolayer L-B films of cadmium stearate deposited on a ZnSe plate (transmission) and a silver-evaporated glass slide (RA). Among various kinds of metal salt of stearic acid, cadmium salt is known to have highly ordered and packed film structure,[37] from which a strong anisotropy is expected. The set of spectra clearly indicates that all the bands complementarily appear: weak bands in RA spectrum appear strongly in transmission, and vice versa, as expected. Since the transmission and RA spectrometries have complementary surface selection rules, we can roughly evaluate molecular orientation without calculation. The CH_2 stretching vibration bands and the anti-symmetric COO^- stretching vibration band are suppressed in the RA spectrum, while they appear very strongly in the transmission spectrum. This strongly suggests that the methylene groups in the hydrocarbon chain are oriented nearly parallel to the film surface. This is consistent with that the CH_2 wagging vibration, whose transition moment has a parallel direction to the hydrocarbon chain is recognized in the RA spectrum only. These support a schematic model, in which the hydrocarbon has a perpendicular orientation to the film surface. On the contrary, the symmetric COO^- stretching vibration band appears strongly in the RA spectrum, while it is largely suppressed in the transmission spectrum. Both results for the antisymmetric and symmetric stretching vibration modes for an identical group suggest that the transition moment of the COO^- group has nearly perpendicular orientation to the film surface.

The quantitative analysis for the bands gave clear results, as shown in Table 3.3. Both CH_2 stretching vibration modes have an orientation angle of 85° from the surface normal, which gives a tilt angle of the hydrocarbon chain of 7°.[35] The evaluated results are fairly consistent with the qualitative model discussed above. In the same manner, the antisymmetric and symmetric COO^- stretching vibration modes have angles of 83° and 18° from the surface normal. As a result, a schematic picture of the cadmium stearate molecule can be drawn by using the quantitative values.[35] Therefore, this technique had a strong impact on various researchers who study L-B films.

Schmelzer et al.[50] fabricated an interesting L-B film that consists of 3-thienyl-pentadecanoic acid (3TC15) and distearylviologene (DSV) molecules (Figure 3.27). A mixture of the compounds (2:1 for 3TC15:DSV) in a mixed solution of chloroform and ethanol (3:1) was spread on an aqueous solution of cadmium chloride at pH 6.0, and compressed films were transferred on solid substrates to form 6-monolayer L-B films. They proposed a schematic model of the L-B film as presented in Figure 3.28 after an analysis of π-A isotherm measurements. To investigate the model by spectroscopy, they prepared a similar L-B film, in which 3TC15 was replaced by alkyl-deuterated stearic acid (D18), so that the hydrocarbon parts in DSV and D18 could be discriminated in infrared spectra.

Figure 3.29 presents FT-IR transmission (a) and RA (b) spectra of the L-B film. The spectra are very different from each other, and it is impressive that some bands derived from methyl groups clearly appear in the RA spectrum. This characteristic strongly suggests that both DSV and D18 are almost perpendicularly oriented to the film surface. They employed the method of Umemura et al. for evaluating orientation

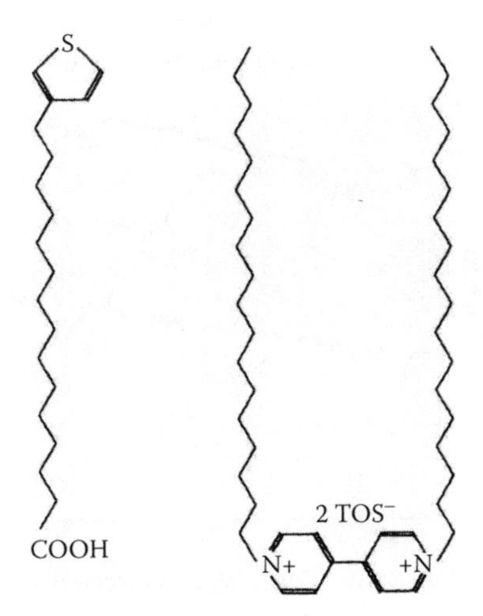

FIGURE 3.27 Schematic representation of 3TC15 (left) and distearylviologene (DSV) (right). [Reprinted with permission from *Thin Sol. Films* 243: 620–624, 1994. Copyright (1994) Elsevier Science.]

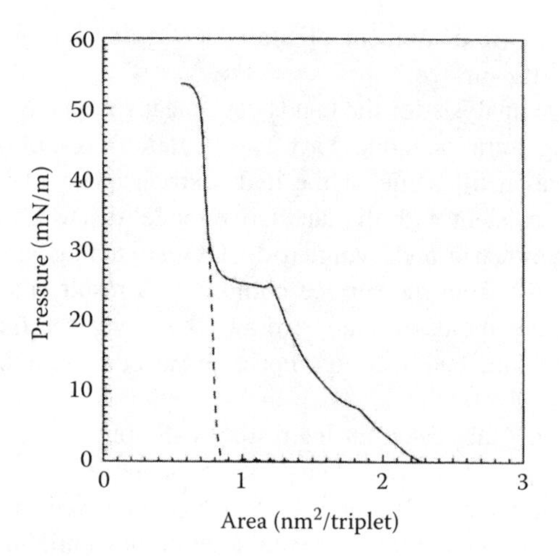

FIGURE 3.28 p-A isotherms of 3TC15-DSV (2:1) at 19°C. The solid line is for the initial compression, while the dashed line is for subsequent expansion and compression cycles. [Reprinted with permission from *Thin Sol. Films* 243: 620–624, 1994. Copyright (1994) Elsevier Science.]

angles, and they found that the tilt angle of the 3TC-DSV L-B film was 7.5° with respect to the surface normal. They also employed the surface plasmon resonance (SPR) technique for evaluating the thickness of the film. The SPR technique is, in principle, a method to measure reflectance as a function of an angle of incidence, and the shift of the angle that gives the minimum of the reflectance is analyzed by the Fresnel equation. Therefore, it is similar to elipsometry. They found that the thickness

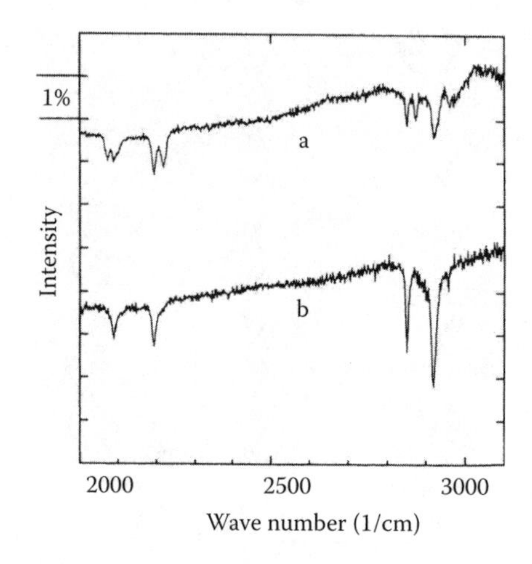

FIGURE 3.29 FT-IR (a) RA and (b) transmission spectra of mixed films of deuterated stearic acid and DSV (2:1). [Reprinted with permission from *Thin Sol. Films* 243, 620–624, 1994. Copyright (1994) Elsevier Science.]

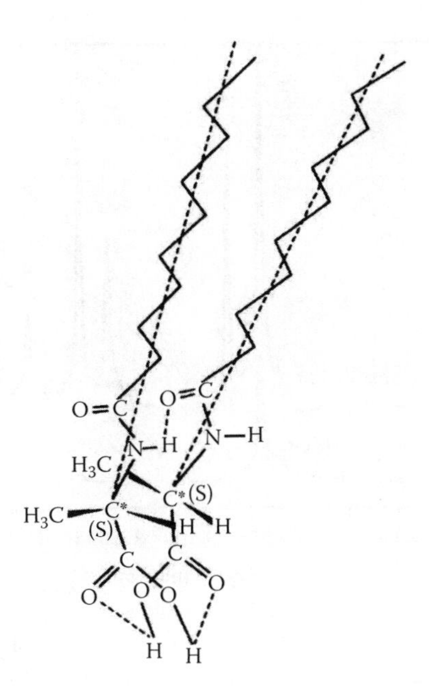

FIGURE 3.30 Pair of *N*-octadecanoyl-L-alanine molecules in a tightly aggregated state. [Reprinted with permission from *Chem. Phys. Lett.* 313: 565–568, 1999. Copyright (1999) Elsevier Science.]

of the L-B film was 3.0 nm, which supported the interdigitated molecular structure. In this manner, the combination technique of transmission and RA spectrometries is a powerful means of discussing fine molecular arrangement in thin layers.

Du and Liang[51] examined the relationship between molecular chirality in an L-B film and the molecular orientation by the combination technique. They synthesized *N*-octadecyl-L-analnine (Figure 3.30), and investigated the structural property of L-B films of the molecule. They prepared the 11-monolayer L-B films on a CaF_2 plate and a silver-evaporated quartz plate for transmission and RA measurements, respectively. The spectra are presented in Figure 3.31; it was found that they were significantly different from each other. The most striking difference is that the amide A (N-H stretch at 3324 cm^{-1}), the amide I (ν(C=O) at 1646 cm^{-1}), and the CH_2 scissoring (singlet, 1472 cm^{-1}) and the C=O stretching vibration bands that arose from the carboxyl group (1705 cm^{-1}) were suppressed or unavailable in the RA spectrum, whereas they were very strong in the transmission spectrum. This suggests that the L-B film has a strong anisotropy, at least uniaxially. Since the N-H stretching vibration mode for a free N-H group is found at 3510 and 3420 cm^{-1},[52] the large, low-wavenumber shift to 3324 cm^{-1} suggests that the N-H group is strongly hydrogen bonded. In general, the hydrogen bonding network helps keep the molecular aggregation highly stable in the two dimensional plane, and the groups in the network tend to have specific orientation in the plane. This is true for this case, and the hydrogen bonding makes the N-H group perfectly parallel to the film surface, which results in the disappearance of the amide A band in the RA spectrum. This speculation is fully consistent with the fact that the amide I band appears at a

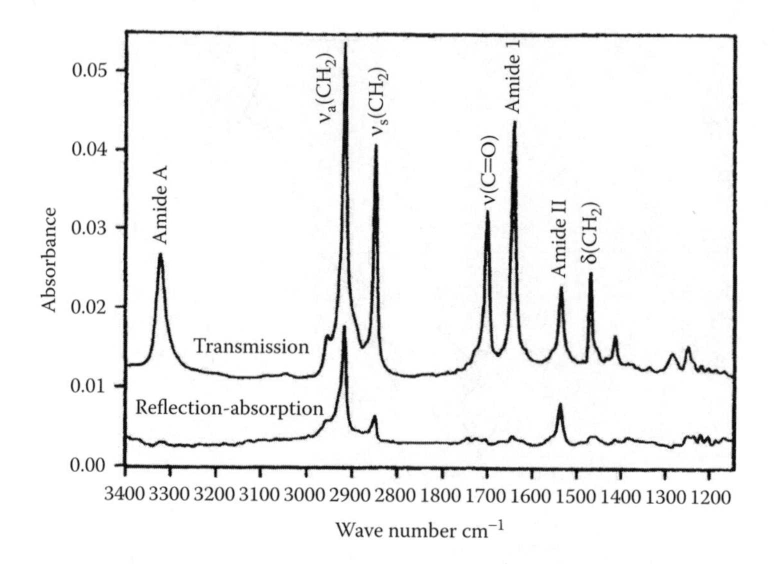

FIGURE 3.31 FT-IR spectra of an 11-monolayer *N*-octadecanoyl-L-alanine L-B film. [Reprinted with permission from *Chem. Phys. Lett.* 313: 565–568, 1999. Copyright (1999) Elsevier Science.]

typically low wavenumber (1646 cm^{-1}) when it is partially ionized by associating with adjacent molecules, and the band also disappears in the RA spectrum. Therefore, the N-H and C=O groups are hydrogen bonded with each other, and they have parallel orientation to the film surface. Du and Liang confirmed this by performing another FT-IR transmission measurement of the same L-B film prepared on a 1 mM $AgNO_3$ aqueous solution, because the disappearance of the C=O band may be caused by forming silver salt after a reaction with the silver substrate. The transmittance spectrum indicated that the silver ion made the ν(C=O) band at 1705 cm^{-1} disappear, but the amide I band was not influenced by the silver ion. Therefore, the parallel orientation of the C=O group has been experimentally confirmed.

They also observed in-plane anisotropy (dichroism) in polarization transmission spectra of the L-B film, which suggested that both head group components and hydrocarbon chains in the film take a biaxial orientation. These highly oriented structures of the molecule in the film were inferred to originate from the chiral centers of the molecule. They conclude that the minimum-energy configuration of a pair of the same enantiomers favors a twisted angle between them, and the twist between the two adjacent molecules gives rise to the chirality of the aggregates.

3.7 APPLICATION STUDIES OF ER SPECTRAL ANALYSES OF L-B AND LANGMUIR FILMS

Infrared ER spectroscopy is still not a familiar method for thin-material analyses in practice, since the analytical technique is a little complicated and it requires precise optical parameters such as complex refractive index and thickness of the film prior to the analysis. Another problem with this technique is that the signal-to-noise ratio becomes poor in comparison to other reflection techniques, since the reflectivity is low especially for the p-polarization near Brewster's angle. Nevertheless, this technique is still very much worth using, since it responds to the molecular orientation sensitively, and both TO and LO modes can be detected simultaneously. Further, the spectral characteristics can be controlled by changing the angle of incidence. It is also of importance to emphasize that many practical applications specifically require this method, due to an optical limitation of the substrate. For example, when the substrate is not transparent to infrared light, and it is not a metal, the FT-IR ER technique is the only choice for measuring spectra of this film on the substrate. The analysis of thin layers on water surface is a typical example, as will be mentioned later.

Some groups tried to employ this technique for practical analysis of thin adsorbates on nonmetallic material, but it was difficult to discuss the spectra quantitatively. One of the first important trials was the study by Mielczarski et al.[53,54] They measured polarized FT-IR ER spectra of ethyl xanthate spontaneously adsorbed on cuprous sulfide at different angles. They employed the following approximation equations instead of using the exact equations (Equations 3.13 to 3.21) for calculation of absorbances by polarization measurements (A_s and A_p).

$$A_s = -\frac{16\pi}{\ln 10}\left[\frac{\cos\theta}{n_3^2 - 1}\right]\frac{n_2 k_2 d_2}{\lambda} \tag{3.22}$$

$$A_{px} = -\frac{16\pi}{\ln 10}\left[\frac{\cos\theta}{\xi_3^2/n_3^4 - \cos^2\theta}\right]\left[-\frac{\xi_3^2}{n_3^4}\right]\frac{n_2 k_2 d_2}{\lambda} \tag{3.23}$$

$$A_{pz} = -\frac{16\pi}{\ln 10}\left[\frac{\cos\theta}{\xi_3^2/n_3^4 - \cos^2\theta}\right]\frac{\sin^2\theta}{\left(n_2^2 + k_2^2\right)^2}\frac{n_2 k_2 d_2}{\lambda} \tag{3.24}$$

Here, A_p is divided into two components, A_{px} and A_{pz}, which correspond to absorbances of a vibrational mode that is perfectly oriented in the in-plane and out-of-plane directions, respectively. The subscripts 1, 2, and 3 represent the air, adsorbed layer, and the substrate phases, respectively. Of note is that only the film layer (second phase) has the absorption index (k_2) and thickness (d_2), and these approximation formula are not applicable to absorbing substrates such as metals. The parameter $\xi_j = (n_3^2 - n_1^2 \sin^2\theta_1)^{1/2}$ is slightly different from Equation 3.13.

They readily discussed the changes in both s- and p-polarization spectra, but the intensity could not be reproduced at all. The discrepancy between the observed and calculated values was probably due to the following reasons: (1) the anisotropic property in the layer was not taken into account, and (2) the absorbing property of the substrate was ignored. Regardless, it was an epoch-making achievement to fully explain the qualitative spectral changes for the first time.

As mentioned before, Hasegawa and Umemura et al. developed an expanded exact theory that can consider optical anisotropy as presented in Equations 3.13 to 3.21. Urai and Itoh et al.[55] employed this anisotropic theory to analyze an iodine

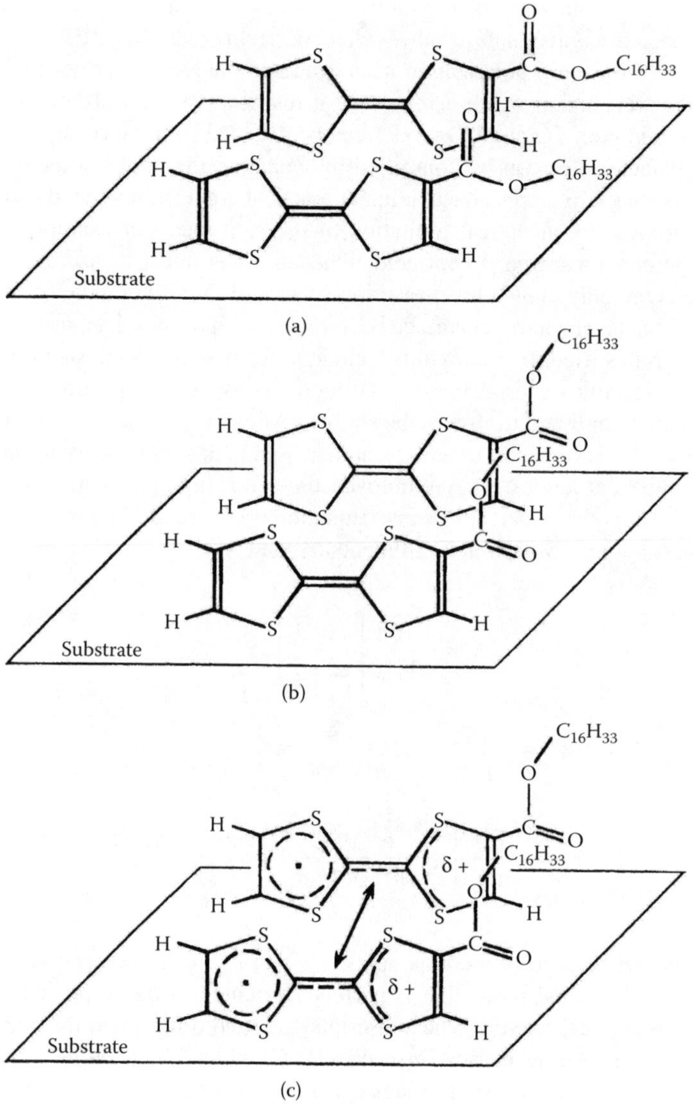

FIGURE 3.32 Schematic representations of the structures of HDTTF in the L-B film before ([a] trans and [b] cis structures) and after (c) iodine doping. [Reprinted with permission from *Langmuir* 14: 4873–4879, 1998. Copyright (1998) American Chemical Society.]

doped conductive L-B film of a tetrathiafulvalene derivative (Figure 3.32). They measured angle-of-incidence-dependent s- and p-polarization FT-IR ER spectra of the L-B film on a silicon wafer, and they found that s-polarization spectra had minute peaks for the $\nu(C_3{=}C_4)$ and $\nu(C_8{=}C_9)$ bands (1570 and 1541 cm^{-1}), while these are clearly observed when an nonoriented sample is measured. Since s-polarization ER spectra respond to surface-parallel vibrational modes only, this evidence suggests that the tetrathiafulvalene ring stands up perpendicularly on the substrate (Figure 3.32).

On the other hand, one of the C=O stretching vibration bands observed at 1701 cm^{-1} appears as a negative band in the p-polarized spectrum when the angle of incidence is small, and the other C=O band is available as a positive band at 1719 cm^{-1}. The latter band is a little shifted in comparison to the corresponding band that appeared in the s-polarization spectrum, but it was attributed to the influence of the adjacent negative band at 1701 cm^{-1}. These results strongly suggest that the L-B film comprises two kinds of molecular species that have different molecular conformations, which are schematically drawn in Figure 3.32a and Figure 3.32b. In Figure 3.32a, the C=O bond has a *trans* conformation relative to the C_3=C_4 bond, whereas it has *cis* conformation in Figure 3.32b. As a result, the C=O group in the *trans* conformer has a nearly perpendicular orientation to the film surface, while the same group in the *cis* conformer has a nearly parallel orientation. They also performed the same analysis for an iodine-doped L-B film, and the molecular stance was found to have the scheme shown in Figure 3.32c.

Although the analytical technique of ER spectra has been recognized as a useful method for a community of surface chemists,[56,57] it has the crucial limitation that it always requires optical parameters. To overcome this limitation, Hasegawa and Theiß et al. have developed a new analytical technique based on the anisotropic optical theory.[58] They introduced complex dielectric-dispersion functions for both IP and OP directions in a film. The dielectric function was chosen by Kim et al. as the oscillator model function,[59] which can take an intermediate state of Gaussian and Lorentzian curves with a switching parameter, α.

$$\chi_j = \frac{4\pi\rho_j \nu_{0j}^2}{\nu_{0j}^2 - \nu^2 - i\tau_j \nu}; \quad \text{where } \tau_j = \gamma_j \exp\left[-\alpha_j \left(\frac{\nu - \nu_{0j}}{\gamma_j}\right)^2\right] \tag{3.25}$$

Here, χ represents an oscillator that corresponds to a band, and ν, ν_0, and γ represent wavenumber, resonance wavenumber, and damping factor, respectively. The numerator in χ is the oscillator strength, which is evaluated by ρ. With this function, all the bands are theoretically constructed, and all the parameters are perturbed, so that the calculated spectra by the function would be converged to the observed ER spectra. Of note is that we have to have at least two independent ER spectra for the convergence, since two independent functions for the IP and OP modes should be converged simultaneously. In their case, the two independent ER spectra of an identical L-B film were obtained by using two different angles of incidence at 25° and 50°.

Figure 3.33 presents two p-polarized ER spectra of five-monolayer cadmium stearate L-B film deposited on a GaAs wafer measured at the two angles. It is found that the spectral shapes are largely different from each other, in response to the angle. Nevertheless, the calculated curves (dashed line) based on an identical set of theoretical functions almost perfectly fit in with the observed spectra. This indicates that all the parameters in the model functions are optimized with respect to the observed ER spectra. In this manner, dispersion curves of dielectric function

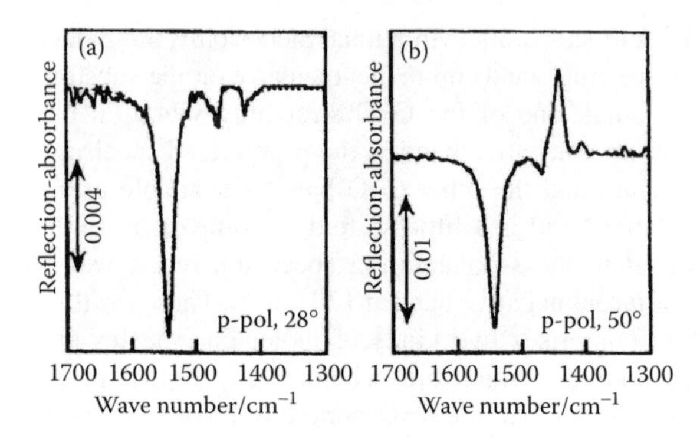

FIGURE 3.33 Simulation spectra (dashed lines) after the convergence of the optimization calculation, which are overlaid on observed p-polarization spectra (solid lines). The calculation was simultaneously performed for the two spectra measured at angles of incidence of (a) 25° and (b) 50°. [Reprinted with permission from *J. Phys. Chem. B* 105: 11178–11185, 2001. Copyright (2001) American Chemical Society.]

have been obtained for both the IP and OP modes. These dispersion functions can directly be converted to refractive-index dispersion curves through a simple relation:

$$\tilde{n}(\nu) = n + ik = \sqrt{\tilde{\varepsilon}(\nu)} \tag{3.26}$$

Figure 3.34 presents the refractive-index dispersions for the IP and OP modes. It is of interest to find that the dispersion curves of absorption index (k) in the IP and

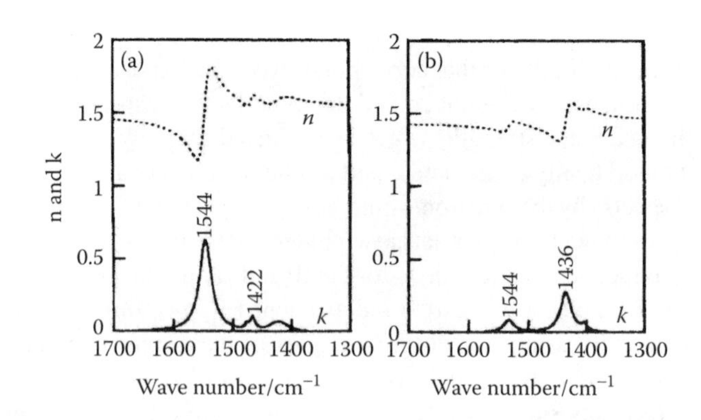

FIGURE 3.34 Refractive-index dispersion curves obtained from the calculated dielectric functions. The real (n) and imaginary (k) parts are represented by the dashed and solid lines, respectively. The surface parallel and normal components are shown in parts (a) and (b), respectively. [Reprinted with permission from *J. Phys. Chem. B* 105: 11178–11185, 2001. Copyright (2001) American Chemical Society.]

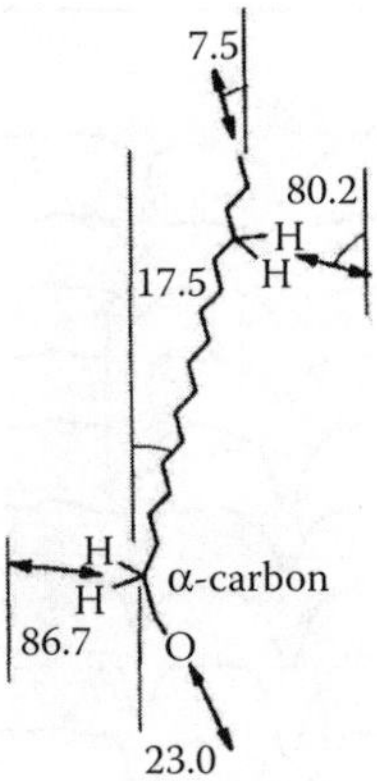

7.5
80.2
H
H
17.5
H
H
α-carbon
86.7
O
23.0

FIGURE 3.35 Evaluated molecular orientation of cadmium stearate molecule in the 5-monolayer L-B film on GaAs. [Reprinted with permission from *J. Phys. Chem. B* 105: 11178–11185, 2001. Copyright (2001) American Chemical Society.]

OP modes are very similar to typical transmission and RA spectra of the film, which proves that the analysis was readily performed.

Another detail to note is that molecular orientation can be calculated by using oscillator strength instead of using absorption index. For details, the reader is referred to Hasegawa et al.[58] With the use of the analytical results, they obtained orientation angles for each normal mode, which includes the CH_2 stretching vibration mode at α-carbon atom in the molecule. Such fine information has never been clarified by other techniques. The estimated molecular structure is drawn in Figure 3.35. The results support previous reports, and finer information has been obtained.

Finally, it would be valuable to add the example of a study of Langmuir (L) monolayer on water surface by means of FT-IR ER spectroscopy. Since water is an extraordinarily strong material for absorbing infrared light, the ER technique, in addition to PM-IRAS and sum-frequency generating (SFG) spectroscopy, is a very important technique for the study of L films.

The analysis of L films by means of the infrared ER technique has been performed by several groups,[60–63] and fundamental techniques have been accomplished thus far. Sakai and Umemura proposed a technique to employ a band-path filter (3500–2000 cm^{-1}) to cut the excess heat energy that makes the L films move on the water surface. With the technique, they obtained clear FT-IR ER spectra of stearic acid and cadmium stearate L films with an increase of the surface pressure. Figure 3.36 presents p-polarization ER spectra of cadmium stearate L film. This experiment was performed on an old FT-IR spectrometer, and the reflectivity of p-polarized light at 38° was very low (less than 1%), but high quality spectra were obtained. In general, it is difficult to collect good p-polarization spectra of L films. With the small angle of incidence in comparison to Brewster's angle (53°), the ER spectra become very similar to transmission spectra,[63] although the sign of the bands are negative, as found in the figure. With these spectra, they evaluated molecular tilt angle, as presented in Figure 3.37. It was found that the stearic acid

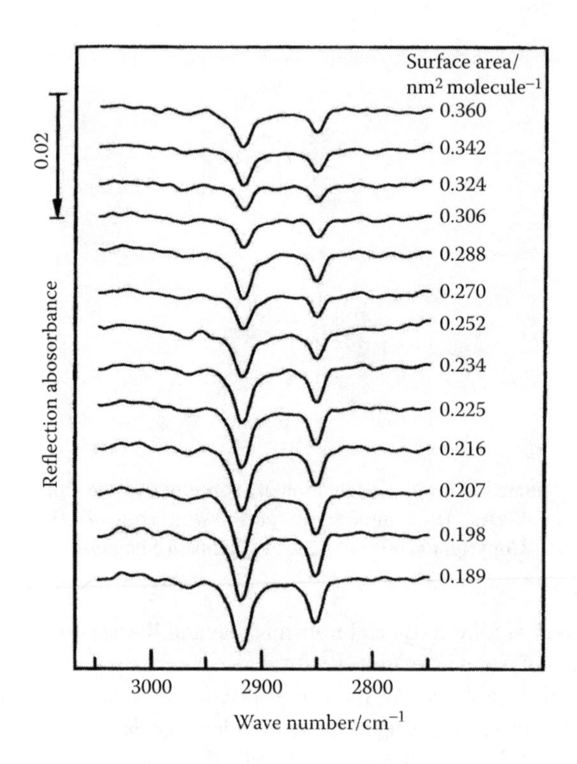

FIGURE 3.36 FT-IR ER spectra of L films of cadmium stearate measured by p-polarized beam at various surface areas. [Reprinted with permission from *Bull. Chem. Soc. Japan* 70: 1027–1032, 1997. Copyright (1997) Chemical Society of Japan.]

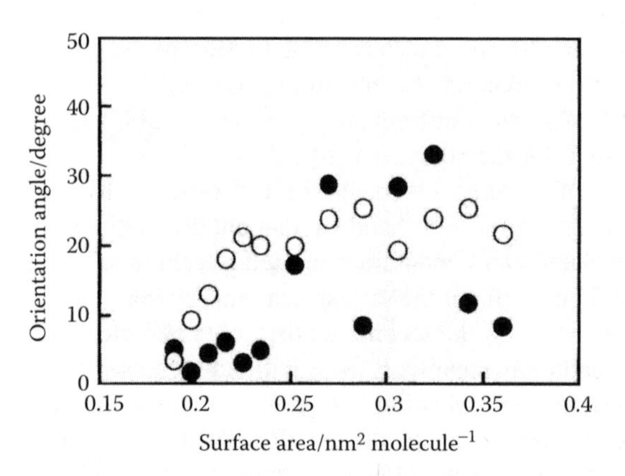

FIGURE 3.37 Molecular orientation in L film of stearic acid (open circle), and cadmium stearate (closed circle) versus surface area. [Reprinted with permission from *Bull. Chem. Soc. Japan* 70: 1027–1032, 1997. Copyright (1997) Chemical Society of Japan.]

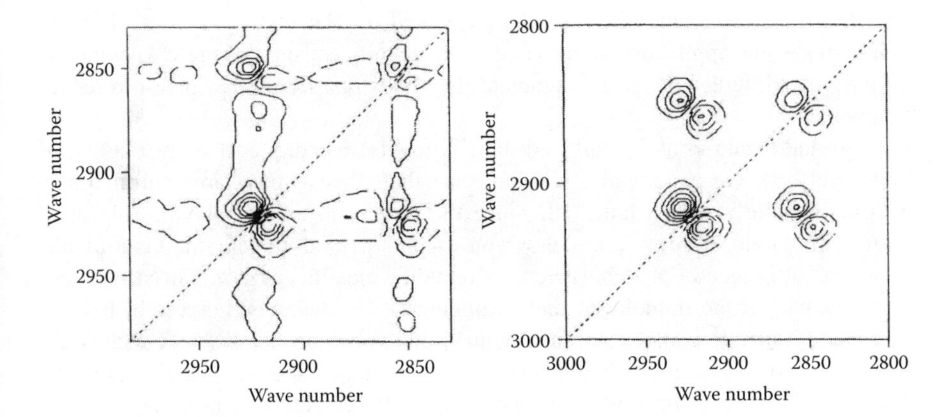

FIGURE 3.38 (a) Experimentally measured and (b) computer simulated 2D-IR asynchronous maps for the low-pressure region subset of spectra collected between 0 and 11 mN m^{-1} of the DPPC monolayer at the air–water interface. [Reprinted with permission from *Appl. Spectrosc.* 54: 965–962, 2000. Copyright (2000) Society for Applied Spectroscopy.]

L film exhibits continuous changes with the surface compression, while the cadmium salt L film presents scattered results for the large surface area. This clearly suggests that the metal salt forms stiff domains on water from the beginning, and the molecular aggregation is very strong, especially for the high-pressure region.

Elmore and Dluhy reported another approach to study a double-alkyl chain molecule, 1,2-dipalmitoyl-sn-glycero-3-phosphocholine (DPPC) L film by means of two-dimensional (2D) correlation analysis.[65] The 2D analytical technique was developed by Noda,[65–67] so that minute spectral changes would clearly be captured. The 2D technique is a powerful way to discuss spectral changes, but it responds to spectral noise too much, particularly for the asynchronous map. Therefore, they measured nonpolarized infrared ER spectra, which is one of the good choices for this reflection technique. They divided the analytical region into two regions: a low-pressure region that corresponded to a liquid expanded phase, and a high-pressure region that corresponded to a liquid condensed phase. They made asynchronous simulation maps using band-shift and no-shift models, and the simulated maps were compared to the 2D maps derived from the observed spectra. One of the comparisons is presented in Figure 3.38 (3A/5A). They clearly showed that the low-pressure region exhibits a process in which nonordered hydrocarbon changes to an ordered structure, and the high-pressure region is a result of band shift. It is of interest to find that the very fine overlapping of bands are clearly separated by the 2D technique, and this analytical technique is useful for monolayer analysis.

3.8 ATR SPECTROSCOPY AND RELATED TECHNIQUES

ATR spectrometry is one of the most practical FT-IR measurement techniques for L-B films. This technique employs internal reflections in a large-refractive-index matter, and the penetrated electric field at a surface of the matter is used for absorption measurements. This is fully explained by Harrick[68] and Mirabella[33] in their books; some application studies and other technique using this optical geometry are introduced here. For the fundamental details of this technique, please refer to the books.

Sato and Ozaki et al.[69] employed the FT-IR ATR technique to analyze a Chl-*a* monolayer on a Ge plate. Chl-*a* is a molecule that consists of a large chromophor (porphyrin moiety) and a long tail. Since the porphyrin moiety plays a role of a hydrophilic group, it forms a monolayer on water at pH 8.0. In general, it is difficult to have reliable spectra of a chelate compound on a metallic surface, since the metal often diverses in the monolayer, and it influences the chelate structure. To have a high-quality spectrum of Chl-*a* on a nonmetallic substrate, FT-IR ATR technique using a Ge prism is a good choice. They readily obtained spectra of the monolayer (Figure 3.39) and cast films of Chl-*a*, and comparative studies were reported in terms of aggregation property.

According to their report, the ATR spectra of the Chl-*a* monolayer were very similar to a spectrum of a solution. Some key bands in the fingerprint region are

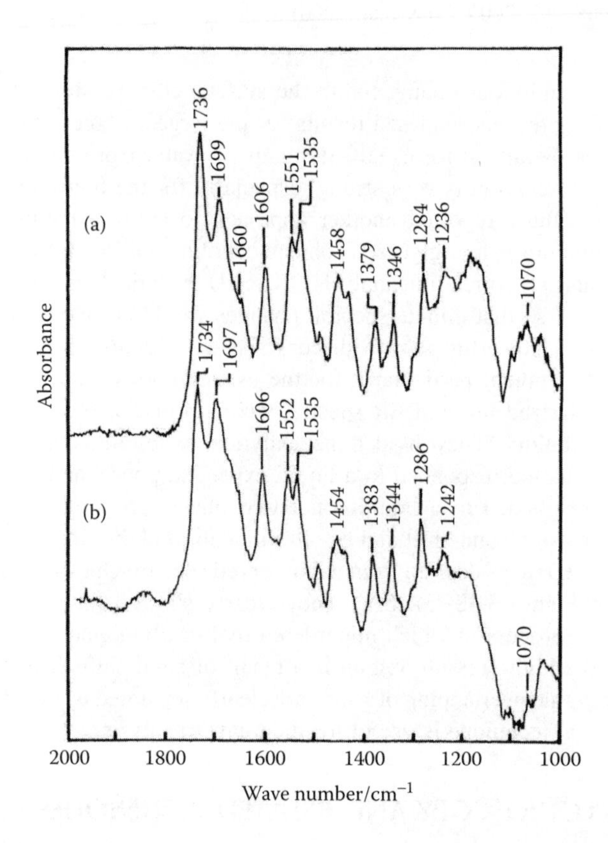

FIGURE 3.39 Infrared ATR spectra of cast films of one-monolayer L-B film of Chl-*a* prepared from diethyl ether (a) and benzene (b) solutions. [Reprinted with permission from *Appl. Spectrosc*. 47: 1509–1512, 1993. Copyright (1993) Society for Applied Spectroscopy.]

known to be correlated to the coordination state of the Mg atom.[70] The similarity suggested that the monolayers comprised five-coordinated monomers, which is quite understandable, since the molecules in a monolayer do not have six-coordinated monomers. On the other hand, transmission spectra of the cast films exhibited strong bands at 1687 cm^{-1} and 1653 cm^{-1} that are assigned to the C=O stretching vibration modes in the free and coordinated 9-keto group. These bands are characteristics of a five-coordinated dimmer species, which is available in a concentrated carbon tetrachloride solution. Furthermore, a very different characteristic was found, in which the free-keto C=O stretching vibration band was suppressed, and the band at 1637 cm^{-1} that was attributed to a C=O stretching vibration mode was incorporated in bonds with a water-coordinated Mg atom (keto C=O…HO(Mg)H…O=C ester). This suggests that the cast film prepared from the water-saturated hexane solution comprises the Chl-*a*-water species and the five-coordinated dimer species. This study reveals that the molecular arrangements in cast films are significantly different from those of L-B films.

ATR spectroscopy is also important because of its optical geometry, which can be used for other applications. One of the most important applications is surface-enhanced infrared reflection–absorption (SEIRA) spectroscopy.[71] A thin metal layer is evaporated on the ATR-prism surface, on which a thin film or molecular adsorbates are deposited. This layer geometry follows the Kretschmann configuration. With this geometry, infrared absorption of the thin layer is largely enhanced. The metal layer for ATR-SEIRA comprises metal particles, and it is known that a continuous metal layer does not contribute to the surface enhancement. SEIRA is often discussed along with surface-enhanced Raman spectroscopy (SERS), which also requires a rough metal surface for a large enhancement. The mechanism of SERS is always discussed in relation to the concept that a dielectric dispersion influenced by the metal topography couples with the incident light to yield a significantly large Raman scattering. This physical model using the concept of surface plasmon has been recognized to be a very comprehensive way to explain SERS. After this success with understanding SERS, SEIRA has also been explained in terms of surface plasmon, due to the metal particles. Nevertheless, it would be more reasonable to consider that the metal particles that have a dielectric property strongly absorb infrared light, through which the adsorbates on the particles give strong infrared absorption bands.[72]

ATR-SEIRA spectroscopy is a very powerful tool for studying electrochemical reaction on an electrode. The metal evaporated surface can be used as a working electrode, and the internal reflection of the infrared light prevents the excess absorption by water. The significantly high sensitivity also contributes to the analysis, since we can reduce the number of scans for FT-IR measurements, which enables us to measure time-resolved spectra. Wan and Osawa et al.[73] measured ATR spectra of adsorbed benzenethiol (BT) molecules on a Au(111) surface in an aqueous solution. One of the spectra is presented in comparison to a transmission spectrum of bulk BT (Figure 3.40). The absence of the S-H stretching vibration band (2568 cm^{-1}) in the SEIRA spectrum indicates that the molecule adsorbs on gold by covalent bonding. The difference of relative band intensities between the two spectra suggests that the molecule has a specific molecular orientation to the surface (Figure 3.41). By

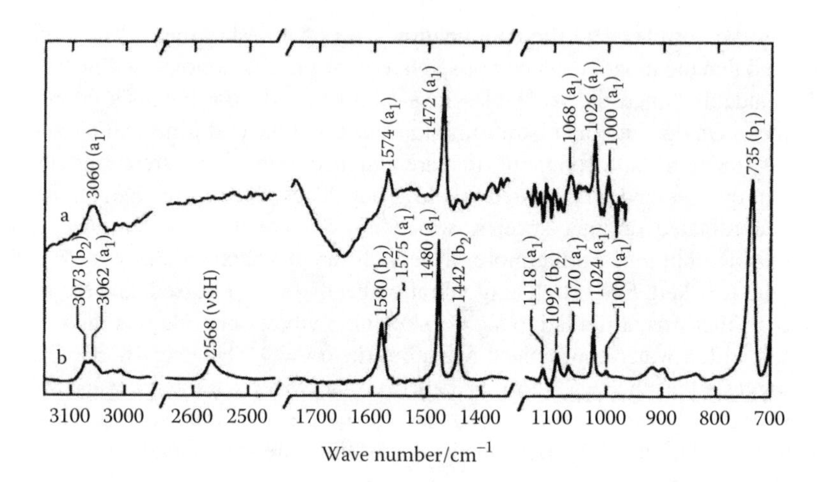

FIGURE 3.40 Representation of the model used in estimating molecular orientation and the directions of dipole moment changes of a_1, b_1, and b_2 modes. [Reprinted with permission from *J. Phys. Chem. B* 104: 3563–3569, 2000. Copyright (2000) American Chemical Society.]

comparison to a transmission SEIRA spectrum of the same molecular species, they estimated the tilt angle θ to be ~30°.

On the other hand, a similar optical geometry is available which consists of prism/air gap/metal phases. This geometry is known as the Otto configuration of ATR. Nagai et al. modified this geometry,[74,75] and they constructed another configuration, such as Ge prism/air gap/Si wafer. They found that the sandwich structure that has a thin air gap enhances a surface perpendicular transition moment significantly, although no metal layer is used. They observed an angle-dependent p-polarization air-gap infrared spectra of an LO mode of silicone oxide (Figure 3.42). They calculated changes of electric-field intensity that depend on the angle of incidence as presented in Figure 3.43 (the thickness was 5 μm). The characteristics of the observed spectra are fully reproduced by the calculation. This technique would be used for practical analyses of L-B films deposited on a semiconductor surface.

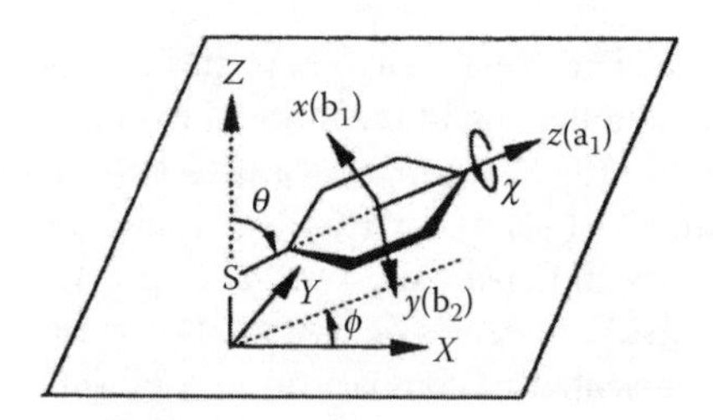

FIGURE 3.41 Comparision of the infrared spectra of BT adsorbed on an Au(111) surface (a) and neat BT (b). [Reprinted with permission from *J. Phys. Chem.* B 104: 3563–3569, 2000. Copyright (2000) American Chemical Society.]

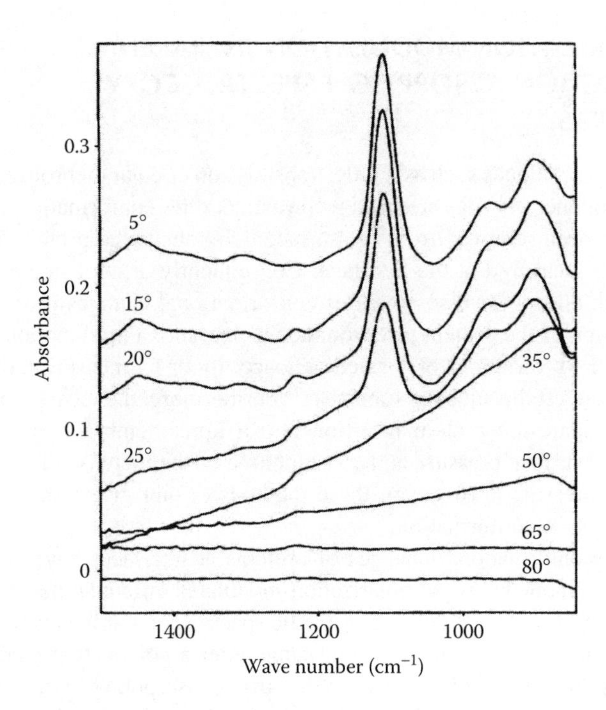

FIGURE 3.42 Air gap ATR spectra of native oxide film on silicone wafer measured as a function of incident angle. Air gap thickness: 5 μm. [Reprinted with permission from *CP430 Fourier Transform Spectroscopy: 11th International Conference*: 581–585, 1997. Copyright (1997) American Institute of Physics.]

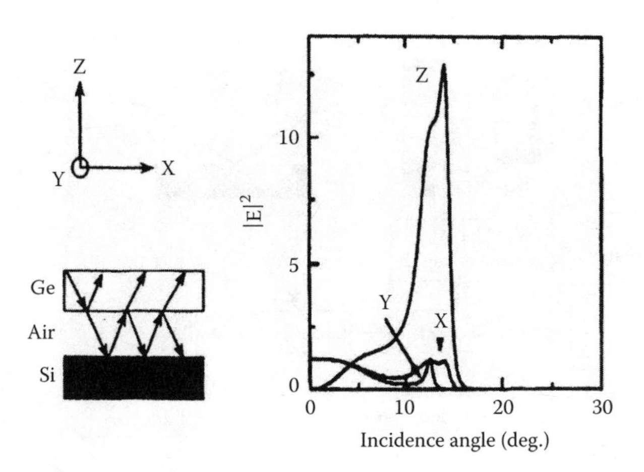

FIGURE 3.43 Calculated results of electric field amplitude generated on the silicon surface by multiple reflections between the prism and silicon.

3.9 POLARIZATION-MODULATION INFRARED REFLECTION ABSORPTION SPECTROSCOPY (PM-IRAS)

Spectroscopic techniques such as FT-IR, transmission circular dichroism, and Nuclear magnetic resonance (NMR) are used to investigate the conformation changes after proteins have been removed from the surface. Obviously, the protein conformation is not directly measured at the interface. Consequently, it is of interest to utilize a method which allows for assessment of conformational changes *in situ*. Until now, the investigation of the protein insertion mechanism into a lipid monolayer has been detected either by variations of surface radioactivity or by measuring the amount of protein that had left the aqueous subphase.[76] Furthermore, the most common method used to investigate the protein insertion into a lipid membrane is measuring the increase in the surface pressure at a constant area or the increase of the surface area at a constant pressure.[77] However, these methods do not give a direct information concerning protein conformation.

Recently, conformational changes of proteins at interfaces have been measured using a new technique known as polarization-modulated infrared reflection–absorption spectroscopy (PM-IRAS).[78] The use of FT-IR spectroscopy at the air–water interface is possible within the entire mid-infrared range after a polarization modulation. This technique has proven to be insensitive to the strong absorption of the water vapor and the isotropic absorption occurring in the sample environment and only weak bands arising from the monolayer are observed. PM-IRAS (Figure 3.44) enables one to characterize *in situ* conformation and orientation of proteins or lipids at the air–water interface.[79–81]

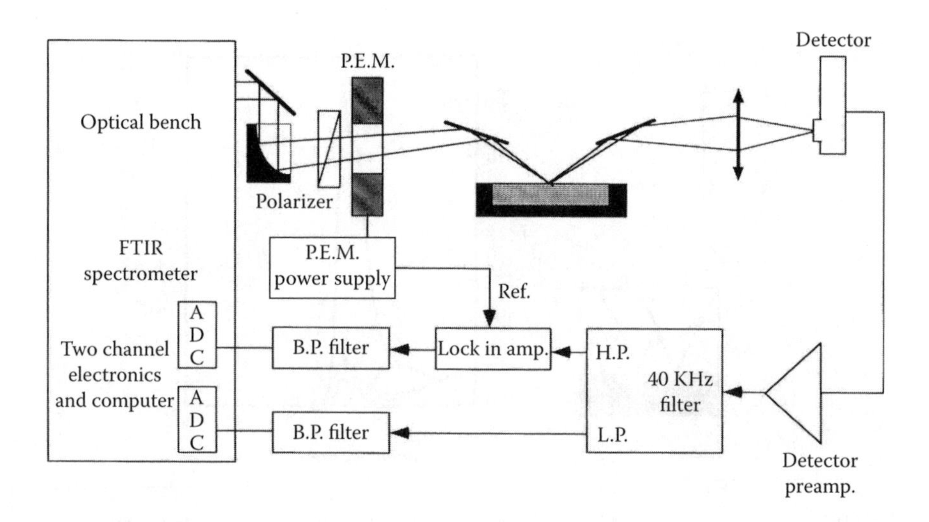

FIGURE 3.44 Schematic diagram of the optical PM-IRAS setup and of two-channel electronic processing.

Although proteins have been spread in monolayers at the air–water interface for decades, no evidence is yet available on the denaturation of their structure. Leblanc's group reported for the first time a study of AChE secondary structure directly at the air–water interface using *in situ* infrared spectroscopy. Based on spectral features of amide I and II bands, they characterized the orientation of α helices and β sheet components relative to the surface as a function of compression and decompression of the monolayer and the surface pressure. They also presented results of an air–water interface interaction between phospholipids and AChE. The adsorption of AChE on phospholipid monolayers, dipalmitoylphosphatidylcholine (DPPC), and dipalmitoyl phosphatidic acid (DPPA) are used as model membranes to mimic the organization of the lipids in biological membranes. The PM-IRAS technique is useful in that it gives simultaneous information on both the enzyme and the phospholipid conformation *in situ.*

3.9.1 Secondary Structure of AChE at the Air–Water Interface Using PMIRRAS

3.9.1.1 Analysis of AChE Secondary Structure at the Air–Water Interface

AChE was spread at the air–water interface at zero surface pressure and with maximum trough area. PM-IRAS spectra were collected during the monolayer compression. Normalized PM-IRAS spectra of the AChE monolayer at different surface pressures are shown in Figure 3.45. Band positions along with the proposed assignments are tabulated in Table 3.4.

In all spectra, two broad bands were observed around 1700–1600 cm^{-1} (amide I) and 1600–1500 cm^{-1} (amide II). Frequencies associated with C-H deformation modes of amino acids side chains vibrations were observed around 1450 cm^{-1}. The amide I contain the most useful information for the analysis of protein secondary structure, and thus is the region of interest. The obtained spectra (Figure 3.45) showed that the positions of the amide I and II bands were similar at the different surface pressures. Careful examination of the amide I band revealed that this band consists mainly of three overlapping bands where two strong absorptions are found at 1655 cm^{-1} and 1630 cm^{-1} and a weaker one around 1696 cm^{-1}. Shoulders were also revealed. The amide II band was centered around 1535 cm^{-1}.

At a surface pressure of 1 mN m^{-1}, the spectrum exhibits an intensive component at 1655 cm^{-1} and shoulders around 1696 cm^{-1} and 1630 cm^{-1}. As the monolayer was compressed, the PM-IRAS signal increased and the shape of the bands became better defined. In fact, the PM-IRAS signal was dependent on the interfacial concentration of the enzyme. The enzyme surface concentration increased with compression, resulting in an increase in signal-to-noise ratio. At 10 mN m^{-1} and 15 mN m^{-1}, the band at 1655 cm^{-1} still remained dominant, but the signal around 1630 cm^{-1} became more prominent. Upon further compression, no changes in the amide I band were observed. However, the intensity of the lower frequency component (1630 cm^{-1}) increased, while the higher frequency band (1655 cm^{-1}) decreased upon further compression. Indeed, at surface pressures of 20 mN m^{-1} and 25 mN m^{-1}, the frequency bands 1630 cm^{-1} and 1696 cm^{-1} were more intense than the one at 1655 cm^{-1}. At 30 mN m^{-1}, the band

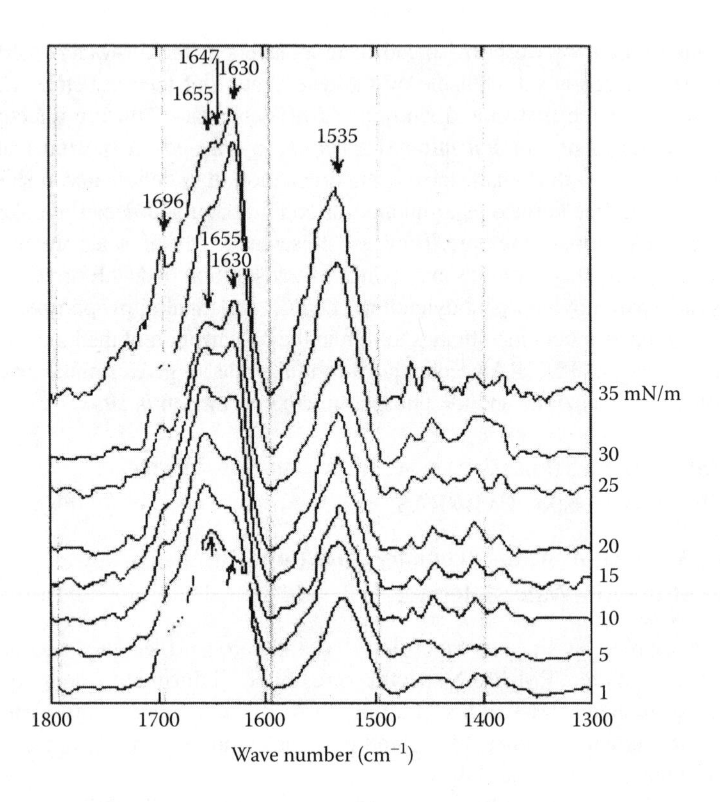

FIGURE 3.45 PM-IRAS spectra of AChE monolayer at the air–water interface at different surface pressures. AChE was spread at nil surface pressure.

at 1655 cm^{-1} was hidden by the low frequency component 1630 cm^{-1} and its sharpness was reduced to a shoulder shown at the same frequency. Further compression led to the formation of a new frequency band at 1647 cm^{-1} at 35 mN m^{-1} (Figure 3.45). On the other hand, the amide II band did not show any significant changes during the compression.

The PM-IRAS spectra of the AChE monolayer exhibited the amide I and II bands, which are consistent with the presence of both α helical and β sheet conformations. Based on both experimental[82–86] and theoretical calculations,[85–88] the 1655 cm^{-1} frequency was assigned to the α helical conformation, while the one at 1630 cm^{-1} was associated with the antiparallel chain pleated sheet conformation (β structure).[87,88] At low surface pressures (1, 5, and 10 mN m^{-1}) (Figure 3.45), α helical structure represented the strongest component and β sheet structure appeared only as a shoulder. The shape of both conformation bands (α helix and β sheet) was better resolved because of the increase of the surface enzyme concentration upon compression. The higher frequency 1696 cm^{-1} was long recognized to be associated with the amide I vibration of the antiparallel chain pleated sheet. In fact, most proteins exhibit an antiparallel pleated sheet with two vibrational modes, a weak band in the region 1690–1696 cm^{-1} together with a well-defined and strong band at 1620–1630 cm^{-1}.[87–89]

Further compression led to the formation of a new band at 1647 cm^{-1}. It is believed that this band is associated with a second population of α helices. In fact, earlier calculations[90] showed a dependency of the length of the α helix on band frequencies. Indeed, an increase in the length of the helix caused a frequency shift to lower wavenumbers, the shape of the spectrum changed gradually, and the frequency splitting increased. A short population of α helices exhibited band at higher frequency, whereas for the longer α helices, the band shifted to lower frequency (by increments of approximately 20 cm^{-1}). Moreover, an X-ray study[91] showed that an AChE monomer consists of 14 α helices in which the length varies from 7 to 35 residues per helix. Therefore, the α helices of the AChE can be grouped into two populations; short α helices with 14 residues or less and longer α helices containing 15 residues or more. In the present work, the two α helix populations were distinguished at higher surface pressures using the PMIRRAS technique. At 35 mN m^{-1}, the enzymes were well packed and the secondary structure components were oriented more vertically so that all the frequencies could be detected at the interface using the PM-IRAS technique.

PM-IRAS spectra collected after the spreading of AChE at a surface pressure of 5 mN m^{-1} are shown in Figure 3.46a and band positions and their proposed assignments are illustrated in Table 3.4. The spectra displayed structural similarity to the ones obtained after spreading at zero surface pressure (Figure 3.45). At 5 mN m^{-1}, the α helix appeared as a strong band at 1655 cm^{-1}, while the β sheet structure was shown as a weak band at 1630 cm^{-1}. At 10 mN m^{-1}, the intensity of the PM-IRAS signal increased and the band at 1640 cm^{-1} became prominent. This frequency component was assigned to unordered structures of the enzyme. A higher frequency band (1696 cm^{-1}) was observed at 25 mN m^{-1} and was associated with the antiparallel pleated sheet.[87] Furthermore, the band at 1630 cm^{-1} became dominant, while the 1655 cm^{-1} band was shown only as a shoulder. The same AChE monolayer was decompressed from 25 to 5 mN m^{-1} and the PMIRRAS spectra were recorded at different surface pressures (Figure 3.46b). The band at 1655 cm^{-1} reappeared with a well-defined form when the monolayer was decompressed to 20 mN m^{-1} while the 1630 cm^{-1} band remained prominent. Upon further decompression, the α helix and β sheet bands were dominant and the unordered structure reappeared at 1640 cm^{-1} at 5 mN m^{-1} and 10 mN m^{-1}. PM-IRAS spectra collected after recompression of the same AChE monolayer to 25 mN m^{-1} exhibited bands associated with the α helix and the β sheet conformations while the frequency 1640 cm^{-1} disappeared. Obviously, the observed behavior of the enzyme secondary structure at the air–water interface is reversible. Indeed, compression and decompression of the monolayer showed that the α helix component was prominent at low surface pressures while the β sheet component was dominant at higher surface pressures.

No significant changes were observed in the amide II band associated with C-N stretching and N-H in-plane bending displacement during the compression. However, spreading at 5 mN m^{-1} caused a splitting of the amide II into two bands at 1523 cm^{-1} and 1535 cm^{-1}. Frequency bands observed in the same region were assigned earlier[90] to random coil and to α helix conformations, respectively. The band at 1535 cm^{-1} may be assigned to random conformation of the enzyme while the band at 1523 cm^{-1} can be attributed to α helical conformation.

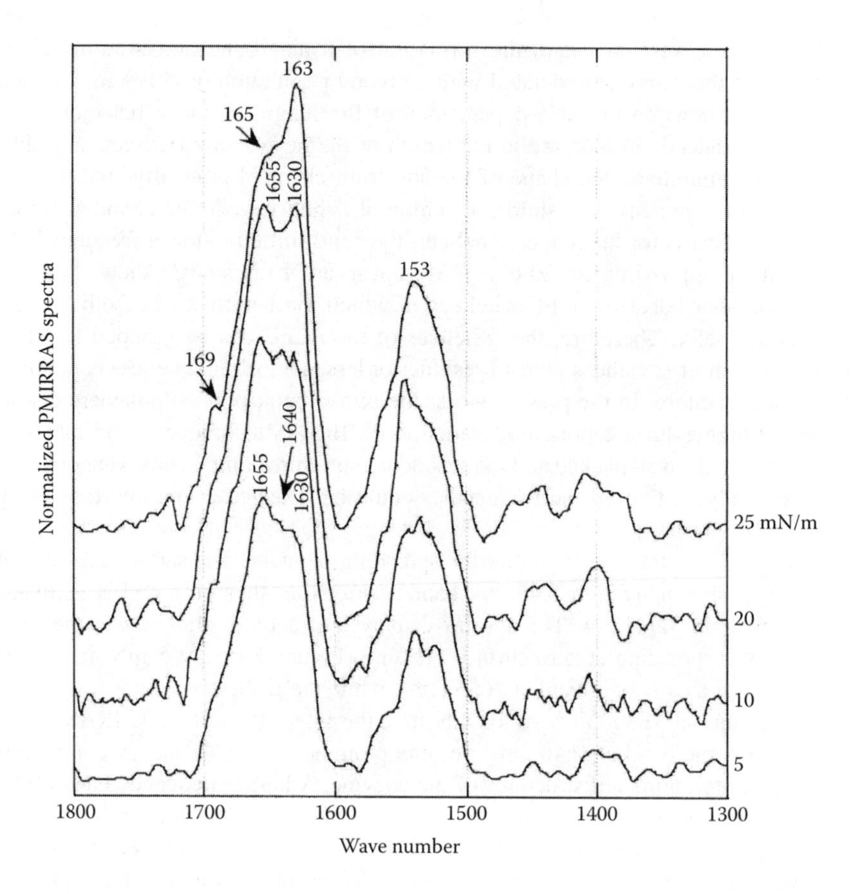

FIGURE 3.46 PMIRRAS spectra of AChE monolayer at the air–water interface during compression. AChE was spread at 5 mN m^{-1}.

The analysis of the amide I and II bands showed no major spectral changes of the enzyme conformation in both the experimental conditions of the AChE spreading at the interface. Application of higher and lower surface pressures did not significantly affect the shape and the position of the amide bands; however, frequencies due to unordered structure appeared only when the enzyme was spread at a high surface pressure and the amide II was split mainly into two bands. The qualitative analysis showed that the secondary structure of AChE monolayer consists mainly of both the α helical and the β sheet conformations. Furthermore, we were able to distinguish between two populations of the α helices (short and longer helix) at the air–water interface using the PM-IRAS technique. This phenomenon was predicted theoretically[90] and was observed experimentally[78] in solution.

Thus, in all PM-IRAS spectra recorded at low surface pressures (1–15 mN m^{-1}), the α helix was dominant while the β sheet structure was less prominent. Upon further compression, the helix band was hidden by the β sheet component, which became more prominent at high surface pressures (20–35 mN m^{-1}). A change in the orientation of the AChE secondary structure components occurred during the compression. Indeed, at zero surface pressure, the spread enzymes have their side chains lying flat at the air–water interface. When the surface pressure increases, the orientation of the polar groups toward the aqueous phase and the nonpolar away from the surface induces changes in the α helix and β sheet orientation at the interface. In fact, computer simulation[77] showed that the α helix was shown as a strong positive band when it was parallel to the water surface. On the other hand, a negative strong band was observed when the helix was perpendicular to the surface. Upon further compression, the helix was lifted toward the air and the intensity of the amide I band associated with the helix decreased while the amide II intensity increased. Based on this theory, we believe that the surface pressure has a significant effect on the orientation of the AChE secondary structure at the air–water interface. Hence, at nil surface pressure, the α helices were lying parallel to the water surface, leading, therefore, to the high signal of the amide I band associated with this conformation. On the other hand, the frequency band assigned to the β sheet conformation was hidden by the α helix band at low surface pressures and was seen only as a weak shoulder. The β sheet component appeared as a strong band during compression, since the helices were oriented more vertically at high surface pressures.

3.9.2 PMIRRAS Study of Phospholipids and Phospholipid–Enzyme Interactions

3.9.2.1 DPPC and DPPA Monolayers

Normalized PM-IRAS spectra of pure DPPC were collected at different surface pressures. The PM-IRAS signal increased as the surface pressure increased and the noise level decreased during the compression due to the increase of the DPPC surface concentration. However, the shape and the position of the bands remained unchanged. At low surface pressure (10 mN m^{-1}), the spectra exhibited six strong bands at 1728, 1468, 1228, 1087, 1058, and 969 cm^{-1}. At higher surface pressures (30 mN m^{-1} and 40 mN m^{-1}), the band position 1728 cm^{-1} shifted to a higher frequency (1735 cm^{-1}). This frequency shift indicated that at low surface pressures, the ester bonds interacted with the water subphase through hydrogen bonding. Upon further compression, there were weak interactions of hydrogen bonding with water molecules and the frequency component associated with the stretching C=O ester bond shifted to 1735 cm^{-1}. The same phenomena was observed recently with the lipid dimyristoylphosphatidylcholine (DMPC).[81] Moreover, the high intensity of this band indicated that the orientation of the carbonyl group was favorably in the plane of the film. On the other hand, the sharp band seen at 1468 cm^{-1} was associated with CH_2 bending mode. Its high intensity suggested a well organization and orientation of the long alkyl chains at the interface in contrast to what was reported recently with the DMPC data,[81] where a large disorder in the hydrocarbon chains was observed.

Asymmetric and symmetric P=O stretching vibrations were revealed at 1228 cm^{-1} and 1088 cm^{-1}, respectively. Similarly, frequency shifts between low and high surface pressure spectra were observed for PO_2^-, which indicate the presence of hydrogen bonding between the phosphate groups and the water molecules at low surface pressures.

The C-O-P stretching vibration modes were observed at 1058 cm^{-1} and the asymmetric stretching modes of the trimethylammonium group $N^{+}(CH_3)_3$ were seen as a sharp band at 969 cm^{-1}. The upward position of this band suggests that the orientation of the transition moment of the C-N stretching vibration was parallel to the water interface.

In the normalized PM-IRAS spectra of DPPA, the PM-IRAS signal increased as the surface pressure increased, due to the increase of the DPPA surface concentration. Similarly, the stretching vibrations of the ester carbonyl group were exhibited as a strong band at 1737 cm^{-1}, and the upward orientation of this band indicates that the transition moment of these vibrations was in the surface plane. The CH_2 bending mode was shown as a sharp band at 1468 cm^{-1} and broad bands were observed around 1106 cm^{-1} and 1020 cm^{-1} and were associated with the symmetric P=O and C-O-P stretching vibrations, respectively. The increase of the intensity and the sharpness of the CH_2 bending mode indicates a good organization of the hydrocarbon chains in the monolayer.

3.9.2.1.1 Phospholipid-AChE Interaction at the Air–Water Interface

The interaction between DPPC and the enzyme was studied at the air–water interface by spreading DPPC as a monolayer and injecting AChE into the subphase. The spectrum of the mixed film exhibited several bands that are associated with the presence of both the phospholipid and the enzyme at the air–water interface. Injection of the enzyme did not induce any changes in the DPPC spectrum. Indeed, bands assigned to the ester carbonyl group C=O, the CH_2 bending mode, the asymmetric P=O, the symmetric P=O, C-O-P, and the asymmetric C-N stretching vibrations were observed at 1733, 1468, 1227, 1085, 1055, and 972 cm^{-1}, respectively. Moreover, the amide I and II bands were observed at the 1750–1600 cm^{-1} and the 1550–1500 cm^{-1} regions, respectively. The main bands of the mixed film were being oriented upward relative to the baseline. This indicates that their transition moments were preferentially in the surface plane. The amide I band was centered around 1630 cm^{-1}, which is assigned to the β sheet strand, while the α helix was shown as a weak shoulder around 1655 cm^{-1}. In contrast to the pure enzyme monolayer, at a surface pressure of 15 mN m^{-1}, the mixed film was well packed and organized and the α helices were being oriented more vertically. Therefore, the β sheet component was much more prominent at this surface pressure. The amide II band was shown as a strong positive band centered at 1535 cm^{-1}. The presence of bands associated with the phospholipid and AChE indicates that the enzyme was incorporated into the lipidic film at the interface. The shape and the position of the amide I and II bands obtained in the mixed lipid–enzyme monolayer suggest that no changes occurred in the AChE secondary structure in comparison with the PMIRRAS spectra of the pure AChE monolayer (Figure 3.45). Furthermore, a similar orientation for the α helix and the β sheet components was observed in the presence of a lipidic film at the interface.

From these results, it can be seen that the polarization modulation technique provides a very sensitive means of measuring the vibrational modes of mixed phospholipids–enzyme monolayer at the interface. This indicates that the enzyme was inserted into the phospholipid film during the compression. Moreover, the orientation or the content of the AChE secondary structure was not influenced by the presence of the phospholipids at the air–water interface.

3.9.2.2 AChE-Substrate Interaction at the Air–Water Interface

The enzyme–substrate interaction was studied at the air–water interface by spreading the enzyme as a monolayer and dissolving the substrate, ATChI, into the subphase. The temperature and the pH of the subphase were 20°C and 8°C, respectively. The amide I and II bands were centered at 1655 cm^{-1} and 1535 cm^{-1}, respectively. At 5 mN m^{-1}, the α helix was seen as a strong band at 1655 cm^{-1}. Higher (1690 cm^{-1}) and lower (1640 cm^{-1}) frequency bands were observed and are assigned to the antiparallel pleated sheet and unordered structure of the enzyme, respectively. At surface pressures of 10 mN m^{-1} and 15 mN m^{-1}, the band at 1655 cm^{-1} shifted to 1658 cm^{-1} and the band at 1630 cm^{-1} became much more prominent. Upon further compression, the β sheet component was dominant and the α helix component was much less prominent. Furthermore, unordered structure of the enzyme was observed around 1640 cm^{-1}. The α-helix component was dominant at low surface pressure and shifted to 1658 cm^{-1} at 10 mN m^{-1} and 20 mN m^{-1}. These frequency shifts were caused by changes in the AChE secondary structure conformation induced by substrate binding during compression.

The band at 1405 cm^{-1} is associated with the stretching vibrations of C-N of the choline moiety and its intensity increases during the compression due to the increase of the surface concentration of the enzyme and the acetylthiocholine. The spectral region 1160–1050 cm^{-1} is mainly attributed to the symmetric stretching vibrations of the C-O-C mode and this indicates that the enzyme was in the acylation form. The interaction of C-C vibrations occurred in this spectral region and made the band stronger. The frequency band 1135 cm^{-1} is associated with the trimethylamine vibrational mode during the substrate binding. On the other hand, the frequency bands 1074 and 1128 cm^{-1} seen only at high surface pressure (25 mN m^{-1}) might be assigned to vibrational modes of the reaction products, thiocholine iodide. In fact, bands in the same region (1080 cm^{-1} and 1130 cm^{-1}) were observed in the IR spectrum of the choline iodide.

The hydrolysis reaction was investigated when the enzyme solution was spread at nil surface pressure. PMIRRAS data indicated the presence of both the substrate and the reaction products at the interface 10 minutes after spreading of the enzyme at high surface pressures. In fact, at 5 mN m^{-1}, the substrate binding and formation of a complex AChE-ACTh is illustrated by C-O-C vibration mode. This is in agreement with the tapping mode atomic force microscopy (TMAFM) images and the UV-Vis results reported recently.[92] TMAFM images of a transferred film 10 minutes after the substrate injection beneath the AChE monolayer indicated the presence of an AChE-ACTh complex. At a surface pressure of 25 mN m^{-1}, frequency bands associated with the thiocholine iodide suggest the presence of a heterogeneous film that consists mainly of free enzymes and reaction products. Moreover, bands (1130 cm^{-1} and 1074 cm^{-1}) arising from the presence of reaction products at the interface were observed at the end on the compression. This agrees with the TMAFM images and UV-Vis data.[92] In fact, the topography of a film transferred 30 minutes after compression of the AChE monolayer showed a heterogeneous surface structure that consisted of both free enzyme and reaction products.

Slight and reversible changes of the enzyme conformation occurred during the compression upon the substrate binding. Indeed, higher frequency components were observed in the amide I band at the beginning of the compression. Moreover, at low surface pressures, the α helix frequency component is seen around 1655 cm^{-1} and shifted to 1659 cm^{-1} upon the substrate binding during the compression. Upon further compression, the components arising from the α helix and β sheet structures reappeared at 1655 cm^{-1} and 1630 cm^{-1}. Using the PM-IRAS technique, we were able to follow the hydrolysis reaction of acetylthiocholine iodide at the air–water interface. In fact, the presence of the substrate binding to the enzyme active site and formation of the reaction products were observed.

Thus, PM-IRAS is proven to be a very sensitive technique for measuring the vibrational modes of the secondary structure of adsorbed AChE at the air-water interface. These data are the first reported clear evidence that not all proteins or enzymes lose their native secondary structure upon spreading at the air–water interface. The shape and the position of the amide I and II bands indicate that AChE consists of both the α helices and β sheets. The content of the amide I was evaluated qualitatively and quantitatively at different surface pressures and surface areas. Application of higher and lower surface pressures did not cause unfolding of the enzyme at the interface; however, the orientation of the α helix and β sheet components at the air–water interface changed upon compression and decompression of the AChE monolayer. Indeed, at low surface pressure, the α helices were parallel to the water surface, while at a high surface pressure, their tilt axis has a tendency to be oriented more vertically to the surface. Furthermore, the incorporation of the AChE into a lipidic membrane did not alter its conformation. Moreover, using the PM-IRAS technique, we were able to follow the progress of the hydrolysis reaction of the acetylthiocholine iodide catalyzed by AChE at the air–water interface. Slight and reversible changes of the AChE conformation occurred upon the acetylthiocholine binding to the AChE active site. The molecular interaction between the organophosphorous and carbamate inhibitors and the AChE showed that only the organophosphate, Paraoxon, caused drastic and irreversible changes in the enzyme secondary structure, while the inhibition by the carbamate was reversible.

3.10 CHEMOMETRIC ANALYSIS FOR SURFACE CHEMISTRY

Chemometrics is a new trend of vibrational spectral analyses that has been developed in recent years. Chemometrics, thus far, has been recognized only as a spectroscopic calibration tool, and the potential for physicochemical discussion has been unknown to most of chemical communities. In recent years, however, a sleeping potential has been revealed for chemometrics to play a role in spectral analyses. Here, only one example, the analysis of L-B film by means of principal component analysis (PCA), is introduced.

PCA is a procedure designed to explain data plots in multidimensional space by producing orthogonal axes sets that are called principal components or factors. A spectrum that consists of N absorbance points such as $(a_1, a_2, \ldots, a_N)$ can be considered a point in N-dimensional space. Therefore, if we measure m spectra, we can plot m points in the space. PCA organizes the movement of the point in the space due to the spectral changes. PCA is formulated by using a spectra matrix, a score, and loading vectors (**A**, **t**, and **p**):

$$\mathbf{A} = \sum_j \mathbf{t}_j \mathbf{p}_j \tag{3.27}$$

The loading vectors are the orthogonal axes set. Since this formulation is a simple mathematical expansion of **A**, the loading vectors have no physical meaning in general. Nevertheless, when a minute chemical species are mixed with a dominant chemical component, the largest spectral changes due to the dominant component are captured by the first loading, and the minor changes due to the minute species

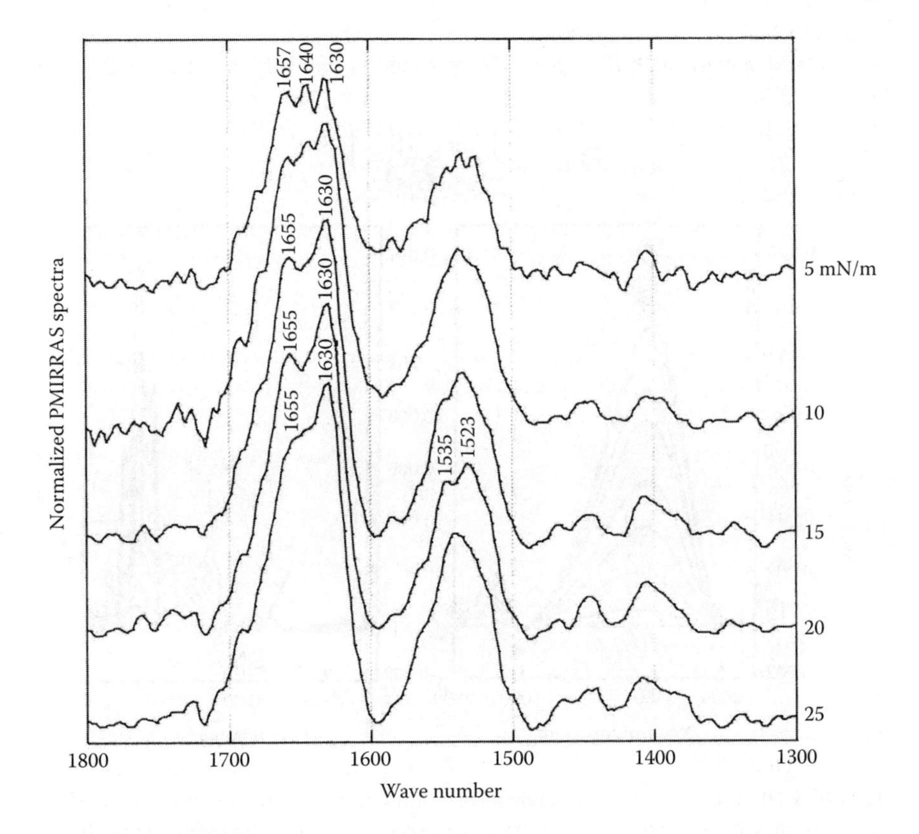

FIGURE 3.47 PMIRRAS spectra of AChE monolayer at the air–water interface during decompression. AChE was spread at 5 mN m^{-1}.

are followed by the second loading.[93,94] In other words, if a mixture is comprised of chemical components with a large concentration ratio, the spectra of the components would be resolved from the mixed spectra.

This technique was first applied to DPPC monolayer L-B film associated with sucrose with various concentrations.[93,94] The raw infrared RA spectra of the L-B film are shown in Figure 3.47. The spectra reflect concentration changes of sucrose, but no more information of interest is available from the raw spectra, since many key bands are overlaid with each other, which makes the analysis of the spectra very difficult. Nonetheless, we can expect that minute information of molecular structure would be extracted from the RA spectra, by use of the characteristic of PCA. Figure 3.48 presents the third loading (third eigenvector) calculated from the RA spectra without any pretreatment. It is striking that the sharp PO_2^- antisymmetric stretching vibration band appears at 1264 cm^{-1}, which is a characteristic band for the strongly hydrated PO_2^- group, and it does not appear in the raw spectra at all. In a similar manner, water-correlated bands appear at 3250 and 1650 cm^{-1}, which are key bands for water with an ice-like structure. Both bands are not observed in the raw RA spectra, either. These results strongly suggest that minute water molecules remain in the L-B film even after drying, and they bind the PO_2^- group in the film through hydrogen bonding.

In this fashion, some chemometric analyses enable us to draw fine information from mixed spectra with no *a priori* knowledge, a fact that is not clear from the

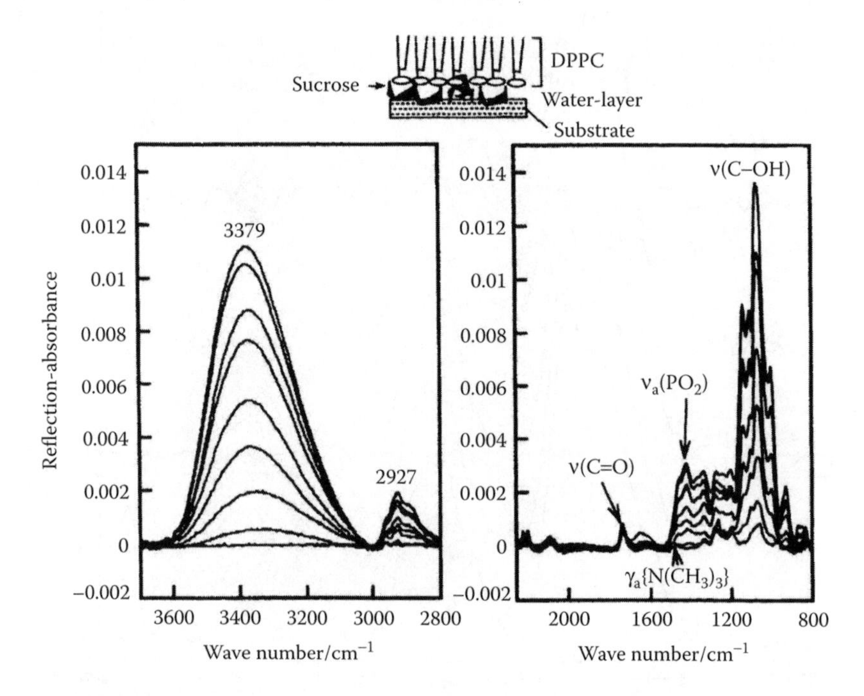

FIGURE 3.48 Sucrose-concentration dependent infrared RA spectra of DPPC-d_{62} L-B film deposited on a gold-evaporated glass slide. The sucrose concentration range is 0–40 mM. [Reprinted with permission from *Anal. Chem.* 75: 3085–3091, 1999. Copyright (1999) American Chemical Society.]

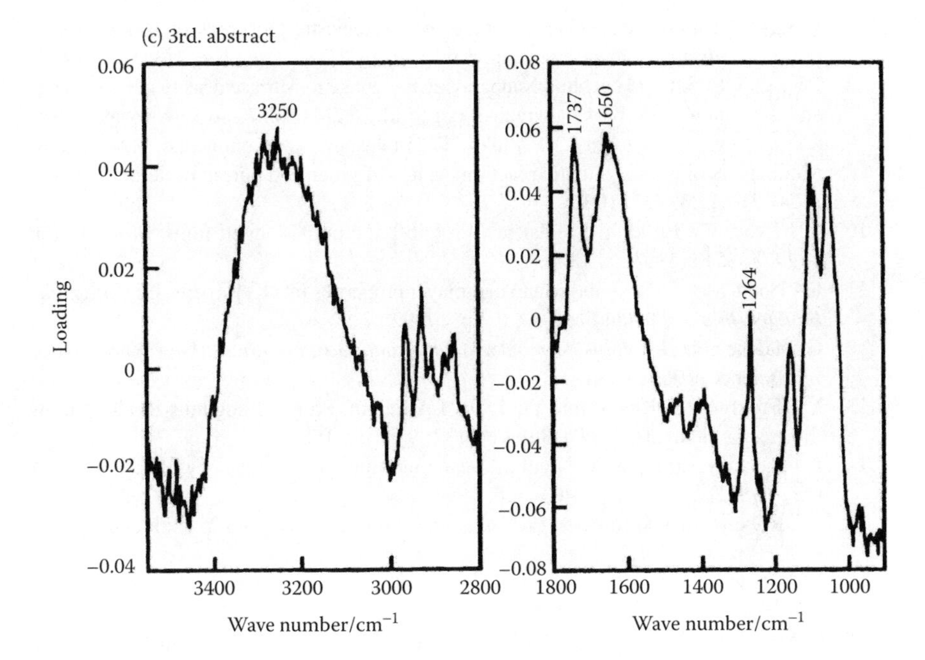

FIGURE 3.49 The third abstract spectrum of the RA spectra in Figure 3.47 yielded by PCA. [Reprinted with permission from *Anal. Chem.* 75: 3085–3091, 1999. Copyright (1999) American Chemical Society.]

raw data. The new characteristic of PCA, which can be used for detection of minute chemical species, is named factor analytical resolution of minute signals (FARMS).

REFERENCES

1. http://west.t.u-tokyo.ac.jp/soft/indexe.html.
2. E Kemnitz, C Werner, SI Trojanov. Structural chemistry of alkaline metal hydrogen sulfates. A review of new structural data. II. Hydrogen bonding systems. *Eur. J. Solid State Inorg. Chem.* 36: 581–596, 1996.
3. RJ Simmonds. *Chemistry of Biomolecules: An Introduction*. North Yorkshire: The Royal Society of Chemistry, 1992, pp. 5–6.
4. BD Ratner, AS Hoffman, FJ Schoen, JE Lemons. *Biomaterials Science: An Introduction to Materials in Medicine*. London: Academic Press, 1996, pp. 11–12.
5. J Rebek Jr. Recognition, Replication and Assembly. In: SM Roberts, Ed. *Molecular Recognition: Chemical and Biological Problems* II. North Yorkshire: The Royal Society of Chemistry, 1992, pp. 65–73.
6. T Hasegawa, J Nishijo, J Umemura, W. Theiß. Simultaneous evaluation of molecular-orientation and optical parameters in ultrathin films by oscillators-model simulation and infrared external-reflection spectrometry. *J. Phys. Chem.* B 105: 11178–11185, 2001.

7. L Shao, NJ Tao, RM Leblanc. Probing the microelastic properties of nanobiological particles with tapping mode atomic force microscopy. *Chem. Phys. Lett.* 273: 37–41, 1997.
8. P Facci, V Erokhin, C Nicolini. Nanogravimetric guage for surface density measurements and deposition analysis of Langmuir-Blodgett films. *Thin Sol. Films* 230: 86–89, 1993.
9. SY Zaitsev, NA Kalabina, VP Zubov, EP Lukashev, AA Kononenko, RA Uphaus. Monolayers of photosynthetic reaction centers of green and purple bacteria. *Thin Sol. Films* 210: 723–725, 1992.
10. YU Lvov, VV Erokhin, SY Zaitsev. Protein Langmuir-Blodgett films. *Biol. Membr.* 4: 1477–1513, 1991.
11. RA Hann. Molecular structure and monolayer properties In: GG Roberts, Ed. *Langmuir-Blodgett Films*. Plenum Press 1990, pp. 17–92.
12. GL Gaines, Jr. *Insoluble Monolayers at Liquid-Gas Interfaces*. New York: Wiley Interscience, 1966.
13. VV Erokhin, RL Kayushina, YU Lvov, LA Feigin. Protein Langmuir-Blodgett films as sensing elements. *Studia Biophys*. 132: 97–104, 1989.
14. R Facci, VV Erokhim., C Nicolini. Scanning tunneling microscopy of reaction centers. *Thin Sol. Films* 243: 403–406, 1994.
15. GG Roberts. In: GG Roberts, Ed. *Langmuir-Blodgett Films*. New York: Plenum Press, 1990, pp. 317–411.
16. RM Stewart. In: GG Roberts, Ed. *Langmuir-Blodgett Films*. New York: Plenum Press, 1990, pp. 273–316.
17. F Mac Ritchie. Air-water interface studies of proteins. *Analyt. Chim. Acta* 249: 241–245, 1991.
18. C Nicolini, M Adami, F Antolini, F Beltram, M Sartore, S Vakula. Biosensors — a step to bioelectronics. *Phys. World* 5: 30–34, 1992.
19. C Nicolini, VV Erokhim, F Anatolini, P Catasti, P Facci. Thermal stability of protein secondary structure in Langmuir-Blodgett films. *Biochim. Biophys. Acta* 1158: 273–278, 1993.
20. G Alegria, PL Dutton. I. Langmuir-Blodgett films of bacterial photosynthetic membranes and isolated reaction centers: preparation, spectrophotometric and electrochemical characterization. *Biochim. Biophys. Acta* 1057: 239–257, 1991.
21. G Alegria, PL Dutton. II. Langmuir-Blodgett films of Rhodopseudomonas viridis reaction center: determination of the order of the hemes in the cytochrome c subunit. Biochim. Biophys. Acta 1057: 258–272, 1991.
22. PK Hansma. *Tunneling Spectroscopy.* New York: Plenum, 1982.
23. Y Yasuda, Y Hirata, H Sugino, M Kumei, M. Hara, J Miyake, M Fujihara. Langmuir-Blodgett films of reaction centers of Rhodopseudomonas viridis: photoelectric properties. *Thin Sol. Films* 210/211: 733–735, 1992.
24. W Kuhlbrandt, DA Wang. 3-dimensional structure of plant light harvesting complex determined by electron crystallography. *Nature* 350: 130–134, 1991.
25. A Ulman. *An Introduction to Ultrathin Organic Films.* San Diego: Academic, 1991.
26. J Umemura. Reflection–absorption spectroscopy of thin films on metallic substrates. In: PR Griffiths, JM Chalmers, Eds. *Handbook of Vibrational Spectroscopy*, Vol. 2. Chichester: John Wiley & Sons, 2002, pp. 982–998.
27. K Yamamoto, H Ishida. Interpretation of reflection and transmission spectra for thin films. *Appl. Spectrosc.* 48: 775–787, 1994.
28. K Yamamoto, H Ishida. Kramers-Kronig analysis applied to reflection-absorption spectroscopy. *Vib. Spectrosc.* 15: 27–36, 1997.
29. PR Griffiths. Beer's Law. In: PR Griffiths, JM Chalmers, Eds. *Handbook of Vibrational Spectroscopy*, Vol. 3. Chichester: John Wiley & Sons, 2002, pp. 2225–2234.

30. M Milosevic, SL Berets. Applications of the theory of optical spectroscopy to numerical simulations. *Appl. Spectrosc.* 47: 566–574, 1993.
31. RL McCreery. *Raman Spectroscopy for Chemical Analysis.* New York: Wiley Interscience, 2000, pp. 35–37.
32. PR Griffiths, JA de Haseth. *Fourier Transform Infrared Spectrometry.* New York: Wiley Interscience, 1986, pp. 213–219.
33. RW Duerst, MD Duerst, WL Stebbings. Transmission infrared spectroscopy. In: FM Mirabella, Ed. *Modern Techniques in Applied Molecular Spectroscopy.* New York: Wiley Interscience, 1998, pp. 11–81.
34. K Kudo. Hikaribussei-Kiso *Introduction to Optical Solid-State Physics*. Tokyo: Ohm-Sha, 1996, pp. 68–70.
35. J Umemura, T Kamata, T Kawai, T Takenaka. Quantitative evaluation of molecular orientation in thin Langmuir-Blodgett films by FT-IR transmission and reflection-absorption spectroscopy. *J. Phys. Chem.* 94: 62–67, 1990.
36. A Dhanabalan, RB Dabke, SN Datta, NP Kumar, SS Major, SS Talwar, AQ Contractor. Preparation and characterization of mixed L-B films of polyaniline and cadmium arachidate. *Thin Sol. Films* 295: 255–259, 1997.
37. T Kamata, J Umemura, T Takenaka. Structure study of Langmuir-Blodgett films of stearic acid and cadmium stearate deposited by different techniques. *C Hem. Lett.* 1231–1234, 1988.
38. J Zhao, Z Wu, J Zhang, T Zhu, A Ulman, Z Liu. Structural characterization of L-B films of a novel azobenzene compound. Mol. *Cryst. Liq. Cryst.* 294: 185–188, 1997.
39. MS Johal, AN Parikh, Y Lee, JL Casson, L Foster, BI Swanson, DW McBranch, DQ Li, JM Robinson. Study of the conformational structure and cluster formation in a Langmuir-Blodgett film using second harmonic generation, second harmonic microscopy, and FTIR spectroscopy. *Langmuir* 15: 1275–1282, 1999.
40. N Johns, AF McKay, RG Sinclair. Band progressions in the infrared spectra of fatty acids and related compounds. *J. Am. Chem. Soc.* 74: 2575–2578, 1952.
41. L Hong-Guo, F Xu-Sheng, X Qing-Bin, W Lei, Y Kong-Zhang. Central metal effect on the organization of porphyrin L-B films. *Thin Sol. Films* 340: 265–270, 1999.
42. M Schmelzer, S Roth, CP Niesert, F Effenberger, R Li. Highly ordered L-B films of DHAP: a donor acceptor substituted polyene. *Thin Sol. Films* 235: 210–214, 1993.
43. S Meucci, G Gabrielli, G Caminati. Comparative investigation of Langmuir-Blodgett films and self-assembled monolayers on metal surfaces. *Mater. Sci. Eng.* C 8–9: 135–143, 1999.
44. T Kawai. Driving force controlling liquid crystals alignment on photochromic polyion complex L-B film. Thin Sol. Films 352: 228–233, 1999.
45. K Ichimura, H Akiyama, K Kudo, N Ishizuki, S Yamamura. Command surfaces 12 [1]. Factors affecting in-plane photoregulation of liquid crystal alignment by surface azobenzenes on a silica substrate. *Liq. Cryst.* 20: 423–435, 1996.
46. K Itadera, K Kudo, S Kuniyoshi, K Tanaka. Charge transfer control in donor molecules adsorbed TCNQ thin films voltage L-B technique. *Synth. Metals* 86: 2261–2262, 1997.
47. T Hasegawa, Y Hatada, J Nishijo, J Umemura, Q Huo, RM Leblanc. Hydrogen bonding network formed between accumulated Langmuir-Blodgett films of barbituric acid and triaminotriazine derivatives. *J. Phys. Chem.* B 103: 7505–7513, 1999.
48. T Hasegawa, S Takeda, A Kawaguchi, J Umemura. Quantitative analysis of uniaxial molecular orientation in Langmuir-Blodgett films by infrared reflection spectroscopy. *Langmuir* 11: 1236–1243, 1995.
49. W Hansen. Electric field produced by the propagation of plane coherent electromagnetic radiation in a stratified medium. *J. Opt. Soc. Am.* 58: 380–390, 1968.

50. M Schmelzer, M Burghard, P Bäuerle, S Roth. Structural information on thiophene Langmuir-Blodgett heterostructures. *Thin Sol. Films* 243: 620–624, 1994.
51. X Du, Y Liang. Well-ordered structure of *N*-octadecanoyl-L-analnine Langmuir-Blodgett film studied by FTIR spectroscopy. *Chem. Phys. Lett.* 313: 565–568, 1999.
52. LJ Bellamy. The Infrared Spectra of Complex Molecules, Vol. 1. London, Chapman and Hall, 1975, pp. 231–330.
53. JA Mielczarski, RH Yoon. Fourier transform infrared external reflection study of molecular orientation in spontaneously adsorbed layers on low-absorption substrates. *J. Phys. Chem.* 93: 2034–2038, 1989.
54. JA Mielczarski, RH Yoon. Spectroscopic studies of the structure of the adsorption layer of thionocarbamate. *Langmuir* 7: 101–108, 1991.
55. Y Urai, C Ohe, K Itoh. Infrared external reflection spectroscopic study on the structures of a conducting Langmuir-Blodgett film of a tetrathiafulvalene derivative. *Langmuir* 14: 4873–4879, 1998.
56. H Sakai, J Umemura. Molecular-orientation change in Langmuir-Blodgett films of stearic acid and cadmium stearate upon surface compression, as studied by infrared external-reflection spectroscopy. *Bull. Chem. Soc. Japan.* 70: 1027–1032, 1997.
57. RA Dluhy, Z Ping, K Faucher, JM Brockman. Infrared spectroscopy of aqueous biophysical monolayers. *Thin Sol. Films* 327–329: 308–314, 1998.
58. T Hasegawa, J Nishijo, J Umemura, W Theiß. Simultaneous evaluation of molecular-orientation and optical parameters in ultrathin films by oscillator-model simulation and infrared external reflection spectroscopy. *J. Phys. Chem. B* 105: 11178–11185, 2001.
59. CC Kim, JW Garland, H Abad, PM Raccah. Modeling the optical dielectric function of semiconductors: extension of the critical-point parabolic-band approximation. *Phys. Rev.* B 45: 11749–11767, 1992.
60. LJ Fina, Y-S Tung. Molecular orientation of monolayers on liquid substrates: optical model and FT-IR methods. *Appl. Spectrosc.* 45: 986–992, 1991.
61. A Gericke, H Hühnerfuss. *In situ* investigation of saturated long-chain fatty acids at the air/water interface by external infrared reflection-absorption spectroscopy. *J. Phys. Chem.* 97: 12899–12908, 1993.
62. RA Dluhy, DG Cornell. In situ measurement of the infrared spectra of insoluble monolayer at the air-water interface. *J. Phys. Chem.* 89: 3195–3197, 1985.
63. H Sakai, J Umemura. Effect of infrared radiation and air flow on fourier transform infrared external reflection spectra of Langmuir monolayers. *Langmuir* 13: 502–505, 1997.
64. DL Elmore, RA Dluhy. Pressure-dependent changes in the infrared c-h vibrations of monolayer films at the air/water inferface revealed by two-dimensional infrared correlation spectroscopy. *Appl. Spectrosc.* 54: 965–962, 2000.
65. I Noda, AE Dowrey, C Marcott, GM Story, Y Ozaki. Generalized two-dimensional correlation spectroscopy. *Appl. Spectrosc.* 54: 236A–246A, 2000.
66. I Noda. General theory of two-dimensional (2D) analysis. In: JM Chalmers, PR Griffiths, Eds. *Handbook of Vibrational Spectroscopy*, Vol. 3. Chichester: John Wiley & Sons, 2002, pp. 2123–2134.
67. Y Ozaki. 2D correlation spectroscopy in vibrational spectroscopy. In: JM Chalmers, PR Griffiths, Eds. *Handbook of Vibrational Spectroscopy*, Vol. 3. Chichester: John Wiley & Sons, 2002, pp. 2135–2172.
68. NJ Harrick. *Internal Reflection Spectroscopy.* New York: Harrick Scientific Corporation, 1987.

69. H Sato, Y Ozaki, K Uehara, T Araki K Iriyama. ATR/FT-IR study of a monolayer film of Chl-*a* on a germanium plate. *Appl. Spectrosc.* 47: 1509–1512., 1993.
70. M Tasumi, M Fujiwara. Vibrational spectroscopy of Chl-s. In: RJH Clark, RE Hester, Eds. *Spectroscopy of Inorganic-Based Materials*. Chichester: John Wiley & Sons, 1987, pp. 407–428.
71. M Osawa. Dynamic process in electrochemical reactions studied by surface-enhanced infrared absorption spectroscopy (SEIRAS). *Bull. Chem. Soc. Japan* 70: 2861–2880, 1997.
72. M Osawa. Private communication.
73. LJ Wan, M Terashima, H Noda, M Osawa. Molecular orientation and ordered structure of benzenethiol adsorbed on gold (111). *J. Phys. Chem.* B 104: 3563–3569, 2000.
74. N Nagai, Y Izumi, H Ishida, Y Suzuki, A Hatta. Infrared surface analysis of semiconductors by a noncontact air gap ATR method. *CP430 Fourier Transform Spectroscopy: 11th International Conference*: 581–585, 1997.
75. M Yoshikawa, N Nagai. Vibrational spectroscopy of carbon and silicon materials. In: PR Griffiths, JM Chalmers, Eds. *Handbook of Vibrational Spectroscopy*, Vol. 4. Chichester: John Wiley & Sons, 2002, pp. 2593–2620.
76. T Wiedmer, U Brodbeck, P Zahler, BW Fulpius, Interactions of acetylcholine receptor and acetylcholinesterase with lipid monolayers. *Biochim. Biophys.* Acta 506: 161–172, 1978.
77. I Cornut, B Desbat, JM Turlet, J Dufourcq. *In situ* study by polarization modulated Fourier transform infrared spectroscopy of the structure and orientation of lipids and amphipathic peptides at the air-water interface. *Biophys. J.* 70: 305–312, 1996.
78. GT Tschelnokow, D Naumann, C Weise, F Hucho. Secondary structure and temperature behavior of acetylcholinesterase. Studies by Fourier-transform infrared spectroscopy. *Eur. J. Biochem.* 213: 1235–1242 1993.
79. T Buffeteau, B Desbat, JM Turlet. Polarization modulation FT-IR spectroscopy of surfaces and ultrathin films — experimental procedure and quantitative analysis. *Appl. Spectrosc.* 45: 380–389, 1991.
80. D Blaudez, JC Buffeteau, JC Cornut, B Desbat, N Escafre, M Pezolet, JM Turlet. Polarization modulation FT-IR spectroscopy of a spread monolayer at the air-water interface. *Appl. Spectrosc.* 47: 869–874, 1993.
81. D Blaudez, JM Turlet, J Dufourq, D Bard, T Buffeteau, B Desbat. Investigations at the air/water interface using polarization modulation IR spectroscopy. *J. Chem. Soc. Faraday Trans.* 92: 525–530, 1996.
82. SN Timasheff, H Susi, LJ Stevens. Infrared spectra and protein conformations in aqueous solutions. II. Survey of globular proteins. *J. Biol. Chem.* 242: 5467–5473, 1967.
83. H Susi, SN Timasheff, L Stevens, Infrared spectra and protein conformations in aqueous solutions. I. The amide I band in water and deuterium oxide solutions. *J. Biol. Chem.* 242: 5460–5466, 1967.
84. H Susi, DM Byler. Protein structure by Fourier transform infrared spectroscopy: second derivative spectra. *Biochem. Biophys. Res. Comm.* 115: 391–397, 1983.
85. L Pauling, RB Corey. Stable configurations of polypeptide chains. Proc. Roy. Soc. (London), B141: 21–33, 1953.
86. S Krimm, Infrared spectra and chain conformation of proteins. *J. Mol. Biol.* 4: 528–540, 1962.
87. T Miyazawa. Perturbation treatment of the characteristic vibrations of polypeptide chains in various configurations. *J. Chem. Phys.* 32: 1647–1652, 1960.

88. T Miyazawa, ER Blout. The infrared spectra of polypeptides in various conformations: amide I and II bands *J. Am. Chem. Soc.* 83, 712–719, 1960.
89. DM Byler, H Susi. Examination of the secondary structure of the proteins by deconvolved FTIR spectra. *Biopolymers* 25: 469–487, 1986.
90. NA Nevskaya, YN Chirgadze, Infrared spectra and resonance interactions of amide-I and II vibrations of alpha.helix. Biopolymers 15: 637–648, 1976.
91. JL Sussman, M Harel, F Frolow, C Oefner, A Goldman, L Toker, I Silman. Atomic structure of acetylcholinestera from *Torpedo california* : aprototypic actylcholine-binding protein *Science* 25: 872–879, 1991.
92. L Dziri, S Boussaad, N Tao, RM Leblanc. Acetylcholinesterase complexation with acetylcholine or organophosphate at the air-aqueous interface: AFM and UV-Vis studies. *Langmuir* 14: 4853–4859, 1998.
93. T Hasegawa, Detection of minute chemical signals by principal component analysis. *Trends Anal. Chem.* 20: 53–64, 2001.
94. T Hasegawa, Detection of minute chemical species by principal-component analysis. *Trends in Anal. Chem.* 75: 3085–3091, 1999.

4 Two-Dimensional Correlation Spectroscopy of Biological and Polymeric Materials

Yukihiro Ozaki and Slobodan Šašic

CONTENTS

4.1 INTRODUCTION

Two-dimensional (2D) correlation spectroscopy is a new spectral analysis method where the spectral intensity is plotted as a function of two independent spectral variables; e.g., wavenumber, wavelength, or frequency [1–4]. Peaks appearing on a 2D spectral plane provide useful information not readily accessible from a conventional one-dimensional spectrum. Figure 4.1a and Figure 4.1b show examples of three- and two-dimensional representations of a 2D vibrational correlation spectrum, respectively [5]. The three-dimensional representation provides the best overall view of the intensity profile of a correlation spectrum while the contour map representation is better for observing the detailed peak shapes and positions. Correlations among bands that belong to the same chemical group, or groups interacting strongly, can be investigated by means of 2D correlation spectroscopy. Figure 4.2 illustrates the general conceptual scheme for obtaining a 2D correlation spectrum. The primary component of the experimental procedure used in 2D correlation spectroscopy is an external perturbation applied to stimulate a system. When an external perturbation is applied to a system, some selective changes in the state, order, or surroundings of constituents occur. The excitation and subsequent relaxation process toward the equilibrium is monitored with an electromagnetic probe. The overall response of the stimulated system to the applied external perturbation leads to distinctive variations in the measured spectrum. The spectral changes are then transformed into 2D spectra by means of a correlation method.

2D correlation spectroscopy based upon the above conception was first proposed by Noda [1–4] in 1986 and extended by the same author to generalized 2D correlation spectroscopy in 1993 [6–8]. In first-generation 2D IR correlation spectroscopy, simple cross-correlation analysis was applied to sinusoidally varying dynamic IR signals to obtain a set of 2D IR correlation spectra [1–4]. This type of 2D IR correlation spectroscopy was successful in the investigations of systems stimulated by a small amplitude mechanical or electrical perturbation. One can find many examples of the applications of 2D IR correlation spectroscopy in the studies of polymers and liquid crystals [9–20]. However, this previously developed approach had one major disadvantage; the time-dependent behavior (i.e., waveform) of dynamic spectral intensity variations had to be a simple sinusoid in order effectively to employ the original data analysis scheme.

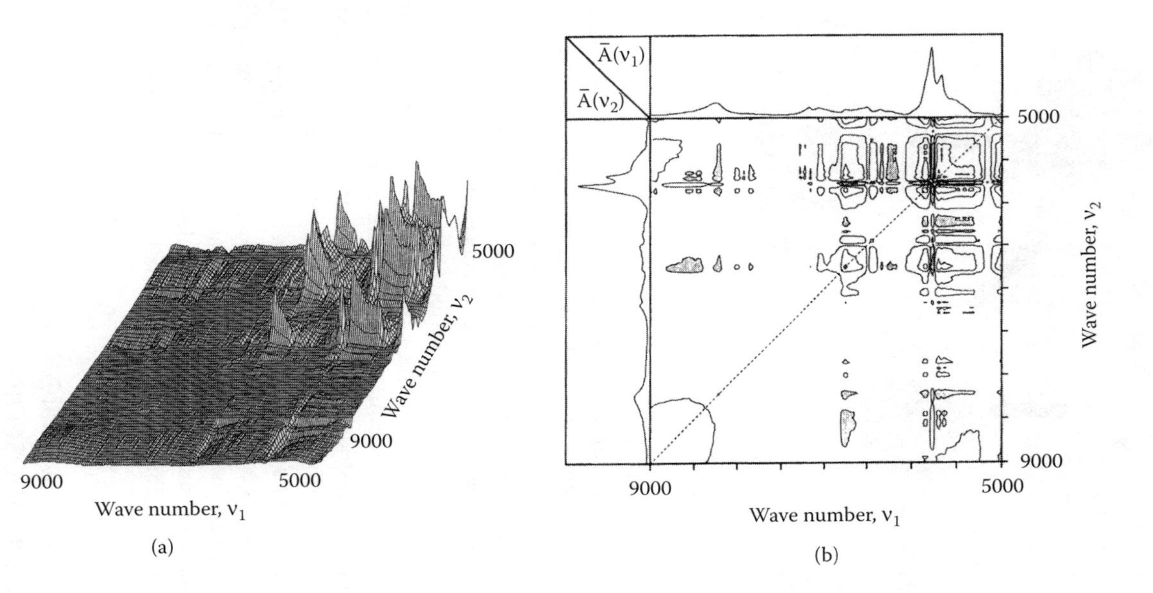

FIGURE 4.1 Examples of (a) three- and (b) two-dimensional representations of a 2D vibrational correlation spectrum. The spectrum was generated from time-dependent FT-NIR spectra in the 9000–5000 cm^{-1} region of Nylon 12 obtained from 30°C to 150°C. [Reproduced from Y Ozaki, I Noda. *J. NIR Spectrosc* 4: 85–99, 1996. Copyright (1997) American Chemical Society.]

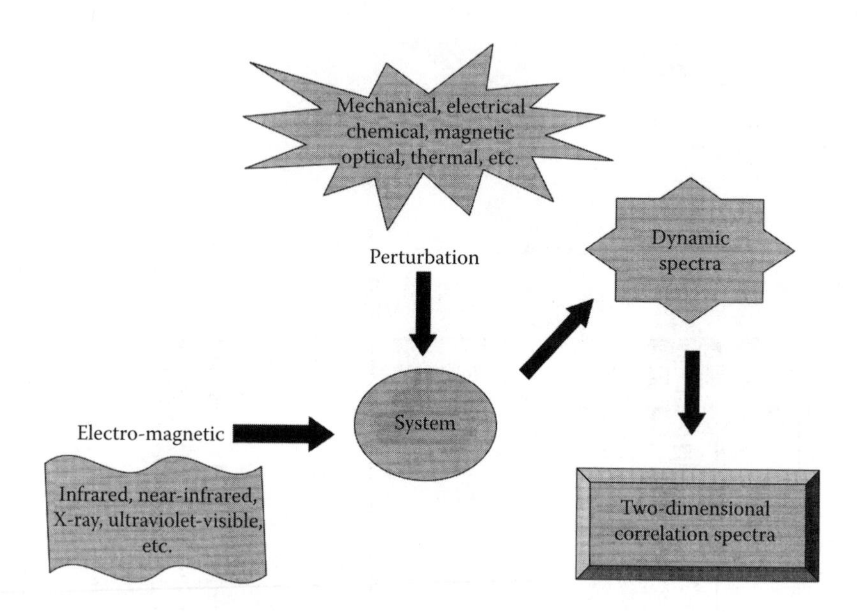

FIGURE 4.2 General conceptual scheme for obtaining a 2D correlation spectrum.

In generalized 2D correlation spectroscopy, a more generally applicable, yet reasonably simple, mathematical formalism is employed to construct 2D correlation spectra from any transient or time-resolved spectra having an arbitrary waveform. The possible variety of external perturbations is virtually limitless [6–8]. The spectral data for 2D correlation analysis may be collected either as a direct function for the perturbation variable itself, such as temperature, pressure, or concentration, or as a function of the secondary consequences induced by the perturbation, such as time-dependent spectral changes by the application of a stimulus. Moreover, extension to other areas of spectroscopy such as near infrared (NIR), Raman, fluorescence, and ultraviolet-visible (UV-Vis) has become quite straightforward. Generalized 2D correlation spectroscopy has opened up the possibility of introducing the powerful and versatile capability of 2D correlation analysis to much wider ranges of applications [21–80].

The purpose of this chapter is to describe the principle, properties, and applications of generalized 2D correlation spectroscopy. The examples of the applications are selected from IR and NIR studies of basic molecules, polymers, biological materials, and others. New possibilities of 2D correlation spectroscopy are also demonstrated in this chapter.

4.2 PRINCIPLE OF TWO-DIMENSIONAL CORRELATION SPECTROSCOPY

4.2.1 Mathematical Background

The basic idea governing 2D correlation spectroscopy is rather simple. 2D correlation is nothing but a quantitative comparison of spectral intensity variations observed at two different spectral variables over some finite observation interval.

Detailed mathematical procedure to generate 2D correlation spectra was provided by Noda et al. [6–8]. Thus, in this chapter only the essence of the procedure is described. In the 2D correlation spectroscopy a series of dynamic spectra are calculated before the correlation analysis.

4.2.1.1 Dynamic Spectra

For a spectral intensity variation $y(\nu,t)$ observed as a function of a spectral variable ν during an interval of some additional external variable t between $T_{\min}$ and $T_{\max}$ (although the external variable t can be any measure of physical quantity such as time, temperature, pressure, or concentration, hereafter t will be referred to as *time* for convenience), the dynamic spectrum $y(\nu,t)$ is defined as

$$\tilde{y}(\nu,t) = \begin{cases} y(\nu,t) - \bar{y}(\nu) & \text{for } T_{\min} \le t \le T_{\max} \\ 0 & \text{otherwise} \end{cases} \tag{4.1}$$

where $\bar{y}$ is the reference spectrum of the system. The selection of a reference spectrum is not strict, but in most cases one can set $\bar{y}(\nu)$ to be the time-averaged spectrum, as defined by the following equation.

$$\bar{y}(\nu) = \frac{1}{T} \int_{-T/2}^{T/2} y(\nu,t)dt. \tag{4.2}$$

A different type of reference spectrum may be selected by choosing a spectrum obtained at some fixed reference point. It is also possible to set the reference spectrum to be zero.

Let us show one example of dynamic spectra. Figure 4.3a displays temperature-dependent NIR spectra in the 6000–5500 cm^{-1} region of Nylon 12, measured over a temperature range of 30°C to 150°C, and Figure 4.3b presents dynamic NIR spectra calculated from the spectra in Figure 4.3a [5]. In this case, temperature was used as the external perturbation. Each dynamic spectrum can be regarded as a difference spectrum between the individual NIR spectrum and a preselected reference spectrum. In the present case, the reference spectrum was set to be the average of all spectral data. Comparison between Figure 4.3a and Figure 4.3b reveals that the dynamic spectra are a powerful means of exploring temperature-dependent intensity variations of NIR spectra of Nylon 12 [5].

4.2.1.2 Generalized 2D Correlation Spectrum

A formal definition of the generalized 2D correlation spectrum $X(\nu_1,\nu_2)$ is given by

$$X(\nu_1,\nu_2) = \Phi(\nu_1,\nu_2) + i\,\Psi(\nu_1,\nu_2) = \frac{1}{\pi\,(T_{\max} - T_{\min})} \int_0^\infty \tilde{Y}_1(\omega)\cdot\tilde{Y}_2^*(\omega)\,d\omega \tag{4.3}$$

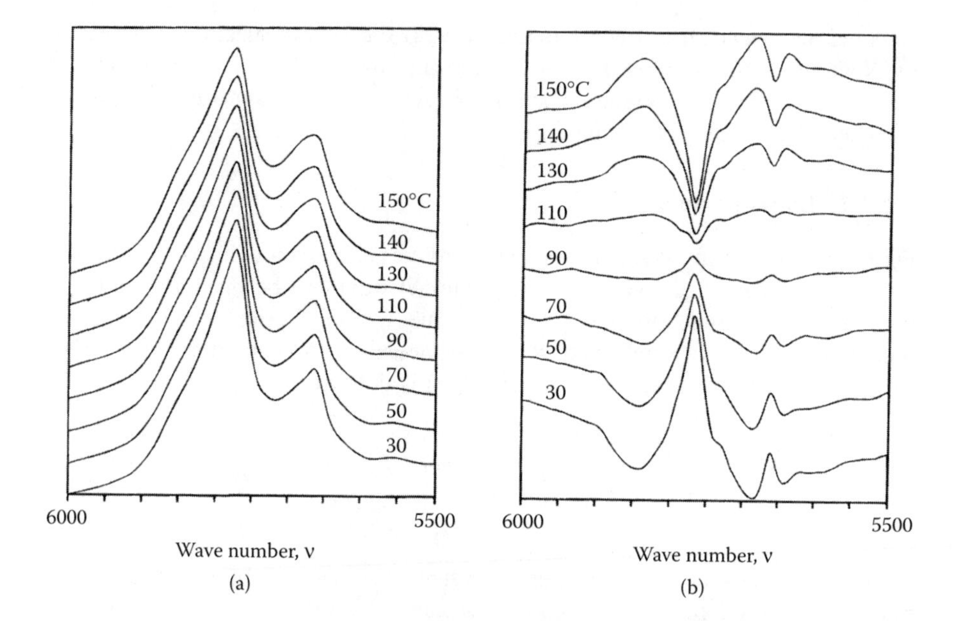

FIGURE 4.3 (a) Temperature-dependent NIR spectra in the 6000–5500 cm^{-1} region of Nylon 12 measured over a temperature range of 30°C to 150°C. (b) Dynamic NIR spectra calculated from the spectra shown in (a). [Reproduced from Y Ozaki, I Noda. *J. NIR Spectrosc* 4: 85–99, 1996 with permission. Copyright (1997) American Chemical Society.]

The 2D correlation spectrum comprises two orthogonal components, $\Phi(\nu_1,\nu_2)$ and $\Psi(\nu_1,\nu_2)$, known, respectively, as the synchronous and asynchronous 2D correlation intensities [6–8]. They represent, respectively, the overall similarities and differences of the time-dependent behavior of spectral intensity variations measured at two distinct spectral variables, ν_1 and ν_2, during the observation period between T_{min} and T_{max}. The term $\tilde{Y}_1(\omega)$ is the forward Fourier transform of the spectral intensity variations $\tilde{y}(\nu_1,t)$ observed at some spectral variable ν_1.

$$\tilde{Y}_1(\omega) = \int_{-\infty}^{\infty} \tilde{y}(\nu_1,t)e^{-i\omega t}dt = \tilde{Y}_1^{Re}(\omega) + i\tilde{Y}_1^{Im}(\omega) \tag{4.4}$$

where $\tilde{Y}_1^{Re}(\omega)$ and $\tilde{Y}_1^{Im}(\omega)$ are, respectively, the real and the imaginary components of the Fourier transform. The Fourier frequency ω represents the individual frequency component of the variation of $\tilde{y}(\nu_1,t)$ measured along the variable t. Likewise, the conjugate of the Fourier transform $\tilde{Y}_2^*(\omega)$ of spectral intensity variations $\tilde{y}(\nu_2,t)$ observed at spectral variable ν_2 is given by

$$\tilde{Y}_2^*(\omega) = \int_{-\infty}^{\infty} \tilde{y}(\nu_2,t)e^{+i\omega t}dt = \tilde{Y}_2^{Re}(\omega) - i\tilde{Y}_2^{Im}(\omega) \tag{4.5}$$

The 2D correlation spectrum can be expressed more simply as

$$X(\nu_1, \nu_2) = \left\langle \tilde{y}(\nu_1, t) \cdot \tilde{y}(\nu_1, t') \right\rangle \tag{4.6}$$

Equation 4.6 clearly shows that $X(\nu_1, \nu_2)$ is the measure of a functional comparison of spectral intensity variations $\tilde{y}(\nu_1, t)$ measured at different spectral variables, ν_1 and ν_2, during a fixed interval of the external variable t.

4.2.1.3 Hetero-Spectral Correlation

If there is any commonality between the response patterns of system constituents monitored by two different techniques under the same perturbation, it may be possible to detect the correlation even between the different classes of spectral signals [6–8]. The hetero-spectral 2D correlation is given by

$$X(\mu_1, \nu_2) = \left\langle \tilde{x}(\mu_1, t) \cdot \tilde{y}(\nu_2, t') \right\rangle \tag{4.7}$$

where $\tilde{x}(\mu_1, t)$ and $\tilde{y}(\nu_2, t')$ are dynamic spectra measured by different techniques.

4.2.1.4 Computation of Asynchronous Spectra

The computation of a synchronous spectrum is relatively simple; it could be obtained by a numerical method without involving Fourier transformation of data. In contrast, the computation of an asynchronous spectrum is rather cumbersome. Recently, Noda [7,8] proposed to use the discrete Hilbert transform algorithm as a computationally efficient method for the calculation of asynchronous spectra. This method is given by

$$\Psi(\nu_1, \nu_2) = \frac{1}{m-1} \sum_{j=1}^{m} \tilde{y}_j(\nu_1) \cdot \sum_{k=1}^{m} N_{jk} \cdot \tilde{y}_k(\nu_2) \tag{4.8}$$

The term N_{jk} corresponds to the jth row and kth column element of the discrete Hilbert-Noda transformation matrix given by

$$N_{jk} = \begin{cases} 0 & \text{if } j = k \\ 1/\pi(k-j) & \text{otherwise} \end{cases} \tag{4.9}$$

4.2.1.5 Representation of 2D Correlation Spectra by Linear Algebra

Recently, Šašic et al. [81, 82] reported that 2D correlation spectra can be represented by linear algebra. In general, a data matrix X composed of mean-centered spectra can be represented by the following equation;

$$X = CP \tag{4.10}$$

where C $(w \times n)$ is a matrix of the n pure spectra (w points) and P $(n \times s)$ is a matrix of the n concentration profiles (s points). Every row of the data matrix can be viewed as a vector of the spectral intensity changes at a given wavenumber. Thus, one has a total of w such vectors with s coordinates. If one wants to compare these vectors — i.e., to examine correlation between intensity changes for any particular pair of wavenumbers ν_1 and ν_2 — then one should multiply X by X^T and form so-called rows cross product matrix Z (covariance matrix). Z is a square matrix of w x w:

$$Z = 1/(s-1)XX^T \tag{4.11}$$

Elements of Z show similarity or dissimilarity between the intensity variations for the pair of wavenumbers ν_1 and ν_2. Note that Z holds all the features of a synchronous spectrum, and, in fact, it is a synchronous spectrum calculated by generalized 2D correlation spectroscopy [81,82]. It is of importance to recognize the relation between the synchronous spectrum and the cross product matrix because the latter is a well-known conception in the field of classic correlation analysis. In this way, the synchronous spectrum can be considered as a table of the correlation coefficients between dynamic vectors of a given spectral matrix [81,82].

An asynchronous spectrum can be expressed by a correlation between dynamic vectors of the data matrix and those of Hilbert orthogonalized X^T.

$$Z = 1/(s-1)XHX^T \tag{4.12}$$

Here, H is the Hilbert-Noda transformation matrix.

4.2.2 Synchronous and Asynchronous Spectra

The generalized 2D correlation spectra consist of synchronous and asynchronous spectra [6–8]. Examples of the synchronous and asynchronous 2D correlation spectra are shown in Figure 4.4a and Figure 4.4b, respectively. A synchronous spectrum is symmetric with respect to a diagonal line corresponding to spectral coordinates, $\nu_1 = \nu_2$. The intensities of peaks located at diagonal positions correspond to the autocorrelation function of spectral intensity variations observed during a period T. Those peaks are therefore referred to as autopeaks. The intensities of autopeaks represent the overall extent of dynamic fluctuations of spectral signals. Cross peaks located at the off-diagonal positions of a synchronous spectrum represent the simultaneous changes of spectral signals at two different wavenumbers. If the sign of a cross peak is positive, it means that spectral intensities at corresponding wavenumbers are either increasing or decreasing together. If the sign is negative, it is indicated that one spectral intensity is increasing while the other is decreasing.

An asynchronous 2D correlation spectrum, which consists exclusively of off-diagonal cross peaks, provides information complementary to the synchronous spectrum. Asynchronous cross peaks develop only if the basic trends of dynamic spectral variations observed at two different wavenumbers of the cross peaks are dissimilar.

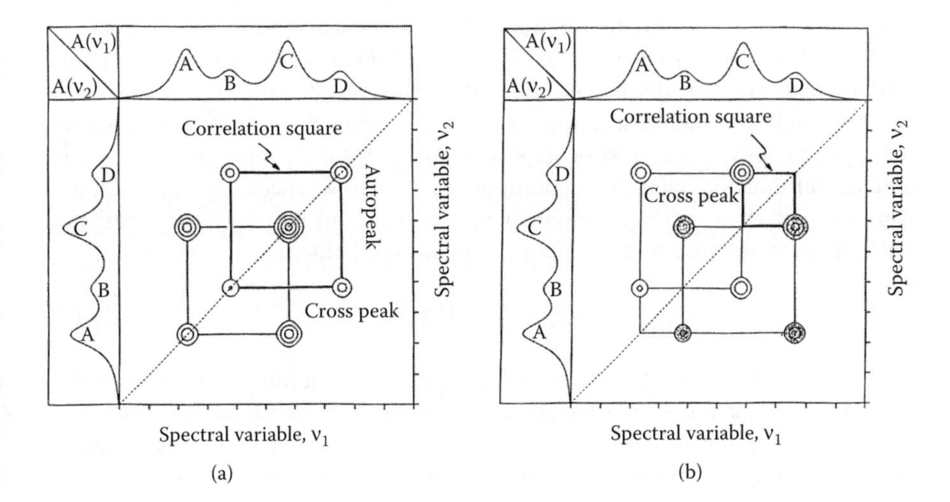

FIGURE 4.4 Example of the (a) synchronous and (b) asynchronous 2D correlation spectra.

The sign of an asynchronous cross peak can be either positive or negative. If the intensity change at ν_1 takes place predominantly before that at ν_2, the sign is positive. On the other hand, if the change occurs after ν_2, the sign becomes negative. It must be kept in mind that this rule is reversed if $\Phi(\nu_1, \nu_2) < 0$.

In general, the asynchronous spectra are a more useful analytical tool than the synchronous spectra because of the following two features. One is that the asynchronous spectra are more powerful in band deconvolution. Another is that they allow one to probe the specific order of the spectral intensity changes occurring during the measurement or value of controlling the variable affecting the spectrum. The usefulness of asynchronous spectra in generalized 2D correlation spectroscopy comes from the fact that all the linearity among the data is zeroed, greatly simplifying the covariance spectrum.

4.3 RECENT PROGRESS IN THEORETICAL ASPECTS OF TWO-DIMENSIONAL CORRELATION SPECTROSCOPY

Šašic et al. [82–84] have recently proposed two new possibilities of 2D correlation spectroscopy. One is sample-sample 2D correlation spectroscopy [82,83] and the other is statistical 2D correlation spectroscopy [84]. These two correlation spectroscopies have extended the potential of 2D correlation spectroscopy.

4.3.1 Sample-Sample 2D Correlation Spectroscopy

In sample-sample correlation spectroscopy [82], generalized 2D correlation maps having sample axes are created instead of generating 2D maps with the variable (wavenumber, wavelength, etc.) axes. In the usual generalized 2D correlation spectroscopy (variable-variable or wavenumber-wavenumber correlation spectroscopy) a correlation between bands is discussed, while in sample-sample correlation spectroscopy, one can discuss the concentration dynamics directly. The idea of sample-sample correlation spectroscopy was born from the idea that synchronous and asynchronous spectra can be expressed using linear algebra. The data matrix contains information about the concentrations as well as spectra (Equation 4.10). Thus, one can show a correlation between concentrations by use of the covariance matrix Z, as in the case of the variable-variable correlation:

$$\mathbf{Z} = 1/(w - 1)\mathbf{X}^T\mathbf{X} \tag{4.13}$$

In this way two types of covariance matrices can be formed [82]. The matrix generated by Equation 4.11 is the spectral cross-product matrix and by its decomposition one can obtain information about the spectral features of components of samples (loadings), while the cross-product matrix obtained by Equation 4.13 provides knowledge about concentration dynamics of components (scores). Each matrix shows a different aspect of the species, spectra, and concentrations examined. The cross-product matrix calculated by Equation 4.13 gives rise to a 2D synchronous correlation spectrum with the samples on both axes, and each point in the 2D map represents a correlation between the concentrations of a given pair of samples, s_i and s_j [82].

The corresponding asynchronous sample-sample correlation spectrum is given by

$$\mathbf{Z} = 1/(w - 1)\mathbf{X}^T\mathbf{H}\mathbf{X} \tag{4.14}$$

In sample-sample correlation spectroscopy the spectra are considered as the intensity changes with the samples, while ordinary spectra are supposed to be spectral responses. Sample-sample correlation spectroscopy offers possibilities of analyzing the concentration changes of the species as a function of external perturbation. The results obtained from sample-sample correlation analysis and those from variable-variable correlation analysis are often complementary each other because the two kinds of analyses depict two essential quantities of any spectral system. Therefore, the sample-sample approach is particularly powerful in combination with the variable-variable approach.

4.3.2 Statistical 2D Correlation Spectroscopy

Newly proposed statistical 2D spectroscopy is different from generalized 2D correlation spectroscopy in that the former abstracts spectral features by pretreatment and by 2D maps that are limited in the figure range from 1 to −1 [84]. Although in this aspect statistical 2D correlation spectroscopy is not similar to generalized 2D correlation spectroscopy, its results can be interpreted in an analogous way to those obtained by generalized 2D spectroscopy. The most significant advantage of statistical 2D correlation spectroscopy is that the calculation of a 2D map is made very easily by using only the *corrcoef* command from Matlab, and the results are purely mathematical in nature, thereby eliminating any subjective involvement of an experimenter; while the inherent weakness of the method lies in its sensitivity to noise.

The idea of statistical 2D correlation spectroscopy came from the original approach by Barton et al. [85]. Their approach is mainly concerned with finding correlations among bands of spectra measured by different spectroscopic techniques. Šašic and Ozaki [84] presented several improvements concerning with objects and targets of correlation analysis as well as relatively simple linear algebra presentation that the methodology utilizes.

Statistical 2D correlation spectra display correlations between the spectra and concentration profiles. To obtain the correlation coefficients one must pretreat the experimental data, since the correlation coefficients present the cosines of the angles between unit-length vectors. The most common way to scale the data to bring them to these forms is auto-scaling; that is

$$d'_{ij} = \frac{(d_{ij} - m_j)}{s_j} \tag{4.15}$$

where d'_{ij} represents the auto-scaled element of jth column of D, m_j is the mean of jth column and s_j is the standard deviation of the jth column. Auto-scaling equalizes variances of all the variables and transforms the spectra into the shapes that are visually very far from the common spectral shapes.

Since the auto-scaling makes all the variances comparable and limits the vectors to the unit lengths, all possible scalar products between these vectors take the values between 1 and −1 [84]. These two figures mean perfect correlation, while 0 corresponds to the absence of correlation. The correlation coefficients can be directly transformed into angles. If variable-variable correlation maps are considered, then correlation coefficients (or angles) between concentration profiles are measured. On the other hand, if sample-sample correlation maps are concerned, then measured are the similarities of the auto-scaled spectra. As all the correlation coefficients are displayed in one map, one must choose the ranges of the highest and smallest coefficients to be investigated. The usefulness of asynchronous spectra in generalized 2D correlation spectroscopy comes from the fact that all the linearity among the data is zeroed, greatly simplifying the covariance spectrum. Since the band intensities are of no importance in statistical 2D correlation, the need for the special attention paid to the minor spectral changes through the asynchronous spectrum no longer exists and the 2D maps herein do not have the asynchronous components. In fact, the asynchronicity does not exist as individual conception in statistical 2D correlation spectroscopy [84].

4.4 APPLICATIONS OF TWO-DIMENSIONAL CORRELATION SPECTROSCOPY

Generalized 2D correlation spectroscopy has proven to be a very effective way to accentuate subtle features of spectral changes induced by external perturbation. The technique is extremely versatile and flexible. One can find a great number of articles that report applications of generalized 2D correlation spectroscopy [21–80]. Selected examples of the applications described below are concerned with 2D-IR and 2D-NIR correlation spectroscopy studies of polymers, proteins, and others. Applications of statistical 2D correlation spectroscopy and sample-sample correlation spectroscopy will also be discussed in Section 4.4.4.

4.4.1 Applications to Polymer Research

2D correlation spectroscopy in vibrational spectroscopy started with polymer research. Noda et al. [1–4,10] applied 2D-IR correlation spectroscopy to polymers stimulated by a small-amplitude oscillatory mechanical perturbation. Since that time, particularly after the introduction of generalized 2D correlation spectroscopy, various kinds of polymers including copolymers and polymer blends have been investigated by 2D-IR [7,36–39], 2D-NIR [26,53,54,58,59], 2D Raman [64,65], and 2D hetero-spectral correlation [66,68,69]. Information provided by 2D correlation spectroscopy studies on polymers may be summarized as follows: (1) molecular conformation and configuration; (2) inter- or intramolecular interactions between functional groups; (3) relative reorientation directions and the order of realignment sequence of submolecular units; (4) relation between local dynamics of side groups and polymer main chain; (5) depth profile of polymers and microscopic spatial distribution of submolecular components of polymers. In this section, three examples concerning 2D-IR studies of polymers will be discussed.

4.4.1.1 Premelting Behavior of Nylon 12 Studied by 2D-NIR Correlation Spectroscopy

Thermal behavior and hydrogen bonds of polyamides have been studied extensively by the use of IR and NIR spectroscopy [5,86]. However, the detailed processes of temperature-induced structural variations have not been explored well. In an IR spectrum the stretching band due to the free NH group is extremely weak and overlaps with the much stronger broad feature arising from the hydrogen-bonded species [86]. In contrast, one can observe bands due to the free NH group clearly in a NIR spectrum because of the large anharmonicity constant for the free NH group [5].

Ozaki et al. [5] applied 2D Fourier transform (FT) NIR correlation spectroscopy to study the premelting behavior and hydrogen bonds of Nylon 12. Figure 4.5 displays NIR spectra in the 9000–5000 cm^{-1} region of Nylon 12 measured over a temperature range of 30°C to 150°C where gradual weakening of inter- or intramolecular associative interactions and decrease of local order leading to the eventual fusion of Nylon 12 crystals are observed [5]. It is well known that Nylon 12 exhibits the melt temperature well above 170°C. Strong bands in the 5900–5500 cm^{-1} region are due to the first overtones of the CH_2 stretching modes of Nylon 12 while their second overtones are observed weakly in the region of 8800–8100 cm^{-1}. A group of bands in the 7300–6900 cm^{-1} region arise from combination bands of CH_2 vibrations. Bands in the 6800–6550 cm^{-1} and 6500–6100 cm^{-1} regions are assignable

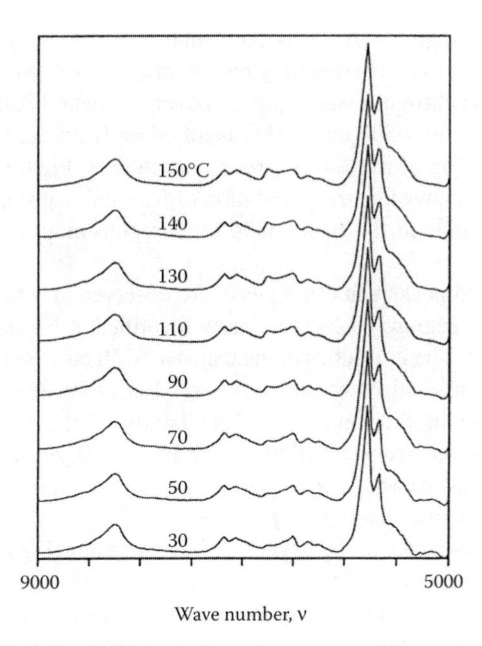

FIGURE 4.5 NIR spectra in the 9000–5000 cm^{-1} region of Nylon 12 measured over a temperature range of 30°C to 150°C. [Reproduced from Y Ozaki, I Noda. *J. NIR Spectrosc* 4: 85–99, 1996 with permission. Copyright (1997) American Chemical Society.]

to the first overtones of stretching modes of free and hydrogen-bonded NH groups of Nylon 12, respectively. Since the amide groups of Nylon 12 can assume several hydrogen-bonded forms as shown in Figure 4.6, the 6800–6100 cm^{-1} region should be very complicated.

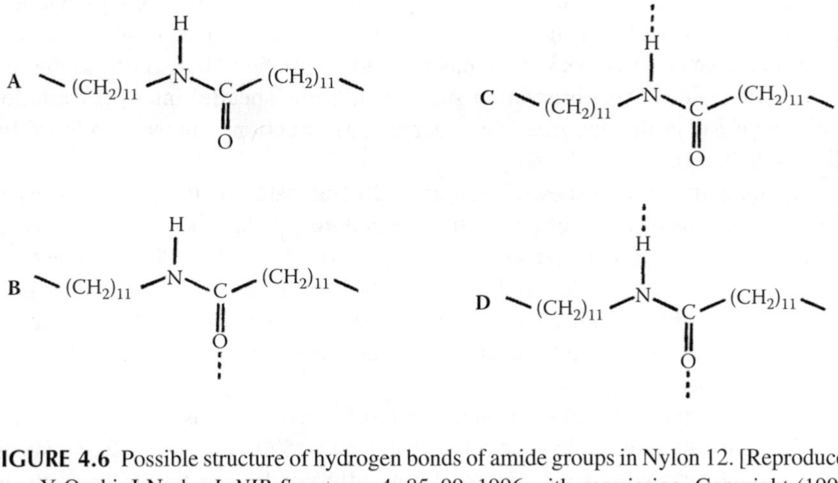

A $\diagdown(CH_2)_{11}-N(H)-C(=O)-(CH_2)_{11}\diagdown$

B $\diagdown(CH_2)_{11}-N(H)-C(=O\cdots)-(CH_2)_{11}\diagdown$

C $\diagdown(CH_2)_{11}-N(H\cdots)-C(=O)-(CH_2)_{11}\diagdown$

D $\diagdown(CH_2)_{11}-N(H\cdots)-C(=O\cdots)-(CH_2)_{11}\diagdown$

FIGURE 4.6 Possible structure of hydrogen bonds of amide groups in Nylon 12. [Reproduced from Y Ozaki, I Noda. *J. NIR Spectrosc* 4: 85–99, 1996 with permission. Copyright (1997) American Chemical Society.]

Figure 4.7a and Figure 4.7b show the synchronous and asynchronous 2D-NIR spectra of Nylon 12 in the NH-stretching region, respectively [5]. The most dominant autopeak of the synchronous spectrum is observed near 6750 cm^{-1} with broad extension down to about 6500 cm^{-1}. This band arises from the first overtone of the NH stretching vibration of free amide group (structure A, Figure 4.6). The autopeak at this wavenumber shows that the population of free amide group is increasing with temperature. The bands in the 6700–6550 cm^{-1} region may be due to structure B (Figure 4.6).

Several other autopeaks and cross peaks are observed at different coordinates in the 6800–6150 cm^{-1} region; peaks are clearly identified at 6490, 6440, 6350, 6270, 6220, and 6170 cm^{-1}. The intensities of the bands at 6750 cm^{-1} and 6440 cm^{-1} increase with temperature while all other bands observed at 6490, 6350, 6270, 6220, and 6170 cm^{-1} decrease in their intensity. The bands in the 6500–6400 cm^{-1} and 6380–6100 cm^{-1} regions are ascribed to structure C and D, respectively. The appearances of several bands in each region suggest that there are a variety of environments or structures for structure C and D [5].

Most of the cross peaks in the asynchronous spectrum (Figure 4.7b) are located at the spectral coordinate near 6750 cm^{-1}. This suggests that the temperature dependence of the stretching vibration of free NH group in Nylon 12 is quite different from that of other NH-stretching bands. The signs of cross peaks appearing in the expanded view (not shown here) of the asynchronous spectrum point to the following sequence of events. (1) The intensity of the band at 6440 cm^{-1} due to structure C increases slightly before any other bands show an intensity change as the temperature of the system is raised from 30°C to 150°C. (2) The intensity increase at 6440 cm^{-1} is followed by a marked increase of the intensity around 6750 cm^{-1}. (3) At higher temperatures, the decrease of band intensity at 6330 cm^{-1}, as well as the decreases at 6270 cm^{-1} and 6220 cm^{-1}, are detected.

Figure 4.8a and Figure 4.8b represent the synchronous and asynchronous 2D-NIR spectra of Nylon 12 in the 6000–5500 cm^{-1} region, respectively [5]. Spectral features in this region are mainly due to the first overtone of the CH_2 stretching vibrations. It is noted in the synchronous spectrum that a band at 5770 cm^{-1} shares negative cross peaks with bands at 5840, 5680, and 5640 cm^{-1} and the rest of the NIR bands. This observation suggests that the spectral intensity variation at 5770 cm^{-1} is in the opposite (i.e., decreasing) direction compared to those for other NIR bands.

The dynamic spectra shown in Figure 4.2b elucidate that the peak intensity at 5770 cm^{-1} is steadily decreasing with temperature, while other peaks located at 5840, 5680, and 5640 cm^{-1} are all increasing. Thus, Ozaki et al. [5] concluded that the band at 5770 cm^{-1} is due to the first overtone of a CH_2 stretching vibration of an ordered or highly associated form of Nylon 12, which decreases with temperature. Other NIR bands may be attributed to more disordered forms created by the temperature increase.

It is noted that in the asynchronous spectrum there are cross peaks correlating the 5770 cm^{-1} band with those at 5840, 5680, and 5590 cm^{-1}. This clearly shows that the disappearance of the ordered form corresponding to the decrease in the 5770 cm^{-1} band is not directly coupled with the creation of other components, such

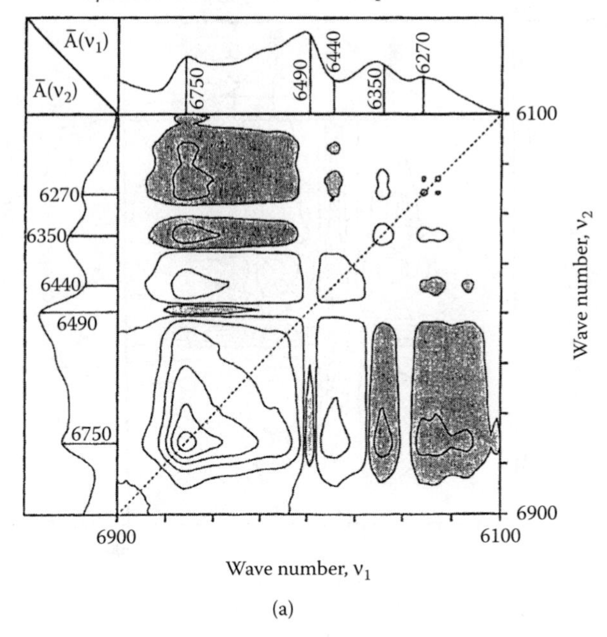

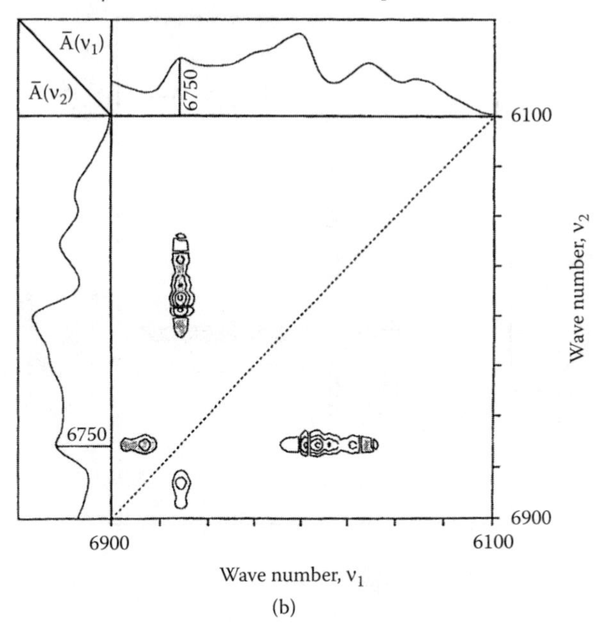

FIGURE 4.7 The (a) synchronous and (b) asynchronous 2D-NIR Spectra of Nylon 12 in the NH-stretching region. [Reproduced from Y Ozaki, I Noda. *J. NIR Spectrosc* 4: 85–99, 1996 with permission. Copyright (1997) American Chemical Society.]

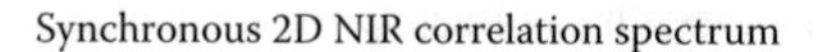

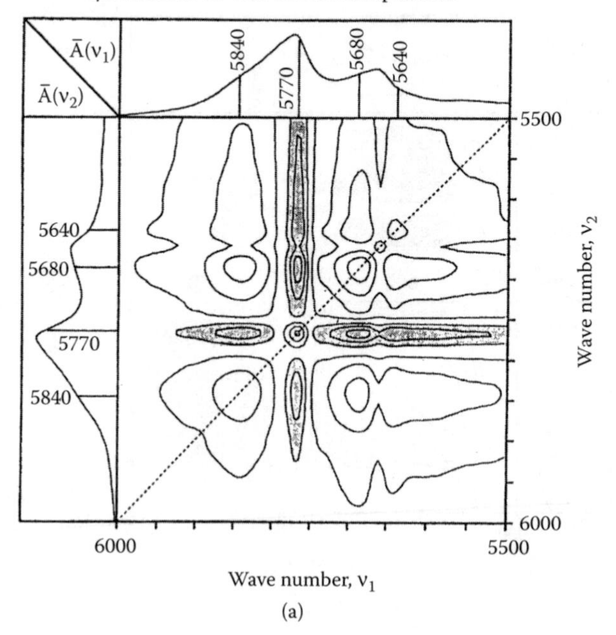

Asynchronous 2D NIR correlation spectrum

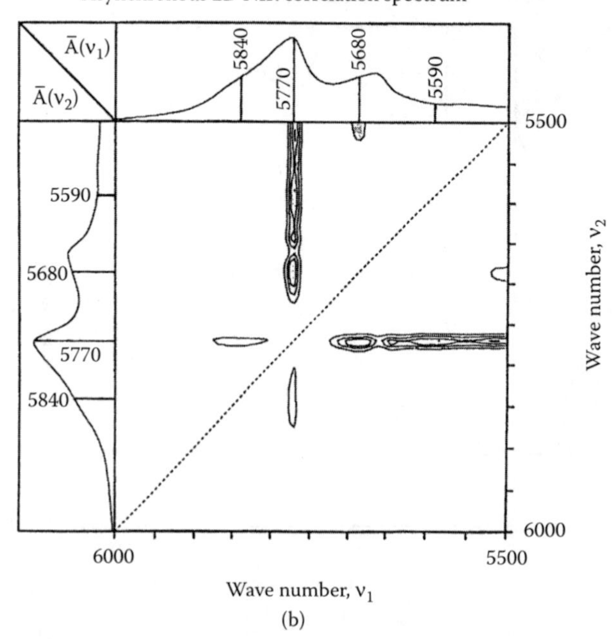

FIGURE 4.8 The (a) synchronous and (b) asynchronous 2D-NIR spectra of Nylon 12 in the 6000–5500 cm^{-1} region. [Reproduced from Y Ozaki, I Noda. *J. NIR Spectrosc* 4: 85–99, 1996 with permission. Copyright (1997) American Chemical Society.]

as those giving rise to the intensity changes in the 5840, 5680, and 5590 cm^{-1} bands, since these events are not occurring simultaneously. The signs of cross peaks at 5840, 5680, and 5590 cm^{-1} indicate that the intensity variations of these bands actually proceed at a lower temperature compared to the onset of the intensity decrease in the 5770 cm^{-1} band. It is very likely that substantial amount of disordered or dissociated components appear as the premelting precursors to the precipitous decrease of ordered components associated with the melting of Nylon 12 occurring at a much higher temperature.

Ozaki et al. [5] also investigated correlations between different spectral regions where signals from different modes of vibrations appear. The following conclusions could be obtained from comparison between the two regions corresponding to the first overtones of CH_2 and NH stretching vibrations and comparison between the two regions containing the first and second overtones of CH_2 stretching modes, respectively. (1) The premelting precursors appear before structure A is created. (2) The bands at 5770 cm^{-1} and 8200 cm^{-1} are probably due to the first and second overtones of antisymmetric CH_2 stretching modes, respectively, of the ordered components.This study has demonstrated the potential of generalized 2D-NIR correlation spectroscopy in investigations of hydrogen bonds of amide groups and conformational changes in the hydrocarbon chains of polymers.

4.4.1.2 Thermally-Induced 2D-IR Spectroscopy of Linear Low-Density Polyethylene

Noda [7] investigated the premelting process of linear low-density polyethylene (LLDPE) by means of 2D-IR spectroscopy. LLDPE is synthesized by copolymerization of ethylene and α-olefins and has a linear molecular structure with short branches. The incorporation of branches reduces the density and crystallinity of polyethylene, while in turn improves flexibility, clarity, and impact strength. IR spectra of semicrystalline LLDPE samples are complicated by the presence of overlapped contributions from coexisting crystalline and amorphous regions.

Figure 4.9 shows temperature-dependent IR spectra of an LLDPE film during the premelting process [7]. The intensity of sharp crystalline bands around 1462 cm^{-1} and 1473 cm^{-1} decreases gradually with temperature. Concomitant with the intensity decrease in the crystalline bands is an increase in the intensity of a much broader band ranging from 1470 cm^{-1} to about 1440 cm^{-1} and associated with the liquid-like amorphous component. The decrease of crystalline bands is accompanied by the subtle development of shoulders, especially around 1465 cm^{-1} and 1455 cm^{-1}, arising presumably from the increase in the contribution from the amorphous component.

Figure 4.10a and Figure 4.10b show 2D-IR synchronous and asynchronous correlation spectra, respectively, generated from the temperature-induced spectral changes in Figure 4.9 [7]. The presence of a pair of positive cross peaks confirms that the simultaneous decreases in intensities of two major crystalline bands at 1473 cm^{-1} and 1462 cm^{-1}. The much broader negative cross peaks are also observed between the crystalline bands and the lower side of the board amorphous band below 1455 cm^{-1}, being consistent with the fact that the intensities of crystalline bands are decreasing, while the amorphous band intensity is increasing. Of note is that even

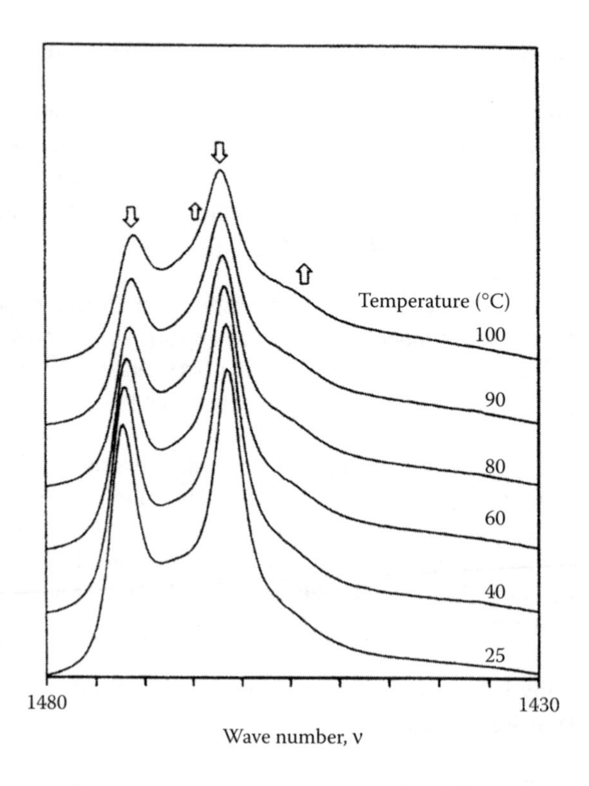

FIGURE 4.9 Temperature-dependent IR spectra of a LLDPE film during the premelting process. [ReproducedI Noda, A E Dowrey, C Marcott, Y Ozaki, G M Story. *Appl Spectrosc* 54: 236A–248A, 2000 with permission. Copyright (2000) Society for Applied Spectroscopy.]

at temperatures well below the T_m (123°C) of LLDPE, some of the ordered crystalline structure is converted into a liquid-like amorphous state, as the sample is heated.

In the asynchronous spectrum (Figure 4.10b), several cross peaks are located around the crystalline bands. The bands near 1473 cm^{-1} and 1462 cm^{-1} are split into two separate bands located around 1473 cm^{-1} and 1471 cm^{-1} and around 1463 cm^{-1} and 1461 cm^{-1}, respectively; the asynchronous spectrum clearly reveals that the increase in the amorphous component does not result in a simultaneous decrease in the entire crystalline component contribution [7]. The signs of asynchronous cross peaks indicate that the intensities of the bands at 1471 cm^{-1} and 1463 cm^{-1} decrease at much higher temperatures than those of the bands at 1473 cm^{-1} and 1461 cm^{-1}. The latter two bands may be due to the polymer chains located within a highly ordered crystalline state where coupled interchain interaction of molecular vibrations is strong. The former two bands, assignable to less coupled vibrations, presumably arise from a different type of crystalline structure. This result indicates the intriguing possibility of the existence of a distinct intermediate crystalline state between the dominant crystalline state with strong interchain vibrational coupling interactions and the totally liquid-like amorphous state, as the semicrystalline polymer system undergoes the premelting process [7].

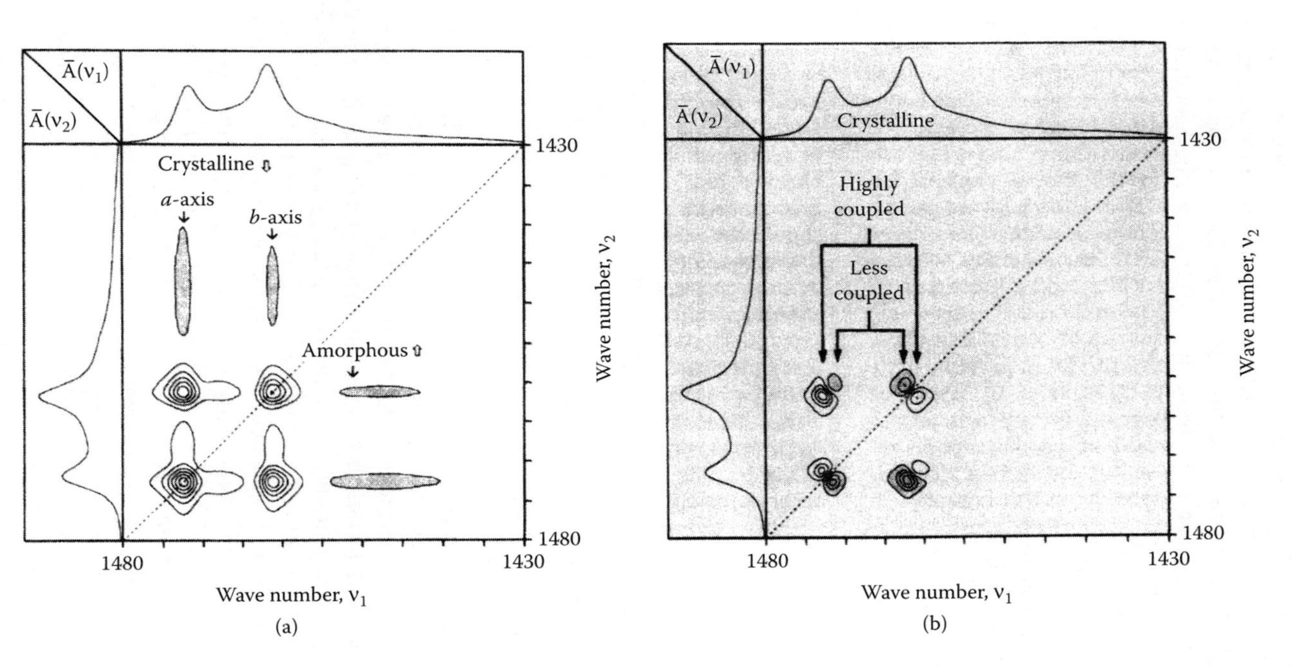

FIGURE 4.10 2D-IR (a) synchronous and (b) asynchronous correlation spectra generated from the temperature-induced spectral changes in Figure 4.9. [Reproduced from I Noda, A E Dowrey, C Marcott, Y Ozaki, G M Story. *Appl Spectrosc* 54: 236A–248A, 2000 with permission. Copyright (2000) Society for Applied Spectroscopy.]

4.4.1.3 2D-IR Study on the Crystallization Process of Poly(ethylene-2,6-naphthalate)

Kimura et al. [39] investigated conformation changes during the crystallization process of poly(ethylene-2,6-naphthalate) (PEN) including the induction period by use of generalized 2D correlation spectroscopy. A broad band due to the C=O stretching vibration and that arising from the naphthalene ring vibration were resolved into several peaks, indicating the existence of different conformations with different coplanarity of the carbonyl group and the naphthalene ring. A difference in conformation change was found between cold crystallization and melt crystallization [39].

PEN has received much attention because it undergoes chain alignment under magnetic fields during melt crystallization. A liquid crystalline-like phase of PEN transiently appearing during crystallization is most likely responsible for the magnetic alignment. Observations by magnetic birefringence and X-ray diffraction indicated that some ordered structure, which is responsible for the magnetic orientation, occurs during the induction period of crystallization. This ordered structure seems to be similar to the nematic-like structure observed during the induction period of cold crystallization of PET near its glass transition temperature.

Figure 4.11 shows synchronous 2D correlation spectra of PEN in the C=O stretching vibration region generated from the spectra obtained during crystallization at 250°C measured (a) from 0 min to 18 min, (b) from 21 min to 39 min, and (c) from 42 min to 60 min, respectively. In Figure 4.11, cross peaks are observed at (1710, 1723), (1710, 1740), and (1723, 1740) cm^{-1}. Kimura et al. [39] attributed the difference in the peak position to the difference in the coplanarity of the C=O group with respect to the naphthalene ring. The more enhanced the coplanarity with respect to the naphthalene ring, the stronger the conjugation becomes, resulting in a lower shift of wavenumber for the C=O group.

In Figure 4.11a small and large intensity changes are observed for the autopeaks at 1710 cm^{-1} and 1723 cm^{-1}, respectively, and cross peaks develop at (1710, 1723)cm^{-1}. These observations lead to the conclusion that the increase in moderately conjugated conformation in the amorphous phase is more significant than the increase in highly conjugated conformation, which is characteristic of the α-crystal form. It seems that the moderately conjugated conformation in the amorphous phase increases before the formation of the α-crystal form. On the other hand, in Figure 4.11b, the change due to crystal growth is more prominent than that due to the moderately conjugated conformation. It is also noted in the same figure that the band due to the moderately conjugated conformation at 1723 cm^{-1} splits into two bands at 1720 cm^{-1} and 1725 cm^{-1}. It can be seen from Figure 4.11c that the decrease in the moderately conjugated conformation is observed during the period of crystallization from 42 min to 60 min.

Figure 4.12 shows synchronous correlation spectra of PEN for the naphthalene ring vibration region [39]. In Figure 4.12, three cross peaks are developed at (1190, 1178), (1190, 1172), and (1190, 1165) cm^{-1}. The band at 1190 cm^{-1} is assigned to the crystalline phase, while the other bands are due to the amorphous phase. The original spectra show that the intensity at 1190 cm^{-1} increases during the crystallization

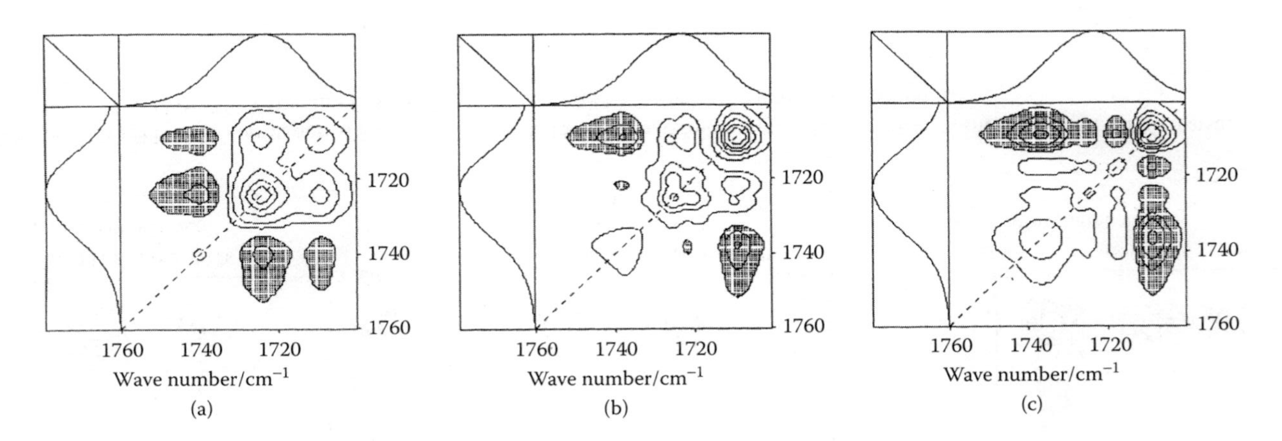

FIGURE 4.11 Synchronous 2D correlation spectra obtained during crystallization at 250°C measured (a) from 0 min to 18 min, (b) from 21 min to 39 min, and (c) from 42 to 60 min. [Reproduced from F Kimura, M Komatsu, T Kimura. *Appl Spectrosc* 54: 974–977, 2000 with permission. Copyright (2000) Society of Applied Spectroscopy.]

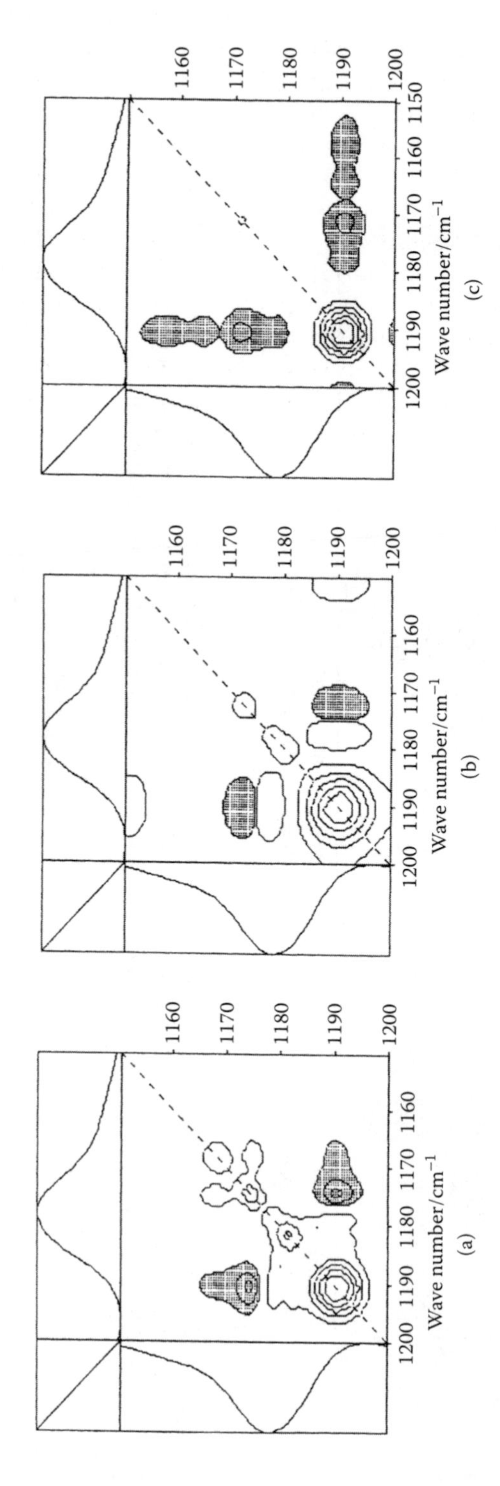

FIGURE 4.12 Synchronous correlation spectra in the naphthalene ring vibration region. [Reproduced from F Kimura, M Komatsu, T Kimura. *Appl Spectrosc* 54: 974–977, 2000 with permission. Copyright (2000) Society of Applied Spectroscopy.]

process. Since the sign of the cross at (1190, 1178) cm^{-1} is positive between 0 min and 39 min and negative between 42 min and 60 min, one can conclude that the band intensity at 1178 cm^{-1} increases between 0 min and 39 min and decreases between 42 min and 60 min. On the other hand, the cross peak at (1190, 1172) cm^{-1} exhibits negative sign throughout the observed period, indicating the decrease in the band intensity at 1172 cm^{-1} from 0 min to 60 min. The temporal intensity changes of cross peaks at (1710, 1723) cm^{-1} and (1710, 1740) cm^{-1} observed in Figure 4.11 exhibit close similarity to those of cross peaks at (1190,1178) cm^{-1} and (1190,1172) cm^{-1} in Figure 4.12, respectively. Therefore, the band at 1178 cm^{-1} is assigned to the naphthalene ring vibration of the moderately conjugated conformation, and that at 1172 cm^{-1} to the corresponding vibration of the slightly conjugated conformation.

Figure 4.13a and Figure 4.13b depict synchronous correlation spectra in the C=O stretching vibration region measured during annealing at 245°C for the time periods of 0 min to 9 min and 12 min to 60 min, respectively [39]. It is noted

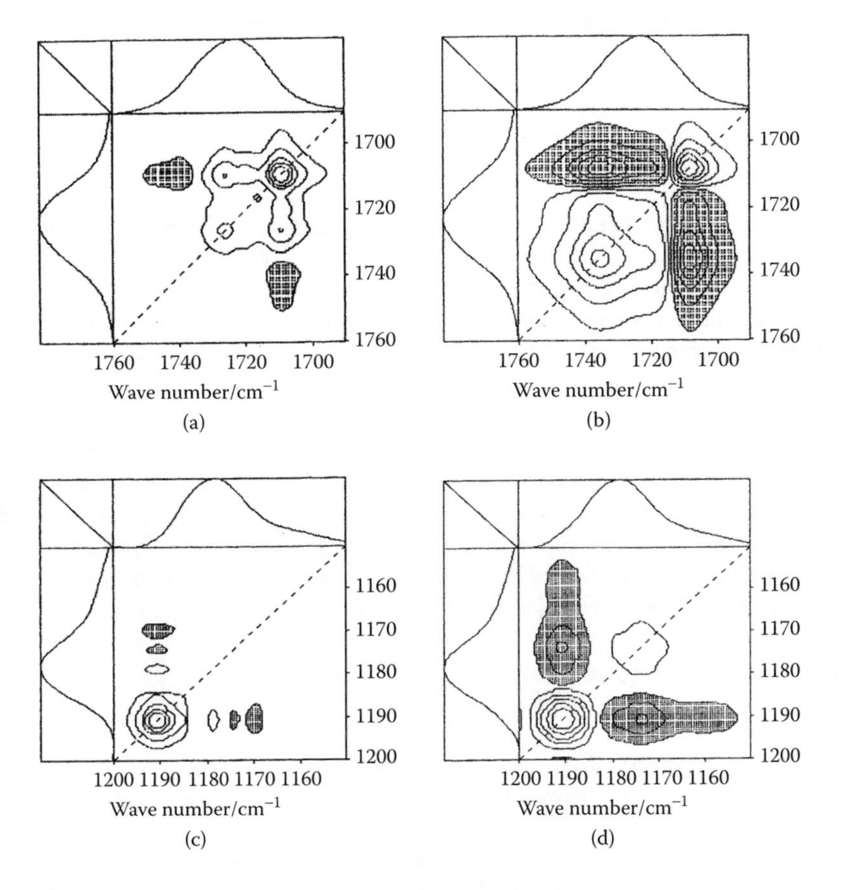

FIGURE 4.13 Synchronous correlation spectra in the C=O stretching vibration region [Reproduced from F Kimura, M Komatsu, T Kimura. *Appl Spectrosc* 54: 974–977, 2000 with permission. Copyright (2000) Society of Applied Spectroscopy.]

from 0 min to 9 min that the intensities of the C=O stretching vibration of moderately conjugated conformation in the amorphous phase (1723 cm^{-1}) as well as that of -crystal form (1710 cm^{-1}) increase. This behavior is similar to that observed in the initial stage of crystallization at 250°C. Figure 4.13 shows that the intensities at 1723 cm^{-1} and 1740 cm^{-1} decrease while the intensity at 1710 cm^{-1} increases. This behavior is also similar to that observed at 250°C in the time period from 42 min to 60 min.

The corresponding synchronous spectra in the naphthalene ring vibration region are shown in Figure 4.13c and Figure 4.13d, respectively. Increase in the intensities at 1178 cm^{-1} and 1190 cm^{-1} (-crystal form) are observed in Figure 4.13c. The pattern of the cross peak in this 2D spectrum is similar to that in Figure 4.12a and Figure 4.12b. The 2D spectrum of the naphthalene ring vibration region (Figure 4.3d) also bears close resemblance to Figure 4.12c.

2D correlation analysis was also employed to analyze a set of transient spectra obtained during the isothermal melt and cold crystallization of PEN [39]. It was found that the conformation changes occurring during crystallization are different for each of the two types of crystallization.

4.4.2 Applications to Biological Compounds

A variety of biological materials and molecules including biomedical samples have been investigated by 2D correlation spectroscopy [21–23]. Among them proteins have been studied most frequently [29,30,33–35,43–45,50,57,60,67,70]. 2D-IR correlation spectroscopy enables the highly overlapped amide I, II, and III bands of proteins to be resolved into component bands ascribed to various secondary structures. Also, it is possible by using 2D correlation spectroscopy to monitor the specific order of the secondary structure changes induced by temperature, possible, pH, and concentration variations. Hydrogen-deuterium (H-D) exchange is also a useful perturbation for exploring protein conformation by means of 2D-IR [30]. 2D-NIR correlation spectroscopy provides new insights into the hydration of proteins [50,57]. Besides proteins, lipids [74], lipid-protein interactions [44], bacteriorhodopsin [75], synthetic and biological apatites [76], stratum corneum of human skin [3], and keratin films from human hair [20] have been investigated by 2D correlation spectroscopy. In Section 4.4.2 the applications of 2D-IR correlation spectroscopy to proteins are described.

4.4.2.1 2D Correlation Infrared Rheo-Optics of *Bombyx mori* Fibroin Film

Sonoyama and Nakano [45] made IR rheo-optical measurements of regenerated *Bombyx mori* silk fibroin film undergoing sinusoidal mechanical strain by use of dynamic step-scan FT-IR spectroscopy combined with software-based dynamic signal processing (DSP). With the DSP technique, dynamic spectra of *Bombyx mori* silk fibroin film with high signal-to-noise ratio (S/N) were successfully obtained in only 30 min. The dynamic spectra revealed that stress-induced dynamic reorientation in fibroin film is mainly localized in the segment with -sheet conformation and is almost synchronous with the applied mechanical strain.

Rheo-optics, a combined analytical tool of spectroscopy and rheology, has been used extensively to explore the relationship between mechanical properties and structures of polymeric materials [87]. IR and NIR rheo-optics have successfully been applied for polymers under sinusoidal external strain because of several instrumental and computational breakthroughs, such as the development of the step-scan interferometer, the introduction of dynamic alignment with the improvement of highly sensitive detectors, and the introduction of 2D correlation analysis.

Figure 4.14 shows dynamic in-phase and quadrature IR spectra of regenerated *Bombyx mori* fibroin film in the silk I state, together with the corresponding static spectrum [45]. The in-phase dynamic spectrum is much more intense than the quadrature one, indicating that stress-induced dynamic structural changes in the regenerated fibroin film respond to the applied strain almost synchronously. The static spectrum shows amide I, II, and III bands at 1659, 1536, and 1238 cm^{-1}, respectively. These

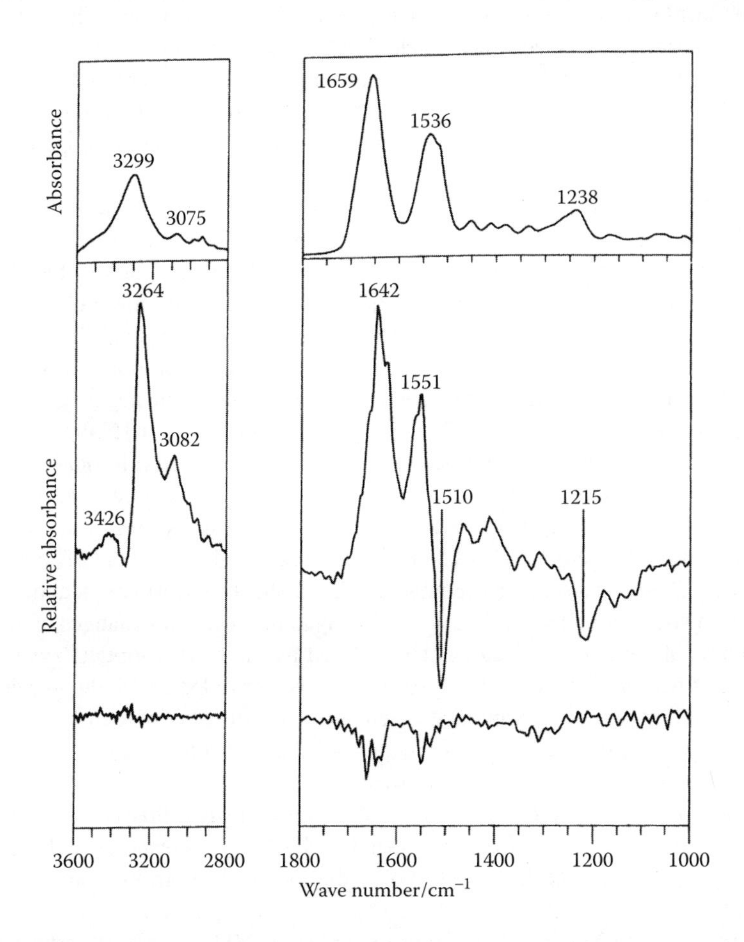

FIGURE 4.14 Dynamic in-phase and quadrature IR spectra of regenerated *Bombyx mori* fibroin film in the silk I state, together with the corresponding static spectrum [Reproduced from M Sonoyama, T Nakano. *Appl Spectrosc* 54: 968–973, 2000 with permission. Copyright (2000) Society of Applied Spectroscopy.]

frequencies suggest that the fibroin assume the helical and disordered structures. In the dynamic spectrum, however, all the amide bands (amide I: 1642 cm^{-1}; amide II: 1551 cm^{-1} and 1512 cm^{-1}; and amide III: 1215 cm^{-1}) are observed at wavenumbers different from those of the corresponding static bands. Of note is that every position of the dynamic bands in the amide I, II, and III regions is in good agreement with that of static features due to -sheet structures in *Bombyx mori* silk fibroin. It is also noted that the dynamic amide bands at 1642 cm^{-1} and 1551 cm^{-1} are positive, while the rest of the dynamic amide bands are negative. The differences in dynamic dichroism of the four bands are also in accordance with the fact that the two positive bands at 1642 cm^{-1} and 1551 cm^{-1}, and the two negative ones at 1512 cm^{-1} and 1215 cm^{-1}, have the same dichroism, respectively, and the dichroism of the former two bands is different from that of the latter two bands. These results led Sonoyama and Nakano [45] to conclude that stress-induced dynamic structural changes in the regenerated silk fibroin film occur mainly in -sheet structures. The fact that the β-sheet structures are more susceptible to mechanical stretching may be related to conformational transformation from the silk I structures abundant in helical and disordered structures to the silk II with -sheet structures upon the strong elongation of silk fibroin, along with the suggestion that sheet structures are mainly responsible for elastic properties of the fibrous protein.

The 2D correlation analysis showed powerful resolution enhancement effect in the amide I band region [45]. A strong autopeak appears at 1642 cm^{-1} in the synchronous spectrum (data not shown). This feature corresponds to the dynamic in-phase band at 1642 cm^{-1}, indicating that stress-induced structural changes occur mainly in -sheet structures. Figure 4.15 shows the asynchronous 2D correlation spectrum of the silk fibroin in the amide I region [45]. In the upper left region of the asynchronous map, the amide I band is separated into four negative correlation peaks at (1661, 1642), (1661, 1615), (1643, 1615), and (1632, 1615) cm^{-1}. The peak at 1642 cm^{-1} corresponds to the band observed in the dynamic in-phase spectrum and in the synchronous correlation spectrum. In addition, the peak at 1661 cm^{-1} arises from the main secondary structure in the regenerated fibroin film. The other two features at 1632 cm^{-1} and 1615 cm^{-1} can be detected only in the 2D correlation spectrum. These peaks are assignable to the β-sheet conformation and tyrosine residues, respectively. It should be noted that dynamic structural changes of aromatic side chains of tyrosine residues can be explored by the IR rheo-optical system. The dynamic structural changes in the tyrosine residues revealed by IR rheo-optics may be an important point to understand mechanical properties of silk fibroin.

The signs of the asynchronous cross peaks indicate that the intensity at 1615 cm^{-1} changes faster, the 1661 cm^{-1} peak moves slower, and the other two features are in the middle range. These dynamic spectral changes suggest that the aromatic side chains of tyrosine residues move first, then the -sheet regions respond, and the helical structures reorient in the last stage, when the mechanical strain is applied to the fibroin film.

Sonoyama and Nakano [45] also investigated the 4000–2800 cm^{-1} region by use of 2D correlation spectroscopy. They reported that the amide A and B bands have a strong synchronous correlation with the amide II mode when silk fibroin undergoes a sinusoidal mechanical strain.

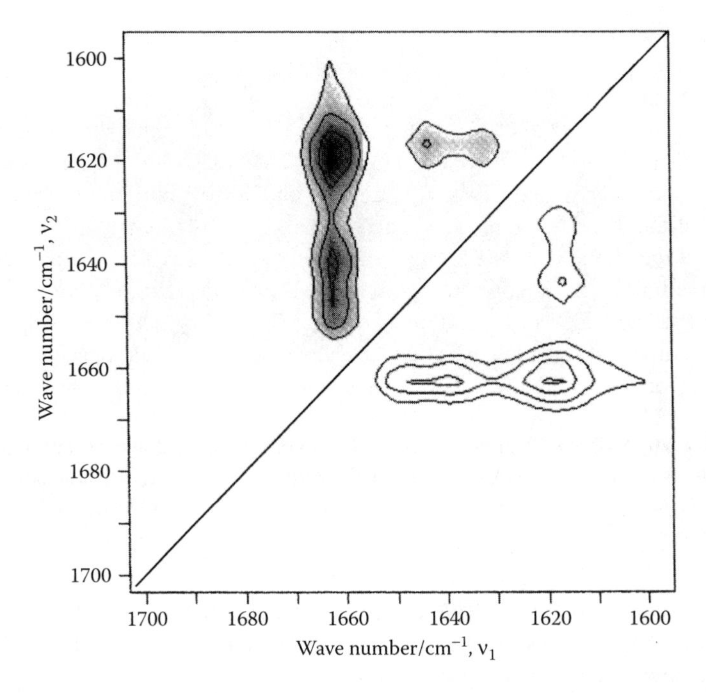

FIGURE 4.15 The asynchronous 2D correlation spectrum of the silk fibroin in the amide I region [Reproduced from M Sonoyama, T Nakano. *Appl Spectrosc* 54: 968–973, 2000 with permission. Copyright (2000) Society of Applied Spectroscopy.]

4.4.2.2 2D Attenuated Total Reflection/IR Spectroscopy Study of Adsorption-Dependent and Concentration-Dependent Structural Changes in β-Lactoglobulin in Buffer Solutions

Czarnick-Matusewicz et al. [43] reported the usefulness of generalized 2D attenuated total reflection (ATR)/IR spectroscopy in the studies of adsorption-dependent and concentration-dependent structural variations of β-lactoglobulin (BLG) in buffer solutions. ATR/IR spectra in the 4000–650 cm^{-1} region were measured for the buffer solutions of BLG with a concentration of 1, 2, 3, 4, and 5 wt% at room temperature. To generate the 2D correlation spectra, the original spectra were subjected to pre-treatment procedures consisting of ATR correction, subtraction of the spectrum of buffer solution, smoothing, and normalization over the concentration. The adsorption-dependent 2D correlation analysis revealed that the interaction between the ATR crystal surface and protein molecules is characterized by pronounced intensity changes of bands due to β-sheet structures buried in a hydrophobic core of BLG. In contrast, the concentration-dependent 2D correlation spectra, which show 10 times more intense features than the adsorption-dependent 2D spectra, reflect changes in various secondary structure elements located in the hydrophilic parts exposed to the solvent.

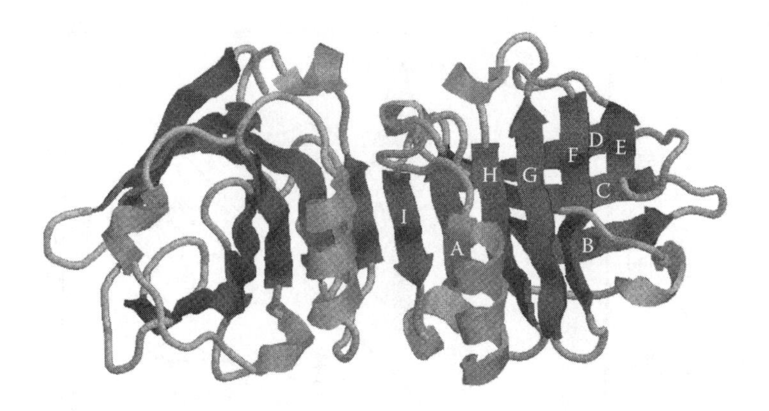

FIGURE 4.16 A ribbon diagram of the single subunit of BLG, drawn from the X-ray data by Brownlow et al. [88] [Reproduced from B Czarnik-Matusewicz, K Murayama, Y Wu, Y Ozaki. *J Phys Chem* B 104: 7803–7811, 2000 with permission. Copyright (2000) American Chemical Society.]

Discussion of the 2D ATR/IR correlation spectra of BLG may be facilitated by briefly looking over the structure of BLG elucidated by X-ray crystallography. BLG, the main component of whey proteins contained in bovine milk, was one of the first proteins explored by X-ray crystallographic analysis [88]. In its mature form BLG is built from 162 amino acid residues, and in an aqueous solution it exists as a dimer of molecular weight of approximately 36 kDa. Figure 4.16 depicts the structure of BLG elucidated by X-ray crystallography [43]. The secondary structure of BLG mainly consists of antiparallel β-sheets. The core of the BLG molecule is made up of a short α-helix segment and eight strands of antiparallel β-sheets, labeled β-A to β-H, which wrap round to form a calyx. The percentages of individual secondary structural components are approximately 35% β-strands, 18% α-helices, 10% 3_{10} helices, and 48% nonperiodic fragments of the BLG main chain. In literature many facts reveal that the secondary structure of BLG is sensitive to the changes in its environment. For example, Qi et al. [89] reported in their study on a thermal denaturation of BLG that the denaturation process strongly depends on the concentration of BLG, particularly in a concentration range below 50 mg/ml.

Figure 4.17 shows ATR/IR spectra in the 4000–650 cm^{-1} region of the buffer solution and the five kinds of BLG solutions. Figure 4.18 displays the spectra of the protein solutions after the four pretreatments [43]. One of the major steps in the preparation of the concentration-dependent spectral data prior to the 2D correlation analysis is normalization.

Figure 4.19 displays a synchronous 2D ATR/IR spectrum generated from time-dependent spectral variations of the 5% BLG solution on the ZnSe surface [43]. The average one-dimensional spectrum (dashed line) and power spectrum along the diagonal line on the 2D map (solid line) are presented on the top of the 2D map. The averaged spectrum is dominated by one broad asymmetrical band with the maximum at 1628 cm^{-1} due to the β-sheet structure, the major structural component of BLG, while the power spectrum consists of a number of well-resolved bands.

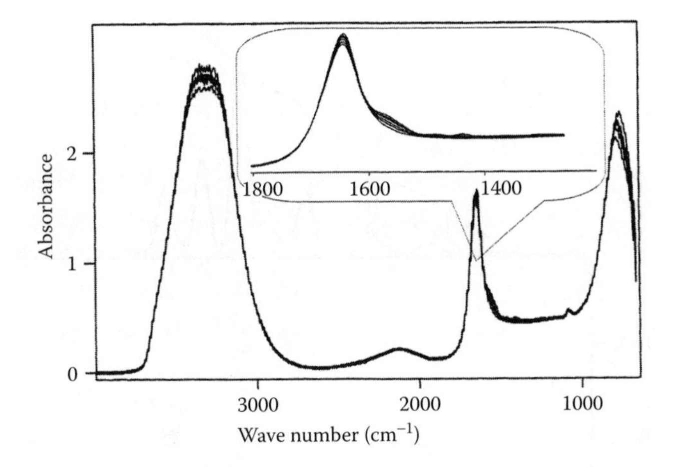

FIGURE 4.17 ATR/IR spectra in the 4000–650 cm^{-1} region of the buffer solution and five kinds of aqueous solutions of BLG with concentrations of 1, 2, 3, 4, and 5 wt%. [Reproduced from B Czarnik-Matusewicz, K Murayama, Y Wu, Y Ozaki. *J Phys Chem* B 104: 7803–7811, 2000 with permission. Copyright (2000) American Chemical Society.]

Note that most of the autopeaks are hardly identified in the original spectrum. Positive cross peaks at (1667, 1638), (1654, 1638), and (1638, 1621) cm^{-1} indicate that the intensity changes correlated by these pairs of wavenumbers are either increasing or decreasing together with respect to the time-dependent adsorption

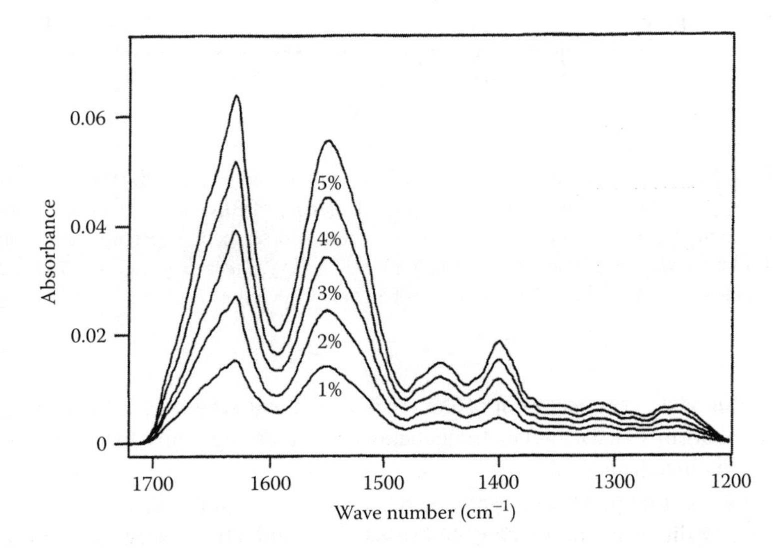

FIGURE 4.18 Concentration-dependent ATR/IR spectral variations in the 1720–1200 cm^{-1} region of the BLG solutions after the four kinds of pretreatments. [Reproduced from B Czarnik-Matusewicz, K Murayama, Y Wu, Y Ozaki. *J Phys Chem* B 104: 7803–7811, 2000 with permission. Copyright (2000) American Chemical Society.]

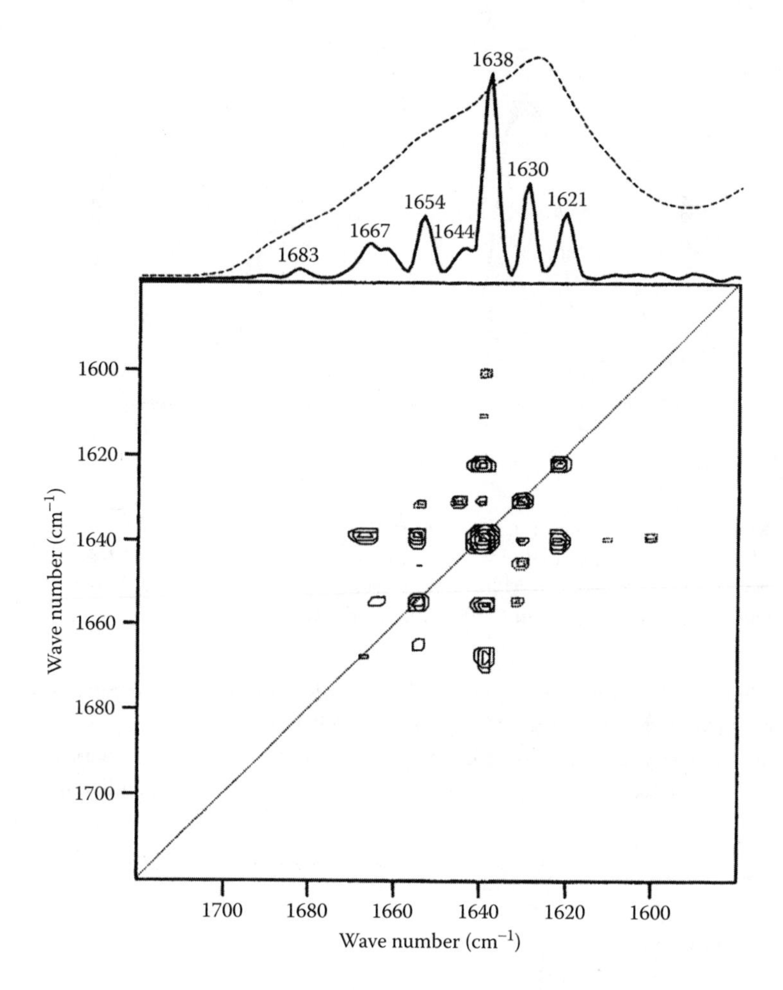

FIGURE 4.19 A synchronous 2D ATR/IR correlation spectrum in the 1720–1580 cm^{-1} region constructed from the adsorption-dependent spectral changes of BLG. A power spectrum (solid line) and an averaged spectrum (dashed line) are presented on the top. [Reproduced from B Czarnik-Matusewicz, K Murayama, Y Wu, Y Ozaki. *J Phys Chem* B 104: 7803–7811, 2000 with permission. Copyright (2000) American Chemical Society.]

process. Negative cross peaks at (1654, 1630) cm^{-1} and (1644, 1630) cm^{-1} suggest that the intensities at the higher frequencies are decreasing while that at the lower frequency is increasing.

The corresponding asynchronous spectrum is presented in Figure 4.20 [43]. On the top of the 2D map a slice spectrum extracted at 1640 cm^{-1} from the asynchronous is shown. The slice spectra may provide information about the separation of overlapping bands and about the order in the intensity changes between one particular band and others. Based upon Noda's rule for the signs of asynchronous cross peaks, Czarnick-Matusewicz et al. [43] proposed the following sequence of intensity

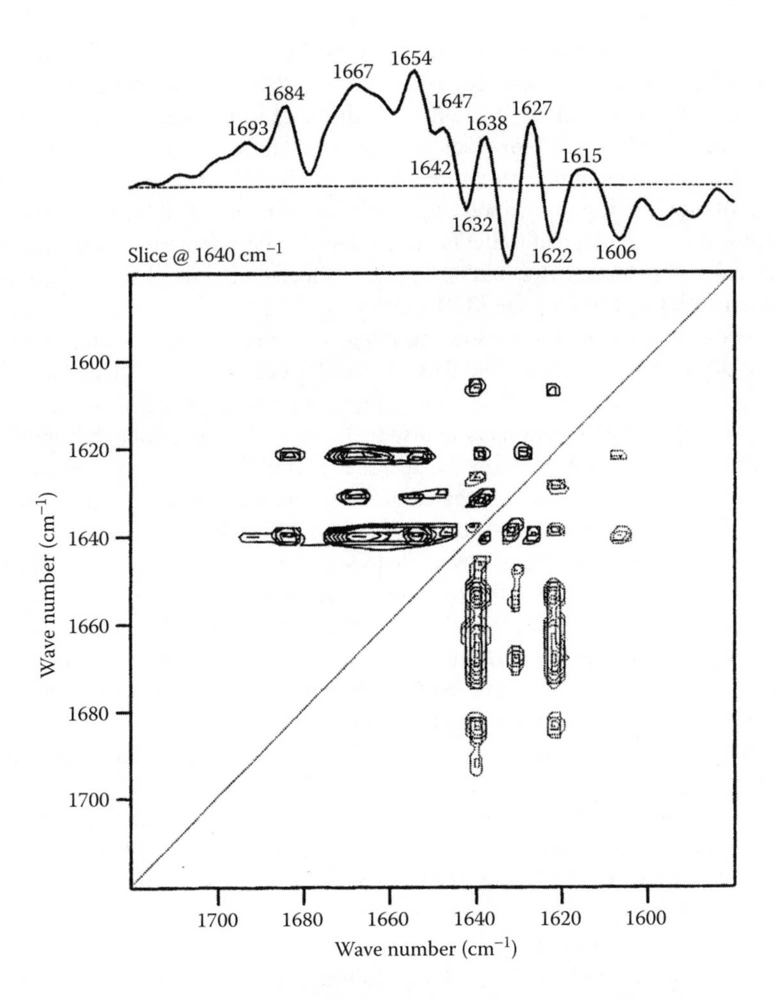

FIGURE 4.20 An asynchronous 2D ATR/IR correlation spectrum in the 1720–1580 cm^{-1} region constructed from the adsorption-dependent spectral changes of BLG. A slice spectrum extracted at 1640 cm^{-1} is presented on the top. [Reproduced from B Czarnik-Matusewicz, K Murayama, Y Wu, Y Ozaki. *J Phys Chem* B 104: 7803–7811, 2000 with permission. Copyright (2000) American Chemical Society.]

changes occurring during the time-dependent adsorption process:

$$(1693 \approx 1684 \approx 1667 \approx 1654 \approx 1647) \rightarrow (1640 \approx 1638)$$
$$\rightarrow (1632 \approx 1627) \rightarrow 1622 \text{ cm}^{-1}$$
$$\text{time of adsorption}$$

According to the infrared studies of BLG in solutions [90–92], the peak at 1638 cm^{-1} is ascribed to a low wavenumber component of antiparallel β-strands creating the characteristic β-sheet motive "buried" in the interior of BLG. The corresponding high wavenumber component is observed at 1693 cm^{-1}. The band at 1630 cm^{-1} is assigned to "exposed" β-strands, i.e., those that are not part of the core of the β-sheets. The peak at 1621 cm^{-1} is attributed to the formation of associated forms with a number of intermolecular hydrogen bonds. The 1644 cm^{-1} component arises from the random coil structures, while the band at 1654 cm^{-1} represents the overlap of signals from the α-helix and random segments. The bands at 1683 cm^{-1} and 1667 cm^{-1} are assigned to the turns connecting the β-strands into the calyx and the 3_{10} helix structures, respectively. The bands located below 1620 cm^{-1} are due to tyrosine (TYR) and tryptophan (TRP) residues of BLG.

The above sequence for the spectral intensity variations suggests that at the first stage of the adsorption process the BLG molecules become more expanded and that, at the expense of the buried β-strands, increases the population of the exposed β-strands [43]. The association could also be facilitated by additional intermolecular hydrogen bonds formed between the residues located in the less ordered parts of BLG. The negative cross peaks detected in the synchronous map (Figure 4.19) reveal that within the evanescent wave, the populations of the intermolecularly hydrogen bonded parts as well as the exposed β-strands, increase with time. On the other hand, the populations of the α-helix, turns, and random coils decrease relatively during the course of the interaction between the ATR prism and protein. This shows that during the adsorption a partly unfolded state of BLG is formed, predominantly due to the increase in the less ordered β-sheet structure controlled by intermolecular hydrogen bonds leading finally into larger oligomeric structures.

Synchronous and asynchronous 2D ATR/IR spectra constructed from the concentration-dependent spectral variations of BLG aqueous solutions is shown in Figure 4.21a and Figure 4.21b, respectively [43]. The power spectra extracted from the two synchronous maps in Figure 4.19 and Figure 4.21 are compared with the same scale in Figure 4.22 [43]. The intensity changes shown by curve (a) are concerned solely with the adsorption process, while those presented by curve (b) are composed of contributions from both the adsorption process and the concentration change in the BLG solutions. It is noted that the intensity variations induced only by the adsorption process are 10 times smaller than those due to concentration as well as adsorption changes. Accordingly, it is very likely that the majority of intensity changes shown by curve (b) originate from the concentration-dependent perturbation. The adsorption-dependent process creates much less significant intensity changes. It is also of note in Figure 4.22 that the nature of the changes, represented by the two spectra, is remarkably different. The strongest autopeaks appear at 1652 cm^{-1} and 1662 cm^{-1} in the power spectrum for the concentration-dependent changes (Figure 4.22). These peaks are assigned to hydrophilic secondary structure elements located on the outer surface of BLG that are built from α-helices and 3_{10}-helices. These secondary structures are less protected from water penetration than the β-strands forming the calyx. A minimal intensity change in the band at 1636 cm^{-1} due to antiparallel buried β-sheets suggests that the β-sheets located inside the hydrophobic core of BLG are less accessible to water molecules than those in α-helical fragments and other hydrophobic, solvent-exposed parts of BLG [43].

The asynchronous 2D correlation spectrum (Figure 4.21b) for concentration-dependent spectral changes provides complementary information to the synchronous

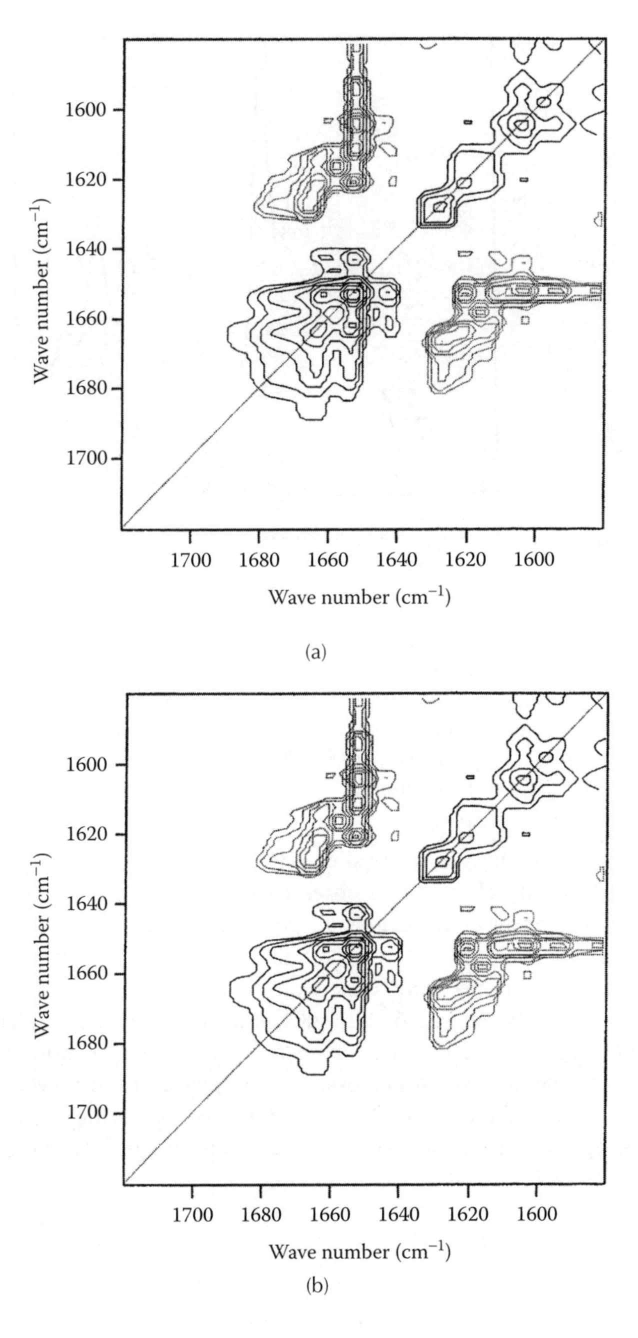

FIGURE 4.21 (a) Synchronous and (b) asynchronous 2D ATR/IR correlation spectra in the 1720–1580 cm^{-1} region constructed from the concentration-dependent spectral changes of BLG. [Reproduced from B Czarnik-Matusewicz, K Murayama, Y Wu, Y Ozaki. *J Phys Chem* B 104: 7803–7811, 2000 with permission. Copyright (2000) American Chemical Society.]

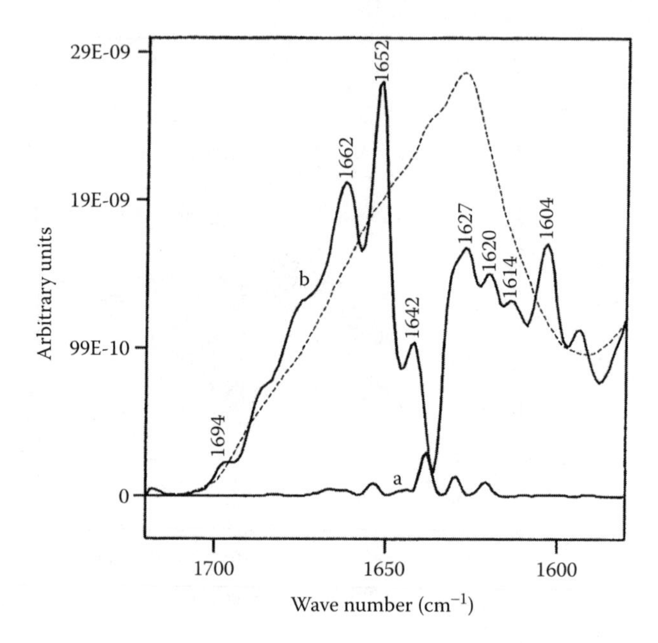

FIGURE 4.22 Power spectra along the diagonal line in the synchronous spectra constructed from the (a) adsorption- and (b) concentration-dependent spectral changes of BLG. The dashed line represents the averaged spectrum calculated from the concentration-dependent spectral set. [Reproduced from B Czarnik-Matusewicz, K Murayama, Y Wu, Y Ozaki. *J Phys Chem* B 104: 7803–7811, 2000 with permission. Copyright (2000) American Chemical Society.]

spectrum. It is important to compare slice spectra extracted at the frequencies where noncorrelated intensity changes are observed individually for the adsorption- and concentration-dependent perturbations, respectively. Such slice spectra extracted at 1640 cm^{-1} and 1665 cm^{-1} from the asynchronous spectra generated from the adsorption-Figure and concentration-Figure dependent spectral changes (Figure 4.20 and Figure 4.21b) are shown in Figure 4.23 [43]. It is noted that the unrelated intensity variations induced by the concentration changes are at least one order higher than those caused by the adsorption process, as in the case of the power spectra. The following sequence of the intensity changes occurring during the concentration change is suggested from the asynchronous map (Figure 4.21b) and the slice spectrum (Figure 4.23).

$$(1645 \approx 1642) \rightarrow (1694 \approx 1683 \approx 1658 \approx 1654) \rightarrow (1673 \approx 1665)\ \text{cm}^{-1}$$
$$(1638 \approx 1633 \approx 1628) \rightarrow 1623 \rightarrow 1616 \rightarrow 1606\ \text{cm}^{-1}$$
concentration

The above sequence reveals that the first event in the secondary structure change occurs in the random coil structure, followed by the changes in the β-sheet, β-turn, and α-helix. The secondary structure variations in another type of β-turn (probably type III) and 3_{10} helix occur as the next event. Changes in both buried and exposed β-strands are behind

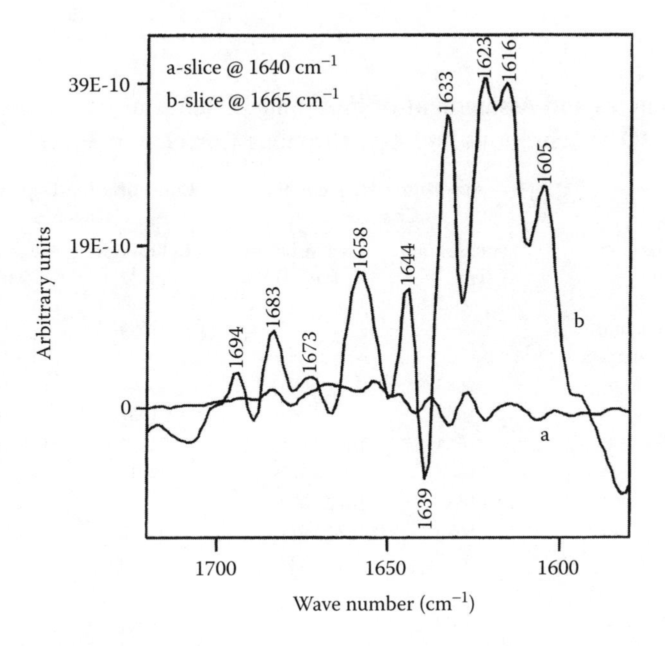

FIGURE 4.23 (a) A slice spectrum extracted along 1640 cm^{-1} in the asynchronous spectrum shown in Figure 4.20, and (b) a slice spectrum extracted along 1665 cm^{-1} in the asynchronous spectrum presented in Figure 4.21b. [Reproduced from B Czarnik-Matusewicz, K Murayama, Y Wu, Y Ozaki. *J Phys Chem* B 104: 7803–7811, 2000 with permission. Copyright (2000) American Chemical Society.]

those due to the structures described above, but ahead of the changes in associated components. The last event in the sequence is structural changes in the side-chains. The minus sign of the synchronous peak around (1652, 1622) cm^{-1} suggests that at the expense of the intramolecular hydrogen bonds located in the disordered/α-helix components (1652 cm^{-1}), the number of intermolecular hydrogen bonds (1622cm^{-1}) increases. The band assignments proposed in this study are summarized in Table 4.1.

In this way, by the use of generalized 2D correlation spectroscopy, Czarnik-Matusewicz et al. [43] have provided new insight into a subject of controversy in the interpretation of ATR/IR spectra of proteins, whether the major intensity variations arise from the concentration changes or the adsorption process on the prism surface. 2D correlation analysis has proved that there are two different mechanisms of secondary structure variations of BLG. In the presence of adsorption the highly hydrophobic β-strands "buried" in the calyx of BLG undergo the main changes. On the other hand, the hydrophilic secondary structural elements are involved in the structural changes caused by the concentration-dependent change.

4.4.3 Other Applications

We have discussed the applications of 2D correlation spectroscopy to polymeric and biological materials. However, a variety of applications of 2D correlation

TABLE 4.1
Frequencies and Assignment of the Amide I Bands of BLG Observed in the 2D Synchronous and Asynchronous Correlation Spectra

Assignment (Secondary Structure)	Adsorption-Dependent Changes		Concentration-Dependenty Changes	
	Synchronous (cm^{-1})	Asynchronous (cm^{-1})	Synchronous (cm^{-1})	Asynchronous (cm^{-1})
Highwavenumber β-sheet component	—	1693	1694	1694
β-turn	1683	1684	—	1683, 1673
3_{10} helix	1667	1667	1662	1665
α-helix	1654	1660,1654	1652	1658,1654
Random coil	1644	1647	1642	1645,1642
β-sheet "buried"	1638	1642,1638	—	1639,1636
β-sheet "exposed"	1630	1632,1627	1627	1633,1628
Aggregated elements by intermolecular hydrogen bonds	1621	1622	1620	1623
Side chains	—	1615, 1606	1614, 1604	1616, 1605

spectroscopy can be found also in liquid crystal research [16,19,21,22,32,42], investigations of self-associated molecules [21,22,46,47,49,51,52], studies of chemical reactions [31,77], and surface and interface studies [72,78]. In this and the next sections, we describe new applications of 2D correlation spectroscopy.

4.4.3.1 2D-IR Correlation Spectroscopy Study on Pressure-Dependent Changes in the C-H Vibrations of Monolayer Films at the Air–Water Interface

Elmore and Dluhy [72] applied 2D correlation analysis to a set of surface pressure-dependent unpolarized IR spectra of a monolayer film of 1,2-dipalmitoyl-*sn*-glycero-3-phosphocholine (DPPC) at the air–water (A–W) interface. They compared the asynchronous 2D-IR spectra calculated from the experimental data with synthetic asynchronous spectra generated by using an overlapped peaks model versus a frequency shifting model. The 2D correlation analysis strongly supports the following conclusions [72]: (1) the low-pressure subset (<11 mN/m), which encompasses both the liquid expanded (LE) and the liquid expanded/liquid condensed (LE/LC) regions of the DPPC monolayer isotherm, is best modeled by two overlapped peaks correlated with ordered and disordered conformational states of the monolayer film; and (2) the high-pressure subset (>11 mN/m), reflecting solely the liquid condensed (LC) phase, is best modeled by a single peak, which undergoes a minor frequency shift, and which may be primarily correlated with gradual packing of the liquid condensed structure.

Recently, the technique of external reflection FT-IR spectroscopy developed. This method could differentiate between ordered and disordered conformations in phospholiped monolayer films at the air–water interface. In particular, the conformation-sensitive C-H stretching vibrations from the lipid hydrocarbon chains could be used to monitor the expanded-to-condensed thermodynamic phase transition of the monolayer. Elmore and Dluhy [72] applied 2D-IR correlation analysis to a set of unpolarized IR spectra of the DPPC monolayer to investigate spectral changes observed in the C-H region. Figure 4.24a shows the unpolarized external reflectance IR spectra in the C-H stretching region of a DPPC monolayer film [72]. To collect these spectra, the monolayer was applied to the surface of the film balance, allowed to equilibrate, and then compressed step-wise over a range of surface pressures from 0.9 mN/m to 50.0 mN/m with spectra measured during the interval between each successive compression. Bands at ~2920 cm^{-1} and ~2850 cm^{-1} are due to the antisymmetric (V_a) and symmetric (V_{as}) CH_2 stretching modes, respectively, and a band at ~2960 cm^{-1} is assigned to the asymmetric CH_3 stretching mode. A slight shoulder on the symmetric CH_2 band arises from the symmetric CH_3 vibration (~2870 cm^{-1}). The intensities of the C-H bands become stronger as the lipid surface density increases with increasing surface pressure and the lipid molecules become more ordered. It is also noted that the V_{aa} CH_2 band shifts from ~2925 cm^{-1} to 2919 cm^{-1} while theV_s CH_2 band shifts from ~2855 cm^{-1} to 2851 cm^{-1}.

Figure 4.24b depicts the spectra of the monolayer after normalization to account for changing surface coverage due to monolayer compression [72]. The normalized spectra clearly reveal that IR intensity increases significantly from ~7 mN/m to ~11 mN/m. This change takes place just after the main transition plateau (LE/LC phase transition region) in the surface pressure molecular area isotherm of the DPPC monolayer and is primarily attributed to the large change in molecular reorientation occurring in this region.

Figure 4.25 and Figure 4.26 show asynchronous 2D correlation maps for the low- and high-pressure region subsets of spectra collected between 0 mN/m and 11 mN/m and between 12 mN/m and 50 mN/m, respectively, of the DPPC monolayer at the A–W interface [72]. The 2D correlation analysis was carried out for both the non-normalized and the normalized IR spectra. A low-pressure region below 11 mN/m corresponds to the liquid expanded (LE) and liquid expanded/liquid condensed (LE/LC) isotherm phases while a high-pressure region greater than 11 mN/m corresponds to the liquid condensed (LC) phase of the DPPC monolayer. In the asynchronous maps for the low-pressure region subset the most dominant cross peaks appear at 2925, 2916, 2855, and 2849 cm^{-1} and are easily assigned to the V_a and V_s CH_2 stretching modes. For the high-pressure region above 11 mN/m the most significant asynchronous cross peaks are observed at 2922, 2915, 2852, and 2949 cm^{-1}. These are also assigned to the V_a and $V_s CH_2$ stretching modes. For the normalized DPPC monolayer spectrum, a cross peak quartet and curved elongations along the diagonal are observed that correspond to a peak shift range of ~2920 cm^{-1} to ~2919 cm^{-1} for the V_a CH_2 band and ~2952 cm^{-1} to ~ 2951 cm^{-1} for the $V_s CH_2$ band (Figure 4.26b). Of note is that a more distinctive pattern in the high-pressure region is observed in the normalized asynchronous plot rather than the non-normalized asynchronous plot.

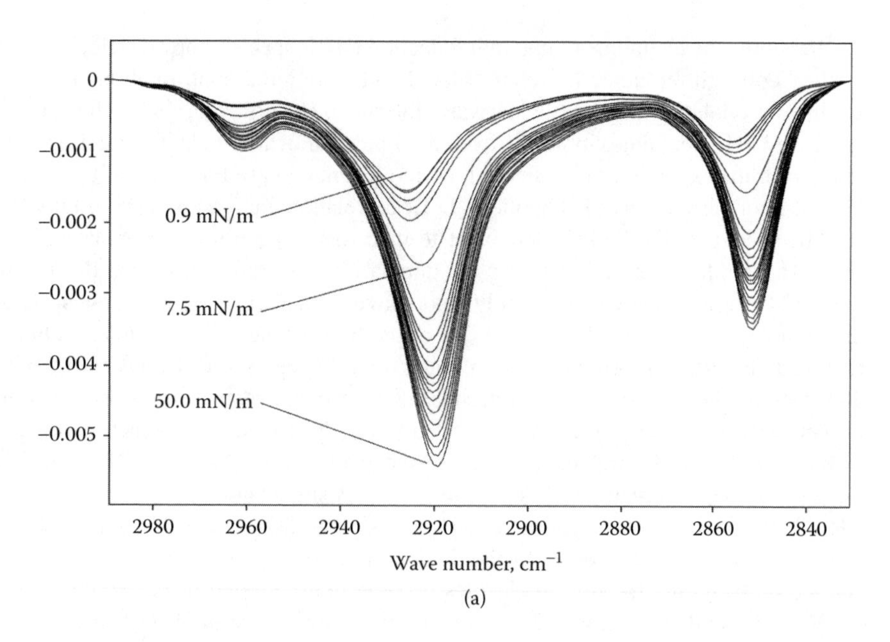

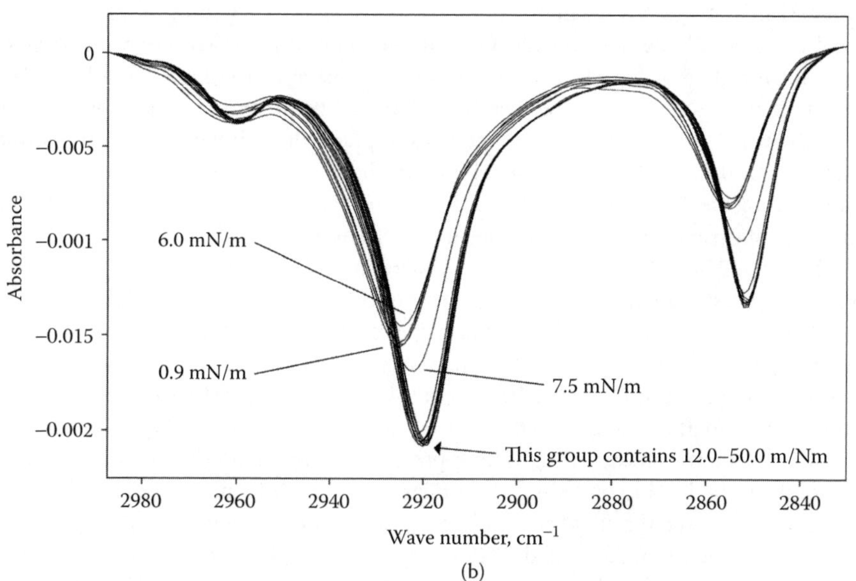

FIGURE 4.24 (a) The unpolarized external reflectance IR spectra in the C-H stretching region of a DPPC monolayer film. (b) The spectra of the monolayer after the normalization to account for changing surface coverage due to monolayer compression. [Reproduced from D L Elmore, R A Dluhy. *Appl Spectrosc* 54: 956–962, 2000 with permission. Copyright (2000) Society of Applied Spectroscopy.]

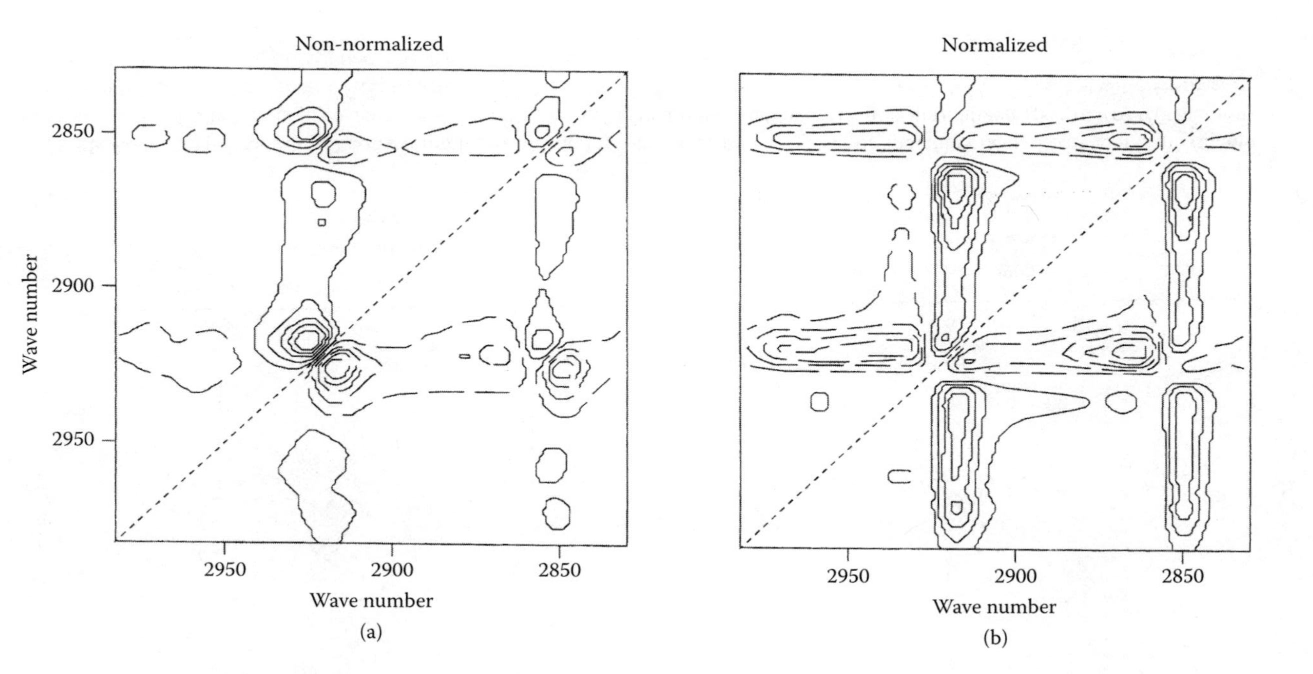

FIGURE 4.25 Asynchronous 2D correlation maps for the low-pressure region subset of spectra collected between 0 mN/m and 11 mN/m of the DPPC monolayer at the A–W interface. (a) Correlations calculated for non-normalized spectra. (b) Correlations calculated for normalized spectra. [Reproduced from D L Elmore, R A Dluhy. *Appl Spectrosc* 54: 956–962, 2000 with permission. Copyright (2000) Society of Applied Spectroscopy.]

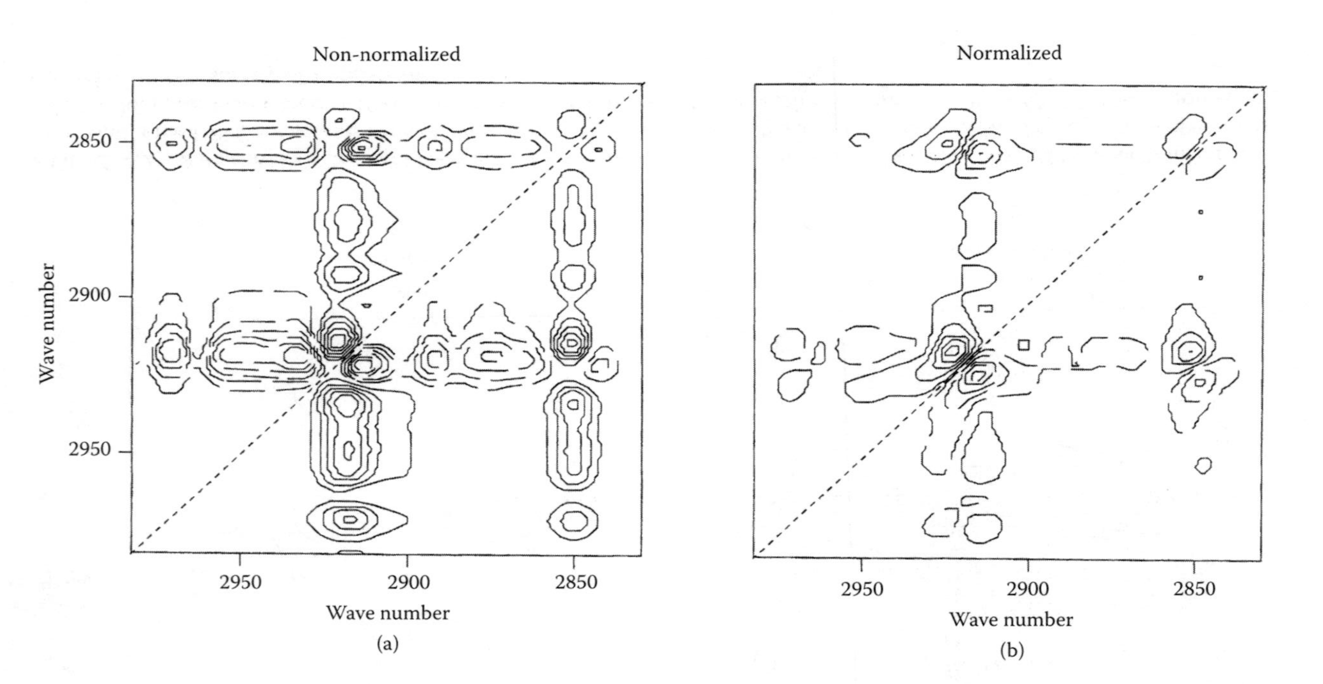

FIGURE 4.26 Asynchronous 2D correlation maps for the high-pressure region subset of spectra collected between 0 mN/m and 11 mN/m of the DPPC monolayer at the A–W interface. (a) Correlations calculated for non-normalized spectra. (b) Correlations calculated for normalized spectra. [Reproduced from D L Elmore, R A Dluhy. *Appl Spectrosc* 54: 956–962, 2000 with permission. Copyright (2000) Society of Applied Spectroscopy.]

To elucidate the causes of the frequency shifts and band splitting in the 2D spectra, synthetic monolayer 2D-IR spectra were calculated for two limiting cases. The first case was concerned with an overlapped peaks model in which an overall vibrational band was calculated as the sum of two individual sub-bands whose frequencies remained constant but whose relative intensities changed through the simulated monolayer transition. The second case was a "frequency shifting" model in which a single band undergoes a simple frequency shift. A detailed description of these models was given in literature [93].

Figure 4.27 shows computer-simulated 2D-IR asynchronous correlation maps for the overlapped peaks model in which two closely related peaks representing ordered and disordered frequencies change relative intensity but do not shift frequency [72]. The simulated asynchronous spectrum shows correlation intensity cross peak doublet, one positive and one negative, with no elongation along the diagonal. On the other hand, the computer simulations for the frequency shifting model shown in Figure 4.28 reveal that a simple frequency shift could be distinguished in the simulated asynchronous spectrum by the presence of a cross peak quartet, two with positive correlation intensities, and two with negative. In addition, a curved elongation of these cross peaks along the diagonal was associated with this frequency shift.

Figure 4.25a clearly shows that CH_2 bands both split into two components in the low-pressure region. The resulting 2D asynchronous map bears close resemblance to the non-normalized overlapped peaks model (Figure 4.25a and Figure 4.27a). Of particular note is that while the non-normalized low-pressure monolayer spectra show characteristic band splitting, the normalized spectra are dominated by noise and baseline fluctuations. This feature is in agreement with the normalized overlapped peaks model.

The high-pressure region yields quite different results. The most notable features in Figure 4.26 are the quartet of cross peaks and curved elongations along the diagonal that are observed in the normalized spectra. These patterns are in good agreement with the frequency shifting model in which a single band shows a shift (Figure 4.26b and Figure 4.28b).

The 2D-IR correlation analysis of DPPC monolayer films at the A–W interface revealed that the low-pressure region associated with the liquid expanded-to-condensed LE/LC phase transition region is characterized by two overlapped peaks in the C-H stretching region [72]. The frequency of these two sub-bands indicated that they are correlated with ordered and disordered alkyl chains in the DPPC monolayer film. The high-pressure region associated with the LC phase is characterized by a single peak that undergoes a frequency shift. The frequency of this latter peak indicates an ordered monolayer film, and the minor frequency shift observed in the high-pressure regions is primarily correlated with a gradual packing of the LC structure.

4.4.3.2 Applications to Electrochemical Reactions

Osawa [31] applied 2D correlation spectroscopy to analyze time-resolved IR spectra for the reduction of HV^{2+} (1,1′-diheptyl-4,4′-bipyridinium or heptyl viologen) to $HV^{\bullet+}$ at an Ag electrode for a potential step from −0.2V to −0.55V shown

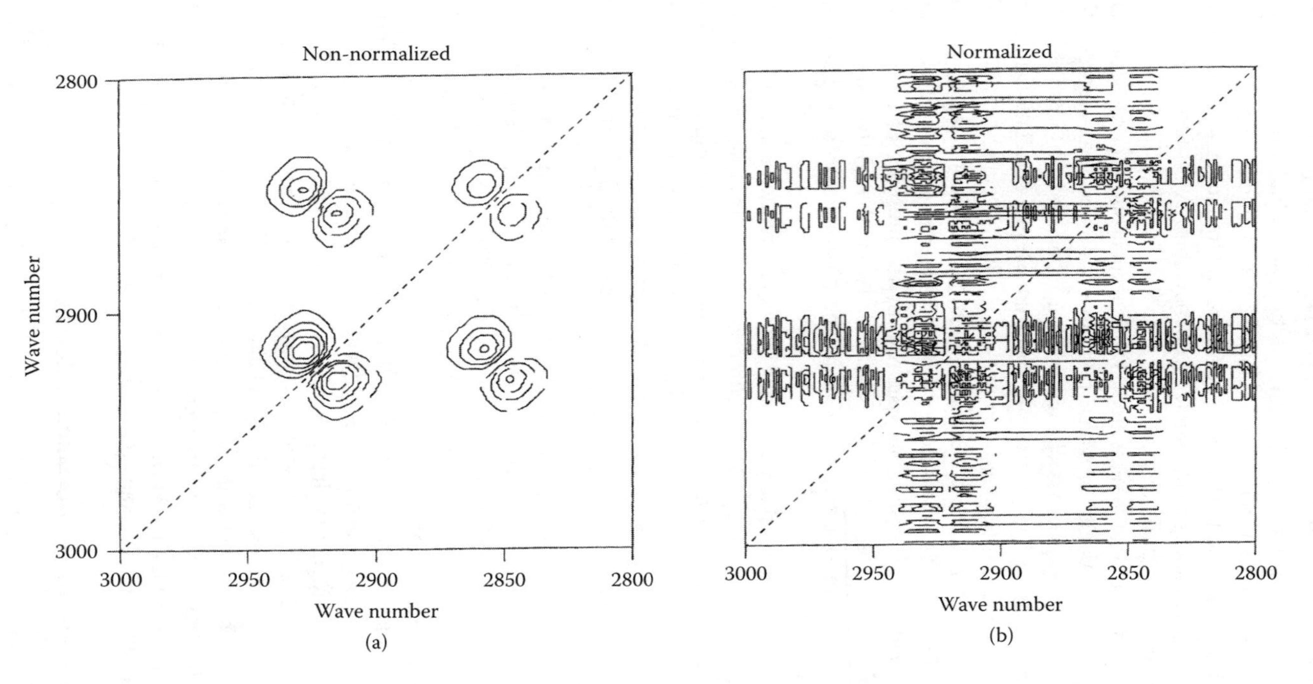

FIGURE 4.27 Computer simulated 2D-IR asynchronous correlation maps for the "overlapped peaks" model. (a) Correlations calculated for non-normalized spectra. (b) Correlations calculated for normalized spectra. [Reproduced from D L Elmore, R A Dluhy. *Appl Spectrosc* 54: 956–962, 2000 with permission. Copyright (2000) Society of Applied Spectroscopy.]

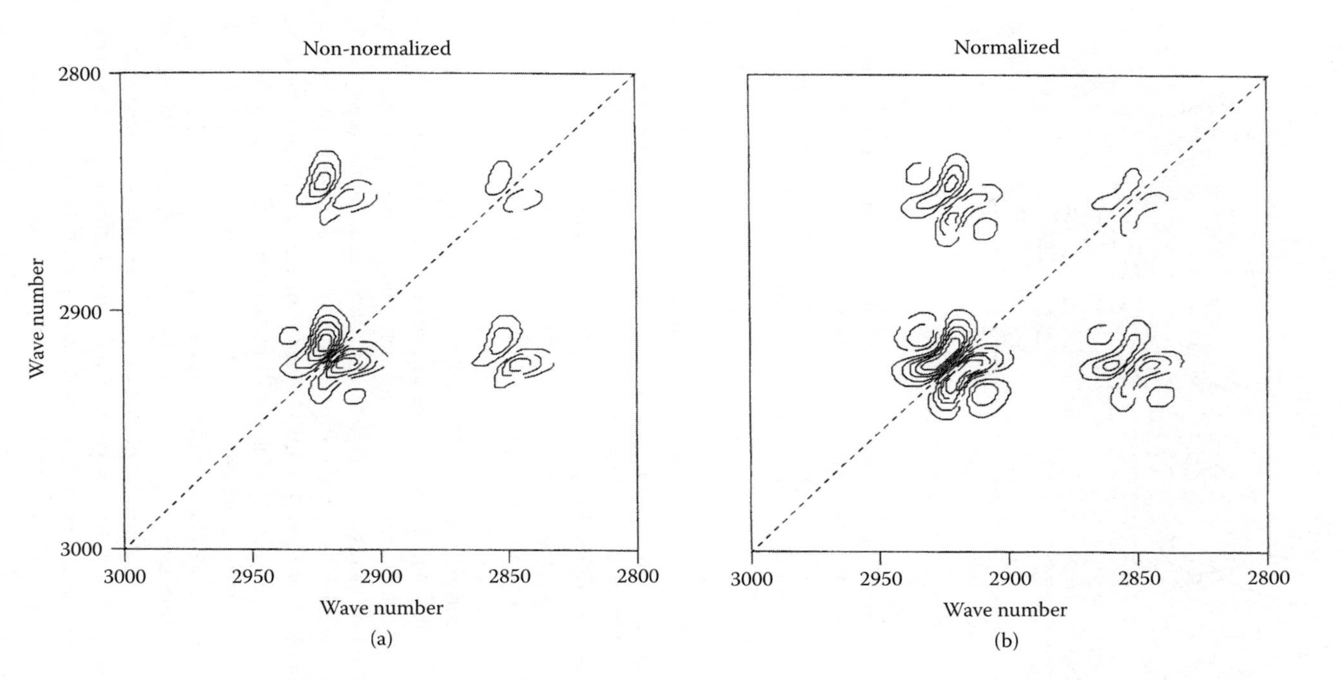

FIGURE 4.28 Computer simulated 2D-IR asynchronous correlation maps for the "frequency shifting" model. (a) Correlations calculated for non-normalized spectra. (b) Correlations calculated for normalized spectra. [Reproduced from D L Elmore, R A Dluhy. *Appl Spectrosc* 54: 956–962, 2000 with permission. Copyright (2000) Society of Applied Spectroscopy.]

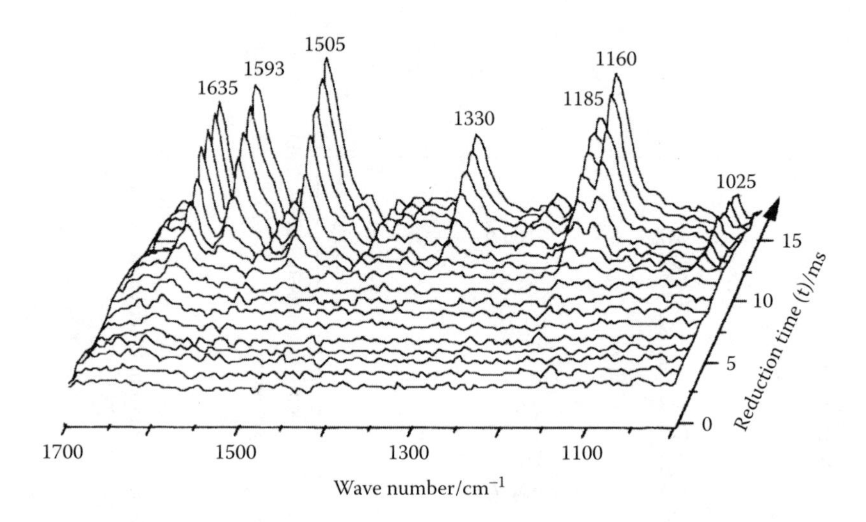

FIGURE 4.29 Time-resolved IR spectra for the reduction of HV^{2+} to $HV^{\cdot+}$ at an Ag electrode for a potential step from –0.2 to –0.55 (versus Ag/Agce). Each spectrum was collected with 100 μs acquisition time, but only the spectra of every 1 ms interval are shown for clarity. [Reproduced from M Osawa. *Bull Chem So Japan* 70: 2861–1880, 1997 with permission. Copyright (1997) Japan Chemical Society.]

in Figure 4.29 [31]. Figure 4.30a and Figure 4.30b show the synchronous and asynchronous 2D correlation spectra constructed from the time-resolved spectra shown in Figure 4.29, respectively [31]. Six autopeaks and cross peaks indicate that all the six bands at 1636, 1593, 1504, 1331, 1184, and 1161 cm^{-1} change their intensities with time simultaneously (namely, as the amount of the radical increase). Since the sign of the cross peaks is positive, the intensities of all the bands are changing in the same direction (increasing).

Asynchronous 2D correlation analysis provides deeper insight into the reaction dynamics. The 1636 cm^{-1} and 1184 cm^{-1} bands, both assigned to the monomeric radical, have no asynchronous correlation with each other, indicating that the intensity changes of these two bands are completely synchronized. On the other hand, it is noted that these two bands have a strong asynchronous correlation with the remaining for major bands at 1593, 1504, 1331, and 1161 cm^{-1} assigned to the dimer. Therefore, it seems that the intensity changes of the monomer bands and the dimer bands do not occur completely in-phase. The sign of the cross peaks suggests that the monomer bands appear earlier than the dimer bands.

It is also worth noting that in the asynchronous spectrum a number of weak cross peaks correlating to the dimer bands appear. The cross peaks indicate that the dimer bands are composed of several components that change their intensities at different rates. This is clearly seen from the spectral region between 1610 cm^{-1} and 1460 cm^{-1} shown in Figure 4.31 [31]. The asynchronous cross peaks along the diagonal line at about 1600 cm^{-1} and 1500 cm^{-1} reveal that these two bands are doublets: 1594 cm^{-1} and 1581 cm^{-1}, and 1508 cm^{-1} and 1497 cm^{-1}. The higher frequency components at 1594 cm^{-1} and 1508 cm^{-1} share cross peaks with the lower

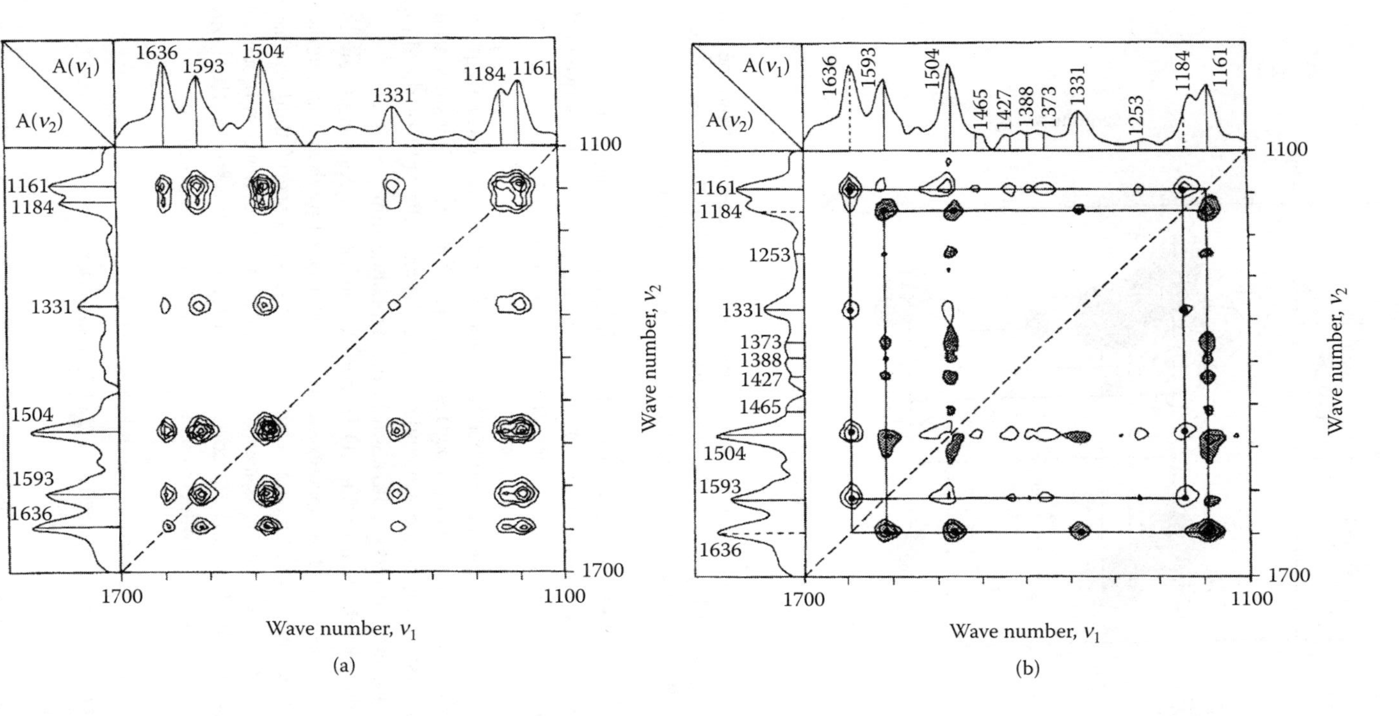

FIGURE 4.30 (a) Synchronous and (b) asynchronous 2D-IR correlation spectra of $HV^{\cdot+}$ generated from the time-resolved spectra shown in Figure 4.29. [Reproduced from M Osawa. *Bull Chem So Japan* 70: 2861–1880, 1997 with permission. Copyright (1997) Japan Chemical Society.]

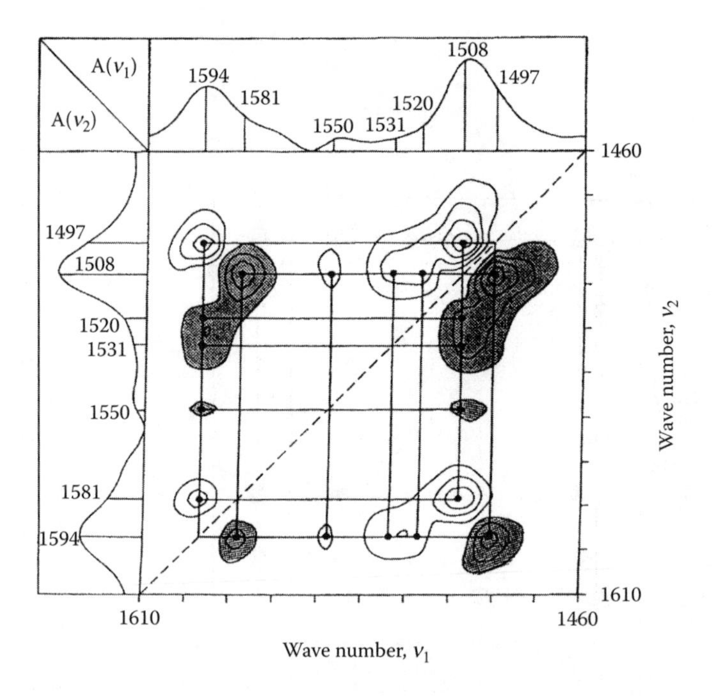

FIGURE 4.31 Asynchronous 2D correlation spectrum of $HV^{\bullet+}$ generated from the time-resolved spectra shown in Figure 4.29. [Reproduced from M Osawa. *Bull Chem So Japan* 70: 2861–1880, 1997 with permission. Copyright (1997) Japan Chemical Society.]

frequency components at 1581 cm^{-1} and 1497 cm^{-1}. On the other hand, no cross peaks are developed between the higher frequency components and between the lower frequency components. Therefore, these four bands are classified into two pairs by their characteristic time-dependent behavior. The sign of the cross peaks indicates that the higher frequency components appear before the lower frequency components. Since the vibrational properties of the $HV^{\bullet+}$ dimer are sensitive to the crystal structure (stacking of the molecules in the crystal), the split of the dimer bands may be ascribed to recrystallization in the deposited film.

The important conclusions in this 2D-IR correlation spectroscopy study are that IR signals arising from different species can be distinguished by their characteristic time-dependent behavior and that one can monitor the sequence of their intensity changes. This sort of information is useful for detailed analysis reaction processes. Ataka and Osawa [73] also used 2D correlation analysis to investigate IR spectra of complex surface reactions collected as a function of applied potential.

4.4.4 Application of Sample-Sample Correlation Spectroscopy and Statistical 2D Correlation Spectroscopy

Sample-sample correlation spectroscopy has been used to analyze temperature-dependent NIR spectra of oleic acid in the pure liquid state [84] and short-wave

NIR spectra of raw milk [94], and statistical 2D correlation spectroscopy has been employed to investigate IR spectra measured during polymerization of bis (hydroxyethl terephthalate) (BHET) [84]. Recently, it has been found that variable-variable and sample-sample generalized and statistical 2D correlation spectroscopies hold considerable promise in investigating chemical reactions [95]. In this section variable-variable and sample-sample statistical 2D correlation spectroscopy studies on the polymerization of BHET are outlined.

Šašic and Ozaki [84] applied statistical 2D correlation spectroscopy to ATR/IR spectra of the polycondensation reaction of BHET that yields poly(ethylene terephthalate) (PET) as a final product and one molecule of ethylene glycol (EG) per each condensation reaction. Figure 4.32a displays the 44 experimental spectra measured during the polymerization process [84]. EG is expelled out of the system, and the spectra contain very weak contribution from EG mainly due to the difference between the rates of EG production and evacuation. Figure 4.32b shows IR spectra of BHET, PET, and EG. The only available spectrum of PET was the spectrum of the cast film, which is not truly identical with the spectrum of PET produced during the reaction, and hence has limited usefulness.

Figure 4.33 illustrates the variable-variable correlation map generated from the spectra shown in Figure 4.32a [84]. Note that almost all the intensity changes perfectly correlate with each other. There is a regular alteration of the fields with figures 1 and –1, which refer to the variable regions where intensities increase or decrease. It can be seen from the correlation map that band intensity changes in the regions of 1170–1134, 1108–1073, and 1015–1009 cm^{-1} take place in the same direction and they occur in the opposite direction to the band intensity changes in the regions of 1129–1117, 1067–1038, 1023–1018, and 1005–947 cm^{-1}. The intensities in the former regions increase during the reaction while those in the latter regions decrease. The boundaries of the regions roughly correspond to the positions of the isosbestic points. There is, however, one exception for the variables in the 1038–1023 cm^{-1} region. The spectral variations in this region do not show high correlation with those in other regions. The weak correlation along the 1038–1023 cm^{-1} region may be due to the presence of EG bands that can contribute evidently only to the 1038–1023 cm^{-1} region. The EG bands disarrange the perfect correlation existing for all other variables. This result is in good agreement with that of principal component analysis (PCA) of mean-centered data, which reveals that the peak at 1030 cm^{-1} is very prominent in PC2.

Figure 4.34 depicts the sample-sample correlation map with correlation coefficients calculated from the spectra in Figure 4.32a [84]. In the present case, the correlation map has time axes. It seems that the concentrations of BHET and PET become equal around 80 min while the eventual influence of the EG concentration vector onto the correlation maps is hard to evaluate. The sample region of 65–85 min shows correlation with all other samples that have correlation coefficients less than 0.7 and higher than –0.7. The values of 0.7 and –0.7 were chosen arbitrarily to indicate noise-free lower boundaries of similarity between the spectra viewed as unit vectors. In general, for highly overlapped spectra sample-sample correlation maps are less sensitive than variable-variable correlation maps simply because the former are calculated from a highly similar array of numbers with no preferences on a given range, as is the case for the 1038–1023 cm^{-1} region in Figure 4.33. Samples at 65 min and

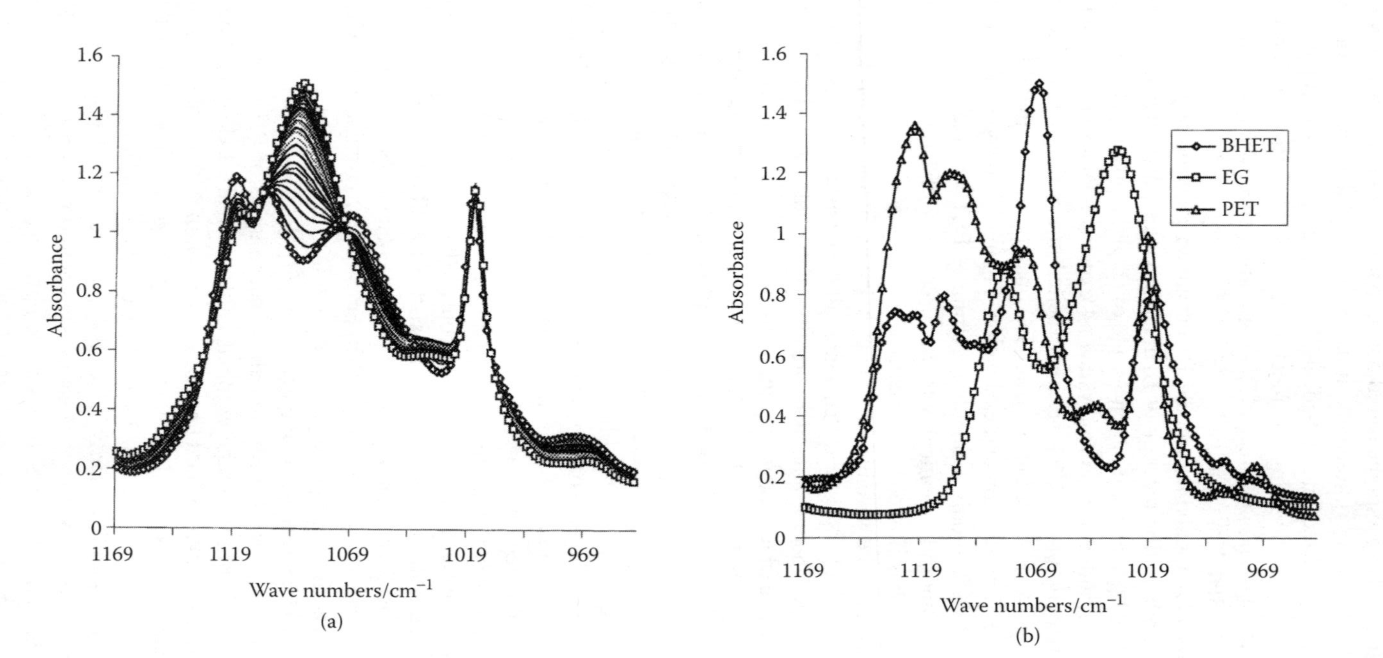

FIGURE 4.32 (a) The 44 experimental IR spectra measured on-line during the polymerization process of BHET. The first spectrum (diamonds) and the last spectrum (squares) are marked. (b) The spectra of BHET, PET, and EG. [Reproduced from Šašic, Y Ozaki. *Anal Chem,* May 2001 with permission. Copyright (2001) American Chemical Society.]

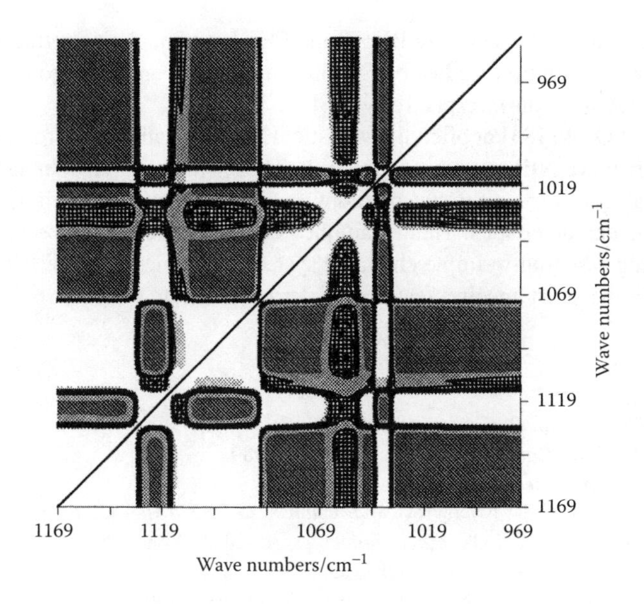

FIGURE 4.33 The variable-variable correlation map generated from the spectra shown in Figure 4.32a. White and light gray: correlation coefficients 0.6 to 1. Gray: correlation coefficients –0.6 to –1. Squares: correlation coefficients 0.4 to –0.4. [Reproduced from Šašic, Y Ozaki. *Anal Chem,* May 2001 with permission. Copyright (2001) American Chemical Society.]

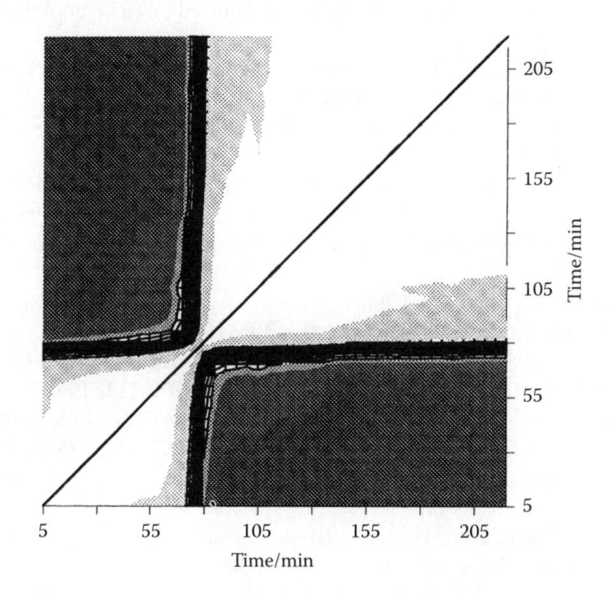

FIGURE 4.34 The sample-sample correlation map generated from the spectra shown in Figure 4.32a. White and light gray: correlation coefficients 0.7 to –1. Gray: correlation coefficients –0.7 to –1. Squares: correlation coefficients 0.6 to –0.6. [Reproduced from Šašic, Y Ozaki. *Anal Chem,* May 2001 with permission. Copyright (2001) American Chemical Society.]

80 min were found as those where BHET and PET have equal concentrations in self-modeling curve resolution studies by Simplisma (65 min) and orthogonal projection approach (OPA) (85 min), respectively [96].

Šašic and Ozaki [84] applied the statistical 2D correlation spectroscopy also to analyze short-wave NIR spectra of raw milk. These studies have demonstrated that the statistical 2D correlation spectroscopy successfully extract their spectral and concentration characteristics. This technique is a fast and reliable screening method for identifying essentially simple chemical systems regardless of their data structure and for outlining most specific features of the multivariate systems.

REFERENCES

1. I Noda. *Bull Am Phys Soc* 31: 520, 1986.
2. I Noda. *J Am Chem Soc* 111: 8116–8118, 1989.
3. I Noda. *Appl Spectrosc* 44: 550–561, 1990.
4. C Marcott, I Noda, A Dowrey. *Anal Chim Acta* 250: 131–143, 1991.
5. Y Ozaki, Y Liu, I Noda. *Macromolecules* 30: 2391–2399, 1997.
6. I Noda. *Appl Spectrosc* 47: 1329–1336, 1993.
7. I Noda, A E Dowrey, C Marcott, Y Ozaki, G M Story. *Appl Spectrosc* 54: 236A–248A, 2000.
8. I Noda, in *Handbook of Vibrational Spectroscopy*, P R Griffiths and J M Chalmers, Eds., Chichester: John Wiley & Sons, 2123, 2002.
9. C Marcott, A E Dowrey, I Noda. *Anal Chem* 66: 1065A–1075A, 1994.
10. I Noda, in *Modern Polymer Spectroscopy*, G Zerbi, ed., Weinheim: Wiley-VCH, 1–32, 1999.
11. I Noda, A E Dowrey, C Marcott. *Appl Spectrosc* 42: 203–216, 1988.
12. I Noda. *Chemtracts-Macromol Chem* 1: 89–106, 1990.
13. P A Palmer, C J Manning, J L Chao, I Noda, A E Dowrey, C Marcott. *Appl Spectrosc* 45: 12–17, 1991.
14. I Noda, A E Dowrey, C Marcott. *Appl Spectrosc* 47: 1317–1323, 1993.
15. C Marcott, A E Dowrey, I Noda. *Appl Spectrosc* 47: 1324–1328, 1993.
16. V G Gregoriou, J L Chao, H Toriumi, R A Palmer. *Chem Phys Lett* 179: 491–496, 1991.
17. R M Dittmer, J L Chao, R A Palmer, in *Photoacoustic and Photothermal Phenomena* III, O Bicanic, Ed., Berlin: Springer, 492–496, 1992.
18. R S Stein, M M Satkowski, I Noda, in *Polymer Blends, Solutions, and Interfaces*, I Noda and D N Rubingh, Eds., New York: Elsevier, 109–131, 1992.
19. T Nakano, T Yokoyama, H Toriumi. *Appl Spectrosc* 47: 1354–1366, 1993.
20. A E Dowrey, G G Hillebrand, I Noda, C Marcott. *SPIE* 1145: 156–157, 1989.
21. Y Ozaki, I Noda. *Two-Dimensional Correlation Spectroscopy*, New York: American Institute of Physics, 2000.
22. Y Ozaki, in *Handbook of Vibrational Spectroscopy*, P R Griffiths and J M Chalmers, Eds., Chichester: John Wiley & Sons, 2135, 2002.
23. Y Ozaki, I Noda, in *Encyclopedia of Analytical Chemistry*, R A Meyers, Ed., Chichester: John Wiley & Sons, 322–340, 2000.
24. M Muler, R Buchet, U P Fringeli. *J Phys Chem* 100: 10810–10825, 1996.
25. I Noda, Y Liu, Y Ozaki. *J Phys Chem* 100: 8665–8673, 1996.
26. I Noda, G M Story, A E Dowrey, R C Reeder, C Marcott. *Macromol Symp* 119: 1–13, 1997.

27. P Streeman. *Appl Spectrosc* 51: 1668–1678, 1997.
28. E Jiang, W J McCarthy, D L Drapcho, A Crocombe. *Appl Spectrosc* 51: 1736–1740, 1997.
29. N L Sefara, N P Magtoto, H H Richardson. *Appl Spectros* 51: 563–540, 1997.
30. A Nabet, M Pezolet. *Appl Spectrosc* 51: 466–469, 1997.
31. M Osawa. *Bull Chem So Japan* 70: 2861–1880, 1997.
32. M A Czarnecki, B Jordanov, S Okretic, H W Siesler. *Appl Spectrosc* 51: 1698–1702, 1997.
33. P Pancoska, J Kubelka, T A Keiderling. *Appl Spectrosc* 53: 655–664, 1999.
34. J Kubelka, P Pancoska, T A Keiderling. *Appl Spectrosc* 53: 666–671, 1999.
35. L Smeller, K Heremans. *Vib Spectrosc* 19: 375–378, 1999.
36. I Noda, G M Story, C Marcott. *Vib Spectrosc* 19: 461–465, 1999.
37. K Nakashima, Y Ren, T Nishioka, N Tsubahara, I Noda, Y Ozaki. *J Phys Chem* B 103: 6704–6712, 1999.
38. C Marcott, A E Dowrey, G M Story, I Noda, in *Two-Dimensional Correlation Spectroscopy*, Y Ozaki and I Noda, Eds., New York: American Institute of Physics, 77–84, 2000.
39. F Kimura, M Komatsu, T Kimura. *Appl Spectrosc* 54: 974–977, 2000.
40. N P Magtoto, N L Sefara, H Richardson. *Appl Spectrosc* 53: 178–183, 1999.
41. M Halttunen, J Tenhunen, T Saarinen, P Stenius. *Vib Spectrosc* 19: 261–269, 1999.
42. Y Nagasaki, T Yoshihara, Y Ozaki. *J Phys Chem* B 104: 2846–2852, 2000.
43. B Czarnik-Matusewicz, K Murayama, Y Wu, Y Ozaki. *J Phys Chem* B 104: 7803–7811, 2000.
44. M-J Paquet, M Auger, M Pezolet, in *Two-Dimensional Correlation Spectroscopy* , Y Ozaki and I Noda, Eds., New York: American Institute of Physics, 103–108, 2000.
45. M Sonoyama, T Nakano. *Appl Spectrosc* 54: 968–973, 2000.
46. I Noda, Y Liu, Y Ozaki, M A Czarnecki. *J Phys Chem* 99: 3068–3073, 1995.
47. Y Liu, Y Ozaki, I Noda. *J Phys Chem* 100: 7326–7332, 1996.
48. Y Ozaki, I Noda. *J. NIR Spectrosc* 4: 85–99, 1996.
49. Y Ozaki, Y Liu, I Noda. *Appl Spectrosc* 51: 526–535, 1997.
50. Y Wang, K Murayama, Y Myojo, R Tsenkova, N Hayashi, Y Ozaki. *J Phys Chem* B 34: 6655–6662, 1998.
51. M A Czarnecki, H Maeda, Y Ozaki, M Suzuki, M Iwahashi. *Appl Spectrosc* 52: 994–1000, 1998.
52. M A Czarnecki, H Maeda, Y Ozaki, M Suzuki, M Iwahashi. *J Phys Chem* 46: 9117–9123, 1998.
53. Y Ren, M Shimoyama, T Ninomiya, K Matsukawa, H Inoue, I Noda, Y Ozaki. *Appl Spectrosc* 53: 919–926, 1999.
54. Y Ren, T Murasaki, K Nakashima, I Noda, Y Ozaki. *J Phys Chem B*, 104, 679–690, 2000.
55. B Czarnik-Matusewicz, K Murayama, R Tsenkova, Y Ozaki. *Appl Spectrosc* 53: 1582–1594, 1999.
56. Y Ren, T Murakami T Nishioka, K Nakashima, I Noda, Y Ozaki. *J Phys Chem B* 104: 679–690, 2000.
57. Y Wu, B Czarnik-Matusewicz, K Murayama, Y Ozaki. *J Phys Chem B* 104: 5840–5847, 2000.
58. P Wu, H W Siesler, in *Two-Dimensional Correlation Spectroscopy* , Y Ozaki, I Noda, Eds., New York: American Institute of Physics, 18–30, 2000.
59. G Lachenal, R Buchet, Y Ren, Y Ozaki, in *Two-Dimensional Correlation Spectroscopy,* Y Ozaki and I Noda, Eds., New York: American Institute of Physics, 223–231, 2000.
60. K Murayama, B Czarnik-Matusewicz, Y Wu, R Tsenkova, Y Ozaki. *Appl Spectrosc* 54: 978–985, 2000.

61. K Ebihara, H Takahashi, I Noda. *Appl Spectrosc* 47: 1343–1344, 1993.
62. T L Gustafson, D L Morris, L A Huston, R M Butler, I Noda. *Time-Res Vib Spectrosc VI,* Springer Proc Phys 74: 131–135 1994.
63. I Noda, Y Liu, Y Ozaki. *J Phys Chem* 100: 8674–8680, 1996.
64. Y Ren, M Shimoyama, T Ninomiya, K Matsukawa, H Inoue, I Noda, Y Ozaki. *J Phys Chem. B* 103: 6475–6483, 1999.
65. Y Ren, T Murakami, K Nakashima, I Noda, Y Ozaki. *Appl Spectrosc* 53: 1582–1594, 1999.
66. M A Czarnecki, P Wu, H W Siesler. *Chem Phys Lett* 283: 326, 1998.
67. C P Schultz, H Fabian, H H Mantsch. *Biospectrosc* 4: 519–524, 1998.
68. A Matsushita, Y Ren, K Matsukawa, H Inoue, Y Minami, I Noda, Y Ozaki. *Vib Spectrosc* 24: 171–180, 2000.
69. Y Ren, A Matsushita, K Matsukawa, H Inoue, Y Minami, I Noda, Y Ozaki. *Vib Spectrosc* 23: 207–218, 2000.
70. Y Jung, B Czarnik-Matusewicz, Y Ozaki. *J Phys Chem B* 104: 7812–7817, 2000.
71. Y Ren, A Matsushita, K Matsukawa, H Inoue, Y Minami, I Noda, Y Ozaki, in *Two-Dimensional Correlation Spectroscopy*, Y. Ozaki and I. Noda, Eds., New York: American Institute of Physics, 250–253, 2000.
72. D L Elmore, R A Dluhy. *Appl Spectrosc* 54: 956–962, 2000.
73. K Ataka, M Osawa. *Langmuir* 14: 951–959, 1998.
74. A Nabet, M Auger, M Pezolet. *Appl Spectrosc* 54: 948–955, 1998.
75. P G H Kosters, A H B deVries, R P H Kooyman. *Appl Spectrosc* 54: 1659–1664, 2000.
76. S J Gadaleta, A Gericke, A L Boskey, R Mendelsohn. *Biospectrosc* 2: 353–359, 1996.
77. S Slobodan, J H Jiang, Y Ozaki. To be published.
78. N P Magtoto, N L Sefara, H H Richardson. *Appl Spectrosc* 53: 178–183, 1999.
79. K Murayama, Y Wu, B Czarnik-Matusewicz, Y Ozaki. J Phys Chem, in press.
80. Y Wu, K Murayama, Y Ozaki. *J Phys Chem*, B 105: 6251–6259, 2001.
81. Šašic, A Muszynski, Y Ozaki. *Appl Spectrosc* 55: 343–349, 2001.
82. Šašic, A Muszynski, Y Ozaki. *J Phys Chem A* 104: 63806387, 2000.
83. Šašic, A Muszynski, Y Ozaki. *J Phys Chem A* 104: 6388–6394, 2000.
84. Šašic, Y Ozaki. *Anal Chem,* 2294–2361, May 2001.
85. F B Barton II, D S Himmelsbach, J H Duckworth, M J Smith. *Appl Spectrosc* 46: 420–429, 1992.
86. M M Coleman, J F Graf, P C Painter. *Specific Interactions and the Miscibility of Polymer Blends*, Lancaster: Technomic Publishing Co., 221–308, 1991.
87. P S Stein. *Polym J* 17: 289–295, 1985.
88. S Browmlow, J H M Cabral, R Cooper, D R Flower, S J Yewdall, I Polikarpor, A C T North, L Sawyer. *Structure* 5: 481–489, 1997.
89. X L Qi, S Brownlow, C Holt, P Sellers. *Biochim Biophys Acta* 1248: 43–48, 1995.
90. A Dong, J Matsuura, S D Allison, E Chresman, M C Manning, J F Carpenter. *Biochemistry* 35: 1450–1456, 1996.
91. A F Allain, P Paquim, M Subirade. *Int J Biol Macromal* 26: 337–343, 1999.
92. G Panick, R Malessa, R Winter. *Biochemistry* 38: 6512–6518, 1999.
93. D L Elmore, R A Dluhy. *Coll Surf A* 171: 225–230, 2000.
94. Šašic, Y Ozaki. *Appl Spectrosc* 55: 163–172, 2001.
95. Šašic, J-H Jiang, H Sato, Y Ozaki. *Analyst* 128: 1097–1103 (2003).
96. Šašic, T Amari, H W Siesler, Y Ozaki. *Appl Spectrosc*, in press.

5 Raman and Mid-Infrared Microspectroscopic Imaging

Rohit Bhargava, Michael D. Schaeberle, and Ira W. Levin

CONTENTS

5.1 INTRODUCTION

Raman and infrared spectroscopy have provided molecular characterizations of complex assemblies spanning the chemical, physical, and biological sciences. Over the past forty years, the development of powerful vibrational microscopy techniques has evolved into the ability to reconstruct a sample's image to allow the visualization of the spatial distribution of chemical components across a specific sample area. Images can be reconstructed from two basic instrumental approaches: in the first configuration, the image, employing a single detector, is assembled from specific, small spatial areas that are relatively coarsely spaced across the sample from which spectral information is acquired. These methods are typically termed point spectroscopic mapping. In the second configuration, spectroscopic information using multiple detectors is acquired over a portion of the observed region of interest defined by an optical image. These techniques are closer to conventional optical, bright-field imaging methods and involve either line scanning or the more recent wide-field, global imaging processes. Since the visualization is accompanied by appropriate magnification for examining a specimen's microscopic structure, these spectroscopic image reconstruction techniques are also variously termed vibrational microscopy, vibrational imaging, or microspectroscopic imaging techniques.[1] Since current instrumentation has advanced to the point where it is nearly as efficient to collect a spectroscopic imaging data set as it is to collect spectra at only several points within a sample of interest,[2] we emphasize in this discussion an examination of the technology and applications of vibrational spectroscopic imaging involving spatially resolved spectral measurements across large sections of the optical field of view. The chapter is divided into two main sections, discussing in detail approaches in which spatially resolved Raman and infrared spectroscopic data sets are acquired.

Since large numbers of vibrational spectra, ranging from several thousand to many million, are acquired in single spectroscopic imaging data sets, data storage, analysis, and representation differs from data recorded by either conventional vibrational spectroscopy or optical imaging methods. Unlike classical spectroscopy, where the recorded data are simply one-dimensional intensity distributions as a function of wavelength (λ), or represented as in optical imaging, in which intensity data are specified over a two-dimensional (x,y) spatial location, vibrational spectroscopic imaging data consist of intensity values that are a function of three dimensions, λ

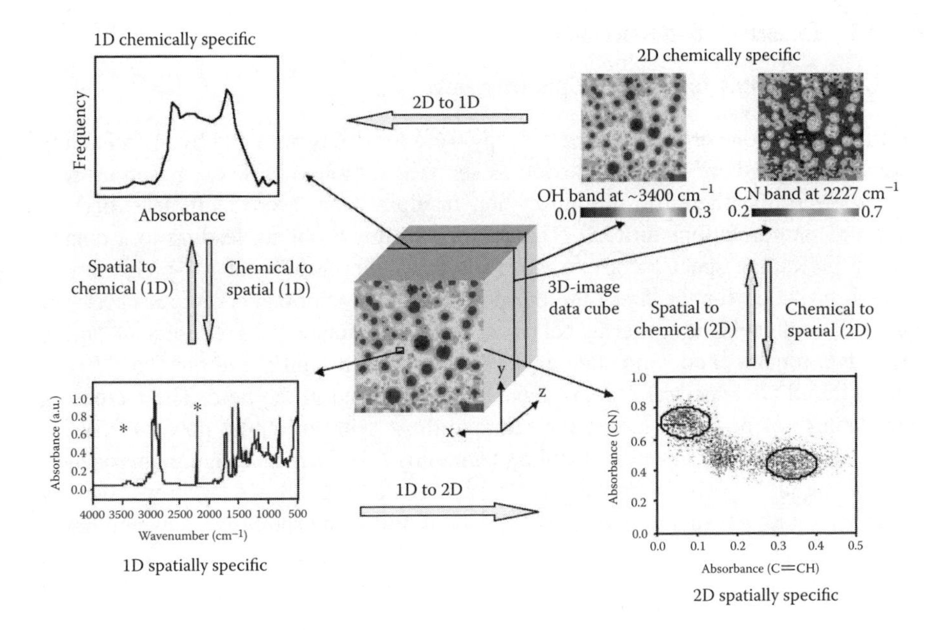

FIGURE 5.1 Visualizations afforded by an imaging data set. (Reproduced from R Bhargava, S-Q Wang, JL Koenig, *Adv. Polym. Sci.*, 163: 137, 2003)

and an *x,y* spatial location, as illustrated in Figure 5.1. Clearly, the entire three-dimensional data set shown in the middle of the figure consists of all the acquired information, but is difficult to visualize in its entirety. By further extracting data from one or two of the data set's three dimensions, relevant information may be obtained. For example, sections of the data set may be visualized either as images at given wavelengths (top, right) or spectra reflecting the chemical composition at specific coordinate locations (bottom, left). A number of other statistical analyses, as, for example, histogram analyses illustrating specific absorbance distributions (top, left) can be constructed from the large number of spectral values accessible from a single data set.

5.2 RAMAN SPECTROSCOPIC IMAGING

Although a wide variety of instrumental configurations are employed for Raman spectroscopic imaging, instrumentation can generally be divided into three main categories based on the mechanism of image formation: (1) scanning techniques,[3–23] (2) image reconstruction methods,[24–32] and (3) global, wide-field imaging strategies.[1,33–59] We will discuss instrumentation reflecting the predominant Raman imaging methods within each category and, using representative examples, will present the advantages and limitations of each method. Since many technological advances have contributed to the feasibility of Raman spectroscopic imaging, it is useful first to discuss some of these enabling technologies.

5.2.1 Enabling Technologies

5.2.1.1 Sources for Raman Spectroscopy

Early applications of Raman spectroscopy were limited historically by the relatively weak broadband arc lamps employed as sources. Although these excitation sources were spectrally filtered to minimize their bandpass, they were inefficient, making spectral interpretations difficult. The solution to this problem, leading to a renaissance in Raman spectroscopy, occurred with the application of gas-phase lasers, which provided for the first time the powerful, monochromatic sources necessary for this weak inelastic scattering behavior. Since then, many types of lasers, including ion, dye, tunable, and solid state devices, and ranging from the ultraviolet (UV) to the infrared (IR) spectral regions, have been employed in the field. The increasing availability of powerful lasers, for example those with an output power of 5W or greater, has been an essential enabling technology for Raman imaging approaches, where the laser power is spread over a large spatial area. Since the power density decreases with the square of the radius of the illuminating spot, high-powered lasers become essential for acquiring images within practical time periods. The power density implications for a variety of techniques have been examined and will be discussed in more detail later in this section.[60–62]

5.2.1.2 Rayleigh Line Rejection

Initial Raman spectroscopy experiments employed either a double or a triple grating monochromator to discriminate Rayleigh scattered laser excitation from the much weaker Raman signal. While this method was effective in suppressing the Rayleigh scattered light, the inherently low throughput of multiple grating monochromators severely limited Raman scattered radiation reaching the detector, therefore leading to time-consuming experiments. Long acquisition times rendered imaging experiments on early instruments impractical, as many days to weeks would have been required to collect the hundreds of individual sampling points needed to reconstruct images at each wavelength. Introduction of wavelength rejection optics, such as dielectric interference and holographic notch filters and beamsplitters, into the excitation/emission radiation paths have allowed incorporation of relatively high throughput single stage dispersive spectrometers. Although dielectric beamsplitters and long pass filters were adequate, their broad cut-off transitions made it difficult to record low wavenumber data (< 400 cm^{-1}) because of the large contribution from scattered Rayleigh radiation in that spectral region.[63] Holographic optics, an alternative to dielectric interference optics, were first employed in the early 1990s. A holographic beamsplitter and notch rejection filter combination later replaced the dielectric interference filter assemblies in many Raman imaging systems and allowed researchers to approach more closely the Rayleigh line than had been previously feasible.[64,65] Recently, the holographic beamsplitter directing excitation radiation toward the sample was replaced by a holographic notch rejection filter (HNRF), which allows data to be acquired within 40–70 cm^{-1} of the Rayleigh line.[64–66] In this case, the HNRF acts as a "mirror" that rejects both the initial laser illumination and the Rayleigh scattered radiation directed toward the microscope, while transmitting the Raman emission.

5.2.1.3 Microscopes

Micro-Raman gained a large impetus with the introduction of the infinity-corrected optical microscope. This new microscope system, unlike conventional fixed tube length models, focuses at infinity, allowing the collimated beam path to be expanded to accommodate the additional optics required for imaging, such as HNRFs, spatial encoding masks, or tunable filters, without degrading the quality of the image. This property allows the microscope system to be adaptable and modular, highly desirable features incorporated into many of the commercially available Raman micro-spectroscopy and imaging instruments.

5.2.1.4 Detectors

Detector technology for Raman imaging has made large advances primarily due to the introduction of silicon charge-coupled device (CCD) detectors[67–69] that provide the sensitive, low noise cameras required for Raman imaging. While the transition from a single point detector to a multichannel detector is required for many of the Raman imaging approaches, the ability to operate in a manner similar to sensitive conventional detectors is not lost. By properly selecting the readout mode, the two-dimensional array can capture spectral information, either as a single spectrum or multiple spectra, as well as images. The ability to employ the same detector for both imaging and macroscopic spectroscopy creates a flexible system that offers instrument manufacturers and users an array of imaging modalities to choose from.

5.2.1.5 Data Handling Systems

An unavoidable consequence of employing rapid multichannel detection is the large size of the data files generated. Typical spectral imaging files can range in size from a few megabytes to several hundred megabytes. This magnitude of data requires high-end computers to process, manipulate, and store. With the advances in computer technology, performance levels have soared and prices have been drastically reduced for these components, making Raman microspectroscopy less expensive and more accessible. An additional area for major advances in recent years is the development of numerical analysis routines[70–72] that benefit from the increased computational power available.

5.2.2 Point Mapping

5.2.2.1 Instrumentation

The simplest Raman microspectroscopic instrumentation, as shown in Figure 5.2a, is the point mapping or point scanning system.[9] Radiation from a laser is focused to define a small spot at the sample. An optical element then gathers, collimates, and transmits the scattered radiation through a filtering optic (typically one or more HNRFs) to remove the Rayleigh scattering contribution, whereupon the filtered radiation is focused onto the slit of a dispersive spectrometer. The spectrum is

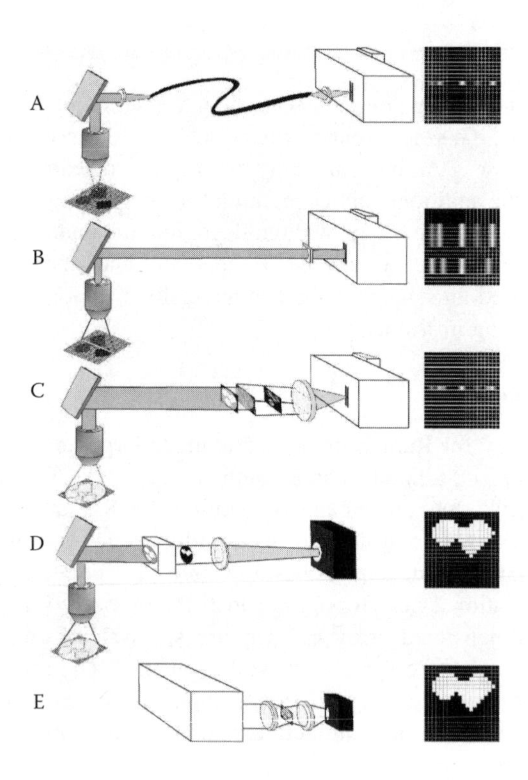

FIGURE 5.2 Various modes of acquiring spatially resolved Raman spectroscopic data include (a) point mapping, (b) line-scan imaging, (c) Hadamard transform imaging, (d) wide-field imaging, and (e) macroscopic imaging. (Reproduced from MD Schaeberle, EN Lewis, IW Levin, *Biological Vibrational Spectroscopic Imaging in Infrared and Raman Spectroscopy of Biological Materials*, H-U Gremlich and B Yan (Eds.), New York, Marcel Dekker, Inc., 2001.)

recorded by either scanning the dispersed wavelengths past a second slit in front of a photomultiplier tube or dispersing the radiation onto a multichannel detector. The result, similar to conventional macroscopic measurements, is a Raman spectrum describing the specific spatial location (x_i,y_i). A motorized sample stage is employed to scan the sample in both the x and y directions and to record a spectrum from every desired location. While a single spectrum provides only a single point from the sample, the series of recorded spectra, along with the spatial locations from where they were collected, can be used to construct a coarse-grained, spatially resolved spectral data set.

A limitation of point mapping conducted in this manner is the lack of specificity of the area being sampled, as the excitation laser beam wake forms a shape similar to an hourglass in the sample. Hence, instead of a single point, dispersed points in the sample above or below the focal plane also contribute to the spectrum collected from a given location. One method to minimize this out-of-focus contribution is a confocal approach, which improves the axial, or depth, resolution of the technique. In this configuration, at least two additional lenses and a pinhole aperture are incorporated into the emission path prior to the detector.[3–6] These additional components limit the depth of field sampled during the experiment by bringing the scattered radiation to an intermediate focus at the aperture. Scattered radiation originating from within the defined depth of field, the "in focus" plane, passes through the aperture's pinhole, while that from regions above or below the plane are almost entirely rejected. The depth of focus and axial resolution can be adjusted simply by changing the size of the pinhole. A lateral resolution of 2 μm and a depth resolution of 4 μm are possible.[9] The spatial resolution can be further improved to 100–200 nm by operating in the near field.[73]

Another confocal approach, which does not require the traditional pinhole aperture, has also been described.[15] In this configuration, two perpendicularly oriented one-dimensional slits, one physical and one virtual, replace the pinhole aperture. The spectrometer's entrance slit is set to a small value, typically 15 μm,[15] and acts as the first slit. The Raman scattered radiation passing through the spectrometer slit is dispersed by the grating and acquired by a multichannel detector. A virtual slit, oriented perpendicular to the spectrometer grating's dispersion direction, is generated electronically by restricting the readout of the multichannel detector to a section of pixels with the highest signal to noise ratio (SNR). In this manner, only the light originating from a given sampling volume is able to reach the area of the detector that contributes to the spectral signal. While confocal point mapping does improve the axial resolution at each point, a large fraction of the Raman scattered radiation is not utilized, decreasing the observed signal and leading to increased experimental times for each sampling point.

5.2.2.2 Applications

Simple instrumentation and low equipment costs make point mapping, whether conventional or confocal, an attractive method for Raman imaging. For example, in the study of biological and polymeric samples Raman analysis can be readily incorporated into existing optical microscopy instrumentation already employed for other techniques. Point mapping has been used to provide insight and understanding into the mechanisms of diffusion in the adhesion materials used in dental composites[16] and bones and to measure a sample's mineral, organic matrix, and protein content.[17] Raman spectroscopy has been employed to examine polymer properties, for example, stress transfer in epoxy composite materials reinforced with high performance polyethlene fibers,[11] domain sizes, relative concentrations of polymer components within domains, distribution of silica fillers and curing agents within polymer blends as functions of both blend composition and history of aging treatments.[8] While standard mapping techniques have been sufficient for several biological and polymeric studies, many complex applications require the improved axial resolution of confocal methodologies. For example, polarized confocal Raman point mapping has been used to qualitatively investigate the molecular orientation of α- and β- transcrystalline regions around poly(ethyleneterephthalate) (PET) fibers embedded in a polypropylene matrix[6] and the microstructure of a two-phase blended polymer.[7] We do not discuss further applications of this methodology as this chapter emphasizes imaging techniques. The interested reader is referred elsewhere.[74]

5.2.3 Line-Scan Mapping

5.2.3.1 Instrumentation

Line scanning, illustrated in Figure 5.2b, is similar to point mapping, except that it employs, for example, either a cylindrical lens,[13,14] a Powell lens,[18–20] or a scanned laser spot[3,12,21] to elongate the excitation laser beam in one dimension to create an illumination line at the sample. The instrumentation is engineered such that the resulting emission line collected by the microscope objective is oriented parallel to the entrance slit of the spectrometer. A line scan system requires a two dimensional detector, such as a CCD detector, to achieve spectral imaging. One dimension of the detector chip is used to record one spatial dimension (the position along the illumination line), while the other is used to capture the dispersed spectrum. Hence, multiple spatial and spectral components are simultaneously recorded. Data from the entire area of interest is acquired by sequentially moving the sample perpendicular to the linear illumination. The advantages and limitations of the various imaging configurations presented here[60–62] depend on a variety of criteria, including the damage threshold of the sample being examined, the achievable laser power, and the time available for data acquisition. Raman line scanning methods offer several advantages over conventional point mapping techniques, specifically including the multichannel detection advantage resulting in lower acquisition times per spatial point and the need to move the sample in only one dimension. Line scanning, however, disperses the laser power over a larger linear area, decreasing the power density. While increasing the power of the laser can correct for the power density loss, this approach is not feasible for some samples, since, for example, biological tissues and many polymeric systems experience photodegradation, suffer thermal damage, or undergo chemical changes at high laser power densities. While the spatial resolution perpendicular to the illumination line remains dependent on the width of the line and the size of the sampling step, the resolution in the dimension along the line is theoretically limited only by diffraction. Practically, the resolution in this dimension depends on the size of the CCD pixels and the magnification of the system, with typically values ranging from 1 μm to 3 μm.[12]

Confocal methodologies, similar to those employed in point mapping, have also been proposed for line scan imaging.[3,12] Current instrumentation employs various implementations of scanning mirrors and apertures to achieve this modality. These methods not only are capable of providing an improved lateral spatial resolution (0.5 μm),[3] but also are capable of achieving an axial resolution of 3 μm,[12] which localizes an analysis to a specific specimen plane.

Although line scanning has several advantages compared to point mapping, it is still plagued by many of the same problems. For example, the need for mechanical movement, whether it is the sample stage or the beam steering optics, can lead to image misregistration and further degrade the spatial resolution of the technique. An additional limitation of line scanning lines arises from the possibility of differing spatial resolution along each axis (x and y) of the image. While this may be advantageous in some experimental studies, it generally renders image interpretation less intuitive. The advantages arising from multichannel detection of the spatial signal compared to point mapping are important, however, as data acquisition times are considerably reduced.

5.2.3.2 Applications

Raman hyperspectral line imaging, often in conjunction with various multivariate processing algorithms,[75,76] has been employed in a variety of studies including model systems,[14,18] polymers,[3,10,12] and biological materials.[19,20,77] Here we discuss some representative examples, emphasizing the insight that may be gained from Raman line-scan mapping in these systems. In polymeric systems, such data are typically employed for correlating the microstructure of a solid polymer or a blend to the material's properties. For example, the crystallinity of syndiotactic polystyrene (sPS)[10] in an injection molded, dumbbell-shaped thin slice (3 mm × 1.5 mm × 0.2 mm) was sectioned perpendicular to the injection flow direction and examined. The relative distributions of crystalline and amorphous sPS could be extracted from the data by employing a factor analysis method.[10] However, the information content from Raman microscopy may be limited. For example, while Raman imaging was able to distinguish local variations in crystallinity, the technique could not be used to quantitate the crystalline content without measuring independently verified crystallinity standards.[10]

Similarly, typical biological analyses using Raman line-scanning involve spatial visualizations of the spectroscopic responses to structural change, as for example, detecting fatigue related microdamage in bone tissue[19,20,77] that may affect the mechanical integrity leading to age-related bone degeneration. In a study employing Powell lens illumination in conjunction with factor analysis algorithms, three distinct regions of tissue, observed with 1.4 μm spatial resolution and originally identified as visibly undamaged, contained microcracks and possessed microdamage.[20] It was discovered that changes occurring in the phosphate ν_1 spectral band of damaged tissue were indicative of different mineral species present in the bone. In undamaged areas, the band appears at approximately 957 cm^{-1}, as expected for carbonated hydroxyapatite bone mineral. In regions of visible microdamage, an additional ν_1 band was present at 963 cm^{-1} and was assigned to the more stoichiometric, less carbonated mineral.[20] Such results allow correlations between microscopic structural development and macroscopic properties. Clearly, the insight gained from simple microscopy techniques will not necessarily provide structural information, while macroscopic Raman spectroscopic imaging techniques may not be sensitive enough to allow localized determinations.

Confocal methodologies, similar to those employed in point mapping, have also been proposed for line imaging.[3,12] These methods provide not only an improved lateral spatial resolution, but also allow analyses to be localized to specific specimen planes. For example, a confocal Raman line imaging system designed for biological applications was able to achieve a lateral resolution of 0.5 μm and an axial resolution of 3.5 μm by retrofitting a conventional confocal microscope with a scanning mirror, slit, and two cylindrical lenses.[3] The system was used to assess the crystallinity across a bone implant interface. In this case, the implant was coated with highly crystalline calcium phosphate apatite. The crystallinity difference between the implant (higher) and the bone (lower) is expected to cause decreased bone bonding and, therefore, lead to an unstable implant.[3] The system was also used to image a polytene chromosome, namely, a large interphase chromosome where the DNA and amino acid content of the band and interband regions were visualized. While variations in content between the two areas were observed, the ratio of DNA to amino acid was found to remain relatively constant.[3] For polymers,[12] a confocal system provides three-dimensional compositional and morphological heterogeneities in polymer blends[12] and the distribution of species in evolving materials as, for example, following unreacted free melamine in cured melamine-formaldehyde resins.[12]

5.2.4 Hadamard Imaging

5.2.4.1 Instrumentation

Image reconstruction methods, such as Hadamard transform (HT) imaging, shown in Figure 5.2c, demonstrate a departure from the conventional point mapping and line scanning strategies. In this approach, epi-illumination is employed to globally excite all the points within the sample area simultaneously.[24–27] The Raman scattered radiation is collimated and directed to a two-dimensional mask, which selectively transmits data to a two-dimensional detector recording one spatial and one spectral dimension, similar to a line scan experiment. For an $N \times N$ matrix, this experimental procedure is repeated $N^2 - 1$ times, each time with a new mask generated by changing the pattern of open (clear) and closed (opaque) areas of the mask according to an HT encoding sequence. The result is a series of simultaneous linear equations that, using an inverse HT, can be used to deconvolve the spectrum at each point in the image. By collecting radiation from many spatial locations simultaneously, HT methods benefit from a multiplexed advantage that is not achievable with standard point mapping or line scanning approaches. However, light rejection at the mask is still substantial and the application of multiple masks remains cumbersome.

Many differing technologies have been used to create masks for HT imaging. Originally, photolithography was used to generate mechanically movable, two-dimensional or rotating masks.[24–29] While this type of mask was adequate, the spatial resolution was limited by the quality of the photolithography process and misregistration errors resulting from the mechanical nature of the mask movement. In an attempt to remove the mechanical errors, a solid-state approach to mask generation employing liquid crystal spatial light modulators (LC-SLMs) was investigated.[30] While the LC-SLM removes the image misregistration errors caused by physically moving the mechanical mask, it is not an ideal technology for HT spectroscopic imaging. The system experiences loss of fidelity and a decrease in the signal-to-noise ratio (SNR) due to light leakage in the closed elements of the LC-SLM and less than 100% transmission in the open elements, respectively. While the SNR can be increased by multiplexing elements, it is still impossible to achieve the truly binary condition (on/off) required for an ideal HT mask, limiting the usefulness and acceptance of the LC-SLM. Recently, digital micromirror arrays (DMAs), originally developed for use in high definition projection systems, are being employed as masks.[31,32] The highly reflective aluminum micromirrors are mounted directly on top of the integrated circuits used to control their positions. DMAs exhibit true binary operation, fast element switching speed (on/off), reliable positioning, variable mask dimensions, and high radiation tolerance, and currently represent the greatest potential for a commercialization of HT spectrometers and imaging systems.

5.2.4.2 Applications

While HT mapping has been successfully implemented in other spectroscopic[78–80] and imaging[81–83] approaches, its application to Raman imaging has, generally, been limited to technology development and academic studies. Many of the studies focus on evaluating mask technology and testing data acquisition/processing algorithms. A variety of samples, such as polymer laminates,[27] titanium dioxide,[27] and naphthalene,[24,31,32] have been employed in these pilot studies to assess the advantages and limitations of these new approaches. In addition, an HT method for obtaining three-dimensional confocal Raman images has also been presented in the literature.[84]

5.2.5 Fiber-Bundle Image Compression

5.2.5.1 Instrumentation

A relatively new imaging methodology that utilizes a fiber optic bundle in the emission path as a bridge between the microscope and the spectrometer is termed fiber-bundle image compression (FIC) and is shown schematically in Figure 5.2d. The sample is globally illuminated and the scattered radiation is collected by the objective and focused via a lens onto the collection end of a coherent fiber optic imaging bundle, which consists of fibers closely packed into a circular cross-section,[35–37,55] while the distal end of the fiber optic has a line configuration. The linear array of fibers acts as the entrance slit when inserted into a spectrometer. A two-dimensional CCD is used to image the dispersed spectra from each of the fibers in the linear array simultaneously. The novel innovation in FIC imaging is that the spatial information in both the x and y dimensions is maintained by mapping the position of each fiber within the collection spot to a corresponding position in the linear array. By utilizing this encoding scheme, the two-dimensional array of spatial information is transformed into a one-dimensional linear array and allows measurement of two spatial dimensions and one spectral dimension in a single acquisition step. The capture of an entire imaging data set in a single camera acquisition provides data collection speeds that are unparalleled in conventional Raman imaging techniques, which require that either the sample or the laser be raster scanned or that the sequential tuning of an imaging filter be implemented (*vide infra*) in order to collect an equivalent data set. The speed also allows large sample areas to be imaged by employing a rastering scheme similar to point mapping, except that in this case, each sampled point consists of multiple (typically 100) individual points. The system is also capable of working in a "survey and zoom" mode, where a low magnification objective is used to scan a large area of the sample at low spatial resolution, and specific small areas are examined at higher spatial resolution using objectives of greater magnification.[37] These features provide FIC with an efficient means to perform Raman hyperspectral imaging on a time scale required by the biomedical and process control communities for real-time analysis.

5.2.5.2 Applications

Several studies demonstrating the feasibility and functionality of FIC Raman imaging have been performed.[35–37,55] Examples from a preliminary study are shown in Figure 5.3 demonstrating Raman FIC's chemical imaging capability for yielding molecular specific images. In this case, individual particles in a mixture of sucrose and D(-)- fructose sugars,[55] similar in appearance in the optical brightfield image (Figure 5.3a), are distinguished by their characteristic spectral features using a spectral angle mapping (SAM) algorithm.[85] In another study, three amino acids — leucine, isoleucine and valine, which vary only slightly in their chemical structure — were imaged.[37] As in the previous example, it is impossible to identify unambiguously the crystals using only brightfield microscopy. These examples demonstrate that the Raman FIC technique has the ability to generate chemically significant images, even from samples with substantial spectral similarities. The rapid data acquisition and portable nature of such an instrument has potential for use in process control environments.

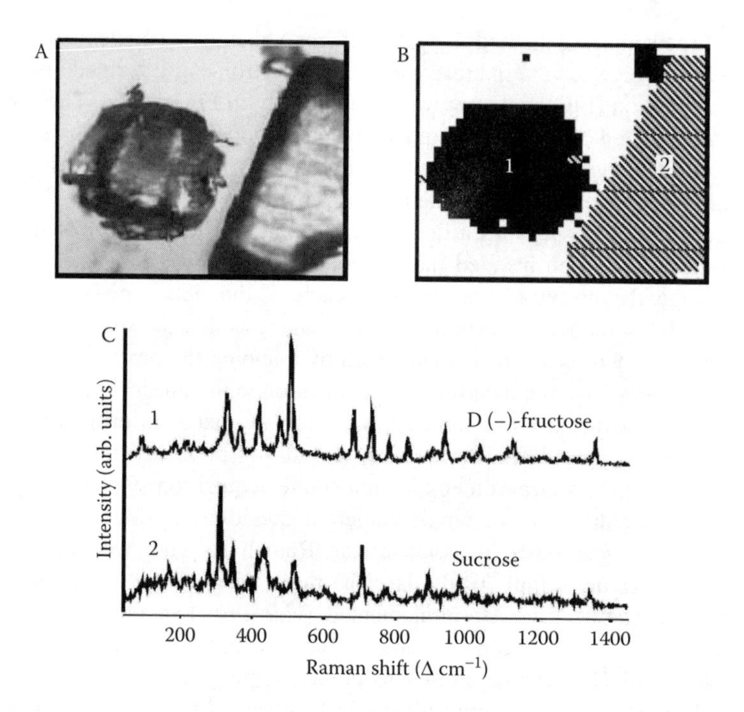

FIGURE 5.3 An example of FIC Raman imaging distinguishing two types of sugars. (Reproduced from AD Gift, J Ma, KS Haber, BL McClain, D Ben-Amotz, *J. Raman Spectrosc.* 30: 757–765, 1999.)

5.2.6 Tunable Filter Imaging

5.2.6.1 Instrumentation

Spectroscopic imaging can also be performed without the use of a conventional dispersive spectrometer by employing a tunable imaging spectrometer.[86] Rotating dielectric filters,[56–58] fixed filter/tunable source,[38,62] acousto-optic tunable filters (AOTFs),[39–46] and liquid crystal tunable filters (LCTFs), based on Fabry-Perot[47] or Lyot/Evans geometries[1,48–54] have all been employed in a variety of Raman hyper-spectral imaging applications. Although the tunable systems' operations are based on different physical or optical interactions, the generalized technique of tunable filter imaging is similar, as shown in Figure 5.2e. Laser excitation is dispersed over a large area of the sample (typically a 200 μm or larger diameter), which allows all points to be excited simultaneously. An image of scattered intensity is filtered by the tunable spectrometer, allowing only a limited spectral bandpass to impinge upon a two-dimensional multichannel detector (CCD) for recording intensity values. The entire spectral dimension is recorded sequentially by tuning the system to specified wavenumber values and collecting an image at each radiation wavelength. This methodology is closest to true imaging in Raman spectroscopy and provides flexibility in the wavelength range, size of image, and time of image acquisition.

5.2.6.1.1 Rotating Dielectric Filters

The rotating dielectric filter[15,57,58] is conceptually the simplest of filtering methods, as it contains no dispersive elements and is capable of providing both spatial and spectral information with a high radiation throughput and inexpensive imaging instrumentation. Based on the average refractive index of the multilayer dielectric filter, the angle between the incident radiation and the normal to the filter can be mathematically equated to the passband wavelength.[57] Since a single dielectric filter covers only a range of ~600–800 cm^{-1}, a series of filters is required to cover a useful Raman spectral region, as for example 3200–400 cm^{-1}. Compared to previously described dispersive techniques, the spectral resolution achieved using dielectric bandpass filters is poor (typically ~20 cm^{-1}).[57] It may be possible to improve the attainable spectral resolution by sacrificing throughput and limiting the achievable overall spectral range of the filter.[57] The practical spatial resolution of the technique when used in conjunction with a longer wavelength laser is about 1 μm, which is not diffraction limited but is adequate for many Raman imaging applications. As with mapping and some Hadamard masking techniques, moving parts of the rotating dielectric filters can cause image misregistration, leading to a degradation of the spatial resolution. In addition, because the circular aperture of the dielectric filter is rotated to become (in projection) an ellipse, the field of view obtained during an experiment varies, increasing the difficulty in data interpretation and image analyses.

5.2.6.1.2 Tunable Laser Systems

Another implementation of narrow bandpass dielectric filters employs a single fixed wavelength filter and tunes the laser source.[38,62] Raman images are focused either onto a fiber optic imaging bundle for delivery to a remote CCD[38] or directly onto a CCD.[62] Using this method, a fixed wavelength filter (or set of filters) is used to select an interval of the spectrum. An image at a particular wavenumber shift is recorded by tuning a near infrared (NIR)[38] or visible[62] laser such that the wavenumber displacement of interest coincides with the interval selected by the dielectric filter. For example, a dye laser tunable from 630–690 nm was used in conjunction with either a 711 nm or 735 nm dielectric filter.[62] The transmission bandwidth of each of the filters was 0.5 nm, which corresponds to ~10 cm^{-1} over the spectral range of the imaging system and provides a spectral resolution sufficient for many applications. The spatial resolution achievable is close to the theoretical diffraction limit. The tunable source, however, adds an additional degree of complexity to the system and increases experimental times considerably. For example, because of the broad tuning range of the source, a single holographic notch rejection filter is not sufficient, and a number of such filters are required, leading to high transmission losses. A dichroic beamsplitter with a sufficiently high cutoff wavelength is also required. The spectral range of such instruments is typically limited by the range of the laser and the properties of the filter. Typically, a range of ~1350 cm^{-1} is available, which can be expanded with the use of additional fixed wavelength dielectric filters.[62] Economic and logistical drawbacks of the system, including the high purchase cost and requirement for trained personnel, limit this method's feasibility mainly to academic environments.

5.2.6.1.3 Solid State Filters

Solid state, electronically tunable filters have also been employed for hyperspectral Raman imaging. For example, a liquid crystal filled dual-stage Fabry-Perot filter (DFPF) with a ~10 cm^{-1} bandpass has been reported.[47] In a typical Fabry-Perot interferometer, two partially transparent optical surfaces are separated by a dielectric spacer, creating a resonant cavity for light filtration. Typically, the interferometer is tuned by varying the refractive index of the dielectric medium or by changing the cavity spacing. Nematic liquid crystals (LCs) are employed as the filling material. A method of electronically tuning the DFPF is achieved by controlling the optical properties of the LC by electro-rheological orientation without inducing any deformations or translations. With appropriate coatings on the optics, the DFPF can be made to operate over the 3000–400 nm range. In conjunction with its spectral resolution of ~10 cm^{-1}, and only a slight deterioration in spatial resolution over a standard imaging system coupled to a relatively high throughput of 60% to 70%, the DFPF is an excellent candidate for Raman imaging.[47] The free spectral range of the DFPF, however, is less than the Stokes-shifted Raman region for many organic compounds, which may limit the device's applicability. In addition, high- and low-pass interference filters may be required to further isolate a single passband interval, further reducing the spectral range of the overall system. A critical limitation is the requirement for strict temperature controls to maintain constant optical properties of the DFPFs. For example, a change of 0.1°C typically introduces an estimated instability in the filter's spectral calibration of ±10 cm^{-1}, severely limiting maintenance-free long term or harsh environmental applications.[47] The AOTF, with its random accessibility, solid state design, high throughput, and fast switching speeds (~1 μsec), has many characteristics that make it a good candidate for Raman imaging. During operation, radio frequencies applied to the piezoelectric tranducer attached at one end of a birefringent crystal are converted into a traveling acoustic wave that moves laterally through the crystal. An acoustic absorber bonded to the distal end of the crystal minimizes reflections that may disturb the primary acoustic waveform. Light incident on the crystal is diffracted by the acoustic pressure wave as it propagates through the phase grating, providing a narrow bandpass of wavelengths diffracted from the crystal at a fixed orientation into both polarization orders. The (+) first-order polarization exits the crystal along the same vector as the propagation of incident light. Changing the frequency applied to the transducer causes light of different wavelengths to be diffracted; these can be subsequently selected by an optical aperture. The undiffracted light and the (–) first-order polarization are rejected by a beam stop. While the minimum bandpass of a particular device is set at the time of construction, it can be increased and "tailored" for a specific application by multiplexing the radio frequency (RF) signals injected into the crystal, allowing a multiplexed image of an entire region.[42] One limitation of the AOTF is its poor spectral resolution (~15 cm^{-1}), although there have been devices presented recently in the literature with much improved spectral[87,88] and spatial[89] resolutions, as well as a polarization-independent AOTF.[90] AOTFs may also lead to a loss of spatial fidelity especially noticeable at high magnification, arising from the wavelength dependence of the diffraction angle of the acoustic interaction in the dimension perpendicular to the dispersion dimension of the device. This one-dimensional loss leads to different spatial resolutions for orthogonal spatial dimensions of the image, making image analysis difficult. By carefully modeling the effect, the shift can be removed with post-acquisition processing.

LCTFs are based on nematic liquid crystal waveplates[91] consisting of a series of cascaded stages based on either the Lyot[92] or Evans[93,94] birefringent filter design. Although the exact design specifics for the two geometries are slightly different, the general principle of operation is similar: Each filter stage consists of a birefringent element and a nematic LC waveplate, which acts as an electronically controlled phase retarder, sandwiched between a pair of parallel linear polarizers. The exit polarizer of one stage acts as the entrance polarizer for the following stage. The birefringent element and the LC waveplate are oriented so that the incident light is normal to their optical axes and rotated 45° relative to the linear polarizers. Linearly polarized incident light is divided into two equal amplitude paths, the ordinary rays (*o-rays*) and the extraordinary rays (*e-rays*) traveling at different phase velocities through the birefringent material. In traversing the stage, *o-rays* and *e-rays* experience a relative optical path difference described by the wavelength dependent retardence of the birefringent element and the LC waveplate. Upon exiting the birefringent element, the *o-rays* and *e-rays* pass to a polarization analyzer which is oriented parallel to the input linear polarizer. Only wavelengths of light that are in phase are transmitted by the linear polarizer and presented to the successive stage, doubling the thickness of the birefringent element. The filter is electronically tuned by changing the potential applied across the liquid crystal waveplates while the spectral resolution and the free spectral range of the device are fixed during manufacturing by the thickest and thinnest birefringent elements, respectively. A spectrometer incorporating the LCTF is compact, contains no moving parts, and offers advantages over other tunable imaging filters. Unlike the AOTF, the LCTF can have near diffraction limited spatial resolution in both dimensions and is capable of providing moderately better spectral resolution (7–12 cm^{-1}).[49,54] The compact nature and in-line beam path of the device allows the LCTF to be easily incorporated into a wide variety of instrumentation ranging from microscopes to airborne macroscopic imaging systems.[1,48–54] One major limitation of LCTF technology for Raman imaging, however, is its low throughput, extending the exposure time required for each image frame and resulting in long experimental times. The extended time period can have adverse effects on studies where slight changes in experimental conditions, such as hydration, have a large impact on the sample.

No single technology satisfies all the criteria of an ideal imaging spectrometer for all applications. The best choice for a particular application will be determined only after a careful consideration of spectral and spatial resolution, throughput, experiment time limitations, and spectral range required. Each of the tunable filter technologies has its advantages and limitations. The best choice for a particular experiment will depend on the desired results and the trade-offs offered by the various technologies. We present some examples next that illustrate the application of filter based Ramana spectroscopic imaging instrumentation.

5.2.6.2 Applications

Wide-field illumination in conjunction with tunable filters and two-dimensional CCDs has proven to be the method of choice for many polymeric and biological Raman chemical imaging applications. Angle tunable dielectric filter imaging systems have been used to access component distribution in polymer blends,[58] to monitor sensitivity and pattern quality of substrates employing multiple-layer Langmuir–Blodgett film as a negative electron beam resist,[59] and to examine polymer lubricant dispersal on metal surfaces.[15,56] The fixed filter/tunable laser method has been implemented in a variety of preliminary biological applications, including investigations of human lymphocytes[33,34,62] and rat eye lenses.[34] While this technique possesses great potential as shown by these studies and especially in its ability to easily perform fluorescent and Raman imaging, the cost and expertise required limit the number of applications. A variation of the fixed filter/tunable laser technique employs a coherent fiber bundle to deliver the Raman image to the CCD.[38] In this case, a tunable NIR laser is coupled via fiber optics to the Raman probe head to image an area approximately 45 by 90 μm at a spatial resolution of ~1 μm.[38] For example, the white light image of *Euglena* shown in Figure 5.4 contains chloroplast, eyespot, nuclei, and reservoir and is well known to be rich in chlorophyll and β–carotene. A corresponding β–carotene Raman image shows that the eyespot and chloroplast are especially rich in this component. The representative spectra confirm that the level of β– carotene decreases away from the eyespot.[38]

Recent developments in the field of Raman chemical imaging have turned their attention to solid state, no-moving-parts imaging spectrometers, such as the dual holographic grating,[95] AOTF,[39–46] and LCTF[47–54] systems. Detection of foreign inclusions in human tissue has become a vigorous area of research. For example, emphasis has been placed on detecting silicone gel inclusions that result from failed breast implants. Although the silicon gel has been the major focus, other foreign components,

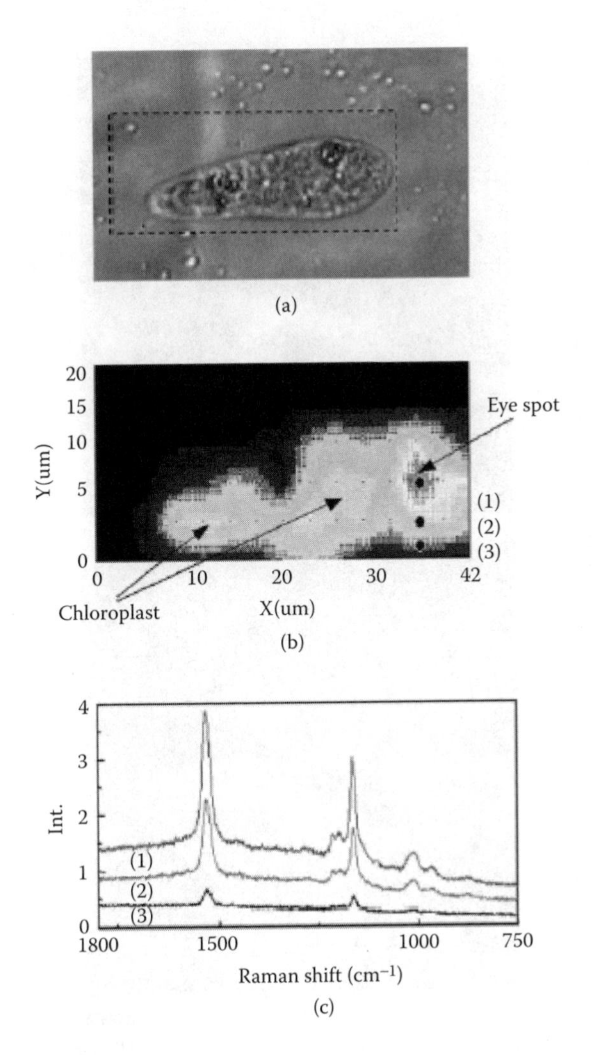

FIGURE 5.4 (a) Mapping area on *Euglena* delineated by a dashed box. (b) Map of the Raman intensity, showing the locations of spectra in (c). (H Sato, T Tanaka, T Ikeda, S Wada, H Tashiro, Y Ozaki, *J. Mol. Structure* 598: 93–96, 2001.)

such as the biocompatible polyester fixative patches that anchor the implant to the chest muscle wall, have been incorporated within the tissue surrounding the implant. Wide-field Raman chemical imaging using a two-dimensional CCD in conjunction with an AOTF was employed to examine foreign inclusions in a sectioned human breast tissue biopsy sample.[44] While it was impossible from the white light image alone to make a definitive determination about the structures observed in the image, the ratiometric Raman image, as shown in Figure 5.5, reveals unambiguously the inclusions as Dacron polyester.[44]

Wide-field LCTF Raman imaging has also been successfully employed in understanding the surface architecture of thermoplastic olefins (TPO) employed in the

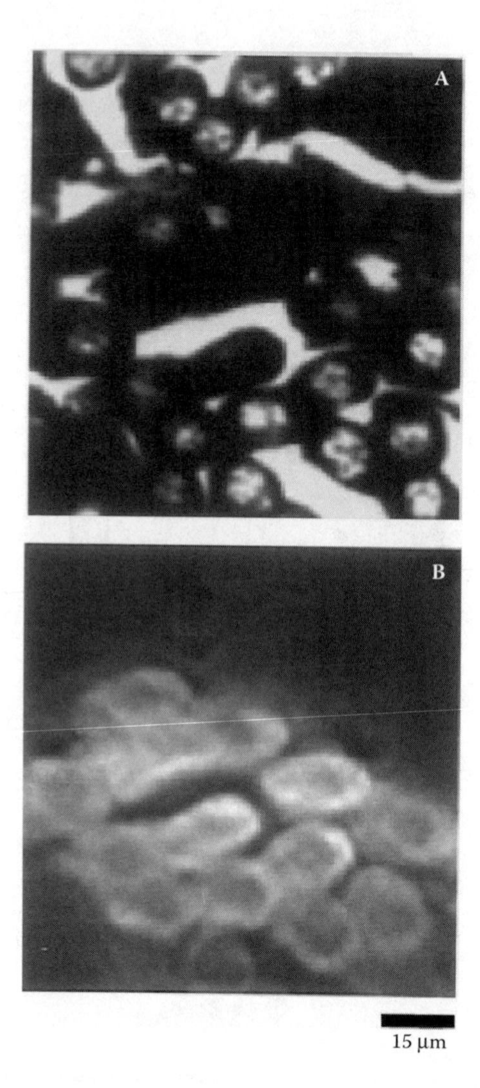

FIGURE 5.5 High-definition images of Dacron polyester in human breast implant capsular tissue: (a) brightfield reflectance image and (b) background ratioed Raman image (1615 cm^{-1}/1670 cm^{-1}); 10 min integration, 20X (NA = 0.46) objective. (MD Schaeberle, VF Kalasinsky, JL Luke, EN Lewis, IW Levin, PJ Treado, *Anal. Chem.* 68: 1829–1833, 1996.)

construction of vehicle bumpers and fascia in the automotive industry.[53] The TPO was created by blending ethylene–propylene rubber (EPR) and polypropylene (PP) with a coating of chlorinated polyolefin (CPO) sprayed onto the molded surface of the TPO part to promote adhesion of the primer and paint to the TPO substrate. A generally accepted model of the CPO-coated injection molded TPO architecture is shown in Figure 5.6. Depending on the exact parameters used during the molding process, various stratified layers are believed to form. The diffusion of the CPO primer through surface PP layers influences critically the adhesion of the CPO to

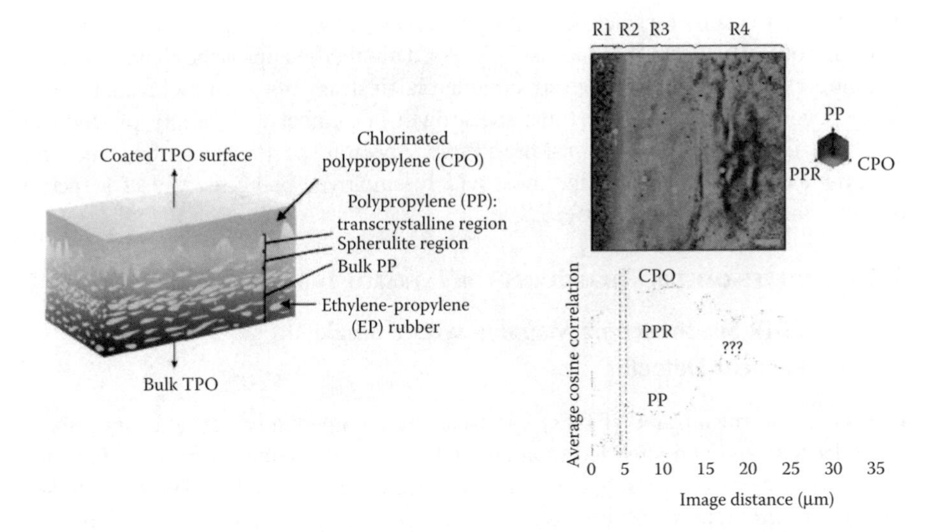

FIGURE 5.6 (Left) Generally accepted model of injection molded TPO after priming with CPO. The CPO-coated TPO is microtomed in cross sections (10 micrometers thick) for examination using Raman chemical imaging. (Right) Cosine correlation analysis of CPO-coated TPO resulting in a (top) RGB composite image showing PP, EPR, and CPO in red, green, and blue, respectively. Color overlap indicates pure component colocalization. The size bar corresponds to 5 μm, and Raman image collection is performed using a 50X objective (NA = 0.80). (Right, bottom) Average correlation scores as a function of sampling depth for CPO, EPR, and PP. The dotted lines emphasize the surface morphology of coated TPO and help illustrate the solvent interaction of the CPO primer with EPR and PP. (HR Morris, JF Turner, B Munro, RA Ryntz, PJ Treado, *Langmuir* 15: 2961–2972, 1999.)

the TPO by chain entanglements. The EPR is believed to reside below the surface and may swell upon application of the CPO solvent, causing the elastomer domains to migrate and combine at the surface. To better understand the structure of this material which critically affects its properties, Raman imaging followed by cosine correlation[72] was employed, to study the architecture of CPO-coated TPO samples. Component distribution in the sample was readily visualized,[53] showing that the outer 2–3 μm was homogeneous but contained greater amounts of strongly colocalized CPO and EPR. CPO formed irregular domains diffusing into the bulk, following closely the distribution of EPR. This also suggested that the EPR provides a flow network for CPO penetration, with the PP acting a foundation for the EPR/CPO entanglement that promotes adhesion to the TPO substrate. Thus, considerable insight into the structure, its formation, and likely effects on performance for multiphase polymers can be obtained from Raman imaging experiments.

5.3 MID-INFRARED SPECTROSCOPIC IMAGING

Infrared spectroscopy, particularly in the mid-infrared spectral region (4000–400 cm^{-1}), provides useful compositional and structural information and has, naturally, become the focus of microspectroscopic examinations. Historically, attempts to harness infrared spectroscopy for microscopic measurements have been successful in various forms for over fifty years.[96] In the last 20 years, a number of approaches have become commonly accepted and numerous commercial instruments are now available or being developed. At the same time, the growth in number and variety of studies involving infrared microscopy has been rapid. Spectral specificity can be achieved by a number of mechanisms, the most widely employed being the use of a spectrometer incorporating an interferometer.

5.3.1 Interferometric Techniques for Infrared Imaging

5.3.1.1 FT-IR Spectroscopic Mapping with a Single Element Detector

Fourier transform infrared (FT-IR) spectroscopy commercially debuted more than three decades ago and soon led to attempts to utilize the same instrumentation for microscopic measurements. Specifically, the advantages afforded by FT-IR spectroscopy of spectral reproducibility and time averaging are particularly useful in examining small spatial regions where optical throughput is low and spectral fidelity is desirable. Infrared interferometers were coupled to infrared microscopes incorporating fast and sensitive cryogenic detectors for commercial availability in the 1980s. These systems and their advanced versions are employed for microscopic infrared analyses at thousands of locations today. Further, the availability of stable detectors and high speed computers with large storage capabilities have allowed for the mapping of millimeter size spatial areas. Due to the sequential nature of the mapping process, however, experimental times are long, thus limiting applicability to experiments where either a small number of discrete measurements are made, data acquisition times are small compared to the time scales of changes in the sample, high spatial resolution is unimportant, or a spectrum from a small sample is required. These qualities have made the single-element microscopy spectrometer the instrument of choice for polymeric applications, for example, but it also has applicability in the biological arena.

5.3.1.1.1 Instrumentation

In single-element microspectroscopic instrumentation, modulated radiation from an interferometer is diverted to a set of optics that condense light, as for example a microscope, as shown in Figure 5.7a, to a small spatial area. Spectral information from a small, specified area of the sample is obtained by restricting the area illuminated by the infrared beam using opaque apertures of controlled size. The collected radiation is then diverted to a sensitive detector. To uniquely identify the area examined, however, a corresponding white light optical image is also required. Clearly, focusing the infrared beam for maximal throughput and minimal dispersion in the sample plane requires the optical and infrared paths be parfocal and collinear. Hence, an optical microscope is integrated into the infrared microscope.

There are important differences between an infrared microscope and one used for optical microscopy or Raman imaging. While optical microscopy and Raman spectroscopic imaging can be carried out by employing high quality, refractive glass optics, infrared microscopes consist of all-reflective optics, since glass does not transmit a

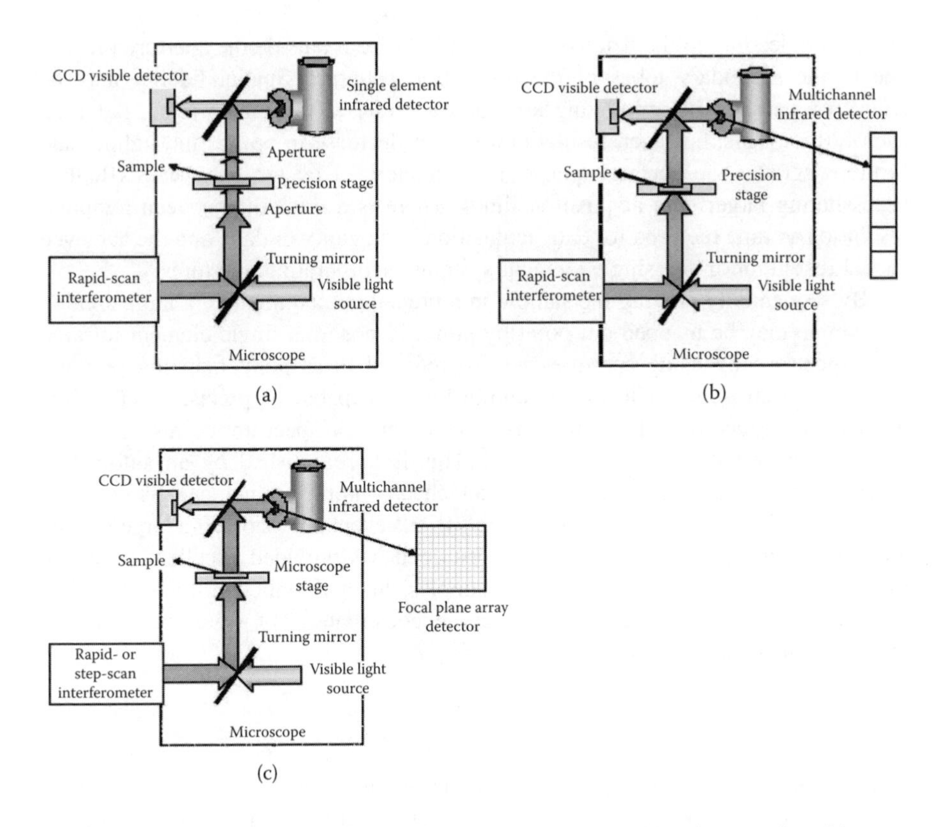

FIGURE 5.7 Modes of conducting FT-IR microspectroscopy: (a) point mapping, (b) raster scanning with a linear array detector, and (c) wide-field multichannel acquisition with a staring focal plane array detector.

large portion of the mid-infrared spectrum (wavelengths longer than ~5 μm). Further, corrected reflective optics allow spectral fidelity to be maintained by minimizing optical aberrations. The modulated radiation is typically derived from an economically constructed rapid-scan Michelson interferometer, although, in principle, any type of radiation modulator may be employed (for example, step-scan interferometers). Radiation corresponding to a small spatial area is allowed through by employing a carbon black coated metal aperture. In some later designs, infrared absorbing glass is used for apertures permitting visible imaging without disturbing the infrared setup. Infrared microscopes using apertures are designed for the largest aperture that may be employed, for example a ~100×100 μm^2 spot. Thus, the spot size at the sample plane allowed by the optics is fixed but the effective area available for light throughput is determined by the area of the aperture opening limiting light utilization efficiency, in some cases to as low as 0.1%.

The use of apertures that decrease the cross-section of radiation at the sample plane also decreases throughput due to diffraction when the aperture is of the same dimension as the wavelength of light (~2–12 μm), limiting the highest spatial resolution achievable. Radiation transmitted through an aperture also undergoes diffraction, leading to the detector sampling light from outside the apertured region due to the secondary lobes of the diffraction pattern.[97] Spatial fidelity may be recovered partially by employing a second aperture in tandem to reject radiation even further. This, however, results in a further decrease in optical throughput and further degrades the spectral quality as characterized by the spectral SNR, thus necessitating larger data acquisition times. There is a trade-off between temporal resolution or time required for data acquisition, the quality of data, and the achieved spatial resolution in the single-element systems incorporating apertures.

By sequentially moving the sample in a predetermined manner, a large area on the sample may be mapped out point by point. Hence, this single-element infrared microspectroscopy using apertures is also referred to as point mapping or point scanning spectroscopy. Clearly, a sample holder capable of precise microscopic movements, a record of the sample movement, and the spectrum corresponding to every spatial point must be maintained. This is accomplished by an automated, programmable microscope stage and an attached computer, which can also be used to control the instrument and synchronize various events in mapping a large sample area. In summary, the instrumentation described has provided excellent results in examining small impurities or defects in samples, but is of limited utility or statistical viability in examining large areas of heterogeneous materials for presence or distribution of chemical species.

5.3.1.1.2 Applications

Discussions on the applications of FT-IR microspectroscopy to biological materials and polymers[98,99] are available.[100] Some representative examples of various applications are presented below to illustrate the application areas of IR microscopy. As explained above, FT-IR microscopy is particularly useful in examining microsamples, additives, contaminants, and microscopically localized degradation products, as the sensitivity of FT-IR spectroscopy to microscopic samples is greatly increased compared to macroscopic measurements. There is an important caveat: the sample area incorporating the defect must be identified first using visible microscopy and only then can it be analyzed using FT-IR microscopy. In contrast, wide field imaging (*vide infra*) may lead to further information that may not be apparent in the optical contrast of the bulk material. If the sample itself represents a small quantity, it has been claimed that FT-IR microspectroscopy can be employed to examine minute quantities down to the nanogram range.[101] The preparation for microscopic quantities of the sample is crucial and many specialized accessories are available for specific tasks. A number of studies have identified gel inclusions in poly(ethylene),[102,103] contaminants on the surface of a semiconductor device,[104] acrylic fiber on a microcircuit die,[105] contaminants in poly(vinyl chloride) (PVC),[106] and a mold release agent on the surface of a polyurethane.[107] The spatial localization of additives,[108] as well as the degradation of polymers[109] indicating the presence of many different types of localized degradation products[110] can also be examined, whether on the surface or in the bulk material after microtoming. Disparate materials like lubricants,[111] pyrolysis products from microsamples, inclusions and growth of foreign materials may be studied.[112] The degradation of a polymeric sample often begins at the surface and proceeds into the bulk. Skin-core oxidation in aged polymers[113] was observed and oxidation depth profiles determined indicating a diffusion-controlled oxidation process.

Single-component or multiple-component fibers[114] or fibers with dyes[115] or pigments[116] have also been detected and analyzed by FT-IR microspectroscopy. Fibers of different materials can be readily differentiated[117] along with fiber blends or fibers in polymeric matrices.[118] However, examinations of thin fibers (> ~10 μm) is complicated by optical effects. Fibers larger than ~30 μm in diameter usually do not require much sample preparation. While the detection and identification of single fibers can be readily accomplished, quantitative analysis of orientation is complex.[118–120] At the interface of polymers and biological systems, FT-IR microscopy can be used to examine the interactions of biological species with polymeric implants,[121] their oxidation and subsequent degradation, which may lead to device failure. In particular, polymer degradation is critical in the area of implanted prosthetic devices,[122] which can be monitored by microscopic measurements. The microstructure of semi-crystalline polymers and blends determines their performance properties. For example, the crystal structure of multi-morphic poly(vinylidene fluoride) (PVDF) has been identified in various phases by FT-IR microscopy. The different types of spherulites seen optically are assigned to different crystal modifications. Further, blends of such polymers also yield information on specific interactions between the blended polymers.[124] The crystalline and amorphous regions of another polymer with potentially wide application, Nafion,[125] were examined. The performance of the fuel cell is critically dependent on the water uptake in the polymer electrolyte, which depends critically on the Nafion structure. Typically, the crystalline regions contain no water, while in the amorphous regions there is a complete proton transfer from the acid to the water molecules, facilitating water retention. Polymer microstructure, which affects critical performance properties, can be routinely determined using FT-IR microscopy. Combined with polarized radiation obtained from the interferometer and a polarizer in the beam path, local composition and relative organization in crystalline polymers provide information complementary to x-ray scattering,[126] DSC,[127] or microscopy studies.[123,128]

Spectroscopically determined intramolecular and intermolecular specific interactions in polymer blends[128] can be correlated to morphology characterized by optical microscopy, since spectral mapping is often insufficient to determine size and shape information. Further, spectral information from small regions may be difficult to reliably correlate with observed polymer behavior. For example, a model polymer blend system[129] was studied spectroscopically in an attempt to characterize its phase diagram. A poor correlation was found between the phase diagram determined in this manner and the phase diagram obtained by optical microscopy, probably due to the large spot size of the infrared beam and the lack of spatial fidelity of apertured spectra. In another case, however, infrared mapping of the sample[123] showed evidence of phase separation based on spectral differences. These examples also illustrate the differences in origin of the contrast mechanisms between optical and infrared microscopies. Although optical microscopy is capable of higher spatial resolution, its discrimination is limited to a difference in the average property of a material, such as the refractive index, unless the system is specially labeled to detect a property involving the behavior of the label. Infrared microspectroscopy derives its contrast mechanism from the intrinsic composition of the material, but suffers from a poorer spatial resolution. A judicious use of the two complementary techniques is often required to achieve a well-defined characterization.

The approach of forming an interface by laminating films of two polymers, allowing them to diffuse, and then microtoming a vertical section to obtain a sample for IR mapping, is a popular method for examining diffusion, a central phenomenon in polymer science. Spectral profiles along the diffusion direction can be obtained and the concentration of the diffusing species directly inferred.[130,131] However, polymer dissolution by small molecular weight solvents has been too rapid to examine using mapping techniques; consequently, the diffusion profiles for very limited low molecular mass substances have been mapped.[132–138] Simultaneous reaction and diffusion is also important in the formation of blend systems, and the competitive effects of the rates of the two processes can be easily carried out using FT-IR microspectroscopy, where the rate of diffusion is visualized by the time evolution of the absorbance (concentration) profiles while the rate of reaction can be monitored as a time evolution of the reactant (or product) absorbance (concentration).[139] Local orientation during formation[140,141] or failure[142] can also be probed.

5.3.1.1.3 Synchrotron Infrared Microscopy

The large collection times and poor SNR characteristics of infrared microscopy using apertures is largely a result of relatively weak light intensity, which is further reduced by apertures. The higher the spatial resolution (or the smaller the aperture opening), the more severe the loss of radiation intensity. A synchrotron source provides radiation throughput that is much greater than conventional infrared sources. Hence, radiation from a synchrotron fed to an interferometer–microscope assembly results in spectral data of considerably higher SNR, while affording higher spatial resolution than that obtained using conventional sources. Typically, the SNR for data obtained using a synchrotron microscope is almost two orders of magnitude higher than that obtained from a standard infrared globar source. Further, an extended spectral range is provided by the synchrotron, which may be useful for some specimens. Apertures as low as $3 \times 3\ \mu m^2$ have been used[143] to image thin, small samples such as single cells.[144] The major impediment in widespread application of the technique has been the requirement for access to a synchrotron, which is uneconomical for most organizations. Thus, studies employing such an instrument are limited.[145] Further, the SNR of data obtained in a given time for a wide field of view using a synchrotron source is comparable to data obtained from focal plane array detectors available today. Hence, synchrotron-based infrared microscopes are not expected to be popular for routine analyses, but will probably maintain an advantage in analyzing very small samples at high spatial resolutions.

5.3.1.1.4 Near-Field Infrared Mapping

Near-field infrared microscopy permits sample examinations at a spatial resolution greater than that achievable using conventional optics (to the diffraction limit of light).[146] The near-field condition is attained by employing specialized radiation collection optics that permit partial conversion of the evanescent fields surrounding high-index optical elements into propagating waves, or by employing near-field apertures. The contrast parameters are different from those in conventional optics as the amount of light propagating from a subwavelength aperture through a flat substrate strongly depends on the spacing of the probe and the sample, easily generating topographic artifacts. The transmitted power, further, is strongly influenced by the refraction index of the sample, resulting in substantial differences between near-field and corresponding far-field spectra. Theoretical models to account and correct for these effects have been developed. This technique, however, is still in developmental stages and will offer limited point mapping capabilities, precluding applications in which data from large spatial areas is required rapidly; however, the achievable spatial resolution is expected to surpass that from any other IR microspectroscopic technique.

5.3.1.1.5 Scanning Probe Microspectroscopy or Photothermal Imaging

Another technique combining scanning probe microscopy at a submicron spatial resolution with an IR interferometer is based on a microthermal analysis approach,[147,148] which permits localized thermal analysis to be combined with surface and subsurface imaging using a scanning probe with a small detection head. In the IR spectroscopic analogue of this system, modulated radiation from a standard interferometer is directed onto a sample and detected by a microprobe.[149,150] The signal from the probe measures resulting temperature fluctuations, thus providing an interferogram analogous to the interferogram normally obtained by direct detection of IR radiation transmitted by a sample. This enables IR microscopy at a spatial resolution which could potentially be as low as tens of nanometers, since the resolution now depends on the probe and not on the system's optics. Once again, sample topographical and thermal properties influence the observed chemical contrast.

5.3.1.2 Raster-Scan FT-IR Imaging Using Multichannel Detectors

While single-element microspectroscopy provides the capability to obtain spectra from small spatial regions, poor SNR characteristics, diffraction effects, and stray light issues resulting from the use of apertures limit the applicability of this point mapping approach. A multichannel detection approach to circumvent some of these issues has recently been implemented,[151] whereby a linear array detector is employed to image an area corresponding to a rectangular spatial area on the sample. The linear array is moved precisely to sequentially image a selected spatial area on the sample. This is referred to as "push broom" mapping or raster scanning. The process is conceptually similar to point-by-point mapping, but takes advantage of the multiple channels of detection. Hence, imaging a large sample area is faster by a factor of m, for a linear array detector containing m elements. Further, spatial resolution is dependent on the relative size of the detector and system magnification, while spatial specificity depends only on the relative position of the array and the sample. The instrument is schematically displayed in Figure 5.7b.

While point mapping detectors are typically 100–250 μm in size, an individual element in a linear array of detectors is of the order of tens of micrometers. Hence, employing a linear array eliminates the need for apertures, as small array detectors directly image different spatial sample regions.[151] For example, a detector pixel 25 μm in size can be operated at 1:1 magnification or 4:1 magnification to provide a 25 μm or a 6.25 μm spatial resolution. Such magnification ratios can be readily achieved by employing available, relatively aberration-free infrared optics. The debilitating effects of apertures when spatial resolution comparable to the wavelength of light is desired are no longer a factor, and consequently the quality of data is higher than that of the single channel detection systems. In addition, the spatial resolution, data quality, and time for data acquisition are no longer coupled as in point mapping methods. The data acquisition time depends solely on the size of the image and quality of data desired, and is less correlated with the spatial resolution, which is determined by the employed optics. The optical and infrared paths, however, are required to be collinear, and the visible image must be referenced to acquire infrared data. A precision stage that reproducibly steps in small increments is required for any mapping sequence larger than that achieved by one row of detectors. Once the sample is manually positioned, a visual image may be obtained by using a visible light camera and moving the precision sample stage. The area from which infrared spectroscopic data is to be acquired can then be marked and acquisition initiated with the interferometer being operated in a continuous scan mode. In combination with high performance multi-channel detectors, this mode combines the most desirable properties of rapid-scan interferometry to yield high quality data. In our experience, the performance of the sample stage, the position of which is critical to prevent misregistration errors, is satisfactory. The system is flexible in that the imaging of virtually any rectangularly defined sample area may be optimally accomplished.

5.3.1.2.1 Applications

The reported applications are few, as the technology is very new. A number of preliminary reports and validation samples highlight the spectral quality and the flexibility of data acquisition. In particular, arbitrarily large images at a high spatial resolution can be acquired. For example, Figure 5.8 shows a large section of skin imaged at a high (6.25 μm) spatial resolution. A typical area using wide-field imaging (*vide infra*) is shown alongside for comparison. Clearly, two-dimensional detectors may also be scanned in a similar manner. However, current detector and computer technology for matching data acquisition, interferometer motion, stage stepping, and detector electronic coupling will likely limit the advantage of two-dimensional detectors operating in a raster scanning mode.

5.3.1.3 Global FT-IR Spectroscopic Imaging

The state of the art in FT-IR microspectroscopic instrumentation is the combination of a focal plane array (FPA) detector and an interferometer,[152,153] as shown in Figure 5.7c. FPA detectors consist of thousands of individual detectors laid out in a two-dimensional grid pattern. An FPA matched to the characteristics of the optical system is capable of imaging the entire field of view afforded by the optics and of utilizing a very large fraction of the infrared radiation spot size at the plane of the sample. The increase in the number of individual detectors compared to a linear

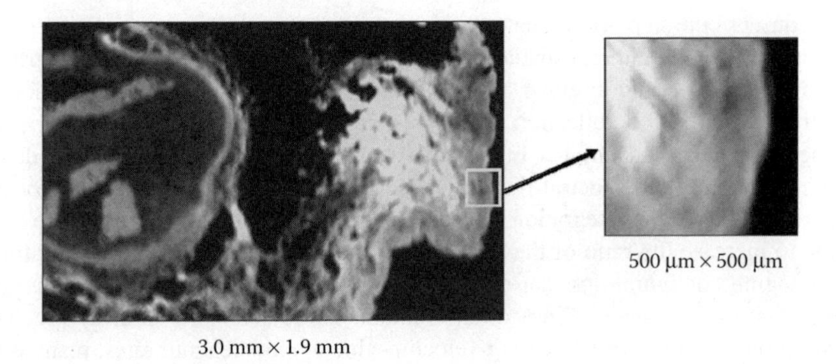

FIGURE 5.8 Raster scanned image of a section of skin demonstrating that large spatial areas, useful for simultaneous morphological and chemical analyses, can be examined without loss in spatial resolution. For comparison, a typical imaging area using two-dimensional detectors is shown.

array results in a correspondingly greater multichannel advantage. For example, a p × p pixel focal plane array detector provides a p^2 time savings compared to a single element detector, and a p^2/m time saving compared to a linear array detector containing m elements. For a 128 × 128 element detector compared to the single element case, the advantage is a factor of 16384, while compared to the linear array detector, the multichannel advantage is a factor of 2048. Further, the two-dimensional detectors are capable of imaging large spatial areas simultaneously without inherent inefficiencies of moving the sample or resetting the interferometer to scan a different area. Due to the considerable reduction in data acquisition times, imaging large areas of static samples is possible, as is the examination of dynamic processes.

Refractive index differences in the sample can be visualized by using a separate optical path or by the brightfield infrared imaging, which contains a scattering as well an absorption/emissive contribution from all infrared wavelengths. The spatial resolution of instrumentation based on FPA detection is determined by the magnification optics (typically 15X) and the size of individual detector elements on the focal plane array (typically 40–60 μm). Although the nominal resolution is determined by the optics and the detector, the resolution limit is usually determined by the diffraction limit for the wavelength of interest. The quality of data and the acquisition time are related, with the minimum acquisition time being on the order of 1 sec.[154] The root-mean-square SNR for data acquired in this short time is typically ~200:1 but can be increased by signal averaging methods.[155] The spatial resolution can be varied from the diffraction limit to a few millimeters by altering the magnification of the instrument.

The first, and to date the most popular, approach to FT-IR microimaging spectrometers incorporates a step-scan interferometer.[156] A step-scan interferometer provides a means to maintain constant optical retardation for a specified, often long, time period. A constant retardation over an extended time period allows for suitable signal averaging and for data storage. Short time delays allow for mirror stabilization at the onset of the step prior to data acquisition. Data acquisition and readout formats may be accessed in either a spatially sequential (rolling) or simultaneous (snapshot) mode across the array. In either case, the signal is integrated for only a fraction of the time required for collection of each frame. The integration time, number of frames co-added, and number of interferometer retardation steps (which is determined by the desired spectral resolution) determine the total time required for the experiment. Since the integration time determines the data quality, efforts have been made to increase the ratio of the integration time to the total data acquisition time.

Imaging configurations that utilize a rapid scan interferometer have been proposed for small arrays.[157] Since a large number of detectors on an array preclude conventional rapid mirror scanning velocities due to slow readout rates, many configurations utilize a slow-speed, continuously scanning mirror. Some manufacturers employ fast step-scan in which the mirror is partially stabilized to achieve the same retardation error as continuous scan spectroscopy; this approach is termed "slow scan". A generalized data acquisition scheme that permits true rapid scan data acquisition for large arrays or higher mirror speed acquisition from small arrays has been proposed,[158] where the integration time of individual frames collected by the FPA detector is negligible on the time scale of the complete interferogram collection. For most FPA detectors available today, however, the motion of the moving mirror does not allow co-addition of frames per interferometer retardation element in the continuous scanning mode. Compared to step-scan data acquisition, rapid scan data collection allows for fast interferogram capture as no time is spent on mirror stabilization; however, the image stored per resolution element is noisier. The error arising from the deviation in mirror position during frame collection is hypothesized to be the next largest contributor of noise compared to the dominant contribution from random detector noise. At present, the advantage of continuous scan interferometry lies in making less expensive instrumentation, compared to step-scan systems, and in increasing data collection efficiency.

5.3.1.3.1 Applications

Unlike single-element mapping devices, the generally poor SNR characteristics of FT-IR imaging have limited its sensitivity. Recently, with improved data acquisition protocols and processing tools, the SNR levels achievable for several minutes of data collection have reached >1500, implying quantification levels of less than 1%. Defects (such as air bubbles) in thin films may be easily visualized. This technique is expected to be popular, for example, in the analysis of polymer inclusions and defects in the semiconductor processing industry. With wide-field imaging techniques, defects that have similar refractive indices within examined material can be readily detected. There has been considerable controversy over the failure of breast implants and the subsequent contamination of breast tissue. Implants typically consist of an elastomeric shell with a silicone filling inside. Over time and due to mechanical failure, the implant contents leak into surrounding tissue. The presence of silicone oil has been readily detected[159] in tissue sections using FT-IR imaging, as shown in Figure 5.9.

Polyvinyl chloride (PVC) is a useful polymer that is becoming less popular due to its degradation behavior. However, blends of PVC with other polymers have the

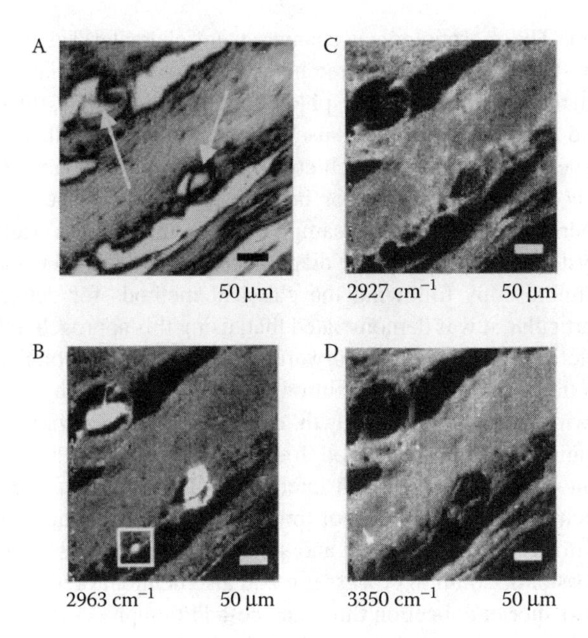

FIGURE 5.9 Si-CH_3 characteristic vibrations were used to provide chemical contrast between the tissue and silicone oil inclusions as small as ~10 μm. Inclusions were also readily observed in cases where optical microscopy contrast was ineffective.

capability to yield composites with greater stability. In one such case, images of phase separated PVC/PMMA blends are reported.[160] Phase separation and concentration quantification can be carried out as a function of starting polymer properties and process conditions although the dispersion sizes determinable by infrared imaging methods are limited to tens of microns. In many polymeric systems, the dispersed phase is smaller than this dimension and the composites are not amenable to FT-IR imaging analyses. Scattering[161] analyses and determinations of relative concentrations of a species corresponding to a given pixel may yield qualitative clues to the morphology of the phase separated material, but cannot yield quantitative morphological information. We anticipate that the spatial distribution of specific interactions in phase separated polymeric systems could also lead to such information, and studies examining these aspects should appear in the near future. Imaging semi-crystalline polymers via their dichroism[162] demonstrates a polymeric spherulitic structure similar to that seen using polarized optical microscopy. The segregation of semi-crystalline blends can also be examined.[162] Components of a laminate, in which individual layers are strongly segregated, were readily identified based on absorbance images from their characteristic vibrational peaks.[163] A polymer–liquid system consisting of phase-separated mixtures of uncured poly(butadiene) and diallyl phthalate were studied to characterize morphology differences before and after the curing process.[164] Optical microscopy of these systems is particularly challenging as the resultant phases have similar refractive indices. Sufficient image contrast was achieved, however, by FT-IR imaging due to the inherent chemical differences manifest in their infrared spectra. The morphological changes were characterized over a period of time and the post-cure sample exhibited homogeneity at the spatial resolution level of the instrument. The properties of phase separated materials, such as polymer dispersed liquid crystals (PDLCs),[165] also depends on the solubility of the blends' constituents. The concentration of each component in phase separated domains can be calculated accurately with an error being estimated using statistical methods. Coupled to thermal control of the sample,[166] this information yields the phase diagram of a system.[167] The final phase diagram corresponded well with that obtained using optical microscopy following the classical methods for determining phase behavior. In particular, it was demonstrated that, using this approach, phase diagrams are readily determined in a straightforward manner that is decoupled from phase separation kinetics. Further, the distribution of surfactants in such systems,[168] which were also shown to retard droplet growth, could also be visualized.

FT-IR imaging presents a method for determining both the spatial and the spectral content of a polymer–solvent interphase. By monitoring the spatial distribution of concentration as a function of time from an initially known state, one can determine the diffusion of a polymer and solvents in contact, as well as the dissolution rate.[169] Faster dissolution processes could also be monitored in real time using a combination of shorter collection times and co-addition processes.[170] The diffusion of liquid crystals into polymers is particularly intriguing given that the liquid crystal may be in an organized state or an isotropic state depending on the temperature. The diffusion of a liquid crystal (5CB) into a PBMA matrix was studied by using the contact method to prepare a gradient.[162] Concentration profiles were obtained as a function of time and temperature. The presence of an anomalous diffusion process was detected. It was shown that fast FT-IR was able to correctly identify the diffusion process as anomalous. As opposed to this, a bulk mass uptake analysis would have led to the conclusion that the process proceeded according to Fick's second law.

The encapsulation of biologically active proteins in sustained release polymer devices determines their function and activity.[171] As the polymeric capsule is prepared, it is subjected to a number of manufacturing processes that may denature the active pharmaceutical agent. In such cases, it is advantageous to have a method available to perform a spectroscopic analysis on the final product without the necessity of extracting the protein[172] and to provide information on the mechanisms of release.[173,174] FT-IR imaging of biological systems has demonstrated a potential to complement other imaging approaches. For biomedical applications, the technique may be used to examine chemical changes due to pathological abnormalities and to follow histological alterations with high accuracy. Nondestructive morphological visualization of chemical composition rapidly provides structural and spatial information at an unprecedented level. Specifically, thousands of spectra routinely acquired in an imaging experiment may be employed for statistically meaningful data analyses, which in the example of biological tissue samples may prove ultimately useful in medical diagnoses. In cases where subtle biochemical differences have been implicated in disease prognosis, mid-IR spectroscopic imaging may provide viable solutions. Since the visualization contrast is dictated by inherent chemical and molecular properties, no external sample treatments, as, for example, the histopathological staining techniques required for optical microscopy, are necessary.

In fact, one of the first demonstrations of FT-IR imaging in ascertaining the distribution of a biological species was that of a lipid [C(16)-lysophosphatidylcholine] in a KBr disk,[175] an analysis that was difficult using optical microscopy alone. A typical example of the type of tissue information that can be retrieved was demonstrated by examining monkey cerebellum sections.[176] Relative lipid to protein distributions allowed easy differentiation of white matter regions relative to gray matter areas. Purkinje cells in rat cerebella, which strongly influence motor coordination and memory processes, were visualized using FT-IR imaging techniques.[177,178] Neuropathologic effects of a genetic lipid storage disease, Niemann-Pick type C (NPC),[179] were distinguishable on the basis of spectral data without the use of external histological staining. Statistical analysis provided a numerical confirmation of these determinations consistent with a significant demyelination within the cerebellum of the NPC mouse. IR spectroscopy has been used for a number of years to characterize mineralized structures in living organisms (notably bone). FT-IR imaging spectroscopy[180,181] of bone allows spatial variations of a number of chemical components to be nondestructively monitored. Correlations in bone between FT-IR imaging and optical microscopy involving chemical composition, regional morphologies and the developmental processes have been made.[182] An index of crystallinity and bone maturity could be determined providing structural information in a nondestructive manner. Analyses of this nature are particularly useful in studying structural modifications in bone, as, for example,[183] those arising in osteoporotic human iliac crest biopsies. These determinations allow correlations to be discerned between degenerative processes and their manifest chemical signatures.[34–38]

5.3.2 Other Infrared Microspectroscopy Approaches

5.3.2.1 Hadamard Transform Infrared Microscopy

Hadamard microscopy involves the use of encoded masks to allow radiation only from specific areas of the sample to be incident on the detector (*vide supra*, in the section on Raman microscopy).[184] By employing a sequence of masks, one may determine the spatial distribution of chemical species over a field of view.[185,186] A form of multichannel advantage is obtained by employing physical masks while retaining the spectral multiplex advantages of the Fourier transform.[187] The technique has not achieved widespread applicability due to the requirements for opaque masks and the large number of computations, which limit the versatility of the technique and do not impart any significant advantages over single element approaches employing apertures. Further, experimental times were large and misregistration occurred while switching masks. In using masks with individual elements, spatial fidelity was compromised by the less than ideal transmission characteristics of the masks. Recently, programmable micro-mirror arrays have been employed for Hadamard transform microscopy.[188]

5.3.2.2 Laser-Diode-Based Imaging

An alternative to using interferometers or filters for spectral discrimination is to employ a source that provides spectral discrimination directly, as in the use of infrared diode lasers. By coupling a laser and an infrared microscope, one can obtain a map of large spatial areas with spatial resolution approaching the diffraction limit.[189] As an extension to this approach, the laser output may be directed onto a small focal plane array detector to provide a multichannel advantage. Although the procedure may allow rapid, high-fidelity monitoring of small spectral regions, only limited spectral ranges are accessible, and the possibility of future multichannel detection will likely be limited by relatively low laser powers.

5.3.2.3 Solid State FILTERING Approaches

The combination of a multichannel detector with a suitable filter, using, for example, infrared AOTFs[190] to restrict radiation to specific wavelengths, can be employed to provide spectral discrimination. By sequentially sampling a wide spectral region in small incremental wavelengths, one may conduct infrared imaging.[191] Solid state approaches provide high spatial fidelity and a rugged system due to the absence of moving parts. While solid state approaches are useful in such settings, interferometer-based systems provide much higher sensitivity and broader spectral coverage, and can be employed to conduct experiments at higher spectral resolutions.

5.4 SUMMARY AND OUTLOOK

The last fifteen years have seen an impressive increase in the number and diversity of types of instrumentation available for Raman and mid-infrared microspectroscopic imaging analyses. The synergistic coupling of instrumentation advances and new applications has provided unprecedented insight into the molecular heterogeneity of materials and their function. The molecular bases of many biological phenomena have been elucidated while providing extraordinary amounts of data at unprecedented rates. There remains, however, a significant need to improve the quality of the acquired data and to develop rapid data processing strategies for effectively extracting and displaying information. As this occurs, we anticipate widespread applications at more sophisticated levels of vibrational microspectroscopic imaging studies in both familiar and unfamiliar fields.

REFERENCES

1. MD Schaeberle, DD Tuschel, PJ Treado, *Appl. Spectrosc.* 55: 257–266, 2001.
2. R Bhargava, IW Levin, *Anal. Chem.* 73: 5157–5167, 2001.
3. CJ de Grauw, C Otto, J Greve, *Appl. Spectrosc.* 51: 1607–1612, 1997.
4. CJH Brenan, IW Hunter, *J. Raman Spectrosc.* 27: 561–570, 1996.
5. NM Sijtsema, SD Wouters, CJ de Grauw, C Otto, J Greve, *Appl. Spectrosc.* 52: 348–355, 1998.
6. MR Frenández, JC Merino, MI Gobernado-Mitre, JM Pastor, *Appl. Spectrosc.* 54: 1105–1113, 2000.
7. P Schmidt, J Kolaík, F Lednick´y, J Dybal, JM Largaron, JM Pastor, *Polymer* 41: 4267–4279, 2000.
8. R Appel, TW Zerda, WH Waddell, *Appl. Spectrosc.* 54: 1559–1566, 2000.

9. M Delhaye, PJ Dhamelincourt, *J. Raman Spectrosc.* 3: 33–43, 1975.
10. SL Zhang, JA Pezzuti, MD Morris, A Appadwedula, CM Hsiung, MA Leugers, D Bank, *Appl. Spectrosc.* 52: 1264–1268, 1998.
11. LCN Boogh, RJ Meier, HH Knausch, BJ Kip, *J. Polymer Sci. Part B: Polymer Phys.* 30: 325–333, 1992.
12. L Markwort, B Kip, *J. Appl. Polymer Sci.* 61: 231–254, 1996.
13. M Bowden, GD Dickson, DJ Gardiner, DJ Wood, *Appl. Spectrosc.* 44: 1679–1684, 1990.
14. M Bowden, DJ Gardiner, G Rice, DL Gerrand, *J. Raman Spectrosc.* 21: 37–41, 1990.
15. KPJ Williams, GD Pitt, BJE Smith, A Whitley, DN Batchelder, IP Hayward, *J. Raman Spectrosc.* 25: 131–138, 1994.
16. DM Wieliczka, MB Kruger, P Spencer, *Appl. Spectrosc.* 51: 1593–1596, 1997.
17. JA Timlin, A Carden, MD Morris, *Appl. Spectrosc.* 53: 1429–1435, 1999.
18. KA Christensen, MD Morris, *Appl. Spectrosc.* 52: 1145–1147, 1998.
19. JA Timlin, A Carden, MD Morris, JF Bonadio, CE Hoffler II, KM Kozoloff, SA Goldstein, *J. Biomed. Opt.* 4: 28–34, 1999.
20. JA Timlin, A Carden, MD Morris, RM Rajachar, DH Kohn, *Anal. Chem.* 72: 2229–2236, 2000.
21. A Feofanov, S Sharonov, P Valisa, E Da Silva, I Nabiev, M Manfait, *Rev. Sci. Instrum.* 66: 3146–3158, 1995.
22. CJH Brenan, IW Hunter, *Appl. Opt.* 33: 7520–7528, 1994.
23. CJH Brenan, IW Hunter, *Appl. Spectrosc.* 49: 1086–1093, 1995.
24. PJ Treado, MD Morris, *Appl. Spectrosc.* 44: 1–4, 1990.
25. PJ Treado, MD Morris, *Appl. Spectrosc.* 43: 190–192, 1989.
26. PJ Treado, MD Morris, *Appl. Spectrosc.* 42: 897–901, 1988.
27. KK Liu, L Chen, R Sheng, MD Morris, *Appl. Spectrosc.* 45: 1717–1720, 1991.
28. P Hansen, J Strong, *Appl. Opt.* 11: 502, 1972.
29. RD Swift, RB Wattson, JA Decker, Jr., R Paganetti, M Herwit, *Appl. Opt.* 11: 1596, 1972.
30. DC Tilotta, RM Hammaker, WG Fateley, *Appl. Spectrosc.* 41: 727–734, 1987.
31. RA DeVerse, RM Hammaker, WG Fateley, *Vib. Spectrosc.* 19: 177–186, 1999.
32. RA DeVerse, RM Hammaker, WG Fateley, *J. Mol. Struct.* 521: 77–88, 2000.
33. GJ Pupples, TC Bakker Schut, NM Sijtsema, M Grond, F Maraboeuf, CG de Grauw, CG, Figdor, J Greve, *J. Mol. Struct.* 347: 477–484, 1995.
34. NM Sijtsema, JJ Duindam, GJ Pupples, C Otto, J Greve, *Appl. Spectrosc.* 50: 545–551, 1996.
35. J Ma, D Ben-Amotz, *Appl. Spectrosc.* 51: 1845–1848, 1997.
36. BL McClain, J Ma, D Ben-Amotz, *Appl. Spectrosc.* 53: 1118–1122, 1999.
37. BL McClain, HG Hedderich, AD Gift, D Zhang, KN Jallad, KS Haber, J Ma, D Ben-Amotz, *Spectroscopy* 15(9): 28–37, 2000.
38. H Sato, T Tanaka, T Ikeda, S Wada, H Tashiro, Y Ozaki, *J. Mol. Struct.* 598: 93–96, 2001.
39. PJ Treado, IW Levin, EN Lewis, *Appl. Spectrosc.* 46: 1211–1216, 1992.
40. PJ Treado, IW Levin, EN Lewis, *Appl. Spectrosc.* 46: 553–559, 1992.
41. EN Lewis, PJ Treado, IW Levin, *Appl. Spectrosc.* 47: 539–543, 1993.
42. MD Schaeberle, JF Turner II, PJ Treado, *Proc. SPIE Int. Soc. Opt. Eng.* 2173: 11–20, 1994.
43. MD Schaeberle, CG Karakatsanis, CJ Lau, PJ Treado, *Anal. Chem.* 67: 4316–4321, 1995.

44. MD Schaeberle, VF Kalasinsky, JL Luke, EN Lewis, IW Levin, PJ Treado, *Anal. Chem.* 68: 1829–1833, 1996.
45. HT Skinner, TF Cooney, SK Sharma, SM Angel, *Appl. Spectrosc.* 50: 1007–1014, 1996.
46. SM Angel, JC Carter, DN Stratis, BJ Marquardt, WE Brewer, *J. Raman Spectrosc.* 30: 795–805, 1999.
47. KA Christensen, NL Bradley, MD Morris, RV Morrison, *Appl. Spectrosc.* 49: 1120–1125, 1995.
48. HR Morris, CC Hoyt, PJ Treado, *Appl. Spectrosc.* 48: 857–866, 1994.
49. HR Morris, CC Hoyt, P Miller, PJ Treado, *Appl. Spectrosc.* 50: 805–811, 1996.
50. HR Morris, B Munroe, RA Ryntz, PJ Treado, *Langmuir* 14: 2426–2434, 1998.
51. NJ Kline, PJ Treado, *J. Raman Spectrosc.* 28: 119–124, 1997.
52. JR Schoonover, F Weesner, GJ Havrilla, M Sparrow, P Treado, *Appl. Spectrosc.* 52: 1505–1514, 1998.
53. HR Morris, JF Turner, B Munro, RA Ryntz, PJ Treado, *Langmuir* 15: 2961–2972, 1999.
54. JF Turner, PJ Treado, Proc. *SPIE, Int. Soc. Opt. Eng.* i. 3061: 280–283, 1997.
55. AD Gift, J Ma, KS Haber, BL McClain, D Ben-Amotz, *J. Raman Spectrosc.* 30: 757–765, 1999.
56. DN Batchelder, C Cheng, GD Pitt, *Adv. Mater.* 3: 566–568, 1991.
57. DN Batchelder, C Cheng, W Müller, BJE Smith, *Makromol. Chem. Macromol. Symp.* 46: 171–179, 1991.
58. A Garton, DN Batchelder, C Cheng, *Appl. Spectrosc.* 47: 922–927, 1993.
59. RV Sudiwala, C Cheng, EG Wilson, DN Batchelder, *Thin Sol. Films* 210–211: 452–454, 1992.
60. D Zhang, JD Hanna, Y Jiang, D Ben-Amotz, *Appl. Spectrosc.* 55: 61–65, 2001.
61. MD Schaeberle, HR Morris, JF Turner II, PJ Treado, *Anal. Chem.* 71: 175A–181A, 1999.
62. GJ Puppels, M Grond, J Greve, *Appl. Spectrosc.* 47: 1256–1267, 1993.
63. F LaPlant, D Ben-Amotz, *Rev. Sci. Instrum.* 66: 3537–3544, 1995.
64. B Yang, MD Morris, H Owen, *Appl. Spectrosc.* 45: 1533–1536, 1991.
65. CL Schoen, SK Sharma, CE Helsley, H Owen, *Appl. Spectrosc.* 47: 305–308, 1991.
66. DN Batchelder, C Cheng, BJE Smith, RJ Chaney, Spectroscopic Apparatus and Methods, U.S. Patent 5,689,333, Renishaw plc, 1997.
67. GR Sims, "Principles of Charge-Transfer Devices," in *Charge Transfer Devices in Spectroscopy*, JV Sweedler, KL Ratzlaff, and MB Denton (Eds.), VCH Publishers: New York, 1994.
68. RL McCreery, "CCD Array Detectors for Multichannel Raman Spectroscopy," in *Charge Transfer Devices in Spectroscopy*, JV Sweedler, KL Ratzlaff, and MB Denton (Eds.), VCH Publishers: New York, 1994.
69. B Chase, "Arrays for Detection Beyond One Micron," in *Charge Transfer Devices in Spectroscopy*, JV Sweedler, KL Ratzlaff, and MB Denton (Eds.), VCH Publishers: New York, 1994.
70. D Zhang, D Ben-Amotz, *Appl. Spectrosc.* 54: 1379–1383, 2000.
71. TT Cai, D Zhang, D Ben-Amotz, *Appl. Spectrosc.* 55: 1124–1130, 2001.
72. JF Turner II, "Chemical Imaging and Spectroscopy Using Tunable Filters: Instrumentation, Methodology and Multivariate Analysis," Ph.D. Dissertation, University of Pittsburgh, 1998.
73. S Weber, DA Smith, DN Batchelder, *Vib. Spectrosc.* 18: 51–59, 1998.
74. D Pappas, BW Smith, JD Winefordner, *Appl. Spectrosc.* Rev. 35, 1–23, 2000.

75. CA Drumm, MD Morris, *Appl. Spectrosc.* 49: 1331–1337, 1995.
76. CA Hayden, MD Morris, *Appl. Spectrosc.* 50: 708–714, 1996.
77. A Carden, MD Morris, *J. Biomed. Opt.* 5: 259–268, 2000.
78. JF Turner II, PJ Treado, *Appl. Spectrosc.* 50: 277–284, 1996.
79. MK Bellamy, AN Mortensen, RM Hammaker, WG Fateley, *Appl. Spectrosc.* 51: 477–486, 1997.
80. HEB da Silva, C Pasquini, *Appl. Spectrosc.* 55: 715–721, 2001.
81. QS Hanley, PJ Verveer, TM Jovin, *Appl. Spectrosc.* 52: 783–789, 1998.
82. QS Hanley, PJ Verveer, TM Jovin, *Appl. Spectrosc.* 53: 1–10, 1999.
83. QS Hanley, DJ Arndt-Jovin, TM Jovin, *Appl. Spectrosc.* 56: 155–166, 2002.
84. A Govil, DM Pallister, MD Morris, *Appl. Spectrosc.* 47: 75–79, 1993.
85. FA Kruse, AB Lefkoff, JW Boardman, KB Heidebrecht, AT Shipiro, PJ Barloon, AFH Goetz, *Remote Sens. Environ.* 44: 145–163, 1993.
86. WL Wolfe, *Introduction to Imaging Spectrometers*, Tutorial Text in Optical Engineering v. TT 25, SPIE Optical Engineering: Bellingham, WA, 1997.
87. N Gupta, NF Fell, Jr., *Talanta* 45: 279–284, 1997.
88. N Gupta, R Dahmani, *Spectrochim. Acta* A 56: 1453–1456, 2000.
89. ES Wachman, WH Niu, DL Farkas, *Appl. Opt.* 35: 5220–5226, 1996.
90. IC Chang, U.S. Patent 6,016,216.
91. PJ Miller, *Metrologia* 28: 145–149, 1991.
92. CR Lyot, *Acad. Sci.* 197: 1593–1596, 1933.
93. JW Evans, *J. Opt. Soc. Am.* 39: 229–242, 1949.
94. JW Evans, *J. Opt. Soc. Am.* 48: 142–145, 1958.
95. DM Pallister, A Govil, MD Morris, WS Colburn, *Appl. Spectrosc.* 48: 1015–1020, 1994.
96. R Barer, ARH Cole, HW Thompson, *Nature* 163: 198, 1949.
97. AJ Sommer, JE Katon, *Appl. Spectrosc.* 45: 1663, 1991.
98. JL Koenig, *Microspectroscopy of Polymers*, ACS: Washington D.C.
99. JE Katon, *Vib. Spectrosc.* 7: 201, 1994.
100. L Rintoul, H Panayiotou, S Kokot, G George, G Cash, R Frost, T Bui, P Fredericks, *Analyst* 123: 571, 1998.
101. R Cournoyer, JC Shearer JC, DH Anderson, *Anal. Chem.* 49: 2275, 1977.
102. MA Hartcock, in *The Design, Sample Handling and Applications of Infrared Microscopes*, PB Rousch (Ed.), American Society for Testing and Materials: Philadelphia, PA, 85–96, 1987.
103. JC Shearer, DC Peters, in *The Design, Sample Handling and Applications of Infrared Microscopes*, PB Rousch (Ed.), American Society for Testing and Materials, Philadelphia, PA, 27–38, 1987.
104. FJ Weesner, RT Carl, RM Boyle, *SPIE* 1575: 486, 1991.
105. KJ Ward, Proc. *SPIE Int. Soc. Opt. Eng.* 1145: 212, 1989.
106. D Garcia, J. Vinyl *Add. Tech.* 3: 126, 1997.
107. MA Hartcock, LA Lentz, BL Davis, K Krishnan, *Appl. Spectrosc.* 40:, 1986.
108. JB Joshi, DE Hirt, *Appl. Spectrosc.* 53: 11, 1999.
109. DE Gavrila, B Gosse, *J. Rad. Nuc. Chem.* 185: 311, 1994.
110. X Jouan, JL Gardette, *Polym. Commun.* 28: 239, 1987.
111. EV Miseo, LW Guilmette, "Industrial Problem Solving by Microscopic Fourier Transform Infrared Spectrophotometry," *The Design, Sample Handling and Applications of Infrared Microscopes*, ASTM STP 949, P.B. Rousch (Ed.), American Society for Testing and Materials: Philadelphia, 97–107, 1987.
112. P Wilhelm, *Micron* 27: 341, 1996.

113. G Ahlblad, D Forsstrom, B Stenberg, B Terselius, T Reitberger, LG Svensson, *Polym. Degrad. Stabil.* 55: 287, 1997.
114. MW Tungol, EG Bartick, A Montaser, 44: 543, 1990.
115. MC Grieve MC, RME Griffin, R Malone, *Sci. Justice* 38: 27, 1998.
116. SP Bouffard, AJ Sommer, JE Katon, S Godber, *Appl. Spectrosc.* 48: 1387, 1994.
117. JE Katon, PL Lang, DW Schiering, JF O'Keefe, "Instrumental And Sampling Factors In Infrared Microspectroscopy," *The Design, Sample Handling and Applications of Infrared Microscopes*, ASTM STP 949, PB Rousch PB (Ed.), American Society for Testing and Materials: Philadelphia, PA, 4–11, 1987.
118. DB, Chase, "Infrared Microscopy: A Single Fiber Technique," in *The Design, Sample Handling and Applications of Infrared Microscopes*, ASTM STP 949, PB Rousch PB (Ed.), American Society for Testing and Materials, Philadelphia, PA, 12–18, 1987.
119. T Hirschfeld, *Appl. Spectrosc.* 39: 424, 1985.
120. LL Cho, JA Reffner, DL Wetzel, *J. Forensic Sci.* 44: 283, 1999.
121. P Vermette, J Thibault, S Levesque, G Laroche, *J. Biomed. Mater. Res.* 48: 660, 1999.
122. L Costa, MP Luda, L Trossarelli, EMB del Prever, M Crova, P Gallinaro, *Biomaterials* 19: 659, 1998.
123. J Kressler, R Schafer, R Thomann, *Appl. Spectrosc.* 52: 1269, 1998.
124. E Benedetti, S Catanorchi, A D'Alessio, P Vergamini, F Ciardelli, M Pracella, *Polym. Int.* 45: 373, 1998.
125. M Ludvigsson, J Lindgren, J Tegenfeldt, *J. Electrochem. Soc.* 147: 1303, 2000.
126. NP Camacho, S Rinnerthaler, EP Paschalis, R Mendelsohn, AL Boskey, P Fratzl, *Bone* 25: 287, 1999.
127. E Benedetti, S Catanorchi, A D′Alessio, P Vergamini, F Ciardelli, M Pracella, *Polym. Int.* 45: 373, 1998.
128. D Dibbern-Brunelli, TDZ Atvars, I Joekes, VC Brabosa, *J. Appl. Polym. Sci.* 69: 645, 1998.
129. R Schafer, J Zimmermann, J Kressler, R Mulhaupt, *Polymer* 38: 3745, 1997.
130. MS High, PC Painter, MM Coleman, *Macromolecules* 25: 797, 1992.
131. JJ Sahlin, NA Peppas, J. Biomat. *Sci. Polym. Ed.* 8: 421, 1997.
132. K Sheu, SJ Huang, JF Johnson, *Polym. Eng. Sci.* 29: 77, 1989.
133. SC Hsu, D Lin-Vien, RN French, *Appl. Spectrosc.* 46: 225, 1992.
134. RE Cameron, MA Jalil, AM Donald, *Macromolecules* 27: 2713, 1994.
135. AM Riquet, N Wolff, S Laoubi, JM Vergnaud, A Feigenbaum, *Food Addit. Contam.* 15: 690, 1998.
136. SR Challa, S-Q Wang, JL Koenig, *Appl. Spectrosc.* 50: 1339, 1996.
137. SR Challa, S-Q Wang, JL Koenig, *Appl. Spectrosc.* 51: 297, 1997.
138. BG Wall, JL Koenig, *Appl. Spectrosc.* 52: 1377, 1998.
139. R Schafer, J Kressler, R Neuber, R Mulhaupt, *Macromolecules* 28: 5037, 1995.
140. M Iijima, S Ukishima, K Iida, Y Takahashi, E Fukada, Jap. *J. Appl. Phys. Lett.* 34: L65, 1995.
141. A Kaito, M Kyotani, K Nakayama, *Polymer* 33: 2672, 1992.
142. J Glime, JL Koenig, *Rubber Chem. Technol.* 73: 47, 2000.
143. N Guilhaumou, P Dumas, GL Carr, GP Williams, *Appl. Spectrosc.* 52: 1029, 1998.
144. DL Wetzel, JA Reffner, GP Williams, *Mikrochim. Acta Suppl.* 14: 353–355, 1997.
145. LE Ocola, F Cerrina, T May, *Appl. Phys. Lett.* 71: 847, 1997.
146. JJ Sahlin, NA Peppas, *J. Appl. Polym. Sci.* 63:103, 1997.
147. A Hammiche, M Reading, HM Pollock, M Song, DJ Hourston, *Rev. Sci. Instrum.* 67: 4268, 1996.

148. HM Pollock, A Hammiche, M Song, DJ Hourston, M Reading, *J. Adhesion* 67: 217, 1998.
149. A Hammiche, HM Pollock, M Reading, M Claybourn, PH Turner, K Jewkes, *Appl. Spectrosc.* 53: 810, 1999.
150. MS Anderson, *Appl. Spectrosc.* 54: 349, 2000.
151. Spotlight FT-IR Imaging System by Perkin-Elmer Company.
152. P Colarusso, LH Kidder, IW Levin, JC Fraser, JF Arens, EN Lewis, *Appl. Spectrosc.* 52: 106A, 1998.
153. EN Lewis, PJ Treado, RC Reeder, GM Story, AE Dowrey, C Marcott, IW Levin, *Anal. Chem.* 67: 3377, 1995.
154. R Bhargava, SW Huffman, S-Q. Wang, IW Levin, Unpublished data.
155. R Bhargava, IW Levin, *Anal. Chem.* 74: 1429–1435, 2002.
156. R Bhargava, IW Levin, *Anal. Chem.* 73: 5157, 2001.
157. CM Snively, S Katzenberger, G Oskarsdottir, J Lauterbach, 24: 1841, 1999.
158. SW Huffman, R Bhargava, IW Levin, *Appl. Spectrosc.* 56: 965, 2002.
159. LH Kidder, VF Kalasinsky, JL Luke, IW Levin, EN Lewis, *Nature Med.* 3: 235, 1997.
160. K Artyushkova, BG Wall, JL Koenig, JE Fulghum, *Appl. Spectrosc.* 54: 1549, 2000.
161. R Bhargava, S-Q Wang, JL Koenig, *Macromolecules* 32: 8989, 1999.
162. CM Snively, JL Koenig, *J. Poly. Sci. Pol. Phys.* 37: 2353, 1999.
163. C Marcott, GM Story, AE Dowrey, RC Reeder, I Noda, *Mikrochim. Acta Suppl.* 14: 157, 1997.
164. SJ Oh, JL Koenig, *Anal. Chem.* 70: 1768, 1998.
165. R Bhargava, S-Q Wang, JL Koenig, *Macromolecules* 32: 2748, 1999.
166. JL Koenig, S-Q Wang, R Bhargava, *Anal. Chem.* 73: 360A, 2001.
167. R Bhargava, S-Q Wang, JL Koenig, *Adv. Polym. Sci.*, 163:137, 2003.
168. JL Koenig, S-Q Wang, R Bhargava, *Inst. Phys. Conf. Ser.* 165: 43, 2000.
169. CM Snively, JL Koenig, *Macromolecules* 31: 3753, 1998.
170. R Bhargava, T Ribar, JL Koenig, *Appl. Spectrosc.* 53: 1313, 1999.
171. M van de Weert, WE Hennink, W Jiskoot, *Pharma. Res.* 17: 1159, 2000.
172. M van der Weert, R van't Hof, J van der Weerd, RMA Heeren, G Posthuma, WE Hennink, DJA Crommelin, *J. Control Release* 68: 31, 2000.
173. S Cohen, T Yoshioka, M Lucarelli, LH Hwang, R Langer, *Pharm. Res.* 8: 713, 1991.
174. H Takahata, EC Lavelle, AGA Coombes, SS Davis, *J. Control. Release* 50: 237, 1998.
175. EN Lewis, IW Levin, *J. Microscop. Soc. Amer.* 1: 35, 1995.
176. EN Lewis, AM Gorbach, C Marcott, IW Levin, *Appl. Spectrosc.* 50: 263, 1996.
177. EN Lewis, LH Kidder, IW Levin, VF Kalasinsky, JP Hanig, DS Lester, *Ann. NY Acad. Sci.* 820: 234, 1997.
178. DS Lester, LH Kidder, IW Levin, EN Lewis, *Cell. Mol. Bio.* 44: 29, 1998.
179. LH Kidder, P Colarusso, SA Stewart, IW Levin, NM Appel, DS Lester, PG Pentchev, EN Lewis, *J. Biomed.* Opt. 4: 7, 1999.
180. R Mendelsohn, A Hasankhani, E DiCarlo, AL Boskey, *Calcif. Tissue Int.* 59: 480, 1996.
181. C Marcott, RC Reeder, EP Paschalis, AL Boskey, R Mendelsohn, *Phosphor. Sulfur* 146: 417, 1999.
182. R Mendelsohn, EP Paschalis, AL Boskey, *J. Biomed. Optics* 4: 14, 1999.
183. R Mendelsohn, EP Paschalis, PJ Sherman, AL Boskey, *Appl. Spectrosc.* 54: 1183, 2000.
184. MK Bellamy, AN Mortensen, EA Orr, TL Marshall, JV Paukstelis, RM Hammaker, WG Fateley, *Mikrochim. Acta Suppl.* 14: 759–761, 1997.
185. WG Fateley, RA Deverse, RM Hammaker, *J. Appl. Polym. Sci.* 70: 1307, 1998.

186. MK Bellamy, AN Mortensen, RM Hammaker, WG Fateley, *Appl. Spectrosc.* 51: 477, 1997.
187. F Zhang, T Gu, *Proc. SPIE* 1205: 150, 1990.
188. RA DeVerse, RM Hammaker, WG Fateley, *Appl. Spectrosc.* 54: 1751, 2000.
189. JA Bailey, RB Dyer, DK Graff, JR Schoonover, *Appl. Spectrosc.* 54: 159, 2000.
190. JM Bennet, *Appl. Opt.* 15: 2705, 1976.
191. EN Lewis, IW Levin, *J. Microscopy Soc. Am.* 1: 35, 1995.

6 Vibrational Circular Dichroism of Biopolymers: Summary of Methods and Applications

Timothy A. Keiderling, Jan Kubelka, and Jovencio Hilario

CONTENTS

6.1 INTRODUCTION

Vibrational circular dichroism (VCD) is a measurement of the differential absorption of left and right circularly polarized light ($\Delta A = A_L - A_R$) by molecular vibrational transitions, typically in the infrared (IR) region of the spectrum [1–10]. As an absorption technique, it samples the same transitions as in IR spectra; however, its intensity (ΔA) must arise from chiral interactions of molecular bonds in an asymmetric molecule, polymer, or medium. For small molecules, the resulting spectral pattern is often a characteristic of the three-dimensional configuration of the atoms bound within the molecule. For biopolymers this "absolute" configuration is not an unresolved question that needs to be answered experimentally. Amino acids in proteins normally all have an L configuration, and the ribose rings in DNA and RNA all have the same configuration. Polysaccharides are more complex, yet absolute configuration is typically not the driving structural issue even for such systems.

In biopolymers, it is the relative stereochemistry of successive residues in a chain that most often dominates structural analysis studies, because this determines the fold characteristics of the chain of repeating subunits. At its simplest, this is termed secondary structure, and relates to the sequential orientation of residues (helical, extended, and so on) in a conformationally uniform segment of the polymer. Since this mutual orientation affects the mechanical coupling of molecular vibrations as well as their associated dipole moments, it impacts both the IR and VCD spectra. The modes important in such coupling-based VCD might be termed the polymer IR chromophores. These modes often differ from those that might be important in the monomer VCD. This provides a separate view of effects from configuration and (polymer) conformation in the spectrum. Consequently, one of the major applications of IR and VCD spectra in biopolymer structural studies has been the determination of average, fractional secondary structure content (most often of peptides and proteins, but occasionally for nucleic acid applications) [10–14]. Contributions to the spectrum from different secondary structural types tend to overlap, so that these studies can typically only yield fractional components (such as % helix), and cannot provide site-specific structural interpretations. IR and VCD have some advantages over other spectral techniques used for secondary structure analyses, but are best used in conjunction with them to exploit complementary sensitivities available with electronic circular dichroism (ECD) in the UV, Raman, nuclear magnetic resonance (NMR), and electron paramagnetic/spin resonance (EPR/ESR). Thus, a brief sketch of those techniques and their strengths for biopolymer studies may help position this review.

The focus on average secondary structure with optical spectroscopic techniques is a consequence of the moderate strengths of the predominant interactions that impact optical spectra, and of the relatively low intrinsic resolution of those spectra, which ordinarily does not provide site-specific information without selective isotopic substitution [15–17]. Such a limit is in contrast to X-ray crystallography or NMR spectroscopy, which naturally yield site-specific structural information due to their very high resolution (at least for NMR). However, though they are invaluable as structural biology techniques, NMR and X-ray structure analyses are slow both in terms of data acquisition and in terms of the complex interpretive process. These techniques are limited to relatively small soluble molecules or crystallizable species, respectively, and they normally require a substantial amount of protein. Furthermore, beyond the interpretive effort required, neither of these measurements lends itself to rapid time-scale measurements, thereby not normally permitting reliable analysis of dynamic structures or of conformations undergoing fast changes. EPR is capable of detecting faster time-scale structural variations and can yield good distance data but requires spin-labeling, which, beyond considerations of the synthetic difficulty, can change the structure. Limited applications of optical spectra for determination of the tertiary structure, or the fold of the secondary structural elements, have appeared. Typically these use fluorescence or near-UV ECD of aromatic residues, but only sense a change in the fold (through change in the chromophore environment or in the distance between chromophores) rather than determine its nature [18,19].

The dominant technique for secondary structure analyses, particularly of peptides and proteins, has been far-UV ECD of π-π^* and n-π^* transitions of the amide group in peptides [20–26] or the bases in nucleic acids [27,28] in the ultraviolet. Its sensitivity to molecular conformation and ability to study relatively small amounts and dilute concentrations of a sample give ECD a distinct advantage. ECD band shapes gain discriminatory capability by their complete sign reversals for selected structural changes. Since these transitions yield unresolved ECD bands, and those corresponding to different conformations are totally overlapped, the usual interpretative methods employ bandshape pattern-recognition-based algorithms that are either qualitative in nature or dependent on a statistical fit to a set of (typically protein) spectra which provide the structural reference set [11,21,22,26].

IR and Raman analyses of secondary structure historically took a different approach [29,30], due to the natural resolution of the spectrum into contributions from vibrational modes characteristic of different bond types in the molecule. The initial focus was on assigning component frequencies to various secondary structural component types, as has been discussed in several reviews [13,17,31–33]. Most effort focused on the mid-IR amide I (C=O stretch) band with additional use of the amide II (N-H deformation plus C-N stretch) in IR spectra and amide III (oppositely phased N-H deformation plus C-N stretch) with Raman spectral methods. In nonaqueous media, the near-IR amide A (N-H stretch) can also be useful.

Since both IR and Raman techniques give rise to single-signed spectral bandshapes that are effectively just the dispersed sum of the contributions from all the component transitions, and those components differ in frequency by only relatively small intervals compared to the bandwidths, the bandshapes for different proteins are very similar. However, due to the high signal-to-noise ratio (SNR) of Fourier transform IR (FT-IR) spectroscopy, in particular, resolution enhancement using second derivative or Fourier self-deconvolution (FSD) techniques [34–36] can lead to added interpretability, but can also be subject to misuse by the novice user [12,13,37]. Solvent as well as secondary structural segment nonuniformity and end effects have an impact on frequencies, dispersing the contribution of each component over a significant spectral range, leading to real difficulties with the simplifying assumptions typically used for band assignments [38–40]. For example, most methods depend on an assumption that the dipole strengths (extinction coefficients) of all the residues, regardless of their conformation, are the same, while they in fact vary [13,41,42].

Nonetheless, surprisingly accurate analyses have appeared. Bandshape-based analyses, similar to those used for ECD, have also been applied to FT-IR and Raman spectra with reasonable success [11,38,43–46].

This contrast of FT-IR and ECD sensitivities to structure, and the desire to give the more highly resolved vibrational band components a conformationally dependent sign variation, led to the development of VCD and its counterpart, Raman optical activity (ROA). Only VCD will be addressed here, but other reviews dealing with ROA are widely available [9,47–50]. The key impetus for moving to the vibrational region of the spectrum is its rich distribution of resolved transitions, which are characteristic of localized parts of the molecule. By probing their chirality, VCD measurements can expose the distinct stereochemical sensitivity of these vibrational modes [8–10,47,49,51–55]. The chromophores needed for VCD are the bonds themselves as sampled by their stretching and bond deformations. Often the most intense transitions correspond to motions of planar (locally achiral) parts of the molecule. Since in a polymer the secondary structure can impose a chiral interaction between these achiral repeating segments, their VCD bandshapes will reflect the polymeric structural character. VCD is to IR what ECD is to UV absorption spectra; both endow the absorption phenomenon with a three-dimensional structural sensitivity to the physical interactions underlying the measurement.

This benefit comes at a cost, manifested as significantly reduced SNR and some theoretical interpretive difficulty for VCD as compared to IR. Developments on the latter front are fast bringing the theoretical capability for prediction of VCD spectra of small molecules to a level that is demonstrably superior to that for ECD spectra [7,56]. We have been able to extend such reliable calculational methods to moderately sized peptides and, with some assumptions, to relatively large peptides [57–59]. However, until now, *ab initio* quantum mechanical methods have generally been of limited use for large molecules, such as proteins and especially nucleic acids. Thus most previous biomolecular applications of VCD have used empirically based analyses [53,55,60]. Experimentally, instrumentation has reached a stage where VCD spectra for most molecular systems of interest can be measured under at least some sampling conditions [3,6,55,61–65]. It is true that most VCD studies of biomolecules in aqueous solutions, naturally the conditions of prime interest, are restricted to the study of relatively high concentration samples, though typically somewhat less than is characteristic of NMR. Often D_2O based solvents are used to improve SNR around 1650 cm^{-1} (location of the peptide amide I and near the main C=O stretch in nucleic acids) and to permit study of more dilute solutions, all at the cost of added complications due to H-D exchange. (By comparison, previous ROA measurements have often demanded even higher concentrations [50,66]).

Sampling conditions for obtaining experimental VCD spectra of protein and peptide samples are similar to those used in FT-IR studies with the important exception that the data are differential spectra of much smaller amplitude (ΔA is the order of 10^{-4} to 10^{-5} times the sample absorbance A) and thus lower SNR. Consequently, to obtain quality VCD spectra, much longer data collection times are required than for FT-IR or ECD, and specially designed instruments are used. Instrumentation and sampling methods for biomolecular VCD are summarized in the next section and are fully discussed in separate reviews [8,9,61–63,67]. Theoretical techniques for simulation of small molecule VCD are also the focus of several previous reviews [5–7,9,47,51,56] and will not be covered in detail here. A brief survey of biopolymer VCD spectral simulation is presented in a section following the experimental methods to illustrate the state of the art as it was as of 2000. While restricted in direct application to biopolymers, recent advances in computational techniques and hardware have made it possible to carry out larger calculations on realistic peptides [59].

On the other hand, qualitative analyses of secondary structure for any size biopolymer can be easily done utilizing the VCD bandshape and its frequency position, assuming there is a dominant uniform structural type [64]. For small, fully solvated biopolymers, the frequency shifts due to the inhomogeneity of the secondary structure are severe problems for frequency-based analyses (FT-IR, Raman). However, the VCD bandshapes are conserved, arising from interactions between local modes, and consequently shift with the absorption bands, permitting relatively simple analyses. In globular proteins, qualitative estimations of structure can be of some interest for determining the dominant fold type, but quantitative estimations of fractional secondary structure content based on empirical spectral analyses are usually of more interest.

Quantitative methods for analysis of protein VCD spectra follow the methods established by ECD analyses [21,22] and employ bandshape techniques referenced to a training set of protein spectra [11,68–75]. In this respect, globular proteins in aqueous solution are assumed, on average, to have their peptide segments in similar environments and of similar lengths, so that the solvent and length effects on the peptide modes will be relatively consistent for the training set and any unknown protein structures studied. Clearly this assumption imposes an intrinsic limit on the accuracy of such methods. However, independent of experimental conditions, the same sort of limit arises from the difficulty in defining the extent of helices or sheets when all the dihedral angles between residues vary from ideal values. Thus precision in reproducing secondary structures obtained from some algorithm for analysis of X-ray structures should not be the ultimate goal; general conformational accuracy is more important.

In summary, unlike ECD, VCD can be used to correlate data for several different spectrally resolved features; and, unlike IR and Raman spectroscopies, each of these features will have at its source a physical dependence on stereochemistry. But from another point of view, use of a combination of these methods in analyzing a biomolecular structure can compensate for the weaknesses of each, providing a balance between accuracy and reliability. The prime questions remaining in the VCD field relate to application and interpretation of the resulting spectra. It is clear that, despite occasionally inflated claims of fundamental advantages of any one technique, perhaps due to marketing of instruments or pressures to gain recognition (or funding) for one's work, real progress in understanding of biomolecular structures will come from combining the information derived from all the data gathered with various techniques [12,70]. In our biomolecular work, different types of spectral data are used to place bounds on or add credence to the reliability of structural inferences that might be drawn from any one method alone.

6.2 EXPERIMENTAL METHODS

Extending optical activity measurements into the mid-IR necessitates special design considerations, in that the rotational strengths of vibrational transitions as detected in VCD are much weaker than those of electronic transitions detected in ECD and that IR sources are weaker and IR detectors noisier than those available in the ultra violet and visible (UV-Vis). Similarly, because VCD is a differential IR technique, its SNR can never approach that of FT-IR, which represents a summed response. Several research groups have developed instrumentation that makes the measurement of VCD reasonably routine over much of the IR region [61–63,76–82]. Commercial FT-IR vendors are now providing high quality VCD accessories [83–92] or, in one instance (Bomem-Biotools), a stand-alone VCD instrument that has now been shown by its users to have an exceptional SNR and baseline characteristics [93,94]; a second one is expected to be marketed soon [91,92]. In this section, general VCD instrument designs are summarized and compared in some detail.

6.2.1 INSTRUMENTATION: FT-VCD VERSUS DISPERSIVE VCD

Available instrumentation makes routine measurement of VCD possible over much of the IR region down to ~700 cm^{-1} on many samples. This low-frequency limit is primarily due to limits on the transmission of optical materials, strength of sources, and sensitivity of detectors. Development of a VCD instrument is normally accomplished by modifying a dispersive IR or an FT-IR spectrometer to incorporate time-varying modulation of the polarization state of the light and a detection scheme for the modulated intensity that results. Various instruments have been described in the literature in detail [4,6,9,61–63] and a detailed review contrasting these designs and detailing the components needed to construct a VCD instrument was published some time ago by one of the present authors [62]. Consequently, while only a survey of these designs is given here, enough detail is presented to distinguish between them and to bring previous descriptions more up to date. Comparison of their efficacy for biopolymer applications will be given in Section 6.2.3.

VCD instruments share several generic elements with other kinds of CD instruments. All current instruments use a broadband light source, typically based on black-body radiation from a ceramic or graphite-based glower (or tungsten in the near-IR), to allow sampling of a spectrum over the IR region. The method chosen for encoding the optical frequencies divides VCD instruments into two styles, dispersive and Fourier transform. Dispersive VCD instruments use a grating monochromator, which must scan through the wavelength spectrum of interest, and record the spectral response sequentially as a function of wavelength. Such an instrument can be optimized for efficient light collection. Resolution is improved by closing the monochromator slits (and degrading SNR), while SNR can be improved by opening them (degrading resolution) and by increasing the instrumental time constant (and thus scanning more slowly) or by averaging repeated scans. Fourier transform VCD (FT-VCD) instruments use a Michelson interferometer that encodes the optical wavenumbers as a function of the moving mirror position, resulting in an interferogram. All frequencies available in the instrument bandpass are sampled simultaneously, thereby gaining efficiency through the multiplex advantage. With FT-VCD, longer mirror travel serves to improve the resolution of spectral features, and the co-addition of repeated scans improves SNR.

Both styles of VCD instrument have their sample compartments modified to allow linear polarization of the light beam. Normally this is done with a wire grid polarizer, which we have obtained, for example, from Cambridge Physical Sciences (manufactured in the U.K. and distributed by Molectron in the U.S.), with CaF_2 or BaF_2 substrates to allow access from the near-IR down to 1100 cm^{-1} and 900 cm^{-1}, respectively, encompassing regions of use for biopolymer studies. For studies of aqueous solution samples, such a substrate matches constraints of the typical sample cell. These polarizers have very high density Al wire arrays (made by a holographic process) which gives an improved polarization ratio at higher wavenumbers. Alternative polarizers with ZnSe and other substrates are also available for wider spectral coverage but have some added reflection loss that is not a problem with low index materials like BaF_2. Sine-wave modulation of polarization between left and right circular polarization states (actually the degree of elliptical polarization varies over the IR region measured) is obtained with a photo-elastic modulator (PEM) which is placed in the optical path directly following the polarizer. Such modulators are available from Hinds Instruments (Hillsboro, OR) based on CaF_2 and ZnSe optical elements, and have modulation frequencies ranging up from 30 kHz (smaller crystals yield higher frequencies but accordingly result in smaller beam apertures, which can be a design limitation). The beam then passes through the sample (about which more detail will be given below) and onto the detector, typically a liquid N_2-cooled, $Hg_{1-x}(Cd)_xTe$ (MCT) photoconducting diode. While these are available from several sources, detectors with both a very high D* and a relatively large area (e.g., 2 mm square) are desirable. (In our experience, very small detectors have undesirable polarization characteristics, but some other designs have used them successfully.)

The electrical signal developed in the MCT detector and preamplifier contains two modulations, one at low frequencies created either by a chopper or the varying interference changes at each wavenumber caused by mirror motion, and another produced by the polarization modulation. These modulation signals are at very different frequencies so that they can be separated by filters into two signal channels to measure the overall transmitted intensity spectrum (I_{trans}) of the instrument and sample via one channel, and the polarization modulation intensity (I_{mod}) resulting from the VCD via the other. In a dispersive instrument, these measurements are realized with separate lock-in amplifiers, one referenced to the chopper and the other to the polarization modulator. For the FT-VCD instrument, the computer directly digitizes and Fourier transforms the I_{trans} signal from a low-pass filtered channel; while a lock-in amplifier referenced to the polarization modulator is used to convert the I_{mod} signal into an interferogram, representing the polarization-modulated signal that the instrument computer can now recognize and transform.

These signals are ratioed to yield the raw VCD signal either before or after A-to-D conversion, depending on the instrumental design. Because VCD is a differential absorbance measurement, $\Delta A = A_L - A_R$ for left/right (L/R) circularly polarized light, it is necessary to ratio the modulated to the transmitted intensity obtain a measure of absorbance and thereby normalize out any dependence on the source and instrument spectral characteristics. In the limit of small ΔA values,

$$I_{mod}/I_{trans} = (1.15\ \Delta A)J_1(\alpha_0)g_I \tag{6.1}$$

where $J_1(\alpha_0)$ is the first order Bessel function at α_0, the maximum phase retardation (expressed as an angle) of the modulator, and g_I is an instrument gain factor. Evaluation of this term and elimination of the gain factor are attained by calibration of the VCD using a pseudo sample, composed of a birefringent plate and a polarizer pair, or by measuring the VCD of a known sample [3,61,62,79,95,96]. Completion of processing of the computer stored VCD spectrum involves calibration, baseline correction and spectral averaging or smoothing, as desired. Optionally, the spectra can be converted to molar quantities; e.g., $\Delta\varepsilon = \Delta A/bc$ where b is the path length in cm and c is the concentration in moles/L. To give these concepts more substance, some details of the dispersive and FT-IR based instruments used in our University of Illinois at Chicago (UIC) laboratory are given below.

6.2.1.1 Dispersive VCD

Our original dispersive instrument is configured around a 1.0 m focal length, ~f/7 monochromator (Jobin-Yvon, ISA) that is illuminated with a home-built, water-cooled carbon rod source [62,81]. A mechanical, rotating wheel chopper (150 Hz) provides the modulation necessary for detecting the instrument transmission with an MCT detector. The monochromator output is filtered with a long wavepass interference filter to allow only first-order diffraction from the grating to pass and uses mirrors to focus the beam achromatically on the sample. A schematic of this dispersive design, which has been the instrument of choice for obtaining most of our biopolymer VCD spectra, is shown in Figure 6.1. An alternate, more compact design — in our case, with a condensed, nearly parallel beam at the sample — has been shown to have some advantages [76,82].

For the mid-IR, a BaF_2 substrate wire grid polarizer (Cambridge Physical Sciences) and an antireflective coated 38 KHz ZnSe PEM (Hinds Instruments) are used. Following the sample, a ZnSe lens focuses the light onto a three-element MCT detector (Infrared Associates) of high D* ($1.3 - 2.7 \times 10^{10}$ cm $W^{-1}Hz^{1/2}$) chosen in terms of size and shape to match the slit image (~2×6 mm) for optimal signal development. This combination permits operation to below ~900 cm^{-1}. (Recently we have installed a new narrower-band MCT detector with a $D^* > 4 \times 10^{10}$ and with a spectral range limited to $\lambda < 8\ \mu$, and have obtained a substantial increase in SNR.) Alternatively, very high sensitivity in the near-IR (~5000–1900 cm^{-1}) is possible with an InSb photovoltaic detector, CaF_2 modulator and lens, and the use of a higher groove density grating optimized for that region. The near-IR overtone region is also straightforwardly accessible with dispersive designs [97–99]. Alternatively, lower wavenumbers (to $\geq$ 600 cm^{-1}) can be accessed with low groove density gratings and different detectors (e.g., As-doped Si, cooled with liquid He) but this typically results in a loss of SNR [62,100]. The ZnSe transmission cutoff poses the current wavelength limit in those experiments. While KRS5 or other materials could permit further penetration of the far-IR, the optical quality (degree of strain) in such materials has so far restricted such

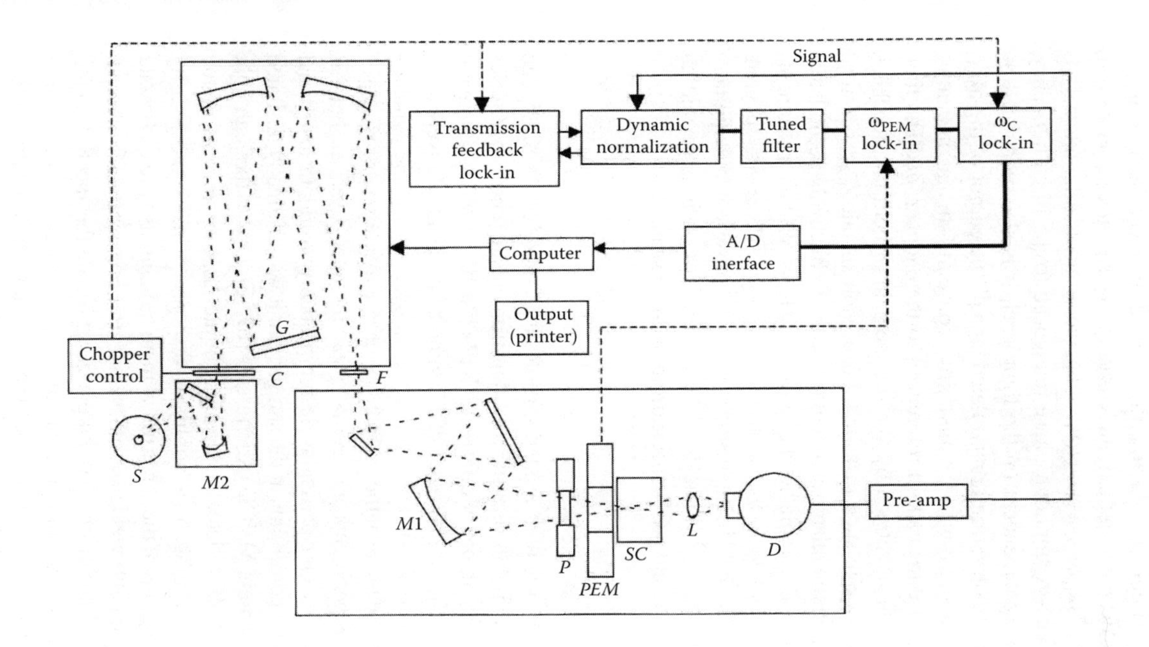

FIGURE 6.1 Schematic diagram of the optical elements (for mid-IR operation) and electronics components of the UIC dispersive VCD spectrometer. S: C-rod light source in cooled housing; M2: focusing mirror (to match monochromator f/#); C: rotating wheel chopper (150 Hz); G: monochromator grating (150 grooves/mm); F: long-wave pass filter; M1: set of mirrors leading to focusing (doubles optical speed and reduces image at sample); P: wire grid polarizer (Al on BaF_2); PEM: photoelastic modulator (ZnSe); SC: sample cell (CaF_2 or BaF_2); L: lens (ZnSe); D: MCT detector (new design, 4 2 × 2 mm elements, stacked). Necessary electronic components are indicated on the upper right: PreAmp consists of 4 preamps matched to each detector element and a summing amplifier; dynamic normalization circuit and tuned L-C filter are homemade, based on a comparator to keep the output signal from the ω_c lockin, (ω_c ~150 Hz) which detects transmission, at ~0.5 V; the ω_{PEM} (~37 kHz) lock-in operates with a minimum time constant so that its output maintains the chopper modulation and can input to the ω_C lock-in. Both of these demodulate the VCD signal, and the A/D convertor does 12 bit digitization and transmits the result every 0.4 nm to a PC-compatible computer for processing. This is significant oversampling, since the resolution at 6 μ is ~30 nm and the normal time constant is 10 sec. As a result, it yields a high-frequency noise element to the VCD spectrum that can be corrected by computational smoothing with no loss of information.

efforts [101]. A modulator capable of using various birefringent optical elements and thus in principle capable of far-IR operation has been built, but initial tests in the mid-IR were not competitive with a PEM on an SNR basis [96,102]. Wide band MCTs (capable of detecting as low as 400 cm^{-1}) have been found to be too noisy for routine VCD spectroscopy in our lab.

To process the signal, a simple lock-in amplifier is used to detect the transmitted intensity proportional to the detector signal (hundreds of mV) developed in phase with the chopping frequency, ω_C. The very weak (μV) polarization modulation intensity is measured with a separate lock-in amplifier, phase referenced to the PEM frequency, ω_{PEM}. Since the VCD signal is dependent on the light-level, it is also modulated by the chopper. Therefore, this polarization demodulated signal (the output of the ω_{PEM} lock-in) is processed with a minimal (τ < ms) time constant, so that it can be demodulated again by using a third lock-in referenced to the chopper, ω_C. The final two signals, V_{mod} and V_{trans}, can be ratioed by varying the gain of both in a feedback circuit incorporating the ω_C lock-in such that the transmission signal, V_{trans}, is forced to be constant. The polarization modulated signal thus becomes effectively normalized, in much the same manner as is accomplished in conventional UV-Vis CD instruments by varying the high voltage applied to the photomultiplier tube which sets its gain. Alternatively, both signals can be simultaneously A-to-D converted and digitally divided in the computer [63,76,82], which has other advantages such as permitting measurement of the absorbance and VCD spectrum in a single scan.

6.2.1.2 FT-VCD

Our FT-IR-VCD spectrometer was built around a BioRad (Cambridge, MA) Digilab FTS-60A FT-IR [62,67,79,80,95], but very successful instruments in other laboratories have been configured around a number of different FT-IRs [78,103–108]. A schematic is illustrated in Figure 6.2. The linear polarizer, PEM, sample, lens, and relatively large area MCT detector, as described above, are contained in an external sample compartment. Use of a nearly collimated beam at the sample was originally chosen for purposes of magnetic VCD (MVCD) experiments and seems to lead to flatter baselines, but this is difficult to test systematically. Our particular design probably results in lost signal and lower SNR due to aperture constraints at the sample, but it was originally designed for high aperture gas phase sample MVCD experiments [67,95]. Due to the high light intensity levels in an FT-IR, MCT detectors easily saturate, as can preamps, if too large a signal is developed, leading to a nonlinear response. This is most easily monitored by checking that regions of the spectrum with no light intensity (e.g.. beyond the detector or filter cutoff) actually yield a zero baseline after Fourier transformation. Optical filters (e.g., 1900 cm^{-1} cutoff low-pass) are used to isolate the spectral region of interest. For aqueous, biological samples, the spectral band pass is additionally strongly limited by the solvent, so that detector saturation is not a significant problem even with high source powers and large apertures. Additionally, use of slightly defocused beams and larger detectors as well as a variable gain preamp can be useful in control of such nonlinearities.

In a rapid scan instrument, the detector signal is processed (Figure 6.2 inset) by a lock-in amplifier referenced to the modulator, and its output signal (processed at a fast time constant, 30–100 μs, which is possible with an SRS 830 lock-in amplifier, Stanford Research Systems, Sunnyvale, CA) forms an ordinary interferogram representing the polarization-modulated signal. This can be Fourier transformed to yield a single beam response spectrum, which, when can be computationally ratioed to the ordinary single beam transmission spectrum yields the raw VCD spectrum. Most rapid scan FT-IR VCD spectra have concentrated on mid-IR bands since the lock-in attenuates the higher frequency near-IR sidebands. Fourier frequencies in rapid scan experiments vary linearly with the mirror speed and optical wavenumber. Slow- or step-scan operation yields better response for higher frequency, near-IR, components of the spectrum [79,101,106–109]. With step-scan operation, removal of the time element results in a simpler phase correction but conversely poses difficulties for measurement of the normal transmission interferogram, since it creates a signal that is effectively DC (modulated only at the step frequency). Phase modulation [106] can overcome this detection problem, but results in reduced SNR for simultaneously measured VCD [79].

The integral of the modulated spectrum is typically very small, since it is often the sum of several positive and negative VCD bands. This results in there being only a very weak center burst in the interferogram. For purposes of interferometer alignment and phase correction, this can pose difficulties [62,79,80,95,110]. Normally, software modifications for transferring phase correction and permitting simultaneous or sequential measurement from two independent detector (I_{trans} and I_{mod}) inputs are required. Unlike the case for FT-IR hardware, where most research grade instruments have sufficient capabilities to be adapted for VCD use, software flexibility should be a central consideration in choosing an instrument for modification to VCD.

Most commercial designs for FT-VCD reflect aspects of instruments developed in academic research labs including that at UIC, described here. An exception is the BOMEM-BioTools (Quebec, Canada and Waconda, IL) design, which is a purpose-built, dedicated mid-IR FT-VCD instrument. All the optical and electronic components of that design were selected to optimize its use for VCD and to provide simple operation and highly reproducible data. Consequently this commercial instrument is less flexible than the UIC or other designs based on an external bench, for example, but has a very high SNR and an exceptionally flat baseline as shown in several recent applications [110–115]. Jasco has also announced (2003) a dedicated VCD instrument. Another variant is available from Bruker Optik (Ettlingen, Germany) which optically consists of an accessory compartment added to a normal FT-IR bench, but computationally collects the transmission and VCD (polarization modulation) interferograms simultaneously in one interleaved data file. The published results with this design are also impressive [83–86,88]. (Data from these instruments for biopolymer samples are compared with data from the UIC dispersive VCD in Section 6.2.3.) Thermo-Nicolet and Digilab market similarly configured, accessory based VCDs, but little has appeared yet in the literature regarding them.

Finally, an extension of the Digilab (Randolph, MA) digital signal processing (DSP) approach to modulated FT-IR applications [116], allows the possibility of detecting polarization modulation in an FT-IR without the use of a lock-in amplifier. This advance is electronics and software based, making use of the conventional external bench approach, but depending on step-scan technology. In the Digilab DSP

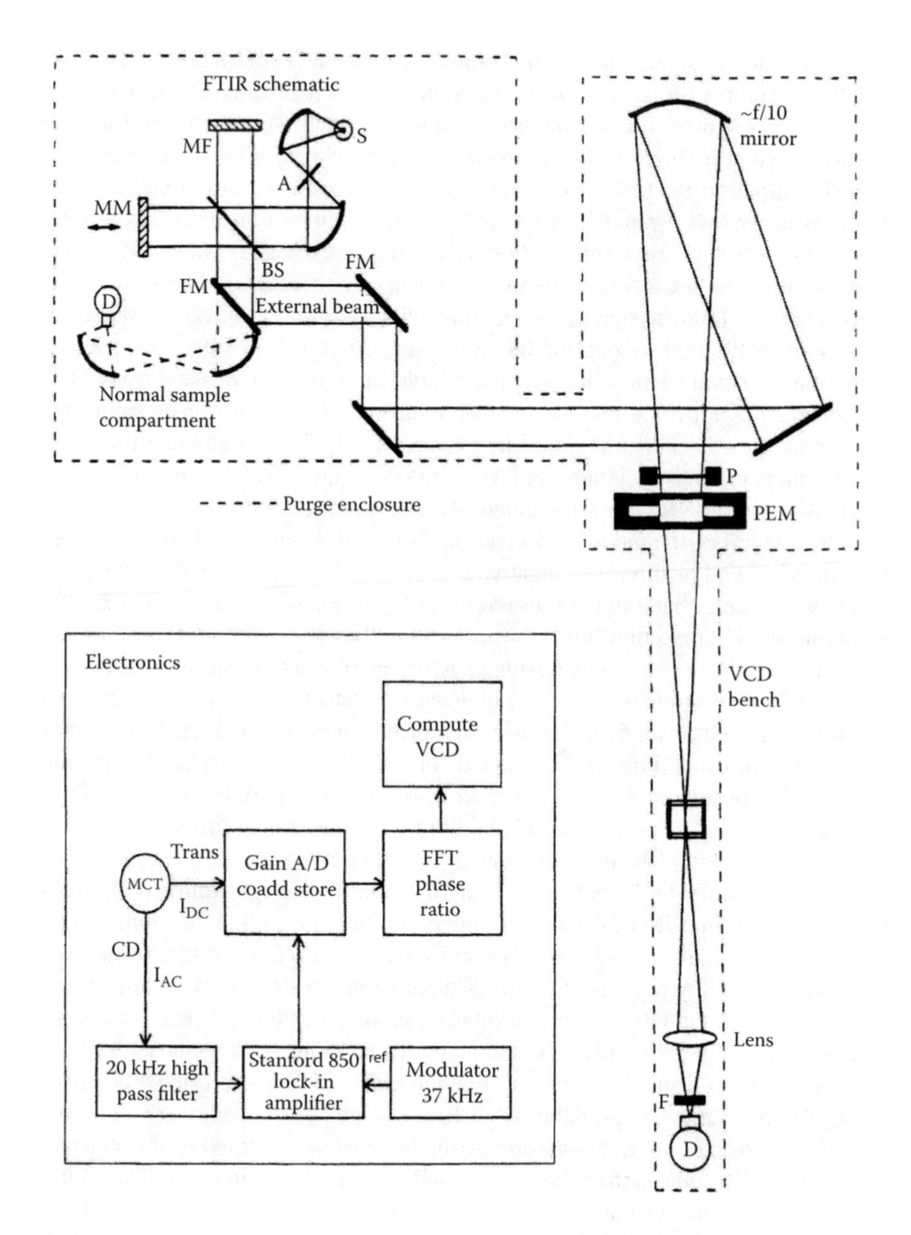

FIGURE 6.2 Schematic diagram of the UIC FT-VCD spectrometer. The components include a Digilab FTS-60A FT-IR with all internal components standard: BS: beam splitter (KBr); FM: flip mirror; MM: moving mirror; MF: fixed mirror; A: aperture; S: high temperature ceramic source. External optics include a long focal length mirror for weak focus (> f/10) at sample; P: grid polarizer; PEM: ZnSe photo-elastic modulator; F: 1900 cm^{-1} cutoff long-wave pass filter; D: detector (MCT) and matched variable gain preamp. This optical design necessarily loses light due to its large image at the sample, but was designed to accommodate a large magnet and minimize the consequences of its stray field for MVCD. Inset: schematic diagram of the electronics components and functions of the UIC FT-VCD instrument.

approach, after the mirror is moved, the whole signal which develops as a function of time is recorded for each step, eventually scanning the whole interferogram. The data at each step is Fourier analyzed to determine the intensities of the oscillating components of the signal that correspond to the applied modulation and associated beat frequencies of interest. Phase modulation generates the I_{trans} signal and polarization modulation the I_{PEM}. After determination of an intensity for each step, the interferograms are transformed, and ratioing again yields the raw VCD signal. While this design was originally developed for linear dichroism studies of polymers, VCD data having high SNR and a close match to published spectra were obtained for small chiral molecules and for ideal biomolecular samples in reasonable measurement times [117, 118]. The restriction to step-scan operation brings some complications, but for typical biological samples, the longer scan times resulting from 1 Hz steps should be comparable to reasonable total measurement times, since extensive signal averaging is needed with conventional designs.

While FT-VCD has many potential advantages for the study of biopolymer systems, in practice, the relatively broad bands and the restriction of data only being measurable in the spectral windows of water can offset the multiplex and throughput advantages of FT-IR. Often this results in dispersive VCD being a more useful method, since it can be used to focus the data collection effort on a single spectral band or narrow region of interest and, with proper optical design, can use relatively wide slits to gain very high throughput thereby offsetting that advantage of FT-IR. Before now, the expected FT-IR advantages had *not* been experimentally realized in terms of enhanced SNR for low resolution biomolecular (aqueous) FT-IR–VCD spectra as compared to what can be measured for a single band with the dispersive instrument over a similar time span [62,79,119]. However, for multiple band measurements, the FT-IR multiplex advantage does save time and improve SNR. Examples of state-of-the-art spectral data for biomolecule samples on each type of instrument are presented in Section 6.2.3. It is clear that the latest FT-IR-based instruments are now very competitive with the older dispersive designs, even for a single band.

Dispersive VCD spectra are often measured with time constants of the order of 10 sec and resolutions of ~10 cm^{-1}, or are averaged over a large number of repeated, faster time-constant scans [76,77]. This means that scanning a typical IR band can take about one half hour or more. FT-IR-measured VCD spectra can sample a much wider spectral region in the same time, but can require extensive averaging over much longer times to match the SNR available using the dispersive instrument for single bands in aqueous phase biopolymers. If, in the end, only one or two adjacent bands are needed for the analysis, much time can be lost with the FT-IR-based technique; but if multiple, widely spaced bands are to be studied, FT-IR-VCD clearly

FIGURE 6.2 (*Continued*) For normal rapid-scan operation, the detector signal is processed through two channels sequentially, with the TRANS (I_{DC}) signal developed in the normal low-pass filtered electronics of the FT-IR and computer. The CD (I_{AC}) signal is developed by use of a high-pass filter (to eliminate the normal interferogram) and a lock-in amplifier to demodulate the PEM frequency signal, which is then amplified, coadded, and fast Fourier transformed (FFT). The ratio of both single beam responses (TRANS and CD) gives the raw VCD. This is normalized by the calibration spectrum (or value), and the final VCD is calculated.

retains its advantage, even for biological samples [79,120,121]. In both cases, VCD scans must be coupled with equally long collections of baseline spectra to correct for instrument and sample induced spectral response. Finally in this comparison of FT-IR and dispersive based VCD instruments, it might be noted that data from the more direct dispersive measurement is intuitively easier to interpret in case something goes wrong, such as excess noise or baseline artifacts coincident with an absorption for which the VCD is sought. Due to the FT process, features in FT VCD corresponding to noise have the same sort of bandshapes as do the desired VCD signals. Each appears at the instrument resolution since, to save time, and enhance SNR, one typically scans the interferogram only out to the resolution needed.

Due to its intrinsically weak signal size, VCD is subject to artifacts which must be corrected by careful baseline subtraction. Though impractical for most biological materials, the best baseline is determined using a racemic sample of the same material, which has an absorbance spectrum and index of refraction identical to that of the chiral sample. However, satisfactory baselines for spectral corrections can often be acquired with carefully aligned instruments by measuring VCD spectra of the same sample cell filled with just solvent. We find this is particularly true of low absorbance aqueous solution samples having very short path length. There exists no completely satisfactory theory of all these artifacts; thus full control of baselines is still an unmet goal. A novel dual modulation method for eliminating birefringence related artifacts, which is most notable as a VCD baseline offset, has been proposed and demonstrated to be effective for specific cases and has been shown to additionally reduce some absorbance artifacts [110,112]. Adding this optical and electronic complexity does have the real promise of eliminating sensitivity to sample-cell and even sample-phase originating birefringence problems. Evidence has developed that collimated or slowly converging beams, few reflections, and uniform detector surfaces give the best baseline results [80,95,107]. Finally, most artifacts can be minimized by careful optical adjustments, and since the alignment is stable, at least in our instruments, resultant baseline characteristics often can be corrected by subtraction.

6.2.2 Biomolecular VCD Sampling Methods

Biomolecular systems are ideally studied to an aqueous environment. This poses difficulties for IR techniques since H_2O has strong fundamental transitions that directly overlap regions of interest such as the N-H and C=O stretches. Consequently, peptide amide I′ (primes indicate N-deuterated amides) VCD at ~1650 cm^{-1}, dominated by the C=O stretch, is normally measured in D_2O. On the other hand, due to their large contributions from the N-H deformations, the amide II at ~1550 cm^{-1} and amide III at ~1300 cm^{-1} are best studied in H_2O to avoid significant wavenumber shifts and alteration in the composition of the modes. These wavenumbers also have little H_2O interference.

Protein samples in D_2O can be prepared at concentrations in the range of 5–50 mg/ml for VCD, depending on the path length and SNR that will be acceptable. An aliquot of the solution (typically 20–30 μl) is placed in a standard demountable cell consisting of a pair of CaF_2 or BaF_2 windows separated by a 25–100 μm spacer (e.g., Teflon or PTFE). For studies in H_2O, concentrations of up to 100 mg/ml (but < 20 μl in volume) and path lengths of 6 μm or less (typically obtained with a mylar spacer) are needed for amide I VCD. Under these conditions, water has an absorbance of ~0.8 at 1650 cm^{-1}. This interference causes a loss in SNR for the protein or peptide VCD, but no increase in artifacts [122]. We have previously found it most useful first to run the H_2O baseline in a refillable cell (Specac Inc., Smyrna, GA), then to remove the solvent completely and refill with sample solution without demounting, thereby maintaining the path length. An alternative cell design is available (BioTools, Waconda, IL) that consists of two CaF_2 plates ground to have a fixed path separation and to be sealed at the edges by contact of the plates. These have proven very useful for biomolecular studies, especially in water where leakage is minimal, but they do require adjustment of the pressure from the holder to achieve a consistent path length for sample and baseline spectra.

Other popular IR sampling techniques such as dried films and mulls of ground powders, while not impossible, are not widely useful for VCD and require great care in preparation. Solid samples, even films, have residual birefringence that can lead to severe VCD artifacts, especially for biomolecular samples. Some film data on peptides have been reported [123–126], but reproducibility and especially interpretability (due to interstrand interaction) were compromised. Spectra of mulls have been measured, but not enough data are available to ascertain reliability. More recent results have suggested that use of the dual modulation technique [110,112] may permit reproducible measurements at VCD for solid-phase samples [127]. On the other hand, ATR cells depend on multiple reflection, which in turn alters the polarization state of the beam. Thus, though a very attractive sampling method for LD studies, ATR is fraught with problems for VCD. It is possible that multiple-modulation techniques may make ATR-VCD feasible at some future date [110]. Gas-phase samples [128–130], of course, have none of these difficulties but at the same time have little biochemical application. Consequently, VCD sampling methods center on use of conventional liquid transmission cells.

Final VCD spectra for biomolecular analyses are obtained by subtraction of a baseline VCD scan from the sample spectrum and by calibration as noted above. An absorbance spectrum of the sample, obtained on the same instrument under identical conditions as the VCD spectra, is useful for interpretive purposes. Additionally, it is important to obtain FT-IR spectra on the same samples at higher resolution and SNR for purposes of comparison, frequency correction (for the dispersive instrument), and resolution enhancement of the absorption spectrum using Fourier self-deconvolution [34]. Ideally, VCD should be plotted in molar units, such as ε and $\Delta\varepsilon$, as is done with ECD measurements. However, since concentration and path lengths are rarely known to sufficient accuracy for these IR-based experiments, VCD spectra of biomolecules are often normalized to the absorbance (as measured on the VCD instrument). Because the absorbance coefficients for different molecules studied will vary, this is only a first order correction for concentration, but is a reasonable method and easy to use. Amides in different conformations [131], coupled to different types of residues and in different solvent environments, will have different molar extinction values.

In our laboratory, ECD spectra are also measured for all peptide and protein samples studied. Normally ECD spectra are obtained for samples in more dilute conditions (0.1–1 mg/ml) and placed in strain-free quartz cells (NSG Precision Cells

Inc., Farmingdale, NY), which can have various path lengths (0.1–10 mm). However, we have also succeeded in measuring ECD down to ~200 nm on the IR or VCD samples themselves, since the CaF_2 windows used for IR transmit into the vacuum UV and the short path lengths can provide adequate UV transmission to 200 nm for protein solutions even at high concentrations [82,132–134]. This approach allows direct comparison of two independent, conformationally sensitive techniques on precisely the same sample. Since relatively small amounts of biopolymer can give rise to significant ECD signals, it is very important to thoroughly clean such sample cells between uses.

6.2.3 Instrument Comparison with Biopolymer Spectra

The characteristics of different VCD spectrometers were discussed in detail in Section 6.2.1. The best way to illustrate the particular features of each design is direct comparison of experimental data measured on the corresponding instruments for the same set of samples under the same conditions. VCD instruments are typically characterized by measuring spectra for standard small chiral molecules, such as α-pinene or camphor, which have relatively large VCD signals. By contrast, proteins and nucleic acids have significantly weaker VCD, which is further complicated by the interference of strong absorption bands due to the aqueous solvent, leading to lower SNR ratio and often substantial baseline artifacts. Therefore, we felt it would be a useful aspect of this review to compare experimental data obtained on several different VCD instruments, not for small molecules, but rather for the polypeptide, protein, and nucleic acid samples of direct interest to the biopolymer focus of this review.

For this purpose, VCD measurements were carried out with the UIC dispersive and FT-VCD instruments in both rapid-scan and DSP modes, and on the commercial, Bomem/BioTools-ChiralIR FT-VCD, as installed at Vanderbilt University in the laboratory of Professor P. L. Polavarapu, and later in the lab of Professor L. A. Nafie at Syracuse University, through their very kind assistance. Additional data on separate samples were obtained on the Bruker IFS-66 with a PMA 35 accessory at Bruker Optik in Ettlingen through the extended cooperation of Dr. Herman Drews and later on a similar instrument installed at the Institute of Chemical Technology, Prague, through the kind cooperation of Professor Marie Urbanova. The sampling conditions and corresponding instrumental measurement parameters are summarized in Table 6.1 and its corresponding footnotes.

Comparison of amide I VCD data on the UIC dispersive spectrometer and the FT-VCD based instruments are shown in Figure 6.3 for poly-L-lysine and in Figure 6.4 and Figure 6.5 for bovine serum albumin (BSA) samples in D_2O. The IR absorption is presented below the VCD for the two instruments with representations of the noise level at the top for each instrument, respectively. All instruments produced the correct (in agreement with previous literature reports) amide I bandshapes: a negatively biased negative couplet for the random-coil form of poly-L-lysine (neutral pH), and a W-shaped contour with a less intense low-frequency negative lobe for the predominantly α-helical BSA. The fundamental difference between the dispersive and FT-VCD spectra is in the nature of the associated spectral noise, as noted in Section 6.2.1. The noise traces are created by subtracting two blocks of data, each

TABLE 6.1
Sampling Conditions and Instrumental Parameters

Samples[a]			Instruments[b]				
				Rapid Scan FT-VCD			DSP FT-VCD
Sample[c]	Conc. [mg/mL.]	Pathlength [μm]	Dispersive UIC[d]	ChiralIR[e]	Bruker[f]	UIC FTS-60A[g]	UIC FTS-60A[g-h]
poly-L-lysine	50	50	X	X	X		
Bovine serum albumin	35	50	X	X	X		
poly-L-proline	25	50	X			X	X
poly(C)•poly(G)	25	50	X			X	X

[a] All samples were dissolved in D_2O. Measurements were done in a home-made brass sample cell that compresses the sample and Taflon spacer between two circular CaF_2 windows. The solvent (D_2O) was measured in a separate cell of the same construction using the same experimental conditions. In all cases, the baseline (solvent) was subtracted from the sample data. The noise traces were calculated from these processed data as one-half of the difference between the two scans (scan blocks).

[b] For detailed description of the instruments, see sec. 2.1.

[c] For each Figure the IR absorption was measured separately for the same sample on a Biorad-Digilab FTS-60 at 8 cm^{-1} resolution with 8-fold zero-filling.

[d] UIC Dispersive conditions: Resolution 8 cm^{-1}, time constant 10s, scan speed 6 cm^{-1}/min, spectral range 1740–1560 cm^{-1}, 2 scans 30 min each, I_{trans} was measured simultaneously as part of the normalization. In the examples shown, BSA data used the new 4-element, restricted range MCT detector. The other were obtained in 2000.

[e] ChiralIR conditions: Resolution 8 cm^{-1}, spectral range restricted by optical filter to 1900–800 cm^{-1}, 2 scan blocks 34 min each. Transmission IR (single beam, I_{trans}) measured before and after each block for about 1 min, no zero filling, cosine apodization, normalization done as part of computation after FFT of each block.

[f] Bruker PMA 35 conditions: Resolution 8 cm^{-1}, 2 scan blocks 30 min each, Backman-Harris apodization with zero filling to 1 cm^{-1} data point density, transmission IR collected simultaneously with VCD and normalization done as part of post-processing.

[g] Measurement conditions with UIC FT-VCD modification of a Biorad-Digilab FTS-60A: resolution 8 cm^{-1}, spectral range filtered: 1900–800 cm^{-1}, total collection time of 68 min. All spectra processed with 8-fold zero-filling. Noise traces obtained from one-half of the difference between two 34 minute blocks.

[h] Due to software limitations, only DSP scans of 8.5 min in length could be obtained (1 sec step. undersampling ratio of 8). 2 scan blocks of 4 scans each (34 min) were measured to produce an equivalent data collect of a 68 min (2, 34 min. blocks) average scan.

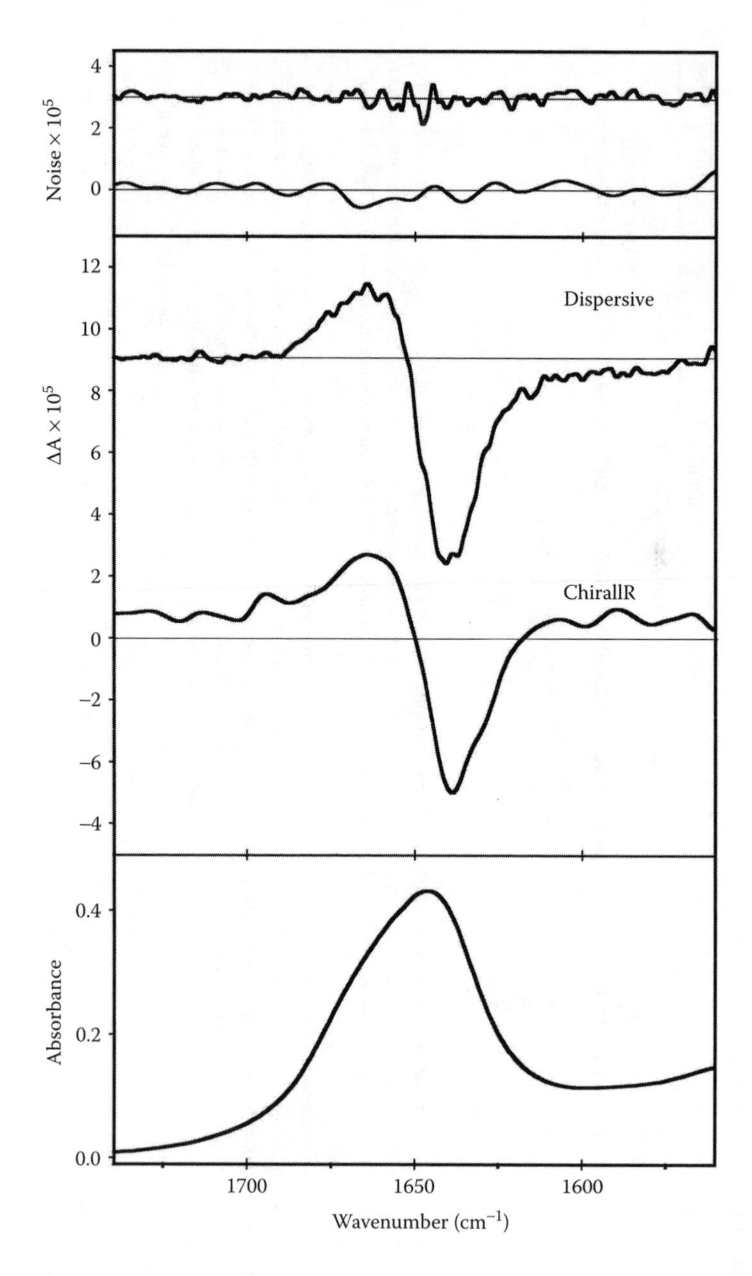

FIGURE 6.3 IR absorbance (bottom), VCD (middle pair of spectra) and noise trace (top pair) spectra for poly-L-lysine in D_2O contrasting dispersive and FT-VCD. The two VCD spectra directly compare, for the same sample, nearly identical time measurements made in 2000 on the then-configured UIC dispersive VCD instrument (middle-top) and on the ChiralIR installed at Vanderbilt (middle-bottom). Noise traces (top) are stacked in the same order and plotted with the same magnitude scale. Parameters for data collection are summarized in Table 6.1.

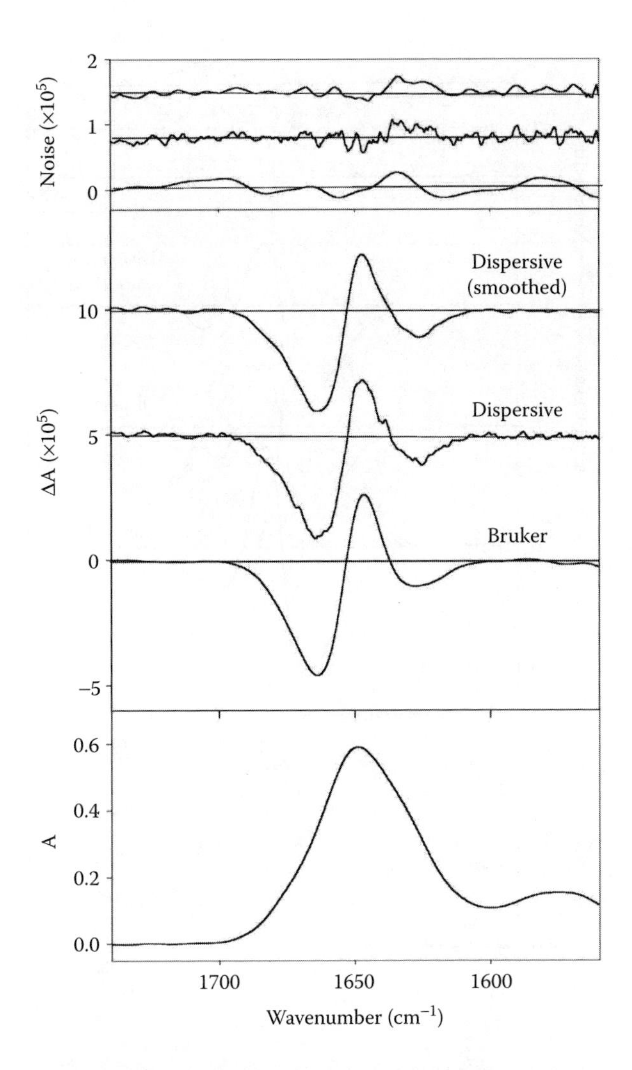

FIGURE 6.4 Comparison of bovine serum albumin (BSA) VCD measured (in 2003) with the Bruker PMA 37 accessory and with the upgraded (new detector) UIC dispersive VCD. The upper VCD spectrum is smoothed to demonstrate the effect of oversampling in our dispersive spectra. All scan times and conditions were the same; identical scans of D_2O were subtracted to correct for the baseline. The new detector, with restricted (< 8 μm) bandpass, results in a significant SNR improvement (~2 times) for VCD. Noise traces are stacked in the same order, but plotted on an expanded scale.

representing one half of the total scans, while for the VCD spectral traces, they are added.

The dispersive VCD traces contain sharp, high-frequency noise (much akin to "pen jitter" but, of course, there is no pen in these totally digital spectra), while the FT-VCD spectra look smoother. This "jitter" comes from our over-sampling the spectra as compared to what is needed to represent the true experimental resolution. These

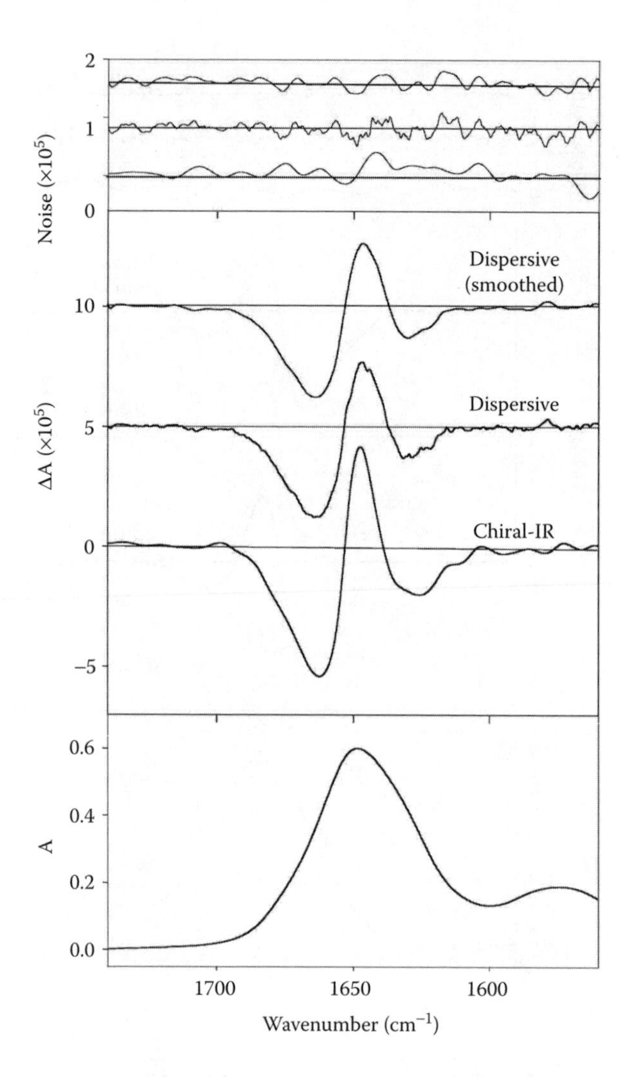

FIGURE 6.5 Comparison of BSA VCD measured (2003) on the Syracuse installed BioTools ChiralIR and the upgraded UIC dispersive VCD, as in Figure 6.4.

data can be computationally smoothed (as shown in Figure 6.4 and Figure 6.5) without loss of resolution to give traces that are more comparable to those obtained with FT-VCD methods. If we compare the noise (the upper set of traces) to the VCD spectra (middle set), it can be seen that the FT-VCD signal oscillations (noise) are significant and of the same width as the spectral features. This is especially apparent for BSA (comparing the bottom and middle traces of Figure 6.4 and Figure 6.5), which has a VCD intensity about two times weaker than that of poly-L-lysine. The width of the features in the FT-VCD noise traces results from the spectral bandwidth being limited by the size of the intereferogram, which was collected to 8 cm^{-1} resolution, the same value as set by the slit width (5 mm) in the dispersive spectrometer. This broad-band nature of spectral noise may in fact be a disadvantage of the FT-VCD, and it is essential that the noise traces be examined carefully in order to correctly interpret the magnitude, and in very difficult cases, the shapes of FT-VCD spectra. By contrast, the 8 cm^{-1} resolution, dispersive VCD is noisy due to significant oversampling (high frequency oscillations) as well as more fundamental uncertainties (low frequency). When it is averaged (smoothed) so as to have the same sort of response as the FT-VCD noise, the baseline fluctuations are directly comparable to the FT-VCD result. This oversampling method has no real cost in terms of dispersive data collection and assures one of sampling the full band profile from which the jitter can be digitally smoothed for presentation, seen in the noise traces (Figure 6.4 and Figure 6.5). Alternatively one could scan faster (thus sampling less often) and average signals from multiple scans [76], as long as the time constant used for the demodulation (lock-in amplifier) do not distort the shape (effectively reduce resolution) obtained with more rapid scanning.

The poly-L-lysine FT-VCD (Figure 6.3) has a positive baseline offset and weak features on both sides of the main couplet. These artifacts may be due to the fact that different cells were used for the sample and baseline (blank) measurement. However, different cells were also used for the dispersive measurement of the sample and of the baseline as well, yet the dispersive VCD has a very flat baseline, with few apparent artifacts. In general we have found FT-VCD instruments to be somewhat more sample cell sensitive than our dispersive instrument, but the dual modulation technique promises a potential correction for this [76]. Noise traces, most evident in the dispersive one, get worse near the absorption band maximum, where the DC signal (i.e., transmission) is small, and division of the modulated signal by the transmission can lead to greater relative errors. Comparison of these test spectra, obtained as just 2 scans at 8 cm^{-1} resolution (middle traces), with more typical biopolymer VCD spectral collections of 12 scans (~6 hours) at 10 cm^{-1} resolution (data not shown) shows these to have increased SNR since the high-frequency noise as well as the lower-frequency baseline oscillations are almost completely averaged out [117]. This small change in resolution has almost no effect on the apparent bandwidth of the spectrum of this aqueous biopolymer (as is the norm). The most evident effect is some loss in intensity for the positive VCD band α-helical proteins at 1645–1650 cm^{-1}, which, lying between two negative bands, will suffer cancellation with loss of resolution. Because the signal level is dependent on the square of the slit width, this slight increase from 8 to 10 cm^{-1} resolution corresponds to nearly a 50% increase in SNR per scan. Such incremental resolution changes are not possible in the BOMEM-BioTools design, nor would they enhance SNR in any significant way. Any throughput advantages gained by opening an aperture incrementally (which may accompany a reduction in resolution) could not easily be used to SNR advantage in a general FT-VCD setup and would be very instrument-specific, at best.

The BSA-based comparisons were made between our recently (2003) enhanced dispersive VCD (narrow band MCT detector) and the Bruker (Figure 6.4) and BioTools (Figure 6.5) instruments. These data are higher in SNR than our earlier tests (2000) and were consistent for several samples but only the BSA results are presented here. The Bruker method of simultaneously collecting transmission and polarization modulation interferograms has advantages for processing and stability. The end result spectra can only sensibly be compared to the dispersive data if we smooth those spectra. While the Bruker resolution is somewhat reduced from the dispersive, as noted above this has no deleterious effects on the spectra. Comparison with the recently remeasured BioTools results shows them to be comparable in quality (Figure 6.5). However, the noise traces for the Bruker and BioTools results do differ in terms of the period of oscillation. While this might reflect resolution, we believe it may be coupled to the apodization favored by each manufacturer. Clearly having such noise components of the same shape as the spectral features can cause confusion.

In summary, FT-VCD methods can be used to measure high quality biopolymer VCD spectra, comparable to and only slightly noisier (if judged on an RMS basis) than that obtained with our (original) dispersive VCD scanned over just the amide I band. Due to the solvent interference, the typical FT-VCD advantage of collecting spectra for whole mid-IR spectral region simultaneously often does not have much practical use for proteins and peptides. FT-IR VCD does not have any SNR advantage compared to dispersive VCD over a narrow spectral region, such as the amide I protein band. On the other hand, one possible disadvantage of FT-VCD is that the noise cannot be easily distinguished from the signal due to their having the same bandwidth. It is important to realize that the dominant noise source in VCD, as in any IR experiment, is detector noise, which is independent of the spectral wavenumber. However, the dispersive experiment is subject to noise due to fluctuation of the sample with time and wavelength as well as errors in wavenumber which impact the scanning and co-addition, which are correspondingly much less of a problem for FT-VCD.

Comparison of the traditional rapid-scan FT-VCD measurement with the novel, lock-in free, step-scan DSP method, was carried out on the UIC FT-VCD instrument. Since this instrument, built a decade ago using a Digilab FTS-60A but with a very large sample focus [80] does not have the same SNR capabilities as the more modern commercial designs (Figure 6.4 and Figure 6.5), samples with larger VCD signals were chosen for this purpose. The rapid-scan and DSP FT-VCD spectra of RNA polynucleotides poly(rC)•poly(rG) as an example is shown in Figure 6.6, and are compared with the base C=O centered VCD measured on the dispersive VCD instrument. As is evident in Figure 6.6 (middle), the VCD bandshapes from both the conventional rapid-scan and the DSP based FT-VCD measurements reflect that obtained on the dispersive instrument (top scan) albeit with some variations in absolute as well as relative intensities. This has been true of a number of different samples for which we have analyzed the DSP FT-VCD method [118]. The initial release of the Digilab DSP software necessary for collection of high frequency (PEM) modulated spectra in DSP mode (DSP v. 3.0) only allows a slowest step speed of 1 Hz. While this has not posed any difficulty when measuring small molecule VCD [118], which can involve measurement of inherently large VCD signals, for biopolymeric systems limited sampling of the signal at each step seems to result in the distortion of the relative intensities of + and – VCD bands. Tests with our instruments and the BioTools (Rina K. Dukor, private communication) instruments, both using the rapid scan method, empirically indicated that averaging more scans before transforming leads to more reliable shapes even if the same total number of scans is averaged in the end. Thus we feel that increased signal sampling over slower steps in the DSP mode potentially may provide enhanced reliability.

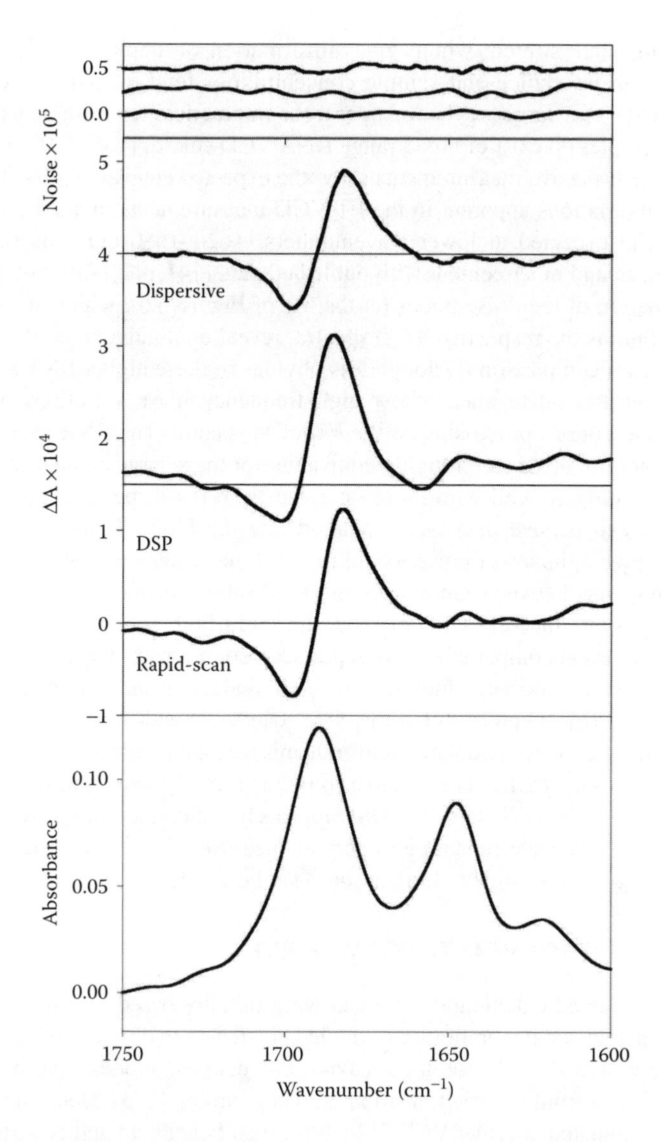

FIGURE 6.6 Comparison of rapid scanning and DSP FT-VCD techniques for the RNA homopolymer duplex poly(rC)•poly(rG) in the C=O base stretching region with a dispersive VCD measurement for reference. IR absorbance (bottom), VCD (middle), and noise traces (top) are compared. VCD and noise traces have the same scale and are stacked in the same order: dispersive, DSP, rapid-scan (from the top).

The RNA homopolymer poly(rC)•poly(rG) adopts a regular A-form (right-handed) double-helix. The VCD signature for poly(C)•poly(G) is that of a positive couplet (–,+ with decreasing frequency) at ~1690 cm^{-1} which arises primarily from coupling of the C=O stretching modes of the stacked planar bases themselves. Other conformationally useful IR/VCD regions can be accessed in nucleic acids, such as the P-O phosphate stretch, which gives information on duplex backbone stereochemistry. Solvent choice and sample concentrations tend to make spectral measurements over very large wavenumber regions impractical for a single experiment, although examples do exist of broad range DNA VCD studies [135–140]. Again, both FT-VCD and dispersive measurements show the expected couplet at ~1690 cm^{-1}, but the baseline variations apparent in the FT-VCD measurements make the very weak negative VCD expected at lower wavenumbers (1620–1660 cm^{-1}, as seen in the dispersive scan and in agreement with published data [141,142]) difficult to discern.

Examination of the noise traces (at the top of Figure 6.6), which are stacked in the same order as the respective VCD spectra, reveal qualitatively similar results as for the previous comparisons. Though less obvious in these higher SNR spectra, the dispersive spectral noise traces show high frequency noise variations, which are masked in the Fourier processing of the FT-VCD spectra. The SNR levels for both FT-VCD scanning modes are roughly equivalent for these large signals and, perhaps surprisingly, compare well with those obtained from the dispersive measurements. (Again, if the dispersive data were sampled like the FT-VCD or smoothed as in Figure 6.4 and 6.5, its lower noise would be much more evident.) More specifically, these comparisons between rapid-scan and DSP modes of measuring FT-VCD for biopolymers prove that the DSP method can yield both spectral bandshapes and spectral noise levels comparable to its rapid-scan counterpart. Thus the DSP's fully digital process (no lock-in amplifier) is both realizable and practical for VCD measurement [118]. It can be noted that step scan VCD with lock-in demodulation, which requires separate modulated and transmission data collections as in conventional rapid scan mode, has been shown to be measurably worse, in terms of SNR, than rapid scan VCD [79]. Thus the DSP approach, which naturally involves simultaneous modulated and transmitted spectral measurements, is the better way to measure step-scan spectra should that approach be needed.

6.3 BIOPOLYMER VCD THEORY SURVEY

Biopolymer-oriented calculations of VCD were initially based on exciton coupling concepts, whereby local vibrations are coupled via transition dipoles yielding a pattern of oppositely signed VCD for the weakly split coupled modes whose frequency dispersion results from electric transition dipolar coupling [143]. Deutsche and Moscowitz first simulated polymer VCD [144,145], then Schellman and coworkers [146] modified the exciton method to simulate VCD for polypeptides in α-helices and β-sheets. Holzwarth and Chabay [147] put forth a dipole coupling (DC) based exciton model for the VCD of a peptide dimer, much as has been used successfully in ECD studies for biopolymers. This model was revived by Diem and coworkers [148,149], who have termed their method the extended coupled oscillator (ECO) model. By use of comparison to more exact theoretical methods [150], DC-based approaches have been shown to be valid only for weakly interacting (non-bonded) high dipolar strength vibrations. Thus, though often invoked for qualitative interpretation of small molecule VCD, the exciton approach actually works best for describing VCD of the largest molecules we have studied, focusing on the characteristic local modes in DNA [148,149,151,152] and is actually less useful for peptides [153].

More accurate means of computing VCD spectra have been developed in the last decade. These involve use of quantum mechanical force fields and *ab initio* calculation of the magnetic and electric transition dipole moments, usually in the form of parameters termed the atomic polar and axial tensors (APT and AAT, respectively). These computations normally involve use of moderately large basis sets and some approximation to represent the magnetic dipole term. In effect, the magnetic field perturbation (MFP) method of Stephens and coworkers [7,51, 154,155] evaluates magnetic transition matrix elements that arise from the perturbation of the ground state wave function by a magnetic field operator. These were initially done at the Hartree-Fock Self Consistent Field (HF-SCF) level and the resultant force fields (FF) were often scaled to get reasonable frequencies. The MFP model is complex to apply to large biomolecules, and the computations become virtually intractable if one attempts to incorporate correlation effects to improve the FF [7,156–158]. Density functional theory (DFT) methods can drastically improve the FF, and hence simulated vibrational frequencies, while keeping the computational effort manageable. However, even DFT computations with reasonable basis sets remain formidable for large biomolecules. Nevertheless, these size-based barriers are dropping fast with continued computational and algorithmic developments [93,114,115,159–163]. We very early reported model HF-SCF calculations for dipeptides [164] as well as, extensions to DFT methods for tripeptides and longer oligopeptides [58,59,165]. To look at even longer peptides, we have exploited a property tensor method to transfer *ab initio* FF, APT, and AAT tensor values obtained from calculations on smaller molecules to simulate spectra for larger polymers [57,153]. These latter approaches have been successfully used to explain spectra for experimentally accessible oligopeptides [59,166–171].

6.4 NUCLEIC ACID VCD SPECTRA AND APPLICATIONS

For nucleic acids, ECD measurements depend primarily on the n-π^* and π-π^* transitions of the bases, which are spread over only a modest range of the near-UV spectrum, resulting in severe overlap of these several broad electronic excitations [172]. Their origin in the highly polarizable π-electron systems of the bases makes them susceptible to significant environmental perturbation, an important consideration for such a complex salt as a nucleic acid polymer. Interactions among these transitions yield information about the nucleotide base stacking convoluted with the local chirality effects of the base–sugar interactions. Spectral transitions centered on other parts of the molecule (e.g., sugar or phosphate groups) are not easily accessed with ECD.

In recent years FT-IR and Raman spectroscopic applications for determining nucleic acid structural variations have proliferated [30,173,174]. In the FT-IR, base deformation modes again overlap in the region from 1800–500 cm^{-1}, but different bases do have some unique absorption patterns. However, due to the ability of vibrational spectroscopic techniques to probe many different types of nuclear motion, FT-IR and Raman can independently sense aspects of the ribose and phosphate backbone conformations by utilizing data from several distinct spectral regions that are relatively independent of base modes.

VCD couples this separate probing of functional groups in the polynucleotide with stereochemical sensitivity to stacking-like interactions, in part originating from dipole coupling. Most DNA and RNA VCD studies have developed empirical correlations of bandshape with helical conformation, but some theoretical modeling has appeared.

Nucleic acid base deformation VCD in the range of 1800–400 cm^{-1}, which is dominated by the C=O and aromatic ring C=N stretch contributions, has normally been measured in D_2O-based solution. On the other hand, phosphate centered modes in the 1250–1000 cm^{-1} region are best studied in H_2O-based solution. Nucleic acid samples in D_2O can be prepared at concentrations in the range of 10–40 mg/ml. At a path of 50 μm, these yield a relatively weak absorbance (≤ 0.1) in the 1700–1600 cm^{-1} region yet still produce acceptable VCD (Figure 6.6), at least for longer duplex samples [175]. In the PO_2^- region, nucleic acids are best studied in H_2O at concentrations of greater than 50 mg/ml and path lengths of 15 μm, which combine to give absorbances of ~0.5 [152]. Identically prepared blank samples of only buffer typically provide adequate VCD baselines, especially for the weakly absorbing base modes.

6.4.1 Empirical Correlation of VCD with Helicity

VCD measurements on RNAs and DNAs in buffered aqueous solution yield quite large signals, in terms of $\Delta A/A$, for a variety of modes. VCD can access the in-plane base deformation modes to study interbase stacking interactions, the phosphate P-O stretches to sense backbone stereochemistry, and coupled C-H or C-O motions to monitor the ribose conformation. The VCD of the sugar-centered modes has proven only marginally useful to date, due to their having little spectral definition in the C-H region, where such characteristic sugar modes are isolated, and due to overlap of the C-O stretches with other nucleic acid modes in the mid-IR. Other diagnostic modes have not been adequately studied to date. ROA studies have also addressed DNA conformations and have utilized sugar-based modes with more success [176,177].

Single stranded RNA samples give rise to a positive VCD couplet in the in-plane base deformation region [175], typically centered over the most intense band lying between 1600–1700 cm^{-1} depending on the base (see Figure 6.6). In most cases, except for A (adenine), this band arises from a C=O stretching mode. However, independent of their detailed sequence, all nucleic acids have a relatively intense band here. Dinucleotides and random copolymers have similar but weaker VCD, while duplex RNAs have similar but more intense patterns. Mononucleotides have little or no detectable VCD in the base deformation or phosphate centered modes, so that the spectra observed for these modes in DNA or RNAs are a direct consequence of their interaction through the polymer or oligomer fold. The generality in VCD bandshape found for simple polynucleotides is a consequence of their dominant right-handed helical twist [175]. For example, the spectrum for a synthetic duplex RNA, such as poly(rG)•poly(rC) (Figure 6.6), is similar to that of a natural t-RNA except in terms of magnitude and temperature dependence [142,178].

Somewhat of an exception to the consistent RNA VCD pattern described above is that of poly(rA)•poly(rU). At 55°C, the spectrum abruptly changes to another complex pattern that can be assigned to formation of a triple helical form, poly(rU) *poly(rA)•poly(rU) [179]. At higher temperatures this complex melts to single strands and the spectrum changes again. This intermediate triple helical VCD spectral form has general characteristics that can be found in all pyrimidine-purine-pyrimidine, A(T)U-based DNA, RNA, or mixed triplexes (Figure 6.7) and provides

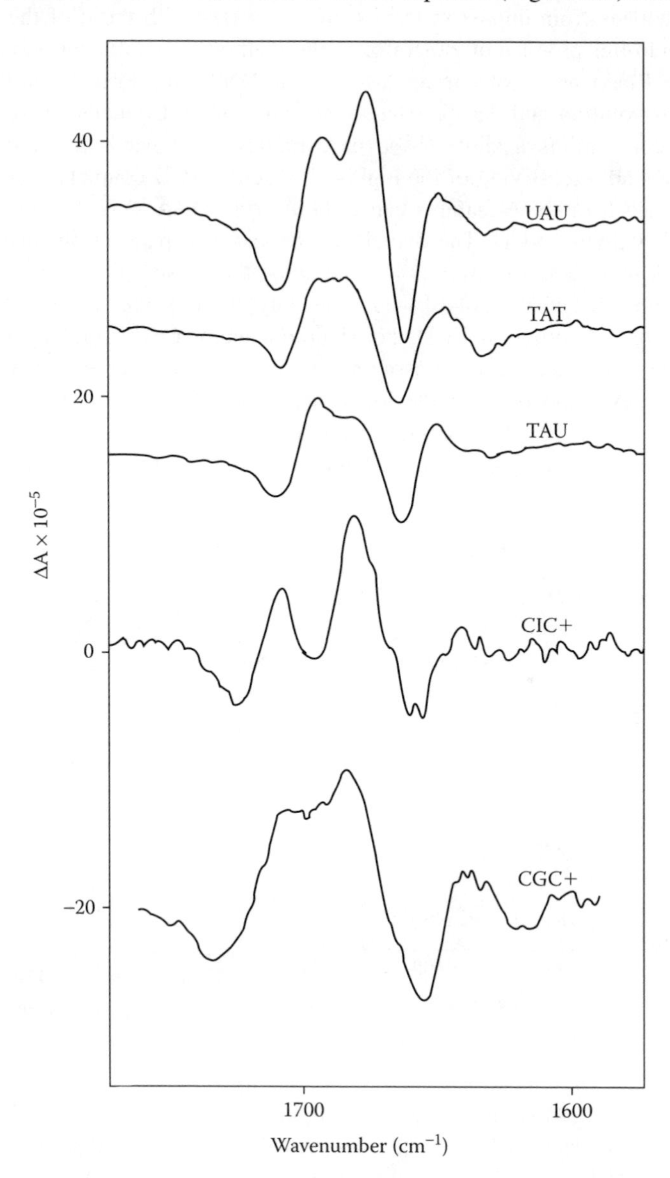

FIGURE 6.7 VCD spectra of triplex DNAs and RNAs composed of different homopolynucleotides for the base-centered modes (C=O, C-N, and ring deformations) of rU∗rA•rU (in D_2O, top); dT∗dA•dT (in D_2O); dA•dT∗rU (in 0.1 M NaCl/D_2O); rI•rC∗rC+ (in D_2O at pD 5.6); and rG•rC∗rC+ (in H_2O at pH 3.9, bottom), where • indicates Watson-Crick pairing and ∗ suggests alternate, presumably Hoogsteen, pairing.

a definitive diagnostic for this triplex conformation [151]. The rG(rI)rC system also can form triplexes but these must be stabilized by ionizing one C strand to form rC•rG*rC+ type triplexes. The pattern seen in the VCD may differ due to the wavenumber variations for the main dipolar transitions of the various base modes, but the changes from duplex to triplex are consistent with those of the dA(dT)rC system and offer promise of generalizing the method for analytical purposes.

While RNAs are mostly in an A-like form, DNAs are normally in the B form in aqueous solution and can be transformed to A form by means of dehydration. The A form exhibits a shift of the base modes to higher wavenumbers and a sharpening and intensifying of the highest frequency VCD couplet, as compared to the B form. All of these features cause the A form DNA VCD to look more like that of RNA duplex VCD. The overall profile and sign patterns are maintained in this B-to-A structural transition [141], including the sensitivities to base content.

The difference in the poly(dA-dT) and poly(dG-dC) type VCD patterns [141] means that base deformation VCD, though dominated by C=O stretching, is sequence dependent and offers a means of assaying DNA base content, at least in a qualitative sense [180]. A comparison of the VCD bandshapes for base modes of DNAs of varying base content, all B form, is given in Figure 6.8a. By contrast to the considerable shape variation for the base centered modes, the PO_2^- modes (Figure 6.8b) are

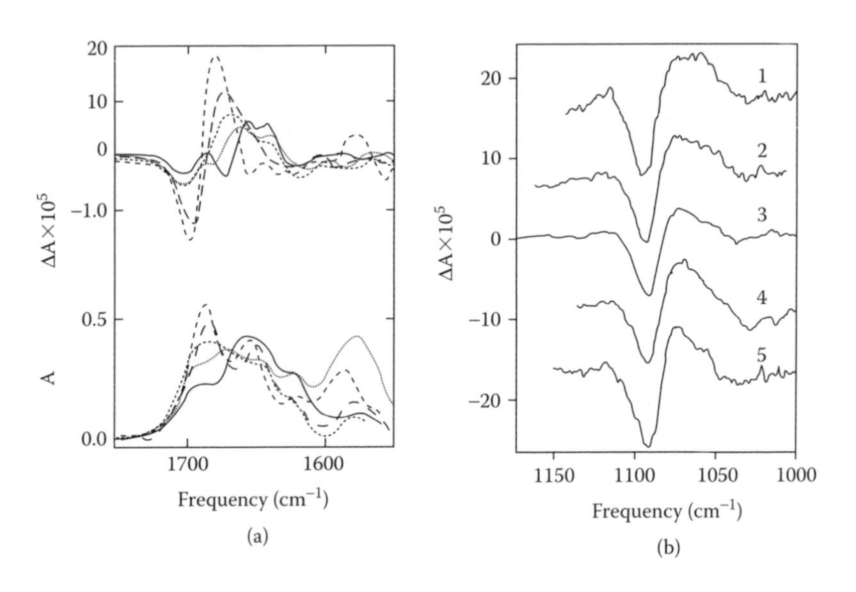

FIGURE 6.8 (a) (Left) Sequence-dependence of the VCD (top) and IR absorption (bottom) of various B-form duplex DNAs in the C=O stretching, base deformation region. (- - - -): poly(dG-dC)•poly(dG-dC); (– – –): *m. lysodeiktus*, 72% GC; (··········): calf thymus, 44% GC; (- · · - · -): *c. perfringens*, 26% GC; (———): poly(dA-dT)•poly(dA-dT). (b) (Right) Sequence-independence of the VCD spectra in the symmetric PO_2^- stretching region for the same B-DNAs: 1: poly(dG-dC)•poly(dG-dC); 2: *m. lysodeiktus*, 72% GC; 3: calf thymus, 44% GC; 4: *c. perfringens*, 26% GC; 5: poly(dA-dT)•poly(dA-dT). Even less variation is evident in the PO_2^- IR absorption.

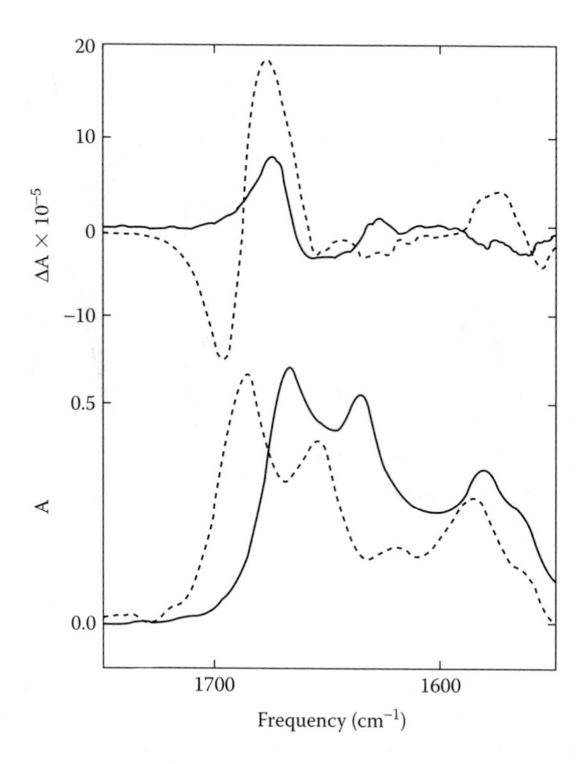

FIGURE 6.9 Comparison of VCD (ΔA, upper) and absorption (A, lower) spectra of the B form (- - -) and Z form (——) of poly(dG-dC)•poly(dG-dC) in the base deformation region. Note that the VCD shifts in frequency with the absorption maximum and changes sign — B(−,+) → Z(+,−) (high to low) — and magnitude for the highest frequency couplet.

virtually constant in form for these same molecules, but do differ between RNA and DNA due to contributions from overlapping ribose modes in RNA that are not present in deoxyribose (DNA) [136,152,181–183].

The VCD spectra of poly(dG-dC) (Figure 6.9) and related DNA oligomers in their B and Z forms have distinctly different bandshapes for the base stretching modes 142,152,178,184]. Both are dominated by a VCD couplet associated with the highest frequency intense base modes, but these IR bands are significantly shifted in frequency and have opposite VCD sign patterns (relative to the absorbance maximum) which reflects the handedness of their duplex helices. While the detailed shape of the Z form base deformation VCD may be dependent on specific conditions used to stabilize that conformation, the generic Z forms resulting from adding either Na^+ or alcohol to induce transitions actually lead to very similar VCD spectra [141]. VCD of Z form DNA is also opposite in sign to that of B form in the symmetric PO_2^- stretching region as seen in Figure 6.10a [142,152,180].

While most nucleic acid studies have been aimed at characterizing the spectra of known structures, some applications for determining unknown structures and determining thermodynamic stability have arisen. The similarity of RNA and DNA

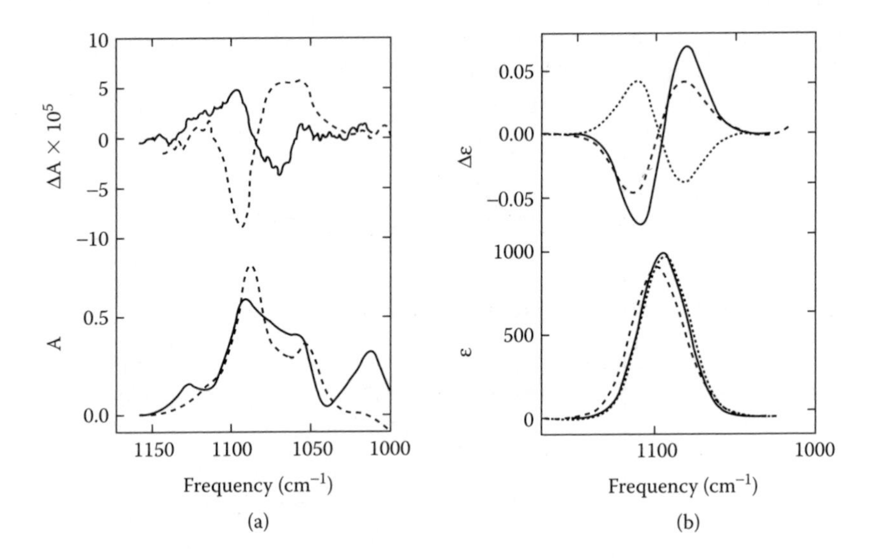

FIGURE 6.10 (a) (Left) Comparison of experimental VCD (top) and IR absorption (bottom) spectra of the B form (- - -) and Z form (——) of poly(dG-dC)•poly(dG-dC) in the symmetric PO_2^- stretch (~1090 cm^{-1}). Note: this region also contains ribose modes, which change with ring conformation change from B to Z (e.g., ~1050 cm^{-1}). (b) (Right) Computed VCD and absorption spectra using a DC model for the A form (- - -), B form (——). and Z form (········) of duplex $(dG\text{-}dC)_5$•$(dG\text{-}dC)_5$ for just the symmetric PO_2^- stretch mode. Spectral intensities are normalized to $\Delta\varepsilon$ and ε per base and reproduced with a Gaussian bandshape with a peak half-width of 20 cm^{-1}, but are best compared with experiment as $\Delta\varepsilon/\varepsilon$, which should equal $\Delta A/A$. In this case the predicted magnitudes are within a factor of 2. No ribose modes are included in the calculational result.

in their VCD spectra of base and phosphate transitions is due to their chirality arising from the local nature of the interactions among oscillators and the independence of that coupling from perturbations by the chiral ribose moiety which differs between RNA and DNA. The bases are spectroscopically the same in both types of molecules, and the helicity is similar; only the spatial relationship of the bases with respect to the axis is different. While the two right handed forms, B and A, give very similar spectra, there is no possibility of confusing them with the spectra of the Z form as has previously happened using ECD [185]. B-form poly(dI-dC) has the opposite near UV ECD pattern to that typical of B form DNA. This discrepancy in the past has led to a series of confused conformational assignments for poly(dI-dC). To clear this up, we have measured the VCD of poly(dI-dC) in both the base stretching and phosphate regions. In both cases a clear spectral pattern results that is consistent with the right-handed B form [142]. These VCD spectra could not result from a Z form. The confusion in ECD results from the bases (I) having slightly different electronic structures. This affects their π-π* transitions but has very little effect on the vibrational transitions, which are the chromophores studied and dipoles coupled in VCD. Other applications have encompassed the previously noted triplex studies which identified structural phase transitions and characterized the single-, double-, and triple-strand equilibria in DNA, RNA, and mixed species of A-T(U) strands [151,179]. Some oligonucleotide studies have established constraints on the possible conformations that can exist for di- and tetra-nucleotides [186]. Finally, the areas of interaction between drugs and DNA and between proteins (or peptides) and DNA remain relatively unexplored by use of VCD measurement, but important progress is being made by the Wieser lab in Calgary using VCD to study metal binding and sequence effects with designed oligonucleotides [136,137,139,140,181–183].

6.4.2 Theoretical Modelling of Nucleic Acid VCD

Such a clear dependence on helicity is a further indication of the role of dipole coupling (DC) in the stacking interaction between bases and the coupling between the phosphate groups. This sort of DC interaction separately leads to the major component observed in the VCD of each group. This is also why model calculations, using the simple coupled oscillator concept as a basis, have had some success in interpreting DNA spectra [142,148,151,152,184,186–188).

The simplest vibrational modes to model with DC interactions are those of the phosphate PO_2^-, which dominate the IR spectrum in the 1250–1050 cm^{-1} region and yield VCD signals characterized by a strong couplet for the symmetric PO_2^- stretch at 1070 cm^{-1} that reflects the helical sense (positive for the B and A forms, negative for the Z form; see Figure 6.10a) [152]. Overlap with ribose modes (locally chiral but uniform in configuration and conformation) can potentially complicate interpretation of this region, but the patterns seen to date are characteristic of the helical conformation. The ribose modes overlapping the PO_2^- modes seem to affect the VCD in a primarily additive manner, implying that their chiral impact, as expected, is highly localized; i.e., they are not strongly coupled to base or phosphate modes. By contrast, the asymmetric PO_2^- stretch at 1250 cm^{-1} has vanishingly weak VCD. The patterns, intensities and relative frequencies seen in the PO_2^- VCD, both for the asymmetric and symmetric stretches, can be quite satisfactorily calculated using the dipole coupling model [142,152,180]. As shown in Figure 6.10b, while A and B forms are not readily distinguished in this manner, Z forms are opposite in sign and reduced in magnitude, much as seen experimentally (Figure 6.10a), and the asymmetric mode at 1250 cm^{-1} (not shown) is much weaker. The large transition dipole moments for these vibrations (involving motions of charged groups) and the weak mechanical coupling of the PO_2^- groups along the backbone make them ideal candidates for a dipole coupling based theory. The important aspect of the PO_2^- modes for applicability of the DC model is their independence from the sequence. These modes sense the helical twist of the backbone directly and are all identical for each nucleotide, thus providing a specific probe of helicity.

The DNA or RNA base modes themselves are more difficult to model with DC based methods. Initially, the problem is one of localizing the normal mode that gives rise to the observed dipole transition. The C=O stretch is the most intense and highest frequency local contribution, but since base deformation modes are combinations of several local bond stretches, all having substantial double bond character, the mode's location (center) and orientation are typically difficult to specify with any assurance of accuracy. DC models adequately represent poly(dG-dC) base mode

VCD spectra but do not do well for dA-dT (or rA-rU) due to the large variance in mode structure between the bases (A having no C=O groups and T or U having two each). Diem and coworkers have attempted to compensate for this by extending the DC approach to multiple nondegenerate oscillators [184,188]. We have tried to model triplex VCD with the same model [151], but were led to what now appears to be a very unlikely structural conclusion. Subsequently an extension of the DeVoe theory model of ECD [189] was applied to DNA VCD with some success [190,191]. These latter parameterized calculations showed that the triplex spectra could be explained with a combination of Watson-Crick and Hoogsteen base pairing. Efforts to extend quantum mechanical determinations of FF, APT, and AAT values computed for segments of DNAs to entire oligomers have provided new insight into the nature of the vibrational modes and their local coupling [135]. In such a model, longer range couplings are effectively the same as in an extended DC model. Here the problem of which interactions to compute at the DFT level and which to leave at the approximate DC level is significant and remains the subject of future studies.

6.5 PEPTIDE VCD STUDIES

6.5.1 Empirical Correlation with Secondary Structure

Proteins are essentially heteropolymers of α-amino acids; hence, it is logical to use peptides as models for fundamental structural units in proteins. Since several polypeptides have well-defined structures, which have been established for several decades, they provided the first entrée of VCD in the biopolymer field. This approach has a long-established tradition for developing the basis of IR and ECD applications for protein secondary structure studies. Some α-helical polypeptides are soluble in nonaqueous solvents, which provided the conditions for early measurement of VCD for several amide transitions [192–194]. For polypeptides having a β-sheet conformation, extensive solution studies are generally not possible due to solubility problems [124–126]. On the other hand, there are polar polypeptides that undergo transitions from coil forms to helical or sheet-like forms in aqueous solution under changes of pH or salt concentration [133,195,196]. More recently, partly in response to the interest in amyloid and prion diseases, several model β-sheet forming peptides have been prepared [197–201] and VCD studies on them have appeared or are underway [111,168,169,202,203].

Aqueous solutions, which are most appropriate for modeling biochemical problems, restrict VCD studies of particular amide bands due to strong solvent absorption interference. Unfortunately, the high concentrations typically needed to obtain enough absorbance for good SNR amide I VCD and yet have a short enough pathlength to obtain sufficient transmitted intensity in H_2O solution are not compatible with maintaining many peptides as uniformly structured monomers, nor with some protein systems, due to complications of aggregation. Consequently, peptide VCD studies focusing on the amide I′ vibration (C=O stretch) have often been carried out in D_2O, after exchange of most of the amide (N-H) protons. In D_2O the amide II′ is strongly shifted and altered in character, and, due to solvent interference, the amide A′ (N-D stretch) and III' vibrations are not detectable with VCD.

Right-handed α-helices yield VCD consisting of a negative couplet in the amide A VCD (~3300 cm^{-1}), a positive couplet (+ then –, with increasing frequency) in the amide I (~1655 cm^{-1}), a negative band lying lower in frequency than the absorption maximum in the amide II (~1550 cm^{-1}), and net positive VCD in the general amide III (1350-1250 cm^{-1}) regions. The prime example of this behavior is found for poly-γ-benzyl-L-glutamate [120,192] which, due to its very long persistence length, has one of the highest intensity and narrowest band width α-helical VCD we have measured. The same (though weaker and broader) bandshape patterns are found for numerous systems composed of α-helical segments of varying lengths [59,120,133, 166,171,192–194,204–206], even including highly helical proteins such as myoglobin, hemoglobin, and albumin (for a comparison of VCD for a highly helical protein in H_2O and D_2O see Figure 6.11 [10,55,72,119,122]). Deuteration of the amide N-H (normally for D_2O studies) changes the shape of the α-helical amide I VCD to a three peaked (–,+,–) pattern (amide I′) [120,122,133,166,192, 193,204,207] and shifts the negative amide II VCD to below 1450 cm^{-1} (amide II′) [193,208].

The β-sheet and coil forms have been shown to have distinctly different amide I′ VCD spectra from that of the α-helix (see Figure 6.12) and are also consistent for a variety of polypeptides [133,195,196,204,209,210]. Widely split amide I′ absorbance features at ~1615 cm^{-1} and 1690 cm^{-1} are characteristic of an antiparallel β-sheet conformation and normally indicate aggregation [12,13,167,211,212]. The weak negative VCD pattern that results in such polypeptides probably relates more

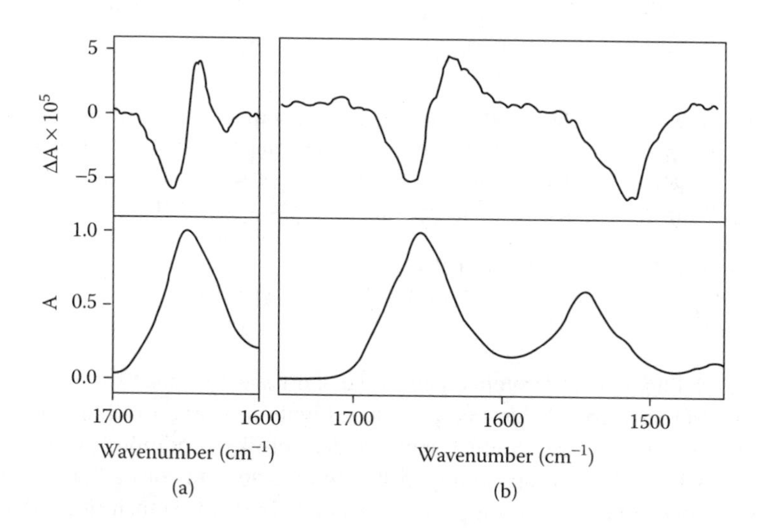

FIGURE 6.11 VCD (top) and IR absorption (bottom) spectra of the highly helical protein bovine serum albumin, for the amide I′ (N-deuterated) band in D_2O (left) and the amide I and II bands in H_2O (right). These VCD show the highly characteristic amide I positive couplet pattern (which becomes –,+,– for I′) and the amide II negative VCD shifted down in wavenumber from the absorbance peak. In a soluble polypeptide of uniform α-helical structure (which would have a negligible fraction of turns and loops) the amide I band would be sharper and more intense (by a factor of ~2) than the amide II, but the same shapes would be preserved (see, for example, Figure 6.15).

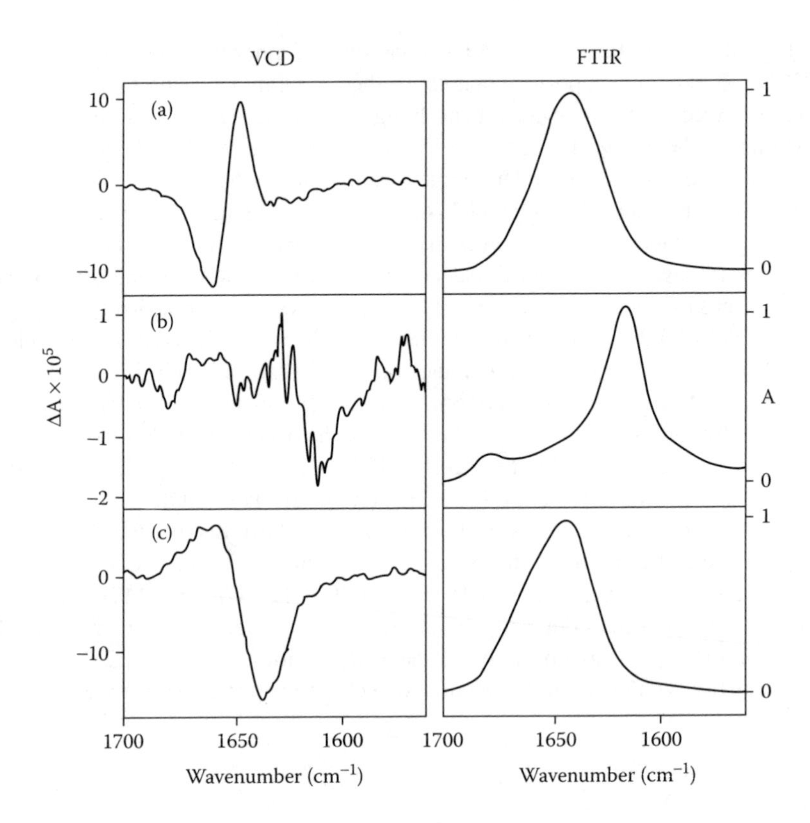

FIGURE 6.12 Comparison of VCD (left) and IR absorption (right) spectra of synthetic polypeptides in α-helical (a, top), β-sheet (b, middle), and random coil (c, bottom) conformations. Biopolymers used to obtain these secondary structures are as follows: (a) poly-(LKKL); (b) poly-(LK) in high salt; and (c) poly-K with neutral pH, where L is L-leu and K is L-lys. All samples were dissolved directly in D_2O with or without added salt. The very weak VCD (10X scale change) for poly(LK) (b) is sensitive to preparation conditions and most likely represents an extensive antiparallel β-sheet aggregate.

to denatured aggregated proteins than to the short sheet segments seen in globular proteins. We have found that this pattern is highly sensitive to sample conditions and can develop a strong couplet shape under gel-like conditions [133,213]. Oligopeptides that adopt an apparently β-structure in non-aqueous solution evidence an absorbance maximum at a higher frequency (~1635 cm^{-1}) than in these β-sheet aggregates and have a VCD whose shape is variable [111,126,167,169,203,214]. On the other hand, both polypeptide and protein β-sheet structures give rise to medium to weak amide II VCD with a negative couplet shape and (for proteins) yield negative amide III VCD [73,122,215].

A number of oligopeptide VCD studies [132,171,216–218] have established that the 3_{10} helix (Figure 6.13e) has VCD of the same sign pattern as the α-helix (Figure 6.13a), but is distinguishable from it by the weaker couplet shape with a more positive bias for the 3_{10} amide I VCD with respect to its amide II, while the opposite

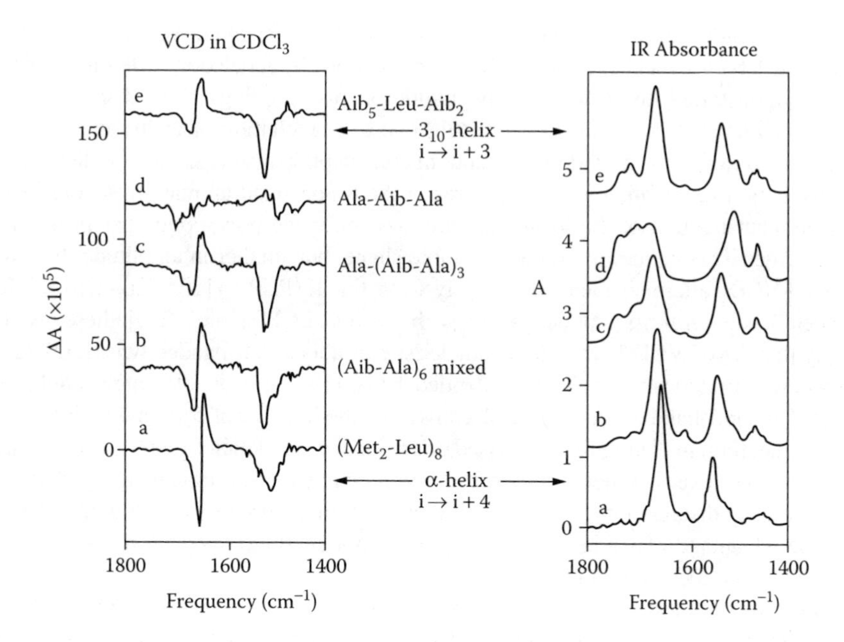

FIGURE 6.13 Comparison of VCD (left) and IR absorption (right) spectra of α-helical [L-Met_2-L-Leu]$_6$ (a, bottom) and 3_{10}-helical Aib_2-L-Leu-Aib_5 (e, top) oligo-peptides both dissolved in $CDCl_3$. Aib (or αMe-Ala or α,α-di-Me-Gly) stabilizes turn structures and, when populating a high fraction of a sequence (more than ~50%), favors 3_{10} helix formation in short peptides. The tripeptide Ala-Aib-Ala (d) lacks any stable secondary structure and gives little VCD; the heptapeptide Ala-(Aib-Ala)$_3$ (c) is mostly 3_{10} helical; and the dodecapeptide (Aib-Ala)$_6$ (b) is a mixed structure of more α-helical character, which is also true in its crystal form, where it is α-helical from the N terminus through the central residues and has a β bend toward the C terminus.

relative intensity patterns are characteristic for the α-helix. This provides a means of differentiating these two related helical structures and identifying mixed structures, such as would be present in the (Aib-Ala)$_4$ octamer [171,218]. Larger peptides, such as (Aib-Ala)$_6$ (Figure 6.13b) are largely α-helical but are distorted on the C-terminus [219], a pattern we have seen in Ala rich peptides as well [166]. Characterizations of the VCD for other minor secondary structure variants, such as β-bend ribbons, alternate *cis-trans* DL proline oligomers, parallel versus antiparallel strands, and various turns, have also been proposed [126,220–223].

The "random coil" form of polypeptides turns out to have a surprisingly intense negative couplet amide I VCD (at ~1645 cm^{-1}, Figure 6.12c), which is opposite in sign to the α-helix VCD pattern and is insensitive to deuteration [195,204,209, 210,224]. Such a result is much too intense to represent a fully disordered polypeptide whose ϕ,φ angles sample much of conformation space. In fact, the VCD for various coil-like peptides decreases in amplitude as the temperature is increased and as the oligopeptide length is decreased [209,210,225]. It is important to realize that individual isolated amide groups have virtually nondetectable VCD, and that the dominant (short range) VCD mechanism in peptides is coupling through the chain. Very recent studies from our lab on α-helical peptides that were isotopically labeled in different patterns have demonstrated the nature and magnitude of this coupling [170]. If the chain is truly disordered, little intensity can result, due to cancellation of contributions from different conformers. Instead these "random coil" peptide results strongly indicate that conformations populating a restricted area of the Ramachandran map must contribute a significant fraction of the local structure — even if the polypeptide has little long range order. This pattern is exactly the same shape, but smaller in amplitude than the amide I VCD pattern characteristic of poly-L-proline II (PLP II) [224,226–228], a left-handed 3_1 helix of *trans* peptides. Along with previous ECD studies [229], these results imply that "coil" VCD is actually characteristic of a local left-handed twist [209,224] which was previously termed an extended helix [195,229] or alternately could be viewed as a left-hand twisted β-strand. Consequently the VCD of coils and β-structures are similar, though distinguishable based on frequency and detailed pattern. Substantial local structure exists in these random coils as shown by the spectral changes reflecting an increase in disorder obtained by heating or addition of various salts [209,224]. VCD has the advantage of being sensitive to much shorter range interactions, which can highlight such local structures.

Perhaps the most important property of peptide VCD, which was established via oligomer studies, is its general length dependence. A comparison of amide I and II VCD strengths versus chain length for several oligopeptides is shown in Figure 6.14. For example, for a series of 3_{10}-helical oligomers, it was found that the VCD band shape for all three amide bands studied (I, II, and A) was established at least by the $n = 3$ or

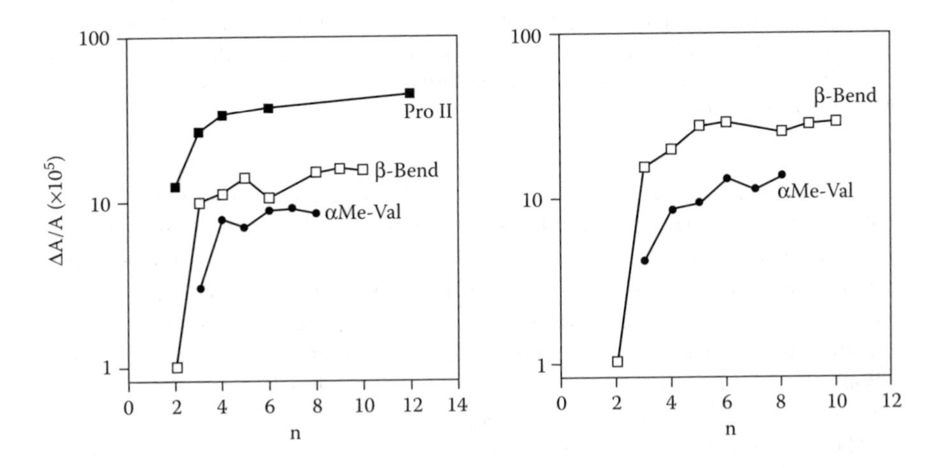

FIGURE 6.14 Comparison of VCD ΔA/A values as a function of oligo-peptide lengths for amide I (left) and amide II (right) bands for peptides of increasing length (n = number of residues) in $(Pro)_n$ (left-hand 3_1 helix), $(Aib\text{-}Pro)_{n/2}$ (β-bend ribbon structure), and $(\alpha Me)Val_n$ (3_{10} helix) oligomers having helical structures. The values show a fast rise to a near constant value for the intrinsic (per residue) VCD (ΔA/A) which indicates a dominance of the VCD by relatively short range interactions. By contrast, ECD values show a slower rise to a stable value with increase in n, due to domination by longer range dipole coupling interactions.

$n = 4$ length [216]. Furthermore, the magnitude of the VCD reached a nearly constant value per subunit by $n = 5$. Similarly, a comparison of the ECD and VCD of Pro_n oligomers showed that ECD ($\Delta\varepsilon$) increased more slowly than the VCD ($\Delta A/A$) with increase in n [53,224]. The shorter range interactions dominate VCD and result in establishing the full signal (normalized for length) for shorter oligomers than with ECD. Oligopeptide ECD primarily arises from through-space, electric dipole coupling, which is long-range in nature, but VCD has a contribution from through-bond mechanical coupling, in addition to a weaker through-space dipolar coupling. The difference between the two is due to the nearly hundredfold smaller dipole moments characteristic of vibrational transitions. Vibrations, which are mechanically coupled, naturally interact over only short ranges so that normally they can be characterized adequately by nearest neighbor residue interactions (aside from inter-residue H-bonds). This characteristic is also borne out in our theoretical studies of small oligomers (see below) [57–59,153,164]. Further proof of this concept of short range dependence is provided by VCD studies of helical peptides containing two (or four) adjacent isotopic labels (^{13}C on the amide C=O) which show a ^{13}C shifted band (~1600 cm^{-1}) with a similar α-helical VCD band pattern as seen (much more intensely) for the bulk of the oligopeptide (^{12}C at ~1640 cm^{-1}) [59,166,230]. The VCD sign pattern was shown to depend on the sign of the coupling constant which changed between sequential and alternate sites resulting in their VCD changing sign as well [170]. Separating such labels reduces the VCD intensity and can change its sign due to the fall-off in coupling and switching between mechanical and dipolar interaction as residues become separated in the peptide chain. The dipole interaction (often termed transition dipole coupling) falls off much more slowly with such separation.

Because of this length dependence, short peptide fragments that have a stable conformation can give rise to a substantial contribution to the observed VCD in a mixed structure. For example, the VCD of *β*-turns may be substantial, as we have seen in model peptides [169], and could provide a means of distinguishing between the various turn-types or of identifying their presence in a larger peptide structure. The short Aib peptide results provide examples of type III *β*-turn VCD [216]. Alternatively, cyclic peptides have been used to study type I and II turns; and those turns, from the results of Diem and coworkers [222,223], are implied to have unique VCD band shapes. Our recent studies on *β*-hairpins have identified unique IR bands associated with turns and correlated them to strong VCD features [169].

6.5.2 Peptide VCD Applications

An important aspect of VCD is its resolution of contributions from the amide group (the repeating aspect) from those originating in the side chains or other parts of the molecule. This was vital in determining the secondary structure type and handedness of several peptides containing aromatic amino acids and non-proteinic residues with chromophoric side chains [204,231–234]. A practical issue arose in this regard with respect to a series of peptides containing *β*-substituted aryl-alanines in various sequences which were used as model for electron transfer. Understanding the secondary structure was central to determination of the separation of donor-acceptor residues. With VCD it was easy to establish in those cases an α-helical conformation

[235], which in turn defined the donor–acceptor distance by use of standard molecular bonds and angles. More recently, we were able to establish the helical handedness of and propose a structure for related short oligopeptides containing $C^{\alpha,\alpha}$-disubstituted residues bridged by binaphthyl and biphenyl side chains [233,236].

The bulk of the examples presented in the previous section and in the following deal with oligo- and polypeptides that establish a conventional secondary structure. A number of early and some continuing studies have also looked at very short peptides, including individual amino acids as well as di- and tripeptides [232, 237–242], and attempted to derive structural insight from their VCD [64]. Freedman and Nafie and also Diem have utilized VCD of alanine and related amino acids and dipeptides to deduce intramolecular H-bond patterns [232,237]. Often, structure in such small peptides can be stabilized by internal H-bonds in a nonpolar solvent [243]. In addition, VCD of tripodal peptides has been analyzed with DC [244]. VCD has also been used to complement Raman and IR component intensity measurements that used to determine (ϕ, ψ) angles in model tripeptides (having just two amide groups) [245]. Finally, several reports of VCD of cyclic peptide have appeared [53,111,169, 222,246], but in general these do not have bandshapes similar to those presented above and must be theoretically analyzed or empirically interpreted by mapping onto known structures.

The sensitivity to handedness and to 3_{10}-helix formation has been used in a series of studies on peptides containing α-Me substituted residues. The usual clear preference of L-amino acids to form right-handed helices can be distorted for residues with α-substitution. Blocked (α-Me)Phe tetra- and penta-peptides were shown to form left-handed 3_{10}-like helices in $CDCl_3$ solution [247] in contrast to the right-handed forms seen for (α-Me)Val, Aib (i.e., (α-Me)Ala), and Iva (isovaline, which has both α-Me and α–Et substitution) [132,217]. Recent VCD studies of a series of peptides based on α,α-biphenyl bridged residues showed the opposite chirality, helical handedness, if C-terminated or N-terminated with an L-Val residue [248]. VCD studies of dehydro phenyl alanine containing peptides also show a tendency to form 3_{10}-helical conformations [64,249].

VCD was used to identify an intermediate during the mutarotation of poly-L-proline from the PLP I to PLP II conformation [228]. An intermediate structure developed a unique IR band and was followed kinetically until it decayed. This intermediate formation could not be detected with ECD, whose analysis had erroneously implied that there was only a two-state transition. The key to uncovering the intermediate was utilizing the frequency resolution of VCD coupled with its short-range conformational sensitivity. While the PLP I and PLP II VCD bandshapes are only marginally distinct, they undergo quite a significant solvent dependent frequency shift that allows detection of the intermediate. In this case, the magnitude of the signal corresponding to the intermediate (which is due to a junction region between the two forms) rose and fell during the transition and was quite substantial in intensity. This has a mechanistic consequence, since the time-dependent intensity was consistent only with development of several junctions in the polymer chain, and stood in contrast to a previously proposed zipper-like mechanism. Detection of this junction between left- and right-handed helical forms was enabled by a previous study of DL alternate proline oligomers which could sample both helicities and establish characteristics for *cis-trans* junctions [221]. Furthermore, the preference of these DL oligomers for a particular helical handedness was determined by VCD to depend on the chirality of the C-terminal residue. While very long DL-Pro polymers have equal populations of optical enantiomers, short oligomers appear to prefer a right-handed helical form for L-residues at the C-terminus (unblocked) [221]. This same sort of selectiveness due to a chiral residue on the terminus occurred in our α,α-disubstituted residue studies [236].

The ability of VCD to distinguish between 3_{10}- and α-helices, in combination with its short range sensitivity, was utilized to solve a lingering problem in the literature based in part on over-reliance on FT-IR frequencies for conformational interpretation. Alanine rich peptides such as Ac-(AAKAA)$_n$-GY-NH_2 have a high propensity for helix formation at low temperatures, even in H_2O, for peptides with more than about 16 residues. Their amide I′ frequencies occur at <1640 cm^{-1} in D_2O, a fact which was originally used to support EPR-based analyses that a 3_{10}-helical structure was formed or had a significant contribution to the conformation [250–252]. Interpretation of the structural impact of this 3_{10}-component has varied as more data developed [253,254]. VCD spectra clearly demonstrate that at high concentration in H_2O both the $n = 3,4$ peptides (17-mer and 22-mer, respectively) are predominantly α-helical, and at lower concentration in D_2O, the $n = 4$ peptide

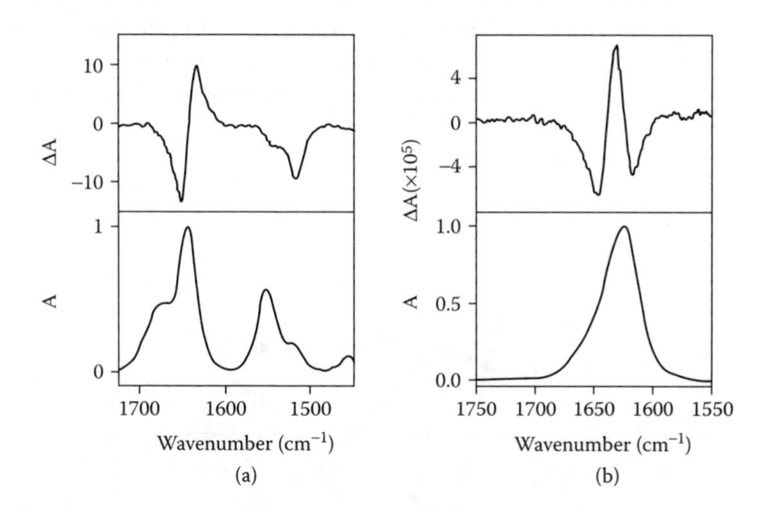

FIGURE 6.15 VCD (top) and IR absorption (bottom) spectra of highly helical alanine-rich peptides for the amide I and II bands in H_2O (a, left) and for the amide I′ in D_2O (b, right). The peptide on the left (a) is Ac-$(AAKAA)_4$ -GY-NH_2 and that on the right (b) is Ac-$(AAAAK)_3$ AAAA-H-NH_2. The IR shoulder at 1672 cm^{-1} are due to a TFA impurity resulting from the solid phase synthesis. While a common problem in interpreting FT-IR spectra of synthetic peptides, it has little effect upon the VCD. Note the low frequency of the IR amide I absorbance maximum in both (~1645 cm^{-1} H_2O and 1637 cm^{-1} D_2O) which has caused misinterpretation of structure in previous papers. By comparison to Figure 6.11, the common α-helical VCD pattern remains evident in both, and, in this case (by reference to Figure 6.13), eliminates the possibility of significant 3_{10} formation and confirms the α-helical character.

is largely α-helical while $n = 3$ is a mixed structure (Figure 6.15) [206]. Certainly the 3_{10} conformation could be involved in the shorter peptide's α-helix to coil transition, but its contribution for longer oligomers at low temperature must be minor. Study of the temperature variation of the spectra of these samples and factor analysis [255] of their resulting bandshapes demonstrated that the VCD, due to its short range length dependence, was able to detect, a third spectral component [206]. This represented an intermediate structure that could be attributed to growth and decay of the junctions between the central helical portion of the molecule and the steadily fraying ends that were coil-like in VCD (Figure 6.16). In such mixed structures, a 3_{10}-component could certainly contribute, especially in a dynamic equilibrium, but could not dominate the structure.

VCD studies [166] of isotopically labeled molecules of a similar sequence have confirmed our initial analysis. The frequency-shifted ^{13}C labeled residues have an amide I' VCD shape indicating a structure variation that depends on the position of the labels in the sequence. VCD for Ac-AAAA(KAAAA)$_3$Y-NH$_2$ labeled with ^{13}C selectivity on the four Ala residues at the N-terminus has a positive couplet for the ^{13}C=O band at ~1600 cm^{-1} and another for the ^{12}C=O band at ~1640 cm^{-1} (Figure 6.17a). However, when labeled at the four C-terminal Ala residues, only a weak negative shoulder is seen for the ^{13}C component indicating those C-terminal residues are frayed or disordered. At a higher temperature, the pattern changes to one characteristic of a coil in both the ^{12}C and ^{13}C residues (Figure 6.17b) that can be observed to develop during the thermal unfolding (Figure 6.18). Analysis of VCD frequency shifts and factor analyses of the bandshapes demonstrate that the ends unfold at lower temperatures than do the middle

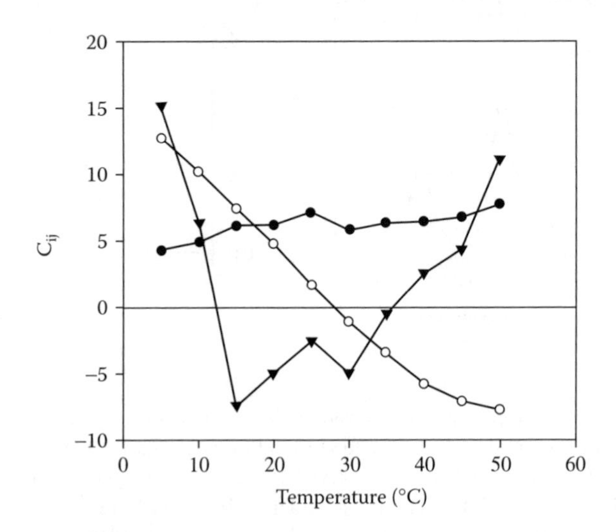

FIGURE 6.16 Variation of spectral component (factor) loadings with increase in temperature for the amide I VCD of Ac-(AAKAA)$_4$-GY-NH$_2$ in D$_2$O. Closed circles represent the loadings for the first factor (indicating a nearly constant contribution from the average spectral contribution), open circles are the loadings for the second factor (tracing out the loss of helical contribution), and the inverted triangles are the loadings of the third factor (accounting for the rise and fall of the spectral contribution from the helix–coil junctions).

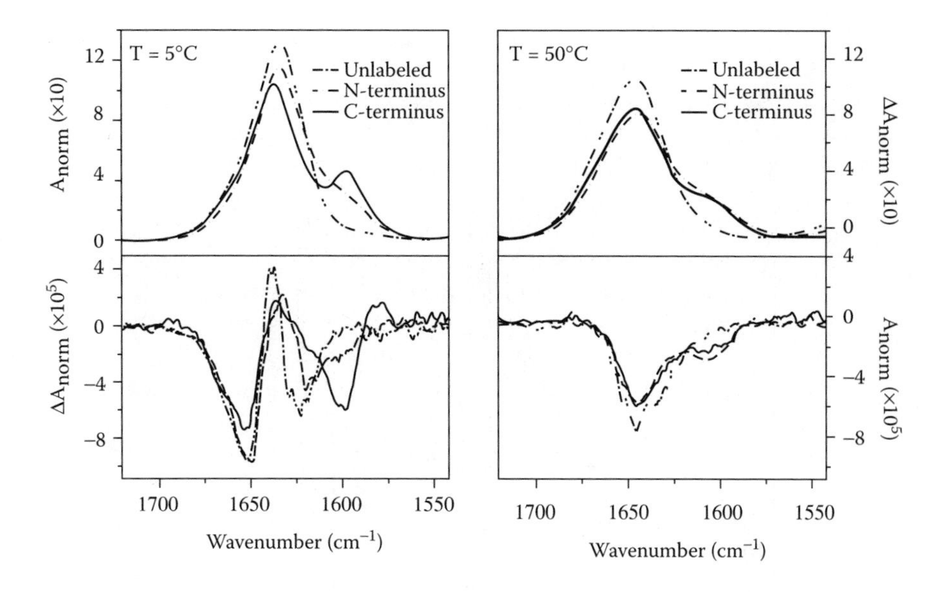

FIGURE 6.17 IR absorption (top) and VCD (bottom) amide I spectra for Ac-(AAAAK)$_3$AAAAY-NH$_2$ in D_2O for the unlabeled (dashed dot) and for the isotopically labeled variants (4 Ala residues with ^{13}C=O) on the N terminal (——) and C terminal (------) A_4 sequence at (left) 5°C and (right) 50°C. The N terminal VCD results show a clear α-helical ^{13}C contribution (~1600 cm^{-1}) at 5°C with the same sign pattern as the ^{12}C VCD (~1640 cm^{-1}), while the C terminal one is coil-like. However, at 50°C both are alike, with just a broad negative band in the ^{13}C region (as compared to the unlabeled species) indicating a full coil form.

residues and that the C-terminal residues are substantially unwound even at low temperatures [166]. In Ac-blocked peptides, the N terminal residues are largely helical at low temperatures, but when unblocked adopt a coil form like the C-terminus [256]. These temperature variations correlating to helix-coil transitions and the associated isotopic shifts are all supported by *ab initio* based calculations of IR and VCD spectra, as discussed in the next section.

6.5.3 Peptide VCD Theory

As has been well established, the Stephens implementation of the *ab initio* quantum mechanical MFP method [51,93,154–158,257–261] can produce fairly reliable theoretical simulations of the VCD of small molecules. In biopolymer applications, we first used the MFP method at the HF-SCF level (4-31G) for a model containing two peptides, constrained to α-helix, β-sheet, 3_{10} helix, and Pro II (left-handed 3_1) helix φ,ψ torsional angles, with excellent qualitative success [164]. We have repeated those calculations at the DFT (BPW91/6-31G**) level and expanded the basis to pseudo tri-alanines containing three peptide bonds linked by chiral (Me-substituted) C_α centers and then to longer helical peptides [58]. These are more complicated than the dipeptide results but confirm their validity. The computed qualitative features, such as positive couplet amide I for α-helix and negative couplet amide I for 3_1-helix

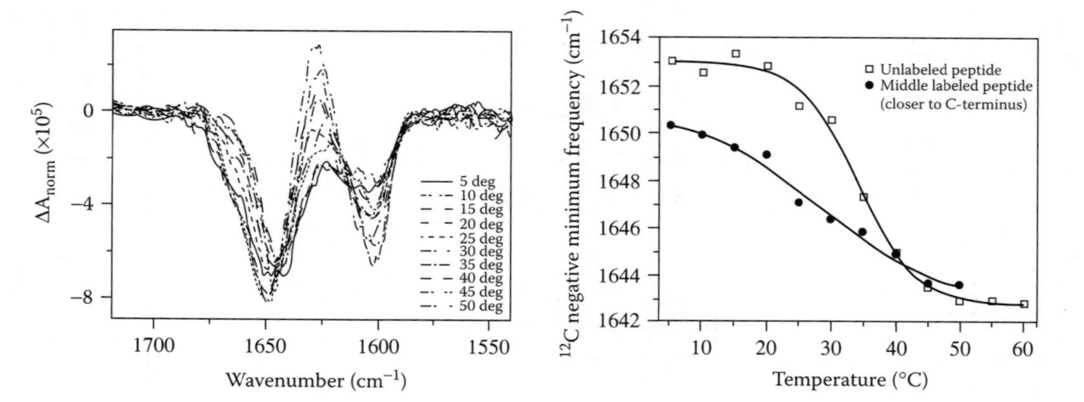

FIGURE 6.18 (Left) Temperature dependence of the amide I′ VCD spectra for the ^{13}C labeled alanine peptide, Ac-(AAAAK)-AAAAK-(AAAAK)$_2$-NH_2. (Right) Plot of the frequencies of the main negative ^{12}C peak versus temperature for the labeled and unlabeled species. The variation indicates that the terminal residues unfold at a lower temperature than do the central ones, since the middle labeled peptide has a higher fraction of ^{12}C residues on the termini and fewer in the middle.

(Pro II-like), and negative amide II for α- and 3_{10}-helix and negative couplet amide II VCD for β-sheet, have all been seen in the experimental data. The short range sensitivity of VCD is strongly supported by the success of this method based on only dipeptide and tripeptide calculations [58,164].

With improved computational facilities and upgraded programs (Gaussian 98) we extended such DFT calculations to peptide segments that include internal hydrogen bonds [59]. Results comparing an α-helical heptapeptide (Ac-A_6-NH-Me), a 3_{10}-helical and 3_1-helical (Pro II-like) pentapeptide (Ac-A_4-NH-Me), and two strands of three peptides each in an antiparallel β-sheet form are shown in Figure 6.19. Though several transitions overlap in the amide I and II regions, the qualitative patterns are still evident and remain in agreement with our experimental results. In these computations a relatively standard (6-31G*) basis set was used that results in too large a splitting between the amide I and II modes (amide I ~100 cm^{-1} too high). We have

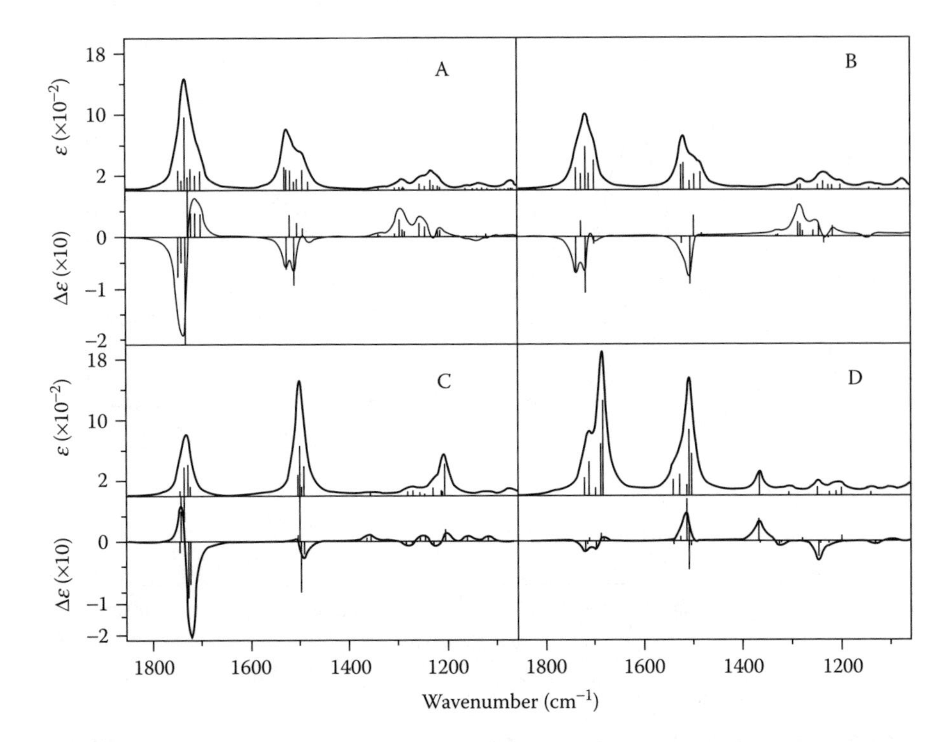

FIGURE 6.19 Comparison of fully *ab initio* DFT simulated IR and VCD spectra predicted for (a) an α-helical heptapeptide (Ac-A_6-NH-Me); (b) a 3_{10}-helical pentapeptide; (c) a 3_1-helical (Pro II-like) pentapeptide (Ac-A_4-NH-Me); and (d) two strands of three peptides each in an antiparallel β-sheet form. Vertical lines indicate the frequencies and the relative D and R values for individual computed normal modes. The spectral region plotted encompasses amide I (computed at ~1700 cm^{-1}), amide II (~1500 cm^{-1}), and amide III (and coupled C_α-H deformation) modes. CH_3 residues of alanine were converted to CD_3 for these computations to reduce spectral interference with amide modes. These calculations form the basis for transfer of property tensors to longer peptides for spectral simulations, as exemplified in Figure 6.20.

identified a modified basis set which can computationally correct this error [262], but the real source of the discrepancy is due to solvent effects, particularly for water. More recently we have developed an empirical correction for the effect of water on the amide I FF and have computed the spectra of N-methyl acetamide and a pentapeptide (β-turn) in explicit water solvent [263]. However, the simulated VCD bandshapes for the amide I and II modes, even with explicit solvent in full *ab initio* IR and VCD simulations, remain qualitatively the same as in Figure 6.19 [169,264,265].

Bour and coworkers have developed a method of transferring both FF parameters and APT and AAT parameters from a small molecule (computed fully *ab initio*) to a larger one that contains elements of the smaller one [57]. Such an approach is ideal for using properties computed for small oligomers to simulate the spectra of large polymers. We have successfully used this method to compute the spectra for oligomers of various sizes (up to ~65 residues in one case [266]). Example results for α-helical and 3_1-helical (Pro II-like) Ac-A_{20}-NH_2 are shown in Figure 6.20,

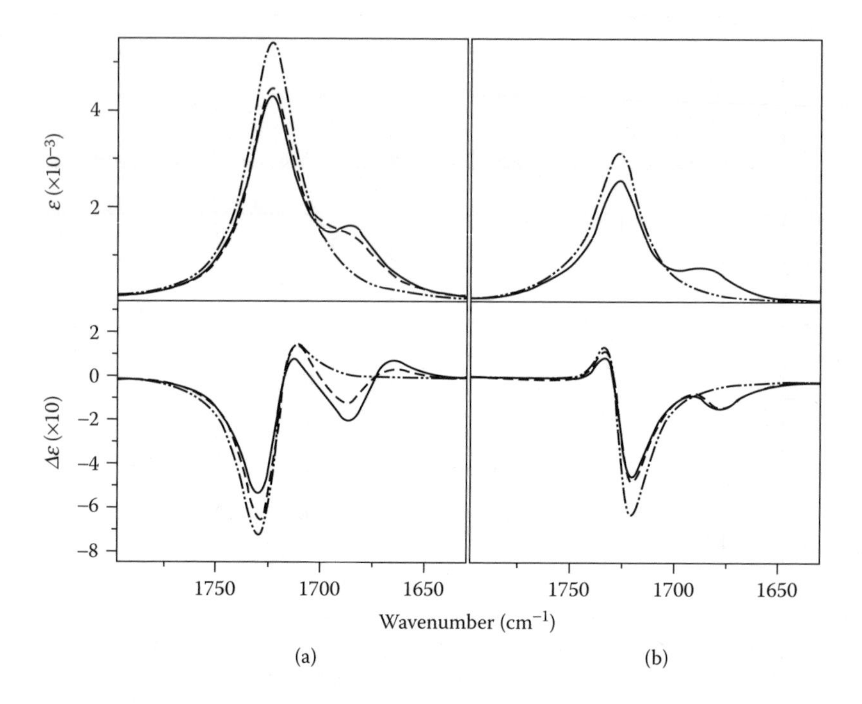

FIGURE 6.20 Simulated IR (top) and VCD (bottom) spectra of ^{13}C isotopically labeled alanine 20-mer peptides in (a) α-helical and (b) Pro II-like helical conformations. Displayed spectra are for unlabeled Ac-A_{20}-NH_2 (— ·· — ··), N-terminus labeled Ac-A*A*A*A*A_{16}-NH_2 (——) and C-terminus labeled Ac-A_{15}A*A*A*A*A-NH_2 (-- –). These spectra can be compared with experimental results in Figure 6.17. The match of the helical simulation and low temperature N-terminal spectra indicate stability for an α-helical conformation on that end experimentally, while the C-terminal match with the coil simulation, even at low temperature, indicates it does not fold to a helix. Note that these simulations are for N-protonated peptides and do not include the lower frequency negative VCD for the Amide I′ mode seen in N-deuterated α-helices (Figure 6.11 and Figure 6.15).

including effects of labeling various sites with ^{13}C on the amide C=O. The experimental results show that just two labeled amide groups are detectable in the amide I IR as a band shifted ~40 cm^{-1} down in frequency [59,230]. We compute such a pattern as well, but would perhaps expect less intensity than observed in the ^{13}C band. For VCD, we compute that the ^{13}C segment, if sequentially labeled, gives rise to the same bandshape pattern as do all the other ^{12}C residues for uniform α-helices or 3_1-helices, whichever one we are simulating. Very recently we have shown that if residues in an α-helix are labeled on alternating positions, the VCD for the ^{13}C amide I band will change sign [170]. Experimentally the VCD of these ^{13}C bands is significantly more intense, relative to the ^{12}C band, than would be expected based on the population of substituted residues for peptides in the α-helical conformation. However, for a coil conformation, the enhancement is minimal.

For β-sheets, Mendelsohn and co-workers [267] have shown even the IR to have striking ^{13}C intensity effects depending on the substitution pattern. Lansbury and coworkers [268,269] and Silva et al. [270] have shown that single substitutions on particular sites can lead to IR enhancements and may be interpreted in terms of the type of β-sheet formed. Our recent *ab initio* based simulations [167] demonstrate that this IR intensity effect is inherent to the aggregates of extended β-strands found in peptide β-sheet models, and arises from in-phase localized vibrations of the labeled residues on the central strands of the multistranded β-sheet. By contrast, isotopic labeling does not dramatically enhance β-sheet VCD in our simulations [262]. Very recent results for β-hairpins have predicted that labeling on opposite, multiparallel strands to form a 14 atom ring yields an intensity enhanced IR absorbance. Experimental realization of this with a 12-residue peptide showed strong VCD effects in the ^{13}C band as well [164,203].

In contrast to all these successes with *ab initio* based MFP modeling, coupled oscillator calculations, which utilize only dipole coupling (DC) to simulate VCD, yield much poorer representations of the qualitative features seen experimentally in peptide VCD [164,170,265,266]. DC has long been proposed as an important factor in peptide vibrational analyses and had been argued to make a critical contribution to the unusual β-sheet spectra [32,267], but our *ab initio* based results appear to encompass most of the characteristic features without resorting to explicit DC (DC is implicit in the short range interactions that are part of the transferred small molecule FF, but no long-range interactions are included). Since the *ab initio* MFP computations of the VCD for just a constrained dipeptide system replicate so many of the observed VCD features, it is clear that the short range contributions represented in such a molecule are very important for understanding the overall VCD spectrum of longer peptides. One might expect that use of DC for long range and MFP for short range interactions might result in a sensible theoretical analysis of peptide and protein VCD. We have carried out such test computations and found small intensity enhancements of ^{13}C bands in helices, but did not recoup the full experimental enhancement. Similar DC extensions of our *ab initio* based simulations of ^{13}C labeled β-sheet models showed a loss of IR absorption intensity, thus moving away from the experimental result [265,272]. DC modeling can be used to fit the spectra for β-sheet peptides if a sufficient number and type of interactions are included as empirical parameters [267].

DC has its most accurate prime biopolymer application in VCD as noted earlier for DNA and RNA simulations of base and PO_2^- modes coupling through space where there are only relatively weak mechanical connections. The stronger through-bond coupling characteristic of the peptide bond through C_α sites demands the higher precision possible with the MFP method in order to describe more localized coupling of molecular vibrations. The property transfer method [57] gives the most useful entrée into large systems under this restriction. Given a sufficient selection of model small molecule calculations, each on a system large enough to develop realistic FF, APT, and AAT parameters for transfer, it is reasonable to propose calculating VCD for large peptides with multiple conformations. The first step in this process was the modeling of hairpin VCD and IR using separate turn and sheet segment calculations for the transfer [169]. Qualitatively correct patterns were obtained, judged by appropriate experimental models. Such efforts might even encompass real protein structures in the near future, and given our recent success with uniform peptides, are likely to be much more accurate than empirically based parameter transfer methods.

6.6 PROTEIN VCD

Proteins are more than just long heteropolymers. Their sequence imparts a specific secondary structure propensity whose understanding is a long-sought goal of biophysical research. Most proteins differ from small peptides in terms of the degree of solvation and the uniformity of the secondary structure segments developed. While a helix in a peptide may terminate in a large number of conformations representing the fraying of that segment, in a protein this termination is likely to be a relatively well-defined or, at least, a conformationally constrained turn or loop sequence. Furthermore the segment lengths in a protein are determined by the entire protein fold rather than by an isolated segment's structural thermodynamic stability. Thus, interpretation of VCD for proteins has largely been dependent on deconvolving spectral data for proteins of known structure and using the understanding (or algorithm) developed to explore structures of unknown proteins. As a check on reasonability, such protein-based protein structure analyses do retain a qualitative dependence on the extensive background of peptide VCD data described above, but they are not constrained to have to have their component spectra match such models. Some deviation is expected since environmental effects on the spectra can be significant, and the globular protein environment for given peptide sequences differs from that of a fully solvated peptide and the effects are somewhat unpredictable at this point. In our laboratory, these interpretations are also constrained to be consistent with ECD and FT-IR data of the same system as is appropriate. These cross-checks help prevent an analysis of a multidimensional structural system from returning nonsensical results, since these spectra have only relatively simple and modestly resolved spectral bandshapes; thus, they must yield limited information content.

6.6.1 Qualitative Spectral Interpretations

In the amide I′ region, if the proteins are first deuterium exchanged before being measured in D_2O or deuterated buffer solution, VCD spectra of the proteins are straightforwardly measurable with path lengths of 25–50 μm and concentrations of 2–5 mg/100 μl [119]. Actual amounts of sample needed are lower since typically only 20–30 μl are needed to fill the cell. Spectra obtained on proteins in D_2O without allowing significant time for complete H-D exchange are of lower quality due to interference from residual HOD and due to the unknown and variable degree of exchange. Amide II and III spectra can be obtained for proteins in H_2O at relatively high concentrations [73,215]. VCD of both the amide I and II bands for a single protein sample in H_2O solution can be measured by using very short path lengths (6 μm), very high concentrations (1 mg/10 μl) and extensive signal averaging (> 20 hrs) (Figure 6.11) [122]. While FT-IR measurements of proteins in H_2O can successfully be made with less sample and concentrations, in order to get adequate SNR VCD for proteins, an absorbance of the order of 0.1 or greater is desirable, which is the source of the above suggested concentrations.

Complete VCD sign pattern inversions and peak frequency variations as large as their band widths can be found if one surveys a broad set of globular protein spectra [11,69,70,72,119]. By comparison, the IR absorption maxima of these same transitions have only modest frequency variations (~20 cm^{-1}) within this same set of proteins. The high degree of bandshape variation seen in VCD is furthermore not seen with ECD [20,24–26,70,71,122,273]. The protein VCD bandshape changes arise from the fact that all types of secondary structure give rise to VCD signals of roughly the same intensity. On the other hand, ECD, particularly in the (historically most heavily studied) $\lambda > 200$ nm part of the UV region, is dominated by the large contribution from the α-helical components.

A selected comparison of normalized bandshapes for the amide I′ VCD, FT-IR, and amide ECD for three proteins in D_2O solution is presented in Figure 6.21. Of these, myoglobin (MYO) has a very high fraction of α-helix, while concanavalin A (CON) has substantial β-sheet component with no helix and triosphosphate isomerase (TPI) has significant contributions of each. These three proteins have ECD spectra similar in shape, the main differences being in intensity (not shown in the figure) and wavelength of zero crossing. CON has a big shift of the positive far UV band and TPI has a broadened negative band as compared to MYO. The near-UV intensity depends strongly on helix content, which is why simple analyses of ECD data in terms of α-helical fractions do so well [20,24–26,70,273].

The FT-IR spectra of these proteins are also quite similar, with the primary difference being a frequency shift of the peak absorbance that roughly correlates with the amount of β-sheet in the protein. Thus CON has a distorted shape with a maximum at ~1630 cm^{-1}, and TPI is broader and slightly lower in frequency from MYO, which is sharper and centered at ~1650 cm^{-1}.

In the VCD spectra of the same three proteins, dramatic changes are seen in the band shapes and significant shifts arise in the frequencies. Together these cause the spectra to exhibit much more sensitivity to the structural variation in these proteins than do either ECD or FT-IR [71,72,119,274]. The highly helical MYO has an amide I' VCD dominated by a positive couplet with a weak negative feature to low energy, closely patterning spectra measured for N-deuterated model α-helical polypeptides (Figure 6.12 and Figure 6.15). By contrast, the CON amide I′ VCD is predominantly negative with the main feature falling between 1630–1640 cm^{-1}, having the same

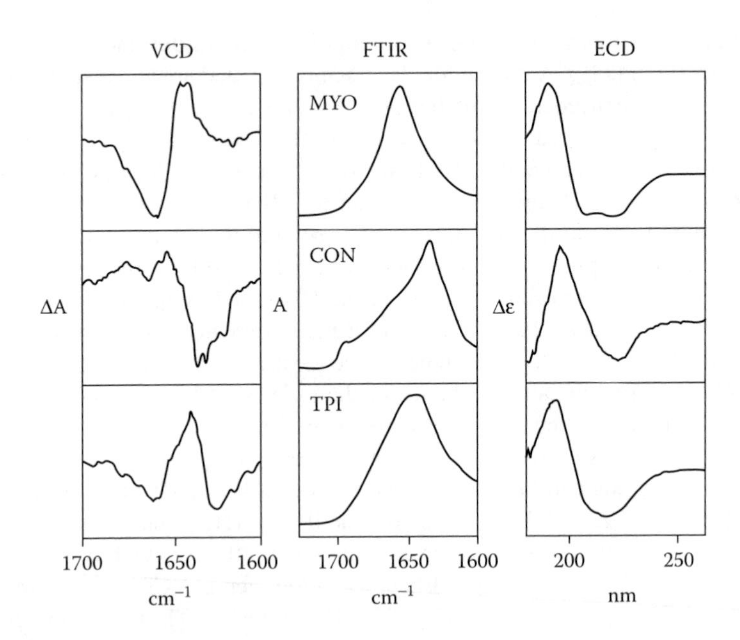

FIGURE 6.21 Comparison of amide I′ VCD (left), FT-IR absorption (middle), and amide ECD (right) spectra of three proteins (in D_2O) with dominant secondary structure contributions from α-helix (myoglobin, top), β-sheet (concanavalin A, middle), and from both helix and sheet (triose phosphate isomerase, bottom) conformations. The comparisons emphasize the relative sensitivity to the differences in secondary structure with the three techniques and the distinct bandshapes developed in VCD for each structural type. Since the spectra are all normalized to the same magnitude, only the bandshapes can be compared. While the actual intensities vary they have no real meaning in this figure.

dominant sign but otherwise not very much like the poly-L-lysine antiparallel β-sheet VCD (Figure 6.12). Globular proteins with a mix of α and β components, such as TPI (Figure 6.21), have amide I′ VCD spectra resembling a linear combination of these two more limiting types. On this qualitative level, the relationship between spectral bandshape and structure is clearer in just the amide I′ VCD spectra than in the amide I′ FT-IR alone, or even in the ECD spectrum measured over the range of 260–180 nm.

Amide II and III VCD for MYO and CON in H_2O are compared in Figure 6.22. Highly α-helical MYO has an intense negative amide II VCD lower in wavenumber than its corresponding absorbance maximum, while CON (β-sheet dominant) has a weaker, negative couplet amide II VCD roughly centered on its absorbance maximum [215]. Proteins with mixed α–β conformations exhibit VCD that is still weaker and representative of a linear combination of these two limiting types. Representative amide III VCDs for these same proteins show an overall opposite sign pattern when averaged over the whole 1400–1100 cm^{-1} region, with helices being net positive and sheets being net negative, especially between 1250 and 1380 cm^{-1} [73]. Amide III VCD is very weak, in part reflecting the very low dipole intensity associated with the amide III mode in the IR. Consequently, its use for structural analysis is limited at best [215,275,276].

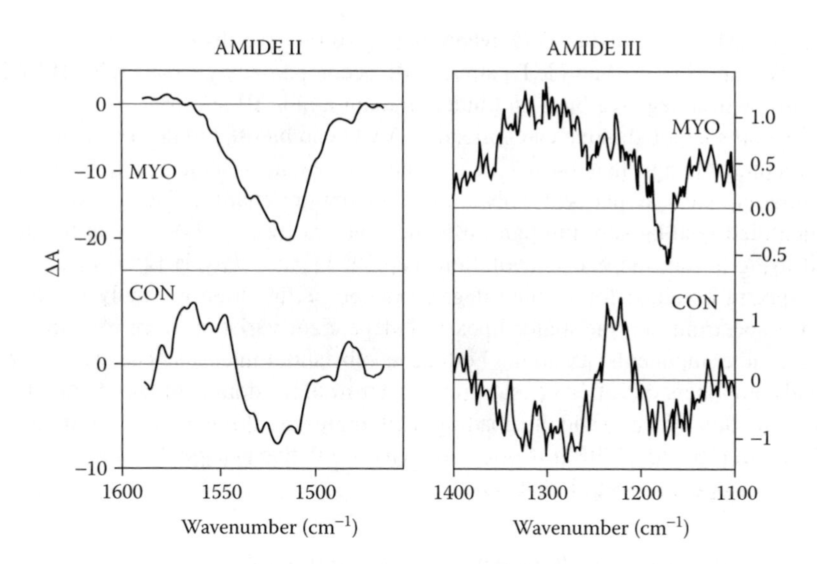

FIGURE 6.22 Comparison of VCD spectra for the amide II (left) and amide III (right) band regions of two proteins with dominant contributions from α-helical (myoglobin, top) and β-sheet (concanavalin A, middle) secondary structure segments. Comparison to the more resolved amide I′ VCD spectra in Figure 6.21 shows the broader and more overlapping nature of the transitions as well as lower SNR in these regions. However, each band retains a distinctive form or sign for a dominant secondary structure type.

While a strength of VCD for conformational analysis is its access to other transitions that can yield added structural selectivity, VCD bandshape patterns vary less in both the amide II and III regions than for amide I′ spectra. The amide II and III bands are broader than the amide I or I′, and all contributions overlap, much as is seen in ECD spectra. This overlap is in contrast to the amide I where frequency as well as bandshape contribute to the analysis. It is even in contrast to Raman spectra of the amide III where each conformer gives rise to bands at characteristic "marker" frequencies [31]. Nonetheless, it is clear that VCD bandshapes, due to their signed nature, have more distinctive character and variation than do the respective FT-IR absorbance bands for each of the accessible protein amide modes (I, II, and III).

Amide I VCD spectra of proteins in H_2O solution have little difference from the D_2O results, except for the predominantly α-helical case where the amide I couplet changes to the amide I′ (–,+,–) bandshape, just as observed for peptides. (Protein and peptide VCD in H_2O and D_2O can be compared by contrasting Figure 6.11 and Figure 6.15, respectively.) The VCD pattern can be used to identify the general secondary structure type. In summary, proteins with a high helical content have an amide I or I′ at ~1650 cm^{-1} dominated by a positive couplet (with a weak low energy negative component in D_2O), an intense negative amide II at ~1530 cm^{-1}, and a net positive amide III at ~1300 cm^{-1}. Those with high sheet content have a negative amide I at ~1630 cm^{-1} with various weak positive features to higher frequency, a negative couplet amide II, and a net negative amide III. Mixed helix–sheet proteins are more easily identified in D_2O, where they give rise to a distinctive, but typically weak, W-shaped (–,+,–) amide I' pattern [60], accompanied by weak amide II VCD, typically with a negative bias, and indeterminant amide III spectra.

Beyond such qualitative categorization, VCD can be used to determine the type of structural change that occurs in a protein when it is subjected to a variable perturbation, such as pH, salt, solvent, or temperature change. These changes can be identified qualitatively through difference spectra [60,277,278] or can be quantified using a bandshape deconvolution method. Factor analysis [255] of a set of such spectra obtained for varying degrees of the perturbation normally yields the average spectrum and the major linearly independent variation from it as the first and second components, assuming both have substantial intensity. The shape of the second component can often be interpreted in terms of the dominant type of structural change in this process and its loading will represent the extent of shift in the equilibrium [166,206,279]. It is this sensitivity to relative change that has long been the *forté* of optical spectral analyses.

6.6.2 Applications of VCD for Protein Structure

A number of applications have been reported that apply VCD based analyses to determination of conformational aspects of specific proteins. An early study of phosvitin [280] was used to correct an ATR–FT-IR data based structure analysis and to show that this highly glycophosphorylated protein went from a coil-like state to an antiparallel β-sheet aggregate at pH 1.6. A study of glucoamylase showed that while glycosylation complicates ECD analysis, it is generally not a problem for VCD, which could be used to determine the thermal denaturation properties of that protein [281]. Similarly a VCD study of two growth hormones was used to eliminate a controversy brought on by misinterpretation of a frequency shift of the FT-IR observed amide I [282]. IR wavenumbers are subject to considerable variation due to solvation and other environmental perturbations, as was dramatically illustrated in our Ala-rich peptide studies and other similar studies [206,252,283].

Analyses of thermally induced Ribonuclease A unfolding [279] have identified a pretransition highlighted by factor analysis of the entire spectrum and structurally defined by comparison of response in separate spectral regions with different techniques. Previous analyses of these thermal transitions based on single frequency intensity variation missed the intermediate. The pretransition is broad, suggesting the unraveling of a helix. Ribonuclease S unfolding results in a broader and weaker pattern with a lower transition temperature due to the cut in the sequence at residue 20. Similar studies of ribonuclease T_1 have identified the thermal loss of the major helical component with retention of partial β-strand character at high temperatures [284].

Bovine α-lactalbumin (BLA) is a protein whose structure appears to be unusually malleable and as such has been the focus of many studies of what is termed the molten globule transition [285]. At low pH, BLA, as well as human and other variants, expands and is said to lose tertiary structure. Our VCD analysis showed that, by contrast, it actually gains secondary structure, in particular α-helix content forming a molten globule structure [60]. Furthermore, although its crystal structure is nearly identical to that of hen egg white Lysozyme (HEWL), its spectra (Figure 6.23) — ECD, FT-IR, and VCD — are noticeably different [277]. In fact, only when at a lower pH or (even more so) in the presence of a helix stabilizing solvent such as TFE or propanol [277], do the spectra of BLA (and other LA proteins as well) begin to resemble those of an (α, β) protein and thus resemble the HEWL spectra in detail. It appears that BLA has a dynamically fluctuating structure in aqueous solution whose conformation samples many local minima, and only under the influence of some perturbation does it find a structural minimum like that of HEWL. This dynamic nature is not evident in the crystal structure and evidences its biggest impact on the VCD spectra. In some cases the helices in BLA may be more 3_{10}-like and thus yield more detectable differences in VCD than seen in other techniques. Such a structural flexibility has a functional component in that BLA binds to β-galactosyl transferase, thereby altering its receptor specificity and enabling the synthesis of lactose in the milk of lactating mammals.

Cytochrome c (cyt c) is a small globular protein with a bound heme group and a dominant helical bundle fold. It has been the target of many protein folding studies for its structure, misfolded or folding intermediate states, and the various spectroscopic probes that provide information on its structural state [286–291]. In our thermal stability tests of acid denatured cyt c, we found that the thermal changes in VCD distinguished between the various forms better than FT-IR or ECD [292]. In particular, the VCD showed the acid-denatured state to have a residual helical component that was reversibly lost on heating. The NaCl induced molten globule state was seen to have both a significant helix fraction and a disordered component, which distinguishes it from the native state (Figure 6.24). The comparison of FT-IR, VCD, ECD, fluorescence and UV-Vis absorbance allows one to isolate various components of the folding change under these different perturbations.

Human chorionic gonadotropin (hCG) protein consists of two subunits. The β-subunit has three hairpin segments that were thought to fold in an initial step (framework model) and then collapse to form the active state [293]. Misfolds are thought to correlate with tumor function. By studying each of the hairpins individually with ECD, FT-IR, and VCD, we have been able to show that the stability of the β form of these peptides outside the protein is not a property of the sequence alone but is dependent on high concentration or can be induced by the presence of a charged micellar environment (e.g., with SDS) [214]. VCD was used to distinguish between the extended (PLP II-like) structure of the sequence corresponding to hairpin H2-β and the antiparallel β-form that could form with the hairpin H1-β and H3-β sequences [214]. The latter (H3-β) was fully dependent on external interactions for forming secondary structure [214], while the former (H1-β) was a stable hairpin form under all conditions studied [134]. In all of these studies, correlation of VCD, FT-IR, and ECD data over a wide concentration range was essential to deriving a reliable basis for interpretation of folding.

A common feature of misfolding is aggregation to form extended antiparallel β-sheet structures. These differ from the twisted β-sheets seen in globular proteins. An example is concanavalin A, which has relatively flat sheets in a tetrameric structure at neutral pH, but is monomeric at low pH. When heated, the con A aggregates, as seen in the IR band growing in at < 1620 cm^{-1} (Figure 6.25) and the shift in the

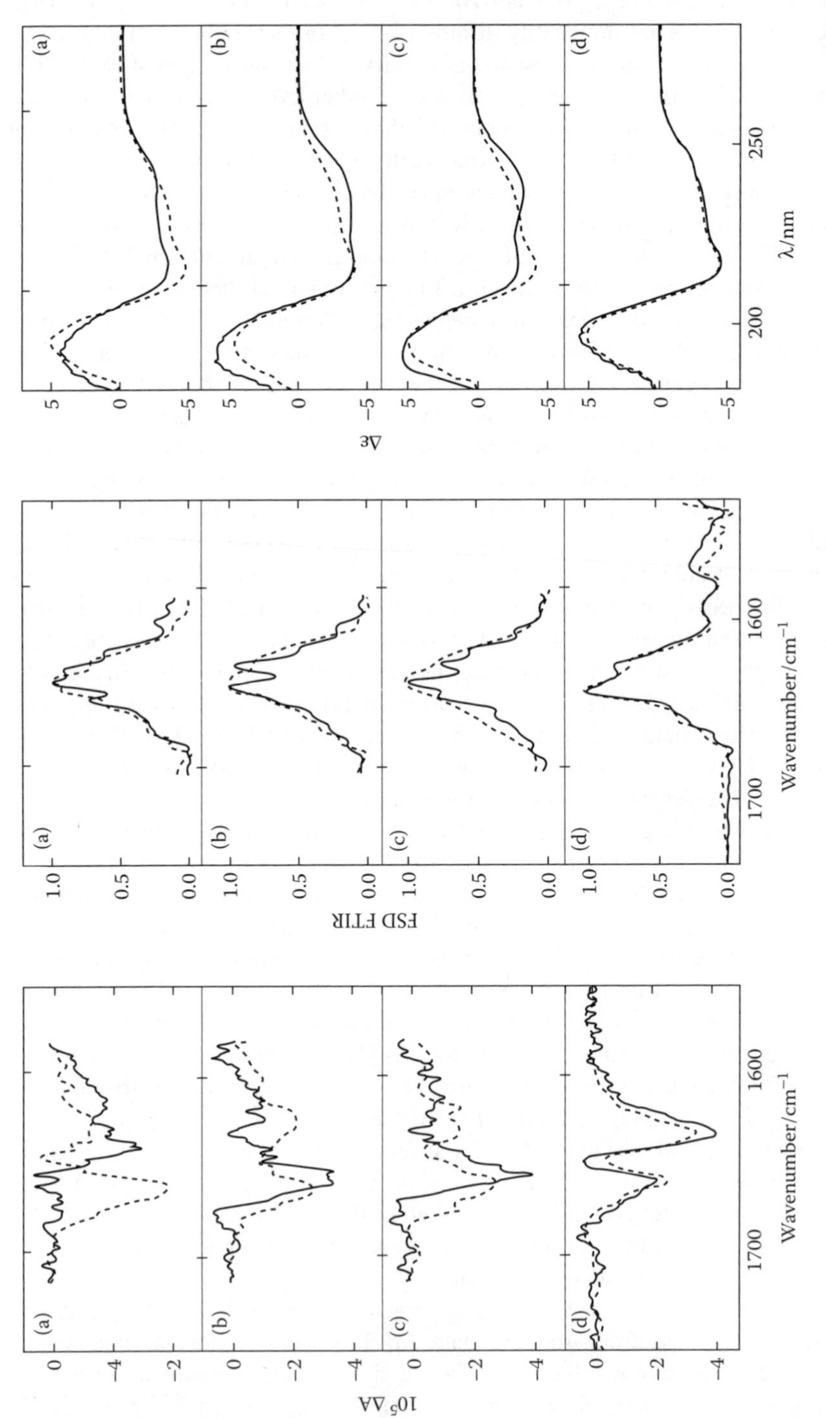

FIGURE 6.23

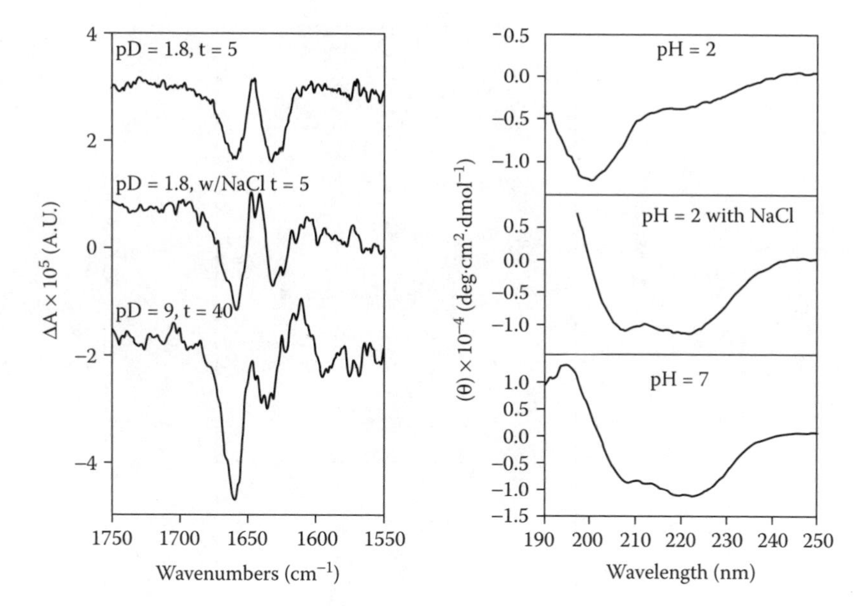

FIGURE 6.24 Comparison of cytochrome c VCD (left) and ECD (right) for native (bottom, neutral pH), molten globule (middle, high salt, low pH) and unfolded (top, low pH) states. The VCD shows residual helix in the unfolded state and discriminates better among them.

VCD shape [294]. VCD is not the method of choice for extended -sheet detection, but loss of VCD intensity coupled to growth of the very low wavenumber amide I band is a useful combined tool.

6.6.3 Protein Secondary Structure Quantitative Analyses

Analysis of protein VCD spectra in terms of the fractional components (FCs) of their secondary structure has centered on the use of the principal component method

FIGURE 6.23 (*Continued*) (Left) Comparison of the amide I′ VCD spectra of (a) human lactalbumin (hLA), (b) bovine lactalbumin (BLA), and (c) goat lactalbumin (GLA), with that of (d) hen egg white lysozyme (HEWL) in D_2O-phosphate buffer at pH 7 (———) and pH 2 (- - - - -). (Middle) Amide I′ deconvolved FT-IR absorption spectra and (right) ECD spectra in the far-UV amide absorption for the same four proteins, again at pH 7 (———) and pH 2 (- - - - -). The variations show that the LA proteins, whose crystal structure (hLA) nearly overlaps that of HEWL, give distinctly different spectra at pH ~7 which should be the native state (VCD having the most extreme variations and ECD being the most uniform among the species). However, at pH ~2, which yields the molten globule forms of LA, the spectra of the LA proteins change and become more similar to each other, while HEWL changes very little. This implies the LA native state has more flexibility than suggested by the crystal structure. Due to their faster time scales for sensing structure, optical spectra can monitor this fluxional variation.

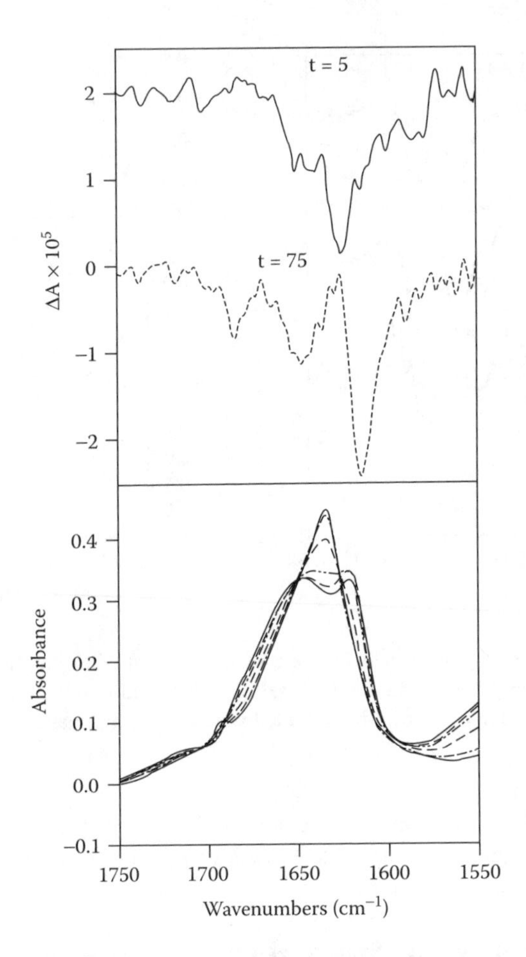

FIGURE 6.25 Comparison of concanavalin A spectra at low pH with increase in temperature. The two VCD (top) show a change from β sheet in the globular form (5°C) to the aggregated one (75°C) while the progression of IR (bottom) show the typical aggregation feature growing in at ~1620 cm^{-1}.

of factor analysis (PC-FA) [255,295,296] to characterize the protein spectra in terms of a weighted combination of a relatively small number of coefficients [11,69,70,72]. Such an approach has much in common with the methods used for determining protein secondary structure from ECD data [20–22,24,26,68,70,71,273,297,298] and methods sometimes used for FT-IR spectral interpretation [11,38,43–45].

In normal principal component analysis, the first spectral component (or factor) represents the most common elements of all the experimental spectra in the training set. The second component represents the major deviations in the set from the most common spectral elements. Each successive component then becomes less significant, eventually representing just the noise contributions. These components are constructed to be orthogonal to each other in a function space spanned by the original basis set of spectra, much as would be found in an eigenvector analysis. Due to the mathematical projection of the first components on the most significant spectral contribution to the entire set of spectra and the construction of the remaining components to be orthogonal to it, the orientations of these components (i.e., the composition of the eigenvectors) are dependent on the makeup of the training set of spectra used. The original spectra θ_i can be represented as a linear combination of factors (or spectral components) Φ_j and their loadings α_{ij}, as

$$\theta_i = \Sigma_j \alpha_{ij}\Phi_j \tag{6.2}$$

While formally this is just a transformation from one basis to another, a compact representation can be obtained by truncating the sum over j to save just the significant spectral components or those containing structural information. This eliminates those Φ_j values that correspond primarily to noise.

For example, the experimental amide I′ VCD (D_2O) spectra for up to 28 proteins [72] yield 6 subspectral components of significance which can be used to reconstruct any spectrum to useful accuracy. For the amide I+II VCD (H_2O) more components are needed [11], and, for very high SNR ratio FT-IR data, even more non-negligible components can be defined [11,43,45,74], while for ECD spectra 5 components seem to be sufficient [68,71]. The loadings of the spectral components corresponding to each protein provide a compact set of values characterizing each protein's VCD spectral bandshape. Regression relationships can then be established between the loadings and structural parameters of interest. The number of significant components indicates the maximum number of interpretable structural contributions. However, these factors may not have a simple relationship to structure, since they are the result of a projection out of the training set (experimental spectra), which can be more or less complete and whose balance is set by the user. In particular, since their orientation depends on the original input set of protein spectra, if that is biased, the component spectra will reflect the bias and the resultant loadings will depend on it [11,21,22,70,298,299]. Thus, attempts to use either a targeted or a broadly based training set of spectra for these analyses will result in structural projections that differ in detail. Either approach might be valid for the specific questions being tested, but they will not be the same. Seeking an optimal method of deriving the spectra–structure relationship for protein VCD (as well as ECD, FT-IR, and Raman spectra) has been a major research effort in our lab in the past [11,69–74,208,284,300–302].

The method we have adopted subjects whichever kind of spectra set is being analyzed (e.g., VCD, ECD, or FT-IR) to PC-FA. This step must include spectra for both the training set of known structure proteins and for the unknown ones to be analyzed. Regression analysis using the loadings (α_{ij}) of the spectra and the FC_ς^i values for the i proteins in the set, ς structural types (e.g., α-helix and β-sheet) and j subspectral coefficients derived from PC-FA is carried out by a complete search for relationships between all possible combinations of j factor loadings, α_{ij}, with the FC_ς^i sets for each ς type. This extensive search determines which coefficients have a statistically significant dependence on structure. To carry out this procedure, the loadings are first fit to the FC_ς values and the resulting relationships are evaluated based on the goodness of fit between the computed $(FC_\varsigma^i)^s$, from spectra, and the known value $(FC_\varsigma^i)^x$ from the X-ray crystal structure. This is then expanded for multiple linear regression relationships for more than one loading by considering all pairs — e.g., α_{ij} and α_{ik}, triples, and so on. There is no *a priori* reason to treat them as sequential groups [45,297], since any one of the loadings may have the most significant dependence on any particular FC_ς value.

We test all possible regression relations by a one-left-out scheme, systematically taking one protein out from the set, redoing the regression [11,69,70], and predicting (not fitting) the structure parameters for the left-out protein. By treating each protein in the set as left out, the resulting prediction errors can be evaluated in terms of a standard deviation of prediction for the entire set. It is this quality factor which then guides selection of the loadings to be included in the final predictive regression algorithm to be used with unknowns. This method is denoted the restricted multiple regression (RMR) approach [11,69,70].

Other methods can be developed for VCD, and numerous methods have been developed for ECD and FT-IR studies [26,74,208,303], but the RMR rather simply focuses on the parts of the spectrum with the maximal structure sensitivity. For example, in amide III studies, we developed an alternate projection-based similarity method which essentially predicts the relative amount of helix and sheet, assuming they are reciprocally related based on similarity of the unknown spectrum with that of the standard [73]. The RMR method has been tested extensively with protein amide I and II spectra measured in D_2O and H_2O with FT-IR and VCD as well as for ECD data and has been found to work best when data from two independent spectral regions, IR and UV, are used in a combined analysis to predict FC values. For example, VCD analysis was better for β-sheet, but ECD was much better for α-helix [70,71] so that combining them gives the best of both methods. Nonetheless, even with multiple data sets, our RMR method shows that the most reliable prediction comes with relatively few loadings used in the regression-developed algorithm for prediction. This is true for FT-IR, VCD, and most especially for ECD data [11,70].

The most far-reaching observation made was that the best secondary structure predictions are found to be correlated with only a few coefficients [11,69,70]. This is in contrast to fitting spectra to structure, where more spectral coefficients are always better, but are not always meaningful. By contrast, prediction as a criterion has a real meaning in terms of the ultimate utility of the algorithm for analysis of unknowns. Since a number of other methods use all significant spectral coefficients [20,43,68], it is important to recognize this fundamental difference with the RMR method. Some recent ECD-based analyses have taken alternate restrictive approaches, complementary to our RMR method, of selectively limiting the training set, cutting off the number of loadings or optimizing the method for self-consistent prediction behavior [45,297–299]. The interdependence of helix and sheet content, previously identified using a neural network study of crystal structure data for 192 non-homologous structures in the protein data bank (PDB) [55,300], and recently extended to ~450 proteins [75] can obscure reliability in the development of such an algorithm. One might think the success of a method implies that it measures two structural properties through their spectral dependencies, while, if the structural elements are correlated, it may sense only one at a physically reliable level. While this might seem to be just a problem of choosing the best training set, we have shown that changing the training set and modifying the spectral region can modify the dependence on different loadings (as would a rotation of the vector orientation), but in all cases, only a few loadings are needed for the most reliable predictions of protein secondary structure. This conclusion is borne out by a number of independent observations [22,45,297]. It also is eminently reasonable. Many factors contribute to the properties of a spectral transition; conformation in terms of ϕ, ψ angles of the peptide chain is only one of them.

Combining the amide I′ and II VCD somewhat improves the predictive ability of the VCD data, making the sheet predictions clearly better than those for the ECD data [70]. Use of the H_2O-based data, for the amide I+II VCD, gave about the same or somewhat worse results [11]. However, a combination of the long-range dependence of ECD with the short-range sensitivity of vibrational spectra (FT-IR or VCD) is the key to better prediction. More importantly, the combined analysis gains stability and results in a significant reduction in the errors associated with outliers. In other words, it eliminates very poor prediction for specific unknowns. Other researchers using independent methods have found similar improvements for combining ECD and FT-IR data, thereby also sensing long- and short-range interactions [44,45]. The best of both methods results in superior α-helix prediction, dependent mostly on ECD, combined with the special β-sheet sensitivity of VCD [11,70].

The standard FC^x values are determined from the X-ray structures using the Kabsch and Sander DSSP method [304] which is a standard component of the PDB. RMR calculations based on the Levitt and Greer [305], King and Johnson [306], and Frishman and Argos [307] algorithms for secondary structure determination give similar, but slightly worse, predictive errors (Table 6.2) [70,71,308]. Thus the limits we find cannot be attributed to choosing the wrong algorithm for secondary structure interpretation. This is consistent with these algorithms being linearly interdependent as illustrated in Figure 6.26 [300,308]. It should be clear that the distortions of helices and sheets in a globular protein mean that there are many residues that cannot be uniquely ascribed to helix, sheet, turn, etc. We feel that it is this ambiguity and nonideality of secondary structure segments that imposes a fundamental limitation on the accuracy of determination of the average secondary structure content by optical spectra.

When a protein is normal — i.e., like the others in the set — the ECD does well by utilizing the standard helix-sheet interrelationship from the training set, but the β-sheet prediction is based on the excellent sensitivity of ECD to helix content only, not on its actually sensing the sheet content independently [70,75]. When a protein structure deviates from this relationship, erratic sheet predictions can result. In fact, ECD predicts sheet content with less error based on helix-sheet correlation than by spectral bandshape alone (Table 6.3) [70]. The field is rife with examples of strange predictions of secondary structure from ECD alone. Certainly others would evolve from FT-IR and VCD, if they were also used alone. It is the combination of ECD and VCD that provides some protection in this regard. Since the best RMR predictions of average secondary structure necessarily neglect some significant components of the spectral data, there must be more information content potentially available in the optical spectra, especially ECD and VCD [11,69,70]. It is natural to assume that the short range dependent techniques such as VCD will sense the distortions characteristic of the conformation of residues at the ends of uniform

TABLE 6.2
Relative Standard Deviations[a] of Secondary Structure Prediction with ECD and FT-IR H-D Exchange Data Using Different Algorithms[b] for Structures

		ECD Data Set		
Algorithm	**Helix**	**Sheet**	**Turns**	**Other**
DSSP[c]	9.3 (1)	20.6 (1)	18.6 (1)	11.5 (3)
XTLsstr[d]	10.7 (1)	18.1 (2)	23.0 (1)	15.6 (1)
STRIDE[e]	10.2 (1)	18.5 (1)	21.4 (1)	22.0 (2)
		H/D Exchange FT-IR Data Set		
Dataset	**Helix**	**Sheet**	**Turns**	**Other**
DSSP[c]	7.1 (5)	13.6 (6)	16.7 (3)	14.1 (5)
XTLsstr[d]	12.1 (5)	14.5 (6)	22.6 (1)	11.9 (6)
STRIDE[e]	8.1 (5)	12.6 (7)	18.1 (3)	15.2 (6)

[a] Calculated for a given component as the percentage of its total variation in the training set, represented by the standard deviation of prediction of that component.
[b] The numbers in parenthesis designate the number of component loadings used for the optimum model (lowest prediction error) for that FC component.
[c] Kabsch and Sander, 1983 [304].
[d] King and Johnson, 1999 [306].
[e] Frishman and Argos, 1995 [307].

segments of secondary structure in a different manner than does ECD with its longer range sensitivity.

To this end we have developed a method to predict the number of segments of uniform structure [helices, sheet strands, and so on] in a globular protein, which is equivalent to the average length, once one knows the FC values and the size of the protein [302]. Our method centers on the characterization of a matrix descriptor of the segment distribution and uses neural network analyses to develop a predictive algorithm from a training set of spectral data. Other labs have taken the simpler approach of predicting a fraction of distorted helix and sheet and assigning that fraction to a fixed number of residues per segment to attain a similar end [309]. Since the fraction of distorted helices is actually correlated to the fraction of helical residues [75], it is not clear if this simpler approach gives more information than assuming an average length for helical segments in globular proteins.

Determination of the FC values, using the FA–RMR methods described previously, is complemented by a segment determination and therefore contains new structural information beyond the average secondary structure. Our method has been applied to the VCD and FT-IR studies of the thermal denaturation of ribonuclease T_1, demonstrating that with increasing temperature the helix segment is lost before the sheet segments [284]. It was also used to determine the number of intramembrane helices in aquaporin using FT-IR data [310].

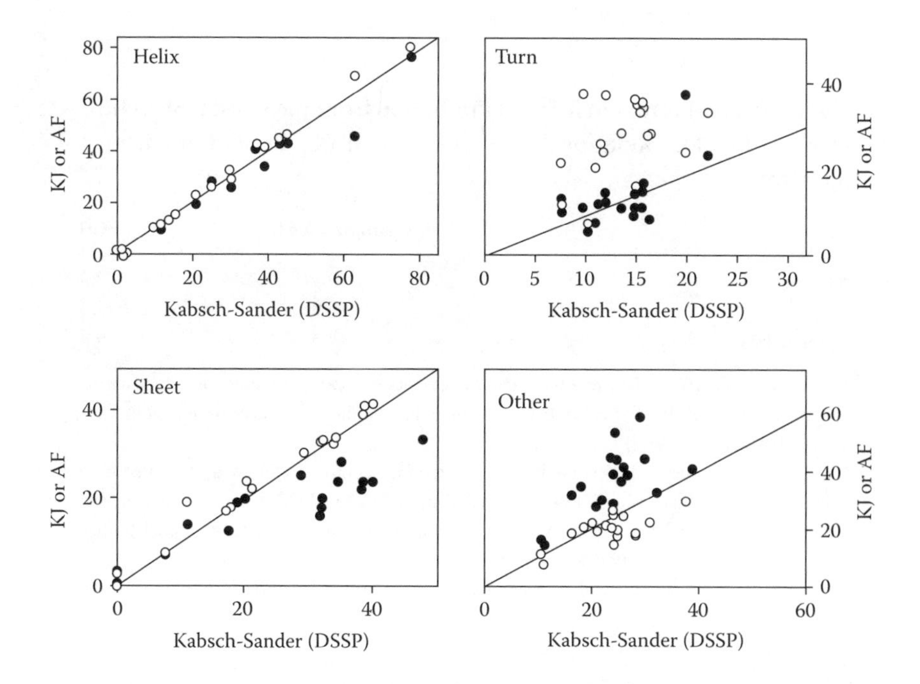

FIGURE 6.26 Plots of the fractional secondary structure composition of our training set proteins as determined with the King and Johnson (XTLsstr, filled circles) or Argos and Frishman (STRIDE, open circles) algorithms, as compared with the Kabsch-Sander (DSSP) algorithm that was used in our best FA-RMR analysis. Clearly helix fraction is about the same with all three, while the STRIDE algorithm predicts the same and XTLsstr algorithm predicts less sheet content for more β-structured proteins. STRIDE predicts more (and essentially uncorrelated with DSSP) turn content while XTLsstr predicts about the same values. Adjusting the sheet content is the key to the improved ECD-based β-sheet predictions with the King and Johnson (XTLsstr) method.

We have shown that FT-IR data, incorporating systematic hydrogen-deuterium (H-D) isotope exchange, can be used to generate even better predicted FC values than possible with conventional ECD or VCD (Table 6.4) [208]. The perturbation of the spectral data set appears to distinguish helix and turn contributions and leads to a more stable analysis. Extension of this isotope approach to include kinetic H-D exchange data and VCD methods may yield even further improvement [311]. This same data set has also been used to predict fold types, at least at the level of intersegment contact maps, thereby yielding tertiary structure information [308].

In summary, the key to the utilization of VCD or any spectroscopic technique for quantitative structure analysis is to establish reliability. We have sought to test thoroughly several different algorithms for interpretation of data to find their most useful realm of application. Several other groups have taken on the same sort of challenge [21,22,26,45,312]. These tests demand use of consistent data, access to several spectroscopic probes, and application of a systematic analysis. To make ours and others tests easier, we have made available several sets of systematically obtained ECD, VCD,

TABLE 6.3
Comparison of Errors in β-Sheet Prediction from the Crystal Structure Derived FC_β-FC_α Relationship with Errors in FC_β Predictions from Spectra[a]

	X-Ray	Amide I′ VCD	ECD
σ[b]	5.8	9.5	8.3
σ_{rel}[c]	12.2	19.9	17.4
σ_{rel} Spectra[a]	—	17.3	19.9

[a] Pancoska et al, 1995 (70) - relative standard deviations of *β*-sheet prediction directly from spectra, using FA/RMR. For ECD these are worse than predicting *β* sheet from α helix (line above), but better for VCD.

[b] Standard deviations of *β*-sheet predictions for training set proteins that was calculated from helix content, obtained from X-ray or predicted from VCD or ECD spectra as indicated in the column headings. Clearly knowing the precise helix content gives better predicted *β*-sheet values than use of spectrally determined values.

[c] Relative standard deviation of best predictions of Kabsch-Sander *β*-sheet fractions from helix content.

FT-IR and Raman spectral data on our website (http://www.chem.uic.edu/takgroup). In one test we have used a neural network to correlate ECD and VCD spectra for the protein training set discussed above [274]. The ECD could be predicted with reasonable accuracy from the amide I' VCD, but the reverse was not true. Thus, despite its added noise, the VCD was shown to have a higher information content which undoubtedly arises from its higher resolution. In an alternate approach to the use of multiple spectroscopic techniques for secondary structure analysis, we have used two-dimensional correlation methods to show which spectral regions are dependent on a given secondary structure type by treating the α-helix or β-sheet content of proteins in a training set as the perturbation variable [301,313,314]. Hetero-correlation between different types of spectra — e.g., ECD with IR and Raman, IR with Raman, VCD

TABLE 6.4
Standard Deviations of Prediction of Secondary Structure Values with Different Spectral Datasets[a]

Dataset	Helix	Sheet	Turns	Bends	Other
ECD	7.2 (1)	9.8 (1)	3.9 (1)	4.0 (2)	4.5 (3)
FT-IR, H_2O	11.1 (4)	5.7 (6)	3.4 (2)	3.7 (6)	5.5 (4)
FT-IR, D_2O	9.6 (8)	6.2 (6)	3.4 (1)	3.2 (7)	6.1 (6)
H/D exchange	5.5 (5)	6.5 (6)	3.5 (3)	4.3 (4)	5.5 (5)

[a] The numbers in parenthesis designate the number of spectral component loadings used for the optimum model for prediction of that structural component. All secondary structures determined from X-ray crystal structures with the DSSP algorithm [304].

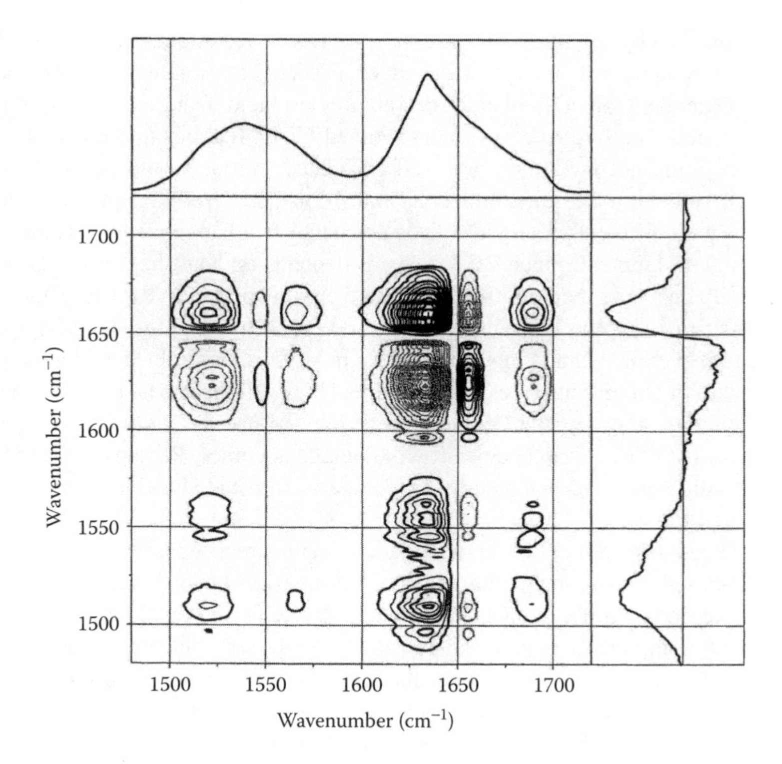

FIGURE 6.27 Representation of the hetero-correlation of VCD (amide I and II) spectra and FT-IR spectra calculated with a novel variation of the generalized two-dimensional spectral correlation method of Noda for the training set of 23 proteins (measured in H_2O). The spectra are correlated with respect to their varying fractional content (FC) of α-helix used as the perturbation. Through positive (black) and negative (gray) correlation cross-peaks, IR peaks at ~1630 cm^{-1} and ~1680 cm^{-1} (amide I), and ~1565 cm^{-1} and ~1520 cm^{-1} (amide II) can be assigned β-sheet conformation, while ~1655 cm^{-1} (amide I) and ~1545 cm^{-1} (amide II) can be assigned to α-helical conformation. (1D spectra: FT-IR (top) concanavalin A, low α-helical content; VCD (right) myoglobin, high α-helical content.)

with IR (Figure 6.27), and VCD with Raman — can be used to identify unique bands with their secondary structure source by using the more easily assigned method (e.g., ECD), to assign the more difficult one (Raman). Such an approach can also be used to identify those spectral regions of highest structural sensitivity for later analyses using the methods described above.

6.7 COMPARISON OF TECHNIQUES

This chapter was constructed primarily to illustrate the advantages of VCD as compared with its close optical spectral counterparts, ECD and FT-IR. However, we must state that ECD and FT-IR data have significant SNR advantages over VCD. In addition, ECD has considerable sampling flexibility, allowing study of more dilute solutions with little or no solvent interference. Due to its very high SNR, FT-IR's characteristic modest bandshape variation, perhaps its major limitation, can be greatly enhanced by use of deconvolution or derivative techniques [12,35,36]. However, for most analyses, interpretation of even such high quality FT-IR data still remains dependent on a frequency correlation of band features with secondary structure types. The frequencies of deconvolved FT-IR features and those of VCD, after band fitting deconvolution, are well correlated, as one would expect for phenomena arising from the same molecular transitions [37]. Yet the sign variations in VCD seen with different protein in a given frequency bandpass indicate an ambiguity in structural assignment. Since VCD analysis depends on bandshape correlations, it can partially decouple the vibrational spectra from this problem. Basically, the VCD transitions shift with the IR transitions, but the broad overlapping bandshapes and the resultant overall sign and intensity pattern in VCD is less influenced by the shift, which makes qualitative analyses quite reliable. However, there is no perfect solution. For quantitative analyses, the FA methods developed use the same frequency base for all proteins; thus, a frequency shift will degrade accuracy. Perhaps a future VCD analysis could correct for such shifts by comparison to the absorbance or even the deconvolved FT-IR.

The signed aspect of optical activity data, with its direct dependence on structure, gives CD-based measurements an important dimension beyond frequency assignment or even other differential measurements. In ECD, the resolution of different contributing components is poor, so that the overall sign pattern and intensity are its most useful properties. For proteins, the strong dependence of far-UV ECD bands on the α-helix contribution has been discussed above. Furthermore, contributions from sugar based transitions, in glycosylated proteins, could distort the amide far-UV spectra, and contributions of the aromatic groups do impact the near-UV spectra [18]. Similarly, in DNA, minor modifications of the bases can shift the π-electron states enough to seriously distort components of the spectrum, as was seen for I-C as compared to G-C [185]. Both aspects of confusion are exacerbated by the intrinsically low resolution of ECD. Thus use of VCD and ECD together accesses the benefits of both techniques, despite the limitations of both.

In the end, questions of biological structure are too important and too complex to be addressed by analysis using only a single technique having known (and sometimes hidden) limitations. The virtues of taking data from multiple spectroscopic techniques and finding the structural model that can satisfactorily encompass all the data obtained cannot be overemphasized. Each technique can probe the biomolecular structure with differing physical sensitivity and act as a control on the interpretation of the others. For some questions, VCD may give important insight, and for most questions it can offer critical data, even if only in a confirmatory role, to add to the overall picture of biopolymer conformation and folding.

ACKNOWLEDGMENTS

This work was originally supported by a long-term grant from the National Institutes of Health (GM 30147), for which we were most grateful, and subsequently parts of it were continued under grants from the National Science Foundation (CHE 03-16014), Research Corporation, and Donors of the Petroleum Research Fund administered by the American Chemical Society. Development of the spectral analysis algorithms was aided by an NSF International Cooperation Grant (with Petr Pancoska of Charles University in Prague). For this chapter we are indebted to Professor Prasad Polavarapu, Vanderbilt University, and Professor Laurence Nafie, Syracuse University, for collecting comparative data on their ChiralIR VCD instruments; Dr. Rina Dukor, Vysis Inc., for discussion of the ChiralIR characteristics and for running other test spectra; Professor Marie Urbanova, Technical University Prague, for running test spectra on her Bruker VCD instrument; Dr. Herrmann Drews of Bruker Optik for discussions and running comparative test spectra; and Dr. David Drapcho, BIORAD Digilab Div., for the loan of DSP software for this test and extensive interactions with us in working out the optimal conditions for its use for VCD. Dr. Vladimir Setnicka and Ms. Rong Huang did numerous checks of the comparative instrument data. The findings described here are the result of the dedicated hard work of a number of postdoctoral and graduate student coworkers, who have been cited and referenced, and who moved the field of VCD applications from a small-molecule theoretical curiosity to the level of a biomolecular structural tool. Many of the samples for special applications came to us through generous gifts or collaborations with co-authors from around the world, as cited in the publications. We are most grateful for these contributions.

REFERENCES

1. G. Holzwarth, E.C. Hsu, H.S. Mosher, T.R. Faulkner, A. Moscowitz. *J. Am. Chem. Soc.* 96: 251–252, 1974.
2. L.A. Nafie, J.C. Cheng, P.J. Stephens. *J. Am. Chem. Soc.* 97: 3842–3383, 1975.
3. L.A. Nafie, T.A. Keiderling, P.J. Stephens. *J. Am. Chem. Soc.* 98: 2715–2723, 1976.
4. T.A. Keiderling. *Appl. Spectrosc.* Rev. 17: 189–226, 1981.
5. P.L. Polavarapu. *Vibrational Spectra: Principles and Applications with Emphasis on Optical Activity.* New York: Elsevier, 1998.
6. L.A. Nafie, G.S. Yu, X. Qu, T.B. Freedman. *Faraday Discuss.* 99: 13–34, 1994.
7. P.J. Stephens, F.J. Devlin, C.S. Ashvar, C.F. Chabalowski, M.J. Frisch. *Faraday Discuss.* 99: 103–119, 1994.
8. M. Diem. Vibrat. *Spectra Struct.* 19: 1–54, 1991.
9. L.A. Nafie. *Ann. Rev. Phys. Chem.* 48: 357–386, 1997.
10. T.A. Keiderling. In: *Circular Dichroism: Principles and Applications.* K. Nakanishi, N. Berova, and R.W. Woody, eds. New York: Wiley-VCH, 2000, pp. 621–666.
11. V. Baumruk, P. Pancoska, T.A. Keiderling. *J. Mol. Biol.* 259: 774–791, 1996.
12. W. Surewicz, H.H. Mantsch, D. Chapman. *Biochemistry* 32: 389–394, 1993.
13. M. Jackson, H.H. Mantsch. *Crit. Rev. Biochem. Mol. Biol.* 30: 95–120, 1995.
14. J.L.R. Arrondo, A. Muga, J. Castresana, F.M. Goni. *Prog. Biophys. Mol. Biol.* 59: 23–56, 1992.
15. G.D. Fasman. *Circular Dichroism and the Conformational Analysis of Biomolecules.* New York: Plenum, 1996.
16. H.H. Havel. *Spectroscopic Methods for Determining Protein Structure in Solution.* New York: VCH, 1995.
17. H.H. Mantsch, D. Chapman. *Infrared Spectroscopy of Biomolecules.* Chichester, UK: Wiley-Liss, 1996.
18. R.W. Woody, A.K. Dunker. In: *Circular Dichroism and the Conformational Analysis of Biomolecules.* G.D. Fasman, ed. New York: Plenum Press, 1996, pp. 109–157.

19. E. Haas. In: *Spectroscopic Methods for Determining Protein Structure in Solution.* H.H. Havel, Ed. New York: VCH, 1995, pp. 28–61.
20. W.C. Johnson. *Methods Biochem. Anal.* 31: 61–163, 1985.
21. W.C. Johnson. *Ann. Rev. Biophys. Biophys. Chem.* 17: 145–166, 1988.
22. N. Sreerama, R.W. Woody. *J. Mol. Biol.* 242: 497–507, 1994.
23. R.W. Woody. In: *Circular Dichroism and the Conformational Analysis of Biomolecules.* G.D. Fasman, ed. New York: Plenum Press, 1996, pp. 25–67.
24. J.T. Yang, C.S.C. Wu, H.M. Martinez. *Methods Enzymol.* 130: 208–269, 1986.
25. M. Manning. *J. Pharmaceut. Biomed. Anal.* 7: 1103–1119, 1989.
26. S.Y. Venyaminov, J.T. Yang. In: *Circular Dichroism and the Conformational Analysis of Biomolecules.* G.D. Fasman, ed. New York: Plenum Press, 1996, pp. 69–107.
27. W.C. Johnson. In: *Circular Dichroism and the Conformational Analysis of Biomolecules.* G.D. Fasman, ed. New York: Plenum Press, 1996, pp. 433–468.
28. J.C. Maurizot. In: *Circular Dichroism: Principles and Applications.* 2nd ed. N. Berova, K. Nakanishi, and R.W. Woody, eds. New York: Wiley-VCH, 2000, pp. 719–739.
29. H.H. Mantsch, H.L. Casel, R.N. Jones. In: *Spectroscopy of Biological Systems.* R.J.H. Clark and R.E. Hester, eds. London: Wiley, 1986, pp. 1–46.
30. F.S. Parker. *Applications of Infrared, Raman and Resonance Raman Spectroscopy.* New York: Plenum, 1983.
31. A.T. Tu. *Raman Spectroscopy in Biology.* New York: Wiley, 1982.
32. S. Krimm, J. Bandekar. *Adv. Protein Chem.* 38: 181–364, 1986.
33. S. Krimm. In: *Biological Applications of Raman Spectroscopy, Vol. 1: Raman Spectra and the Conformations of Biological Macromolecules.* T.G. Spiro, Ed. New York: Wiley, 1987, pp. 1–45.
34. J.K. Kauppinen, D.J. Moffatt, H.H. Mantsch, D.G. Cameron. *Appl. Spectrosc.* 35: 271–276, 1981.
35. D.M. Byler, H. Susi. *Biopolymers* 25: 469–487, 1986.
36. W.K. Surewicz, H.H. Mantsch. *Biochem. Biophys. Acta* 952: 115–130, 1988.
37. P. Pancoska, L. Wang, T.A. Keiderling. *Protein Sci.* 2: 411–419, 1993.
38. F. Dousseau, M. Pezolet. *Biochemistry* 29: 8771–8779, 1990.
39. S. Krimm, W.C. Reisdorf, Jr. *Faraday Disc.* 99: 181–194, 1994.
40. N.N. Kalnin, I.A. Baikalov, S.Y. Venyaminov. *Biopolymers* 30: 1273–1280, 1990.
41. S.Y. Venyaminov, N.N. Kalnin. *Biopolymers* 30: 1259–1271, 1990.
42. H.H.J. De Jongh, E. Goormaghtigh, J.M. Ruysschaert. *Anal. Biochem.* 242: 95–103, 1996.
43. D.C. Lee, P.I. Haris, D. Chapman, R.C. Mitchell. *Biochemistry* 29: 9185–9193, 1990.
44. R.W. Sarver, W.C. Kruger. *Anal. Biochem.* 199: 61–67, 1991.
45. R. Pribic, I.H.M. Van Stokkum, D. Chapman, P.I. Haris, M. Bloemendal. *Anal. Biochem.* 214: 366–378, 1993.
46. R.W. Williams, A.K. Dunker. *J. Mol. Biol.* 152: 783–813, 1981.
47. P.L. Polavarapu. *Vibrat. Spectra Struct.* 17B: 319–342, 1989.
48. L.D. Barron, L. Hecht. In: *Biomolecular Spectroscopy, Part B.* R.J.H. Clark and R.E. Hester, Eds. Chichester: Wiley, 1993, pp. 235–266.
49. L.A. Nafie. In: *Modern Nonlinear Optics, Part 3.* M. Evans and S. Kielich, Eds. New York: Wiley, 1994, pp. 105–206.
50. L.D. Barron, L. Hecht, A.D. Bell. In: *Circular Dichroism and the Conformational Analysis of Biomolecules.* G.D. Fasman, Ed. New York: Plenum, 1996, pp. 653–695.
51. P.J. Stephens, M.A. Lowe. *Ann. Rev. Phys. Chem* 36: 213–241, 1985.
52. T.B. Freedman, L.A. Nafie. *Top. Stereochem.* 17: 113–206, 1987.
53. T.A. Keiderling, P. Pancoska. In: *Biomolecular Spectroscopy,* Part B. R.E. Hester and R.J.H. Clarke, eds. Chichester: Wiley, 1993, pp. 267–315.

54. T.A. Keiderling. In: *Spectroscopic Methods for Determining Protein Structure in Solution*. H. Havel, Ed. New York: VCH Publishers, 1995, pp. 163–169.
55. T.A. Keiderling. In: *Circular Dichroism and the Conformational Analysis of Biomolecules*. G.D. Fasman, ed. New York: Plenum, 1996, pp. 555–598.
56. T.B. Freedman, L.A. Nafie. In: *Modern Nonlinear Optics*. M. Evans and S. Kielich, Eds. New York: Wiley, 1994, pp. 207–263.
57. P. Bour, J. Sopkova, L. Bednarova, P. Malon, T.A. Keiderling. *J. Comput. Chem.* 18: 646–659, 1997.
58. P. Bour, J. Kubelka, T.A. Keiderling. *Biopolymers* 53: 380–395, 2000.
59. J. Kubelka, R.A.G.D. Silva, P. Bour, S.M. Decatur, T.A. Keiderling. In: *The Physical Chemistry of Chirality, ACS Symposium Series*. J.M. Hicks, Ed. New York: Oxford University Press, 2002, pp. 50–64.
60. T.A. Keiderling, B. Wang, M. Urbanova, P. Pancoska, R.K. Dukor. *Faraday Discuss.* 99: 263–286, 1994.
61. P.L. Polavarapu. In: *Fourier Transform Infrared Spectroscopy*. J.R. Ferraro and L. Basile, Eds. New York: Academic, 1985, pp. 61–96.
62. T.A. Keiderling. In: *Practical Fourier Transform Infrared Spectroscopy*. K. Krishnan and J.R. Ferraro, Eds. San Diego: Academic Press, 1990, pp. 203–284.
63. M. Diem. In: *Techniques and Instrumentation in Analytical Chemistry*. N. Purdie and H.G. Brittain, Eds. Amsterdam: Elsevier, 1994, pp. 91–130.
64. T.B. Freedman, L.A. Nafie, T.A. Keiderling. *Biopolymers* 37: 265–279, 1995.
65. T.A. Keiderling. In: *Infrared and Raman Spectroscopy of Biological Materials*. B. Yan and H.U. Gremlich, Eds. New York: M. Dekker, 1999, pp. 55–100.
66. L.D. Barron. In: *Circular Dichroism: Principles and Applications*. K. Nakanishi, N. Berova and R.W. Woody, Eds. New York: Wiley-VCH, 2000, pp. 667–701.
67. T.A. Keiderling, S.C. Yasui, P. Malon, P. Pancoska, R.K. Dukor, P.V. Croatto, L. Yang. 7th International Conference on Fourier Transform Spectroscopy, *Proc., SPIE* 1989; Vol. 1145; 57–63.
68. J.P. Hennessey, W.C. Johnson. *Biochemistry* 20: 1085–1094, 1981.
69. P. Pancoska, E. Bitto, V. Janota, T.A. Keiderling. *Faraday Discuss.* 99: 287–310, 1994.
70. P. Pancoska, E. Bitto, V. Janota, M. Urbanova, V.P. Gupta, T.A. Keiderling. *Protein Sci.* 4: 1384–1401, 1995.
71. P. Pancoska, T.A. Keiderling. *Biochemistry* 30: 6885–6895, 1991.
72. P. Pancoska, S.C. Yasui, T.A. Keiderling. *Biochemistry* 30: 5089–5103, 1991.
73. B.I. Baello, P. Pancoska, T.A. Keiderling. *Anal. Biochem.* 250: 212–221, 1997.
74. S. Wi, P. Pancoska, T.A. Keiderling. *Biospectroscopy* 4: 93–106, 1998.
75. P. Pancoska. In: *Encyclopedia of Analytical Chemistry*. R.A. Meyers, Ed. Chichester: John Wiley & Sons Ltd., 2000, pp. 244–383.
76. M. Diem, G.M. Roberts, O. Lee, O. Barlow. *Appl. Spectrosc.* 42: 20–27, 1988.
77. P. Xie, M. Diem. *Appl. Spectrosc.* 50: 675–680, 1996.
78. C.C. Chen, P.L. Polavarapu, S. Weibel. *Appl. Spectrosc.* 48: 1218–1223, 1994.
79. B. Wang, T.A. Keiderling. *Appl. Spectrosc.* 49: 1347–1355, 1995.
80. P. Malon, T.A. Keiderling. *Appl. Spectrosc.* 42: 32–38, 1988.
81. C.N. Su, V. Heintz, T.A. Keiderling. *Chem. Phys. Lett.* 73: 157–159, 1981.
82. G. Yoder. Vibrational circular dichroism of selected model peptides. PhD Thesis, University of Illinois at Chicago, Chicago, Illinois, 1997.
83. V. Setnicka, M. Urbanova, S. Pataridis, V. Kral, K. Volka. *Tetrahedron Asymm.* 13: 2661–2666, 2002.
84. V. Setnicka, M. Urbanova, V. Kral, K. Volka. *Spectrochim. Acta* Part A 58: 2983–2989, 2002.

85. M. Urbanova, V. Setnicka, K. Volka. *Chemicke Listy* 96: 301–304, 2002.
86. P. Bour, H. Navratilova, V. Setnicka, M. Urbanova, K. Volka. *J. Org. Chem.* 67: 161–168, 2002.
87. M. Urbanva, V. Setnicka, V. Kral, K. Volka. *Biopolymers* 60: 307–316, 2001.
88. M. Urbanova, V. Setnicka, K. Volka. *Chirality* 12: 199–203, 2000.
89. T. Kuppens, W. Langenaeker, J.P. Tollenaere, P. Bultnick. *J. Phys. Chem.* A 107: 542–553, 2003.
90. J. Dobler, N. Peters, C. Larsson, A. Bergman, E. Geidel, H. Huhnerfuss. *J. Mol. Struct. (Theochem)* 586: 159–166, 2002.
91. T. Tanaka, K. Inoue, T. Kodama, Y. Kyogoku, T. Hayakawa, H. Sugeta. *Biospectroscopy* 62: 228–234, 2001.
92. S. Shin, A. Hirakawa, Y. Hamada. *Enantiomer* 7: 191–196, 2002.
93. C.S. Ashvar, F.J. Devlin, P.J. Stephens. *J. Am. Chem. Soc.* 121: 2836–2849, 1999.
94. P.K. Bose, P.L. Polavarapu. *Carbohyd. Res.* 322: 135–141, 1999.
95. R.K. Yoo, B. Wang, P.V. Croatto, T.A. Keiderling. *Appl. Spectrosc.* 45: 231–236, 1991.
96. P. Malon, T.A. Keiderling. *Appl. Spectrosc.* 50: 669–674, 1996.
97. T.A. Keiderling, P.J. Stephens. *Chem. Phys. Lett.* 41: 1976.
98. S. Abbate, G. Longhi, S. Boiadjiev, D.A. Lightner, C. Bertucci, P. Salvadori. *Enantiomer* 3: 337–347, 1998.
99. E. Castigliono, F. Lebon, G. Longhi, S. Abbate. *Enantiomer* 7: 161–173, 2002.
100. R. Devlin, P.J. Stephens. *Appl. Spectrosc.* 41: 1142–1144, 1987.
101. M. Niemeyer, G.G. Hoffmann, B. Schrader. *J. Mol. Struct. (Theochem)* 349: 451–454, 1995.
102. P. Malon, T.A. Keiderling. *Appl. Optics* 36: 6141–6148, 1997.
103. C.A. McCoy, J.A. de Haseth. *Appl. Spectrosc.* 42: 336–341, 1988.
104. D. Tsankov, T. Eggimann, H. Weiser. *Appl. Spectrosc.* 49: 132–138, 1995.
105. R.W. Bormett, G.D. Smith, S.A. Asher, D. Barrick, D.M. Kurz. *Faraday Discuss.* 99: 327–339, 1994.
106. C. Marcott, A.E. Dowrey, I. Noda. *Appl. Spectrosc.* 47: 1324–1328, 1993.
107. F. Long, T.B. Freedman, R. Hapanowicz, L.A. Nafie. *Appl. Spectrosc.* 51: 504–507, 1997.
108. F. Long, T.B. Freedman, T.J. Tague, L.A. Nafie. *Appl. Spectrosc.* 51: 508–511, 1997.
109. L.A. Nafie, R.K. Dukor, J.R. Roy, A. Rilling, X.L. Cao, H. Buijs. *Appl. Spectrosc.* 57: 1245–1249, 2003.
110. L.A. Nafie. *Appl. Spectrosc.* 54: 1634–1645, 2000.
111. C. Zhao, P.L. Polavarapu, C. Das, P. Balaram. *J. Am. Chem. Soc.* 122: 8228–8231, 2000.
112. L.A. Nafie, R.K. Dukor. In: *The Physical Chemistry of Chirality, ACS Symposium Series.* J.M. Hicks, Ed. New York: Oxford University Press, 2000, pp. 79–88.
113. C.S. Ashvar, F.J. Devlin, P.J. Stephens. *J. Am. Chem. Soc.* 121: 2836–2849, 2000.
114. A. Aamouche, F.J. Devlin, P.J. Stephens. *J. Am. Chem. Soc.* 122: 2346–2354, 2000.
115. A. Aamouche, F.J. Devlin, P.J. Stephens. *J. Am. Chem. Soc.* 122: 7358–7367, 2000.
116. D.L. Drapcho, R. Curbelo, E.Y. Jiang, R.A. Crocombe, W.J. McCarthy. *Appl. Spectrosc.* 51: 453–460, 1997.
117. J. Hilario, Optical Spectroscopic Investigations of Model Beta-Sheet Peptides. Ph.D. Thesis, University of Illinois at Chicago, Chicago, 2004.
118. J. Hilario, D. Drapcho, R. Curbelo, T.A. Keiderling. *Appl. Spectrosc.* 55: 1435–1447, 2001.
119. P. Pancoska, S.C. Yasui, T.A. Keiderling. *Biochemistry* 28: 5917–5923, 1989.
120. P. Malon, R. Kobrinskaya, T.A. Keiderling. *Biopolymers* 27: 733–746, 1988.
121. C.N. Tam, P. Bour, T.A. Keiderling. *J. Am. Chem. Soc.* 118: 10285–10293, 1996.

122. V. Baumruk, T.A. Keiderling. *J. Am. Chem. Soc.* 115: 6939–6942, 1993.
123. M. Diem, H. Khouri, L.A. Nafie. *J. Am. Chem. Soc.* 101: 6829–6837, 1979.
124. A.C. Sen, T.A. Keiderling. *Biopolymers* 23: 1533–1545, 1984.
125. U. Narayanan, T.A. Keiderling, G.M. Bonora, C. Toniolo. *Biopolymers* 24: 1257–1263, 1985.
126. U. Narayanan, T.A. Keiderling, G.M. Bonora, C. Toniolo. *J. Am. Chem. Soc.* 108: 2431–2437, 1986.
127. D.M. Tigelaar, W. Lee, K.A. Bates, A. Sapringin, V.N. Prigodin, X. Cao, L.A. Nafie, M.S. Platz, A.J. Epstein. *Chem. Mater.* 14: 1430–1438, 2002.
128. B. Wang, R.K. Yoo, P.V. Croatto, T.A. Keiderling. *Chem. Phys. Lett.* 180: 339–343, 1991.
129. B. Wang, P.V. Croatto, R.K. Yoo, T.A. Keiderling. *J. Phys. Chem.* 96: 2422–2429, 1992.
130. C.N. Tam, T.A. Keiderling. *J. Mol. Spectrosc.* 157: 391–401, 1993.
131. S.Y. Venyaminov, N.N. Kalnin. *Biopolymers* 30: 1259–1271, 1990.
132. G. Yoder, A. Polese, R.A.G.D. Silva, F. Formaggio, M. Crisma, Q.B. Broxterman, J. Kamphuis, C. Toniolo, T.A. Keiderling. *J. Am. Chem. Soc.* 119: 10278–10285, 1997.
133. V. Baumruk, D.F. Huo, R.K. Dukor, T.A. Keiderling, D. Lelievre, A. Brack. *Biopolymers* 34: 1115–1121, 1994.
134. R.A.G.D. Silva, S.A. Sherman, F. Perini, E. Bedows, T.A. Keiderling. *J. Am. Chem. Soc.* 122: 8623–8630, 2000.
135. V. Andrushchenko, H. Wieser, P. Bour. *J. Phys. Chem.* B 106: 12623–12634, 2002.
136. V. Andrushchenko, Z. Leonenko, D. Cramb, H. van de Sande, H. Wieser. *Biopolymers* 61: 243–260, 2001.
137. V. Andrushchenko, Y. Blagoi, J.H. van de Sande, H. Wieser. *J. Biomol. Struct. Dyn.* 19: 889–906, 2002.
138. V. Andrushchenko, J.L. McCann, J.H. van de Sande, H. Wieser. *Vib. Spectrosc.* 22: 101–109, 2000.
139. V.V. Andrushchenko, J.H. van de Sande, H. Wieser, S.V. Kornilova, Y.P. Blagoi. *J. Biomol. Struct. Dyn.* 17: 545–560, 1999.
140. V.V. Andrushchenko, J.H. van de Sande, H. Wieser. *Vib. Spectrosc.* 19: 341–345, 1999.
141. L. Wang, T.A. Keiderling. *Biochemistry* 31: 10265–10271, 1992.
142. L. Wang, T.A. Keiderling. *Nucl. Acids Res.* 21: 4127–4132, 1993.
143. I. Tinoco. *Radiation Res.* 20: 133, 1963.
144. C.W. Deutsche, A. Moscowitz. *J. Chem. Phys.* 49: 3257, 1968.
145. C.W. Deutsche, A. Moscowitz. *J. Chem. Phys.* 53: 2630, 1970.
146. J. Snir, R.A. Frankel, J.A. Schellman. *Biopolymers* 14: 173, 1974.
147. G. Holzwarth, I. Chabay. *J. Chem. Phys.* 57: 1632, 1972.
148. M. Gulotta, D.J. Goss, M. Diem. *Biopolymers* 28: 2047–2058, 1989.
149. W. Zhong, M. Gulotta, D.J. Goss, M. Diem. *Biochemistry* 29: 7485–7491, 1990.
150. P. Bour, T.A. Keiderling. *J. Am. Chem. Soc.* 114: 9100–9105, 1992.
151. L.J. Wang, P. Pancoska, T.A. Keiderling. *Biochemistry* 33: 8428–8435, 1994.
152. L.J. Wang, L.G. Yang, T.A. Keiderling. *Biophys. J.* 67: 2460–2467, 1994.
153. P. Bour. Computational study of the vibrational optical activity of amides and peptides. Ph.D. Thesis, Academy of Science of the Czech Republic, Prague, Czech Republic, 1993.
154. P.J. Stephens. *J. Phys. Chem.* 89: 748, 1985.
155. P.J. Stephens. *J. Phys. Chem.* 91: 1712, 1987.
156. F.J. Devlin, P.J. Stephens. *J. Am. Chem. Soc.* 116: 5003, 1994.
157. P.J. Stephens, F.J. Devlin, K.J. Jalkanen. *Chem. Phys. Lett.* 225: 247, 1994.

158. P.J. Stephens, K.J. Jalkanen, F.J. Devlin, C.F. Chabalowski. *J. Phys. Chem.* 97: 6107, 1993.
159. F.J. Devlin, P.J. Stephens, J.R. Cheeseman, M.J. Frisch. *J. Phys. Chem.* 101: 6322–6333, 1997.
160. F.J. Devlin, P.J. Stephens, J.R. Cheeseman, M.J. Frisch. *J. Phys. Chem.* 101: 9912–9924, 1997.
161. C.S. Ashvar, P.J. Stephens, T. Eggimann, H. Wieser. *Tetrahed. Asym.* 9: 1107–1110, 1998.
162. C.S. Ashvar, F.J. Devlin, P.J. Stephens, K.L. Bak, T. Eggimann, H. Wieser. *J. Phys. Chem.* 102: 6842–6857, 1998.
163. A. Aamouche, F.J. Devlin, S.P. *J. J. Org. Chem.* 66: 3671–3677, 2001.
164. P. Bour, T.A. Keiderling. *J. Am. Chem. Soc.* 115: 9602–9607, 1993.
165. P. Bour, J. Kubelka, T.A. Keiderling. *Biopolymers* 65: 45–49, 2002.
166. R.A.G.D. Silva, J. Kubelka, S.M. Decatur, P. Bour, T.A. Keiderling. *Proc. Nat. Acad. Sci. U.S.A.* 97: 8318–8323, 2000.
167. J. Kubelka, T.A. Keiderling. *J. Am. Chem. Soc.* 123: 6142–6150, 2001.
168. J. Hilario, J. Kubelka, F.A. Syud, S.H. Gellman, T.A. Keiderling. *Biopolymers (Biospectroscopy)* 67: 233–236, 2002.
169. J. Hilario, J. Kubelka, T.A. Keiderling. *J. Am. Chem. Soc.* 125: 7562–7574, 2003.
170. R. Huang, J. Kubelka, W. Barber-Armstrong, R.A.G.D. Silva, S.M. Decatur, T.A. Keiderling. *J. Am. Chem. Soc.* 124: 2346–2354, 2003.
171. J. Kubelka, R.A.G.D. Silva, T.A. Keiderling. *J. Am. Chem. Soc.* 124: 5325–5332, 2002.
172. W.C. Johnson. In: *Circular Dichroism: Principles and Applications*. N. Berova, K. Nakanishi, and R.W. Woody, Eds. New York: Wiley-VCH, 2000, pp. 703–718.
173. E. Taillandier, J. Liquier. *Meth. Enzymol.* 211: 307–335, 1992.
174. M. Tsuboi, T. Nishimura, A.Y. Hirakawa, W.L. Peticolas. In: *Biological Applications of Raman Spectroscopy.* T.G. Spiro, Ed. New York: J. Wiley, 1987, pp. 109–179.
175. A. Annamalai, T.A. Keiderling. *J. Am. Chem. Soc.* 109: 3125–3132, 1987.
176. A.F. Bell, L. Hecht, L.D. Barron. *J. Am. Chem. Soc.* 120: 5820–5821, 1998.
177. L.D. Barron, L. Hecht, E.W. Blanch, A.F. Bell. *Prog. Biophys. Mol. Biol.* 73: 1–49, 2000.
178. T.A. Keiderling, P. Pancoska, R.K. Dukor, L. Yang. In: *Biomolecular Spectroscopy.* R.R. Birge and H.H. Mantsch, Eds. 1989, pp. 7–14.
179. L. Yang, T.A. Keiderling. *Biopolymers* 33: 315–327, 1993.
180. L. Wang, L. Yang, T.A. Keiderling. In: *Spectroscopy of Biological Molecules.* R.E. Hester and R.B. Girling, Eds. Cambridge: Royal Society of Chemistry, 1991, pp. 137–138.
181. V. Andrushchenko, H. van de Sande, H. Wieser. *Biopolymers* 69: 529–545, 2003.
182. V. Andrushchenko, J.H. van de Sande, H. Wieser. *Biopolymers* 72: 374–390, 2003.
183. D. Tsankov, B. Kalisch, H.V. de Sande, H. Wieser. *J. Phys. Chem.* B 107: 647964–647985, 2003.
184. T. Xiang, D.J. Goss, M. Diem. *Biophys. J.* 65: 1255–1261, 1993.
185. J.C. Sutherland, K.P. Griffin. *Biopolymers* 22: 1445–1448, 1983.
186. S.S. Birke, M. Moses, M. Gulotta, B. Kagarlovsky, D. Jao, M. Diem. *Biophys. J.* 65: 1262–1271, 1993.
187. W. Zhong, M. Gulotta, D.J. Goss. *Biochemistry* 29: 7485–7491, 1990.
188. S.S. Birke, I. Agbaje, M. Diem. *Biochemistry* 31: 450, 1992.
189. H. DeVoe. *J. Phys. Chem.* 75: 1509–1515, 1971.
190. B.D. Self, D.S. Moore. *Biophys. J.* 73: 339–347, 1997.
191. B.D. Self, D.S. Moore. *Biophys. J.* 74: 2249–2258, 1998.
192. R.D. Singh, T.A. Keiderling. *Biopolymers* 20: 237–240, 1981.

193. A.C. Sen, T.A. Keiderling. *Biopolymers* 23: 1519–1532, 1984.
194. B.B. Lal, L.A. Nafie. *Biopolymers* 21: 2161–2183, 1982.
195. M.G. Paterlini, T.B. Freedman, L.A. Nafie. *Biopolymers* 25: 1751–1765, 1986.
196. S.C. Yasui, T.A. Keiderling. *J. Am. Chem. Soc.* 108: 5576–5581, 1986.
197. E.K. Koepf, H.M. Petrassi, M. Sudol, J.W. Kelly. *Protein Sci.* 8: 841–853, 1999.
198. H.L. Schenck, S.H. Gellman. *J. Am. Chem. Soc.* 120: 4869–4870, 1998.
199. T. Kortemme, M. Ramirez-Alvarado, L. Serrano. *Science* 281: 253–256, 1998.
200. A.G. Cochran, N.J. Skelton, M.A. Starovasnik. *Proc. Nat. Acad. Sci. USA.* 98: 5578–5583, 2001.
201. C.D. Tatko, M.L. Waters. *J. Am. Chem. Soc.* 124: 9372–9373, 2002.
202. R. Huang, V. Setnicka, J. Hilario, P. Bour, T. A. Keiderling, to be submitted for publication.
203. V. Setnicka, R. Huang, C. L. Thomas, M. A. Etienne, R. P. Hammer, T. A. Keiderling, *J. Am. Chem. Soc.* 127: 4992–4993, 2005.
204. S.C. Yasui, T.A. Keiderling. *Biopolymers* 25: 5–15, 1986.
205. S.C. Yasui, T.A. Keiderling, R. Katachai. *Biopolymers* 26: 1407–1412, 1987.
206. G. Yoder, P. Pancoska, T.A. Keiderling. *Biochemistry* 36: 15123–15133, 1997.
207. R.K. Dukor, T.A. Keiderling. In: *Proc. 20th European Peptide Symposium.* E. Bayer and G. Jung, Eds. Berlin: DeGruyter, 1989, pp. 519–521.
208. B. Baello, P. Pancoska, T.A. Keiderling. *Anal. Biochem.* 280: 46–57, 2000.
209. T.A. Keiderling, R.A.G.D. Silva, G. Yoder, R.K. Dukor. *Bioorg. Med. Chem.* 7: 133–141, 1999.
210. T.A. Keiderling, Q. Xu. *Adv. Prot. Chem.* 62: 111–161, 2002.
211. D.M. Byler, J.M. Purcell. SPIE Fourier Transform. *Spectrosc.* 1145: 539–544, 1989.
212. A.H. Clark, D.H.P. Saunderson, A. Sugget. *Int. J. Peptide Res.* 17: 353–364, 1981.
213. J. Kubelka, R. Huang, T.A. Keiderling, J. Phys. Chem. B 109: 8231–8243, 2005.
214. R.A.G.D. Silva, S.A. Sherman, T.A. Keiderling. *Biopolymers* 50: 413–423, 1999.
215. V.P. Gupta, T.A. Keiderling. *Biopolymers* 32: 239–248, 1992.
216. S.C. Yasui, T.A. Keiderling, F. Formaggio, G.M. Bonora, C. Toniolo. *J. Am. Chem. Soc.* 108: 4988–4993, 1986.
217. S.C. Yasui, T.A. Keiderling, G.M. Bonora, C. Toniolo. *Biopolymers* 25: 79–89, 1986.
218. R.A.G.D. Silva, J. Kubelka, T.A. Keiderling. *Biopolymers* 65: 229–243, 2002.
219. V. Pavone, E. Benedetti, B. Di Blasio, C. Pedone, A. Santini, A. Bavoso, C. Toniolo, M. Crisma, L. Sartore. *J. Biomol. Struct. Dyn.* 7: 1321–1331, 1990.
220. G. Yoder, T.A. Keiderling, F. Formaggio, M. Crisma, C. Toniolo. *Biopolymers* 35: 103–111, 1995.
221. W. Mastle, R.K. Dukor, G. Yoder, T.A. Keiderling. *Biopolymers* 36: 623–631, 1995.
222. H.R. Wyssbrod, M. Diem. *Biopolymers* 31: 1237, 1992.
223. P. Xie, Q. Zhou, M. Diem. *Faraday Discuss.* 99: 233–244, 1995.
224. R.K. Dukor, T.A. Keiderling. *Biopolymers* 31: 1747–1761, 1991.
225. R.K. Dukor. Vibrational circular dichroism of selected peptides, polypeptides and proteins. PhD Thesis, University of Illinois at Chicago, Chicago, Illinois, 1991.
226. R. Kobrinskaya, S.C. Yasui, T.A. Keiderling. In: *Peptides, Chemistry and Biology, Proc. 10th American Peptide Symp.* G.R. Marshall, Ed. Leiden: ESCOM, 1988, pp. 65–66.
227. R.K. Dukor, T.A. Keiderling, V. *Gut. Int. J. Pept. Prot. Res.* 38: 198–203, 1991.
228. R.K. Dukor, T.A. Keiderling. *Biospectroscopy* 2: 83–100, 1996.
229. M.L. Tiffany, S. Krimm. *Biopolymers* 11: 2309–2316, 1972.
230. T.A. Keiderling, R.A.G.D. Silva, S.M. Decatur, P. Bour. In: *Spectroscopy of Biological Molecules: New Directions.* J. Greve, G.J. Puppels, and C. Otto, Eds. Dortrecht: Kluwer AP, 1999, pp. 63–64.

231. S.C. Yasui, T.A. Keiderling. In: *Peptides: Chemistry and Biology, Proc. 10th Amer. Pept. Symp*. G.R. Marshall, Ed. Leiden: ESCOM, 1988, pp. 90–92.
232. T.B. Freedman, A.C. Chernovitz, W.M. Zuk, G. Paterlini, L.A. Nafie. *J. Am. Chem. Soc.* 110: 6970–6974, 1988.
233. J.P. Mazaleyrat, K. Wright, A. Gaucher, M. Wakselman, S. Oancea, F. Formaggio, C. Toniolo, V. Setnicka, J. Kapitan, T.A. Keiderling. *Tetrahed. Asym.* 14: 1879–1893, 2003.
234. M. Urbanova, V. Setnicka, V. Kral, K. Volka. *Biopolymers* 60: 307–316, 2001.
235. S.C. Yasui, T.A. Keiderling, M. Sisido. *Macromolecules* 20: 403–2406, 1987.
236. J.-P. Mazaleyrat, K. Wright, A. Gaucher, N. Toulemonde, M. Wakselman, S. Oancea, C. Peggion, F. Formaggio, V. Setnicka, T. A. Keiderling, C. Toniolo, *J. Am. Chem. Soc.* 126: 12874–12879, 2004.
237. M. Diem. *J. Am. Chem. Soc.* 110: 6967–6970, 1988.
238. W.M. Zuk, T.B. Freedman, L.A. Nafie. *Biopolymers* 28: 2025–2044, 1989.
239. A.C. Chernovitz, T.B. Freedman, L.A. Nafie. *Biopolymers* 26: 1879–1900, 1987.
240. G.M. Roberts, O. Lee, J. Calienni, M. Diem. *J. Am. Chem. Soc.* 110: 1749–1752, 1988.
241. M. Miyazawa, Y. Kyogoku, H. Sugeta. *Spectrochim. Acta.* 50: 1505–1511, 1994.
242. M. Diem, O. Lee, G.M. Roberts. *J. Phys. Chem.* 96: 548–554, 1992.
243. O. Lee, G.M. Roberts, M. Diem. *Biopolymers* 28: 1759–1770, 1989.
244. M.G. Paterlini, T.B. Freedman, L.A. Nafie, Y. Tor, A. Shanzer. *Biopolymers* 32: 765–782, 1992.
245. F. Eker, X.L. Cao, L. Nafie, R. Schweitzer-Stenner. *J. Am. Chem. Soc.* 124: 14330–14341, 2002.
246. P. Malon, P. Pirkova, W. Mastle, T.A. Keiderling. In: *Peptides 1992: Proc. 22nd European Peptide Symp.* C. H. Scheider, A. Y. Eberle, Ed. Leiden: Escom, 1993, pp. 499.
247. G. Yoder, T.A. Keiderling, F. Formaggio, M. Crisma. *Tetrahedron Asymm.* 6: 687–690, 1995.
248. C. Toniolo. Unpublished results.
249. M.J. Citra, M.G. Paterlini, T.B. Freedman, A. Fissi, O. Pieroni. *Proc SPIE,* 2089: 478–479, 1993.
250. S.M. Miick, G. Martinez, W.R. Fiori, A.P. Todd, G.L. Millhauser. *Nature* 359: 653–655, 1992.
251. S.M. Miick, K.M. Casteel, G.L. Millhauser. *Biochemistry* 32: 8014–8021, 1993.
252. G. Martinez, G. Millhauser. *J. Struct. Biol.* 114: 23–27, 1995.
253. G.L. Millhauser. *Biochemistry* 34: 3873–3877, 1995.
254. G.L. Millhauser, C.J. Stenland, P. Hanson, K.A. Bolin, F.J.M. Vandeven. *J. Mol. Biol.* 267: 963–974, 1997.
255. E.R. Malinowski. *Factor Analysis in Chemistry.* New York: Wiley, 1991.
256. S.M. Decatur. *Biopolymers* 54: 180–185, 2000.
257. K.J. Jalkanen, P.J. Stephens, R.D. Amos, N.C. Handy. *J. Am. Chem. Soc.* 109: 7193, 1987.
258. K.J. Jalkanen, P.J. Stephens, R.D. Amos, N.C. Handy. *J. Phys. Chem.* 92: 1781, 1988.
259. R.D. Amos, K.J. Jalkanen, P.J. Stephens. *J. Phys. Chem.* 92: 5571, 1988.
260. R.W. Kawiecky, F. Devlin, P.J. Stephens, R.D. Amos. *J. Phys. Chem.* 95: 9817, 1991.
261. P.J. Stephens, F.J. Devlin, C.F. Chabalowski, M.J. Frisch. *J. Phys. Chem.* 98: 11623–11627, 1994.
262. J. Kubelka. and T. A. Keiderling, *J. Phys. Chem. A* 105: 10922–10928, 2001.
263. P. Bour, T.A. Keiderling. *J. Chem. Phys.* 119: 11253–11262, 2003.
264. P. Bour, T.A. Keiderling, *J. Phys. Chem. B*, 109: 5348–5357, 2005.
265. J. Kubelka. IR and VCD spectroscopy of model peptides: theory and experiment. Ph.D. thesis, University of Illinois at Chicago, Chicago, 2002.

266. J. Kubelka, T.A. Keiderling. *J. Am. Chem. Soc.* 123: 12048–12058, 2001.
267. J.W. Brauner, C. Dugan, R. Mendelsohn. *J. Am. Chem. Soc.* 122: 677–683, 2000.
268. K. Halverson, I. Sucholeiki, T.T. Ashburn, R.T. Lansbury. *J. Am. Chem. Soc.* 113: 6701–6703, 1991.
269. P.T. Lansbury Jr., P.R. Costa, J.M. Griffiths, E.J. Simon, M. Auger, K.J. Halverson, D.A. Kocisko, Z.S. Hendsch, T.T. Ashburn, R.G. Spencer, et al. *Nat. Struct. Biol.* 2: 990–998, 1995.
270. R.A. Silva, W. Barber-Armstrong, S.M. Decatur. *J. Am. Chem. Soc.* 125: 13674–13675, 2003.
271. C. L. Thomas; M. A. Etienne; J. Wang; V. Setnicka; T. A. Keiderling; R. P. Hammer., in *Peptide Revolution: Genomics Proteomics, and Therapeutics. Proc. 18th Amer. Pept. Symp.*, M. Chorev, T. K Sawyer, Ed., Dordrecht:Kluwer Academic Publishers, 2004, pp 381–382.
272. J. Kubelka, R. Huang, J. H. Kim, T.A. Keiderling, to be Submitted for publication.
273. C.T. Chang, C.S.C. Wu, J.T. Yang. *Anal. Biochem.* 91: 13–31, 1978.
274. P. Pancoska, V. Janota, T.A. Keiderling. *Appl. Spectrosc.* 50: 658–668, 1996.
275. K. Kaiden, T. Matsui, S. Tanaka. *Appl. Spectrosc.* 41: 180–184, 1987.
276. F. Fu, D.B. DeOliveira, W.R. Trumble, H.K. Sarkar, B.R. Singh. *Appl. Spectrosc.* 48: 1432–1440, 1994.
277. M. Urbanova, R.K. Dukor, P. Pancoska, V.P. Gupta, T.A. Keiderling. *Biochemistry* 30: 10479–10485, 1991.
278. M. Urbanova, T.A. Keiderling, P. Pancoska. *Bioelectrochem. Bioenerg.* 41: 77–80, 1996.
279. S.D. Stelea, P. Pancoska, A. S. Benight, T.A. Keiderling. *Prot. Sci.* 10: 970–978, 2001.
280. S.C. Yasui, P. Pancoska, R.K. Dukor, T.A. Keiderling, V. Renugopalakrishnan, M.J. Glimcher, R.C. Clark. *J. Biol. Chem.* 265: 3780–3783 3788, 1990.
281. M. Urbanova, P. Pancoska, T.A. Keiderling. *Biochim. Biophys. Acta* 1203: 290–294, 1993.
282. R.K. Dukor, P. Pancoska, T.A. Keiderling, S.J. Prestrelski, T. Arakawa. *Arch. Biochem. Biophys.* 298: 678–681, 1992.
283. S. Williams, T.P. Causgrove, R. Gilmanshin, K.S. Fang, R.H. Callender, W.H. Woodruff, R.B. Dyer. *Biochemistry* 35: 691–697, 1996.
284. P. Pancoska, H. Fabian, G. Yoder, V. Baumruk, T.A. Keiderling. *Biochemistry* 35: 13094–13106, 1996.
285. K. Kuwajima. *FASEB J.* 10: 102–109, 1996.
286. W. Colon, H. Roder. *Nature Struct. Biol.* 3: 1019–1025, 1996.
287. G.A. Elove, A.F. Chaffotte, H. Roder, M.E. Goldberg. *Biochemistry* 31: 6876–6883, 1992.
288. Y.W. Bai. *Proc. Natl. Acad. Sci. USA* 96: 477–480, 1999.
289. V.E. Bychkova, A.E. Dujsekina, S.I. Klenin. *Biochemistry* 35: 6058–6063, 1996.
290. Y.O. Kamatari, T. Konno, M. Kataoka, K. Akasaka. *J. Mol. Biol.* 259: 512–523, 1996.
291. S.R. Yeh, S.W. Han, D.L. Rousseau. *Acc. Chem. Res.* 31: 727–736, 1998.
292. Q. Xu, T.A. Keiderling. *Biopolymers (Biospectroscopy)*, 73: 716–726, 2004.
293. R.W. Ruddon, S.A. Sherman, E. Bedows. *Protein Sci.* 5: 1443–1452, 1996.
294. Q. Xu, T.A. Keiderling *(Biochemistry)*, 44: 7976–7987, 2005.
295. R.J. Rummel. *Applied Factor Analysis*. Evanston: Northwestern University Press, 1970.
296. P. Pancoska, I. Fric, K. Blaha. *Collect. Czech. Chem. Commun.* 44: 1296–1312, 1979.
297. I.H.M. Van Stokkum, H.J.W. Spoelder, M. Bloemendal, R. Van Grondelle, F.C.A. Groen. *Anal. Biochem.* 191: 110–118, 1990.
298. N. Sreerama, R.W. Woody. *Anal. Biochem.* 209: 32–44, 1993.

299. P. Manavalan, W.C. Johnson, *Jr. Anal. Biochem.* 167: 76–85, 1987.
300. P. Pancoska, M. Blazek, T.A. Keiderling. *Biochemistry* 31: 10250–10257, 1992.
301. P. Pancoska, J. Kubelka, T.A. Keiderling. *Appl. Spectrosc.* 53: 655–665, 1998.
302. P. Pancoska, V. Janota, T.A. Keiderling. *Anal. Biochem.* 267: 72–83, 1999.
303. N. Sreerama, R.W. Woody. In: *Circular Dichroism: Principles and Applications.* K. Nakanishi, N. Berova, and R.W. Woody, Eds. New York: Wiley-VCH, 2000, pp.
304. W. Kabsch, C. Sander. *Biopolymers* 22: 2577–2637, 1983.
305. M. Levitt, J. Greer. *J. Mol. Biol.* 114: 181–293, 1977.
306. S.M. King, W.C. Johnson. *Proteins* 35: 313–320, 1999.
307. D. Frishman, P. Argos. *Proteins* 23: 566–579, 1995.
308. B. Baello. Protein secondary structure and tertiary structure studies using vibrational spectroscopy and hydrogen exchange. PhD Thesis, University of Illinois at Chicago, Chicago, 2000.
309. N. Sreerama, S.Y. Venyaminov, R.W. Woody. *Protein Sci.* 8: 370–380, 1999.
310. V. Cabiaux, K. Oberg, P. Pancoska, T. Walz, P. Agre, A. Engle. *Biophys. J.* 73: 406–417, 1997.
311. Q. Xu, and T. A. Keiderling, *Macromol. Symp.*, 220: 17–31, 2005.
312. A.J.P. Alix. In: *Biomolecular Structure and Dynamics.* G. Vergoten and T. Theophanides, Eds. Amsterdam: Kluwer AP, 1997, pp. 121–150.
313. J. Kubelka, P. Pancoska, T.A. Keiderling. *Appl. Spectrosc.* 53: 666–671, 1998.
314. J. Kubelka, P. Pancoska, T.A. Keiderling. In: *Spectroscopy of Biological Molecules: New Directions.* J. Greve, G.J. Puppels, and C. Otto, Eds. Dordrecht: Kluwer Academic Publishers, 1999, pp. 67–68.

7 Membrane Receptor–Ligand Interactions Probed by Attenuated Total Reflectance Infrared Difference Spectroscopy

John E. Baenziger, Stephen E. Ryan, and Veronica C. Kane-Dickson

CONTENTS

The abbreviations used are: ATR, attenuated total reflectance; Carb, carbamylcholine; FTIR, Fourier transform infrared; nAChR, nicotinic acetylcholine receptor

7.1 INTRODUCTION

Understanding the functional mechanisms of any biological macromolecule requires the atomic resolution characterization of both its three-dimensional structure and the nature of its conformational changes. While the elucidation of both can be performed for many cytosolic macromolecules, difficulties associated with the application of physical methods to biological membranes still limit our understanding of integral membrane proteins. Fourier transform infrared (FT-IR) spectroscopy is one physical method that is readily applicable to membrane systems. The technique provides information on the frequencies of molecular vibrations, which are exquisitely sensitive to local environment and covalent structure. In particular, FT-IR spectra are very sensitive to the subtle structural alterations that occur in individual residues upon protein conformational change.

The changes in vibrational frequency and intensity that occur upon conversion of a protein from one conformation (state A) to another (state B) often represent less than 0.1% of the total protein spectral intensity at a given frequency. A direct comparison of spectra recorded from a protein in states A and B rarely reveals the vibrational changes that are associated with the conformational transition (Figure 7.1). To detect these subtle differences, the spectrum of state A must be digitally subtracted from the spectrum of state B (or vice versa) to eliminate the vibrational bands from those residues whose structures are unaffected by the change in conformational state. The resulting difference spectrum exhibits the vibrational bands from only those residues whose structures or environments differ between the two states A and B and thus provides a spectral map of the conformational change (Figure 7.1).

Accurate spectral differences can only be measured under conditions where intensity variations from one spectrum to the next (due to thermal fluctuations, changes in sample concentration, etc.) are much less intense than the changes in vibrational intensity that result from the protein conformational change itself. In most cases, this high degree of spectral reproducibility can only be achieved if the conformational change is triggered while the protein of interest remains inside the FT-IR spectrometer inside an infrared sampling device. The ability to cycle a protein repetitively between two conformational states inside the FT-IR spectrometer is also

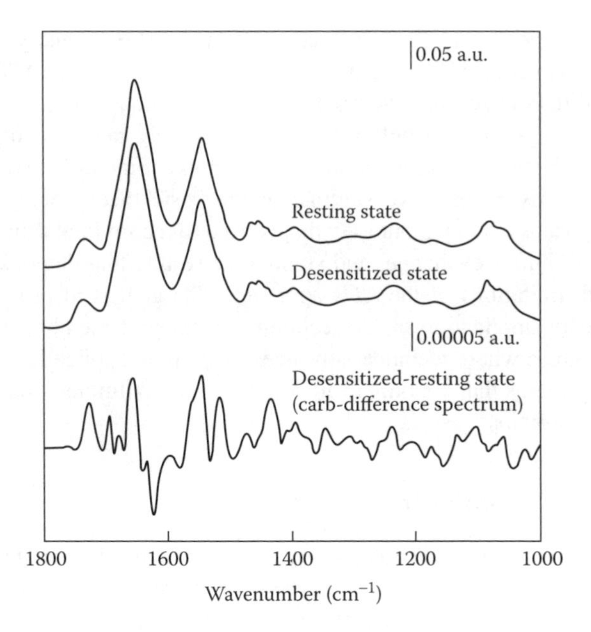

FIGURE 7.1 FT-IR spectra of nAChR membranes recorded using the ATR technique in the presence (desensitized state) or absence (resting state) of the agonist analogue Carb are visually indistinguishable. The vibrational intensity changes associated with Carb binding and the resting-to-desensitized conformational transition cannot be detected against the large background of protein and lipid vibrations that are unaffected by the conformational change. In contrast, the difference between spectra recorded in the resting and desensitized states exhibits a complex pattern of very weak vibrational bands. These vibrational bands reflect the structural changes induced in the nAChR upon the binding of Carb. Note the absorption intensity scale has been increased by a factor of 1000 for the difference spectrum versus the two absorption spectra.

an advantage. Successive difference spectra measured between states A and B can then be signal averaged to achieve a superior signal-to-noise ratio.

FT-IR difference spectroscopy was originally developed to monitor the structural changes that occur in light-activated proteins, such as bacteriorhodopsin and the photosynthetic reaction center. These proteins are particularly well suited for difference spectroscopy because conformational change can be repetitively triggered inside the infrared sampling device with a flash of visible light. Difference spectroscopy has detected changes in both the protonation state and the strength of hydrogen bonding of a number of amino acid side chains upon light activation of both proteins. In many cases, these subtle changes in structure and environment have been monitored in real time using time-resolved FT-IR techniques [1–4].

To apply the difference technique to proteins that lack intrinsic activatable chromophores, a variety of technically innovative methods have since been developed. These methods include (1) light-induced release of an effector ligand from a caged precursor [5], (2) stopped and continuous flow measurements [6], (3) temperature and pressure jump experiments [7,8], (4) equilibrium electrochemistry [9], (5) light-induced photo-reduction [10], and (6) attenuated total reflection (ATR) with buffer exchange [11,29]. For a recent review of the different methodologies, see Reference 13.

The ATR approach with buffer flow is a relatively new and highly versatile method that can be used to probe the molecular details of membrane receptor–ligand interactions. In this chapter, we summarize the basic theory behind ATR FT-IR spectroscopy, discuss the methodology that is used to record ligand-induced spectral differences with buffer exchange, and summarize results that demonstrate both the versatility and limitations of the ATR approach. The goal is to provide a comprehensive guide for application of the technique to integral membrane proteins. We also highlight areas where technical advances may allow application of the method to membrane proteins that are only available in extremely limited quantities, as well as to cytosolic proteins.

7.2 ATR SPECTROSCOPY

The basic theory of ATR spectroscopy is presented here to illustrate how spectra are recorded from biological samples. The theory is necessary to understand how to optimize data acquisition parameters for individual applications, as discussed in Section 7.3. We also summarize the theory behind linear dichroism ATR measurements. Linear dichroism refers to the differential absorption of infrared light that is linearly polarized either parallel or perpendicular to the angle of incidence (see below). Linear dichroism ATR measurements have been used to define the orientations of functional groups relative to the bilayer normal. In theory, changes in functional group orientation, such as the tilting of transmembrane α-helices, during protein conformational change can thus be monitored. Only the basic equations required for the interpretation of linear dichroism ATR difference spectra are presented. A more detailed treatment is found in References 15 to 17.

7.2.1 BACKGROUND THEORY

Infrared light traveling through an infrared transparent material, called an internal reflection element (IRE), undergoes 100% reflection when it strikes the IRE surface at an angle θ above the critical angle θ_c:

$$\theta_c = \sin^{-1} n_{21} \tag{7.1}$$

where $n_{21} = n_2/n_1$, n_1 is the refractive index of the IRE and n_2 is the refractive index of the rarer medium located at the external IRE surface. Superimposition of the incoming and reflected waves yields a standing electromagnetic field within the IRE that is normal to the reflecting surface (Figure 7.2). This electromagnetic field (E_o) also exists in the rarer medium beyond the reflecting interface (E), but its strength decays exponentially according to the following:

$$E = E_o e^{-z/d_p} \tag{7.2}$$

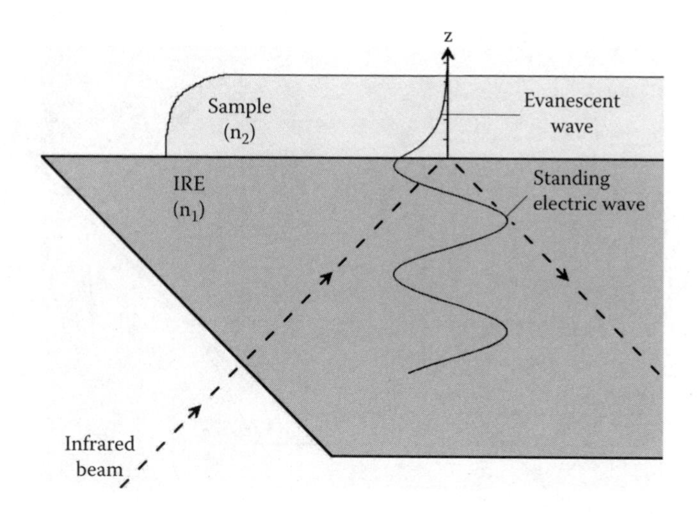

FIGURE 7.2 Schematic diagram of an IRE showing the standing electric wave within the IRE that is created upon superimposition of the incoming and reflected infrared light. The electric field decays exponentially beyond the IRE surface (the evanescent wave). A sample located close to the IRE surface within this exponential decay will absorb the infrared light.

where z is the distance perpendicular to the IRE surface and d_p is the penetration depth of the infrared light defined as follows:

$$d_p = \frac{\lambda / n_1}{2\pi\sqrt{\sin^2\theta - n_{21}^2}} \tag{7.3}$$

The penetration depth of the evanescent wave beyond the surface of an IRE is typically less than 2.0 μm, but is sufficient to interact with a sample in close contact with the IRE surface [14,15].

FT-IR spectra recorded using the ATR technique are essentially equivalent to spectra recorded in the conventional transmission mode, except that the noted dependence of d_p on λ leads to relatively strong absorption of infrared light at lower wavenumbers compared to those observed in transmission spectra. The intensities of individual bands in ATR spectra recorded from membranes oriented parallel to a planar IRE surface can also vary relative to the intensities observed in transmission spectra recorded from isotropic dispersions. Such differences arise because the geometry of the IRE leads to a preferential polarization of those electric field amplitudes of the evanescent wave that are parallel, as opposed to perpendicular, to the IRE surface (see below). Differences in individual band absorption intensity between oriented sample ATR spectra and isotropic sample transmission spectra may reflect an orientational preference of the infrared vibration either parallel or perpendicular to the surface of the IRE.

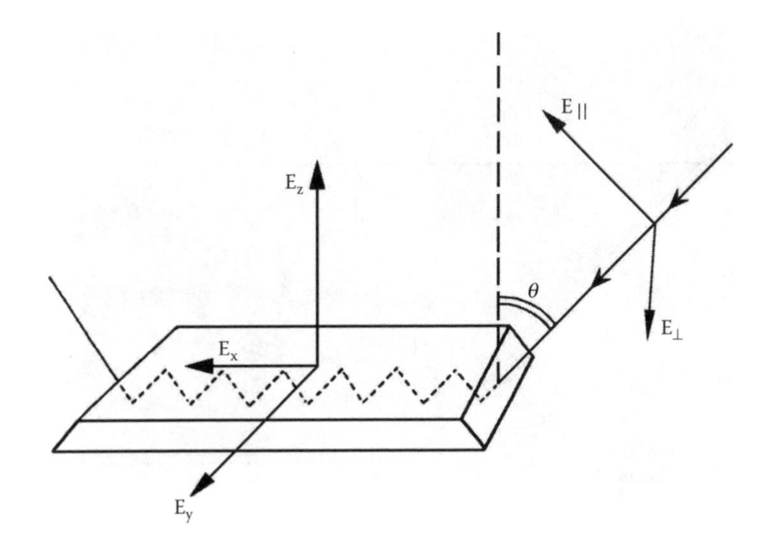

FIGURE 7.3 The electromagnetic field of the infrared light that exists beyond the IRE surface has components in the x, y, and z directions, noted as E_x, E_y, and E_z, respectively. For unpolarized infrared light, there is a net polarization of the infrared light in the plane of the IRE. Infrared light polarized parallel to the plane of incidence ($E_{\}\}}$) has components of its electric field in the x and z directions. Infrared light polarized perpendicular ($E_\perp$) to the plane of incidence has its electric field oriented along the y axis. θ refers to the angle of incidence (see text).

7.2.2 Linear Dichroism

The orientation of a functional group relative to the IRE surface can be determined with linear dichroism measurements [16,17]. The dichroic ratio R of a given absorption band is defined as the ratio of the integrated absorbance intensity obtained with infrared radiation polarized either parallel ($A_\parallel$) or perpendicular ($A_\perp$) to the plane of incidence (Figure 7.3). The plane of incidence is defined by the incoming and reflected infrared beam (xz-plane), where the x-axis denotes the direction of propagation through the IRE and the z-axis is perpendicular to the IRE surface. The angle of incidence θ is defined as the angle between the incoming infrared beam and the z-axis. With the assumption that the membrane components are uniaxially oriented in the xy-plane of the IRE, the dichroism is defined by the following:

$$R = \frac{A_\parallel}{A_\perp} = \frac{E_x^2}{E_y^2} + \frac{E_z^2[f \cdot \cos^2 \gamma + (1/3(1-f)]}{E_y^2[1/2 f \cdot \sin^2 \gamma + (1/3(1-f)]} \tag{7.4}$$

where E_x, E_y, and E_z are the electric field amplitudes in the x, y, z directions, respectively; γ is the angle between the vibrational transition moment and the uniaxial fiber axis (long axis of the molecule); and f is the order parameter describing the average angle β between the fiber axis and the IRE surface/z-axis.

$$f(\beta) = 1/2 \cdot (3\cos^2 \beta - 1) \tag{7.5}$$

To maximize signal strength, ATR difference measurements are usually performed on membrane films that are thicker than d_p. The strengths of the electric field amplitudes E_x, E_y, and E_z are defined for the thick film approximation as follows:

$$E_y = \frac{2\cos\theta}{\left(1-n_{21}^2\right)^{1/2}} \tag{7.6}$$

$$E_x = 2\bullet \frac{\left(\sin^2\theta - n_{21}^2\right)^{1/2} \bullet \cos\theta}{\left(1-n_{21}^2\right)^{1/2}\left[\left(1+n_{21}^2\right)\sin^2\theta - n_{21}^2\right]^{1/2}} \tag{7.7}$$

$$E_z = 2\bullet \frac{\sin\theta \bullet \cos\theta}{\left(1-n_{21}^2\right)^{1/2}\left[\left(1+n_{21}^2\right)\sin^2\theta - n_{21}^2\right]^{1/2}} \tag{7.8}$$

For a germanium IRE ($n_1 = 4$) with an angle of incidence $\theta = 45°$ and a sample refractive index $n_2 = 1.44$, the electric field amplitudes are $E_x = 1.398$, $E_y = 1.516$, and $E_z = 1.625$. For simplicity, we assume that the transition moment of a vibration is coincident with a fiber axis ($\gamma = 0$). The order parameter f can then be defined in terms of the dichroic ratio:

$$f = \frac{R - 2.00}{R + 1.45} \tag{7.9}$$

A molecule that rotates isotropically (e.g., bulk H_2O) exhibits an R value, $R_{\text{isotropic}}$, of 2.00. An R value of 2.00 is also observed if the long axis of the molecule is oriented at the "magic angle" of $\beta = 54.7°$ relative to the IRE surface/z-axis. An R value greater than 2.00 corresponds to an average tilt angle β less than 54.7° relative to the bilayer normal/z-axis. An R value less than 2.00 corresponds to an average tilt angle β greater than 54.7°. In either case, the mosaic spread of lipid multi-bilayers on an IRE surface or the rapid anisotropic motion of the molecule will average the calculated angle β closer to the magic angle (a plot of the dependence of β on R is found in Reference 17). Note that the value of $R_{\text{isotropic}}$ depends on both the refractive index of the IRE and the angle of incidence of the infrared light. The interpretation of linear dichroism difference spectra recorded with IREs made from materials other than germanium or with different angles of incidence requires a re-evaluation of the electric field amplitudes E_x, E_y, and E_z (using the above noted equations) in order to calculate $R_{\text{isotropic}}$.

7.3 TRIGGERING CONFORMATIONAL TRANSITIONS WITH BUFFER FLOW

The limited effective penetration of infrared light into a biological membrane film deposited on an IRE surface allows for the acquisition of spectra in the presence of bulk aqueous solution. Ligands added to the bulk solution percolate through hydrated membrane films and bind to proteins located within the lipid multi-bilayers. In addition, many biological membranes adhere to hydrophilic IREs, thus allowing the acquisition of reproducible spectra in the presence of flowing bulk solution. The ability to record reproducible spectra in the presence of flowing buffer provides a convenient method for triggering ligand-induced conformational change. By alternately flowing buffer either with or without a ligand of interest past the membrane film, one can measure the difference between spectra of the ligand-bound and ligand-free states. Ligand-induced structural change can thus be probed.

We deposit membranes containing the nicotinic acetylcholine receptor (nAChR) on the surface of a germanium IRE and record two spectra in the absence of bound ligand while flowing buffer (wash buffer) at a rate of ~1.5 ml/min past the nAChR film (Figure 7.4). The flowing solution is switched to an identical one containing the ligand of interest (trigger buffer) and a spectrum recorded of the ligand-bound state. The difference between the two ligand-free state spectra (control spectrum) and between the ligand-bound and the ligand-free state spectra (ligand difference spectrum) is calculated and stored. The flowing solution is then switched back to

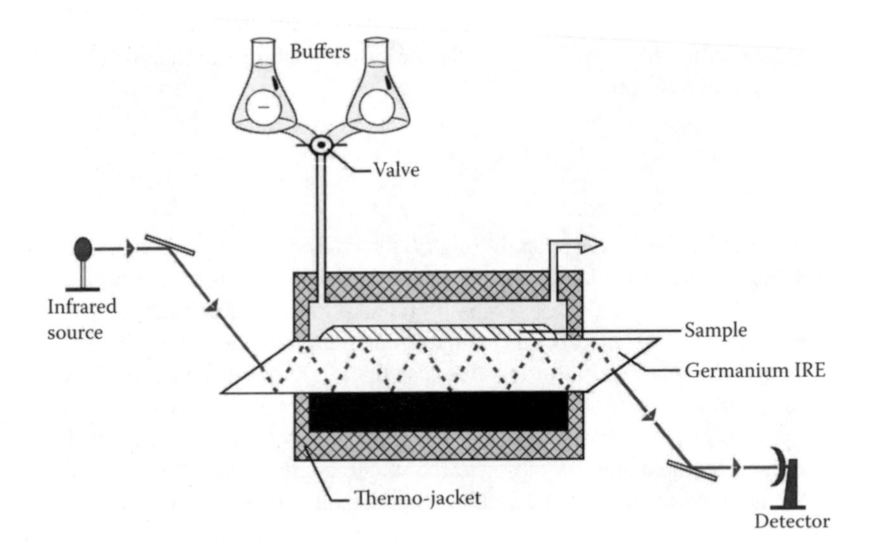

FIGURE 7.4 Schematic diagram of the ATR cell used to record ligand induced spectral differences with buffer flow. The infrared light reflecting off the internal surface of the germanium IRE penetrates into and is absorbed by the sample deposited on the IRE surface. Buffer flows continuously past the membrane sample at a rate of roughly 1.5 ml/min. An electronically controlled valve switches between buffer either with (+: trigger buffer) or without (–: wash buffer) the ligand of interest. The buffer selection is controlled by the spectrometer computer and set according to the spectral acquisition protocol described in the text. The temperature of the ATR cell is maintained constant by flowing buffer from a water bath/circulator rapidly through the two thermo-jackets. The temperature of the wash and trigger buffer flasks are immersed in the working compartment of the same water bath/circulator to maintain a constant temperature (not shown). All tubing from both the buffer solutions and the water bath/circulator to the ATR cell are all placed within a larger insulated tube.

the wash buffer without ligand. After a 20 minute washing period to remove the ligand from the film, the process is repeated many times and individual difference spectra averaged [11,12,18].

The flowing buffer ATR approach is versatile in that the binding of almost any water-soluble small ligand can be studied. Ligand binding can be probed at varying pH and ionic strength. Ligand-induced changes in functional group orientation can be determined by recording difference spectra with linearly polarized infrared light. A number of different IRE materials and dimensions are available, which allows the experimental approach to be optimized for samples with varying quantities of protein (see below). The difference technique, however, requires the formation of a stable membrane film on an IRE surface. Rigorous care must also be taken to avoid variations in flowing buffer temperature, which lead to baseline distortions. The successful application of the ATR method for probing ligand-induced conformational change depends on four interrelated factors: (1) choice of IRE, (2) method of film deposition, (3) formation of stable membrane films on the IRE surface, and (4) choice of appropriate data acquisition conditions.

7.3.1 Choice of Internal Reflection Element

For a typical single bounce IRE, the penetration depth (which corresponds to the effective path length of the ATR "cuvette") is relatively small, leading to weak absorbances and thus potentially weak bands in an ATR difference spectrum. Fortunately, the penetration depth is dependent upon the refractive index of the IRE. Choosing an IRE with a greater penetration depth will increase sample absorption. The effective path length of the ATR cuvette can also be increased by choosing an IRE geometry that allows for multiple internal reflections. The size, geometry, and chemistry of the IRE can thus be optimized for a particular application.

7.3.1.1 IRE Material

Table 7.1 summarizes the most common IRE materials that are available for studies of biological samples. All five transmit light in the mid-infrared region of the electromagnetic spectrum. In cases of unlimited sample size, materials with relatively large penetration depths, such as ZnSe, give rise to the strongest absorption signals and yield the most intense bands in a difference spectrum. If sample size and thus sample thickness on the IRE surface is limited, a Ge IRE is preferable because of the reduced penetration depth. In all cases, sample size and film thickness should match the IRE penetration depth to maximize the sample absorption relative to that of bulk H_2O.

KRS-5 IREs have been used widely for general ATR applications. More recently, ZnSe has replaced KRS-5 as the material of choice because of greater mechanical strength. IREs made from ZnSe are not as easily damaged as those made from the very soft KRS-5 material, although ZnSe has limited use with strong acids and alkalis because prolonged exposure will etch the IRE surface. Complexing agents such as ammonia and EDTA can also form complexes with zinc and will thus erode the IRE. ZnS is slightly harder and more chemically resistant than ZnSe, but has a

TABLE 7.1
Optical and Physical Properties of Different IRE Materials

Material	n_1 (1000 cm^{-1})	d_p[1] (vm)	Useful Spectal Range (cm^{-1})	Incompatible Materials	Cleaning Materials	Properties
Diamond	2.4	1.66	4500–2500, 1,667-33	$K_2Cr_2O_7$,, conc. H_2SO_4	alcohol, acetone, H_2O	non-reactive, pressure and scratch resistant
Germanium (Ge)	4	0.65	5,500–830	hot H_2SO_4 aqua regia[2]	alcohol, acetone, H_2O[4]	hard and brittle, temp sensitive, subject to thermal shock
KRS-5 (eutectic mixture of TlBr and TlI)	2.37	1.73	20,000–400	complexing agents, slightly soluble in H_2O	MEK[3]	deforms under pressure, temp sensitive, toxic
Zinc selenide (ZnSe)	2.4	1.66	20,000–650	acids, strong alkalis	acetone, H_2O, alcohol	withstands limited mechanical and thermal shock
Zinc sulfide (ZnS)	2.2	2.35	17,000–950	acids	acetone, alcohol	comparable to ZnSe, but slightly harder, more chemically resistant

[1] The depth of penetration calculated for each material at 1000 cm^{-1} using a sample refractive index (n_2) of 1.4 and an angle of incidence of 45°.
[2] 3:1 HCl/ HNO_3.
[3] Methylethylketone (2-butanone).
[4] See text.

slightly more limited spectral range. Diamond ATR IREs have excellent physical and chemical properties, but absorb strongly in the 2500 cm^{-1} to 1650 cm^{-1} region. Diamond is an expensive IRE material that has been incorporated into ATR crystals (see below).

The excellent spectral range and high chemical resistance of Ge makes it a popular material for studies of biological membranes. Biological membranes adhere strongly to the very polar germanium IRE surfaces, a prerequisite for recording ATR difference spectra with flowing solution. Ge IREs are easy to clean and their surfaces can be oxidized to enhance film deposition (see below). They tend to be available in the greatest range of geometries and sizes, and also to be the cheapest, although prices vary substantially depending on the supplier. One minor drawback of Ge is that it absorbs visible light in a manner that can deleteriously affect time resolved infrared difference spectra recorded immediately after a laser flash [19]. Conversely, steady-state ATR difference spectra have been reported for bacteriorhodopsin when the protein is continuously illuminated with yellow light [20].

7.3.1.2 IRE Size and Geometry (Optimization of Signal Strength)

The number of reflections N and thus the absorption intensity of a sample is related to the dimensions of the IRE according to the following:

$$N = l / t \tan\theta \tag{7.10}$$

were l is the length of the IRE measured from the center of the entrance aperture to the center of the exit aperture, t is the thickness of the IRE, and θ is the angle of incidence. Longer, thinner IREs with a smaller angle of incidence have a greater number of reflections and thus give rise to stronger absorption bands than shorter, thicker IREs with a larger angle of incidence (Figure 7.5a). For example, the absorption intensities obtained in spectra recorded from various nAChR films deposited on a 50mm × 20 mm × 2 mm 45° Ge IRE (~12 bounces per IRE surface) are compared to the absorption intensities of similar nAChR films formed on a 10 mm × 5 mm × 0.5 mm 45° micro-sampling IRE (~10 bounces per IRE surface) in Figure 7.5b and Figure 7.5c, respectively. The fourfold reduction in thickness of the latter IRE leads to a large number of reflections over a very short distance and thus an absorption intensity comparable to that observed with the larger IRE, although from a much smaller sample [21]. In the extreme case, planar wave guides (very thin, narrow Ge IREs supported by a rigid substrate) have been used to maximize the number of internal reflections over a small sampling area yielding high quality spectra from pico-gram quantities of bacteriorhodopsin as well as from a single frog oocyte [22]. Additional technical advances will further improve sensitivity allowing application of the ATR technique to samples available in very limited amounts.

IREs are commercially available in thicknesses varying from 0.5 mm to 3 mm, in surface dimensions from 10 mm × 5 mm to 50 mm × 20 mm, and with bevel angles of 30°, 45°, and 60°. Some manufacturers supply IREs manufactured with

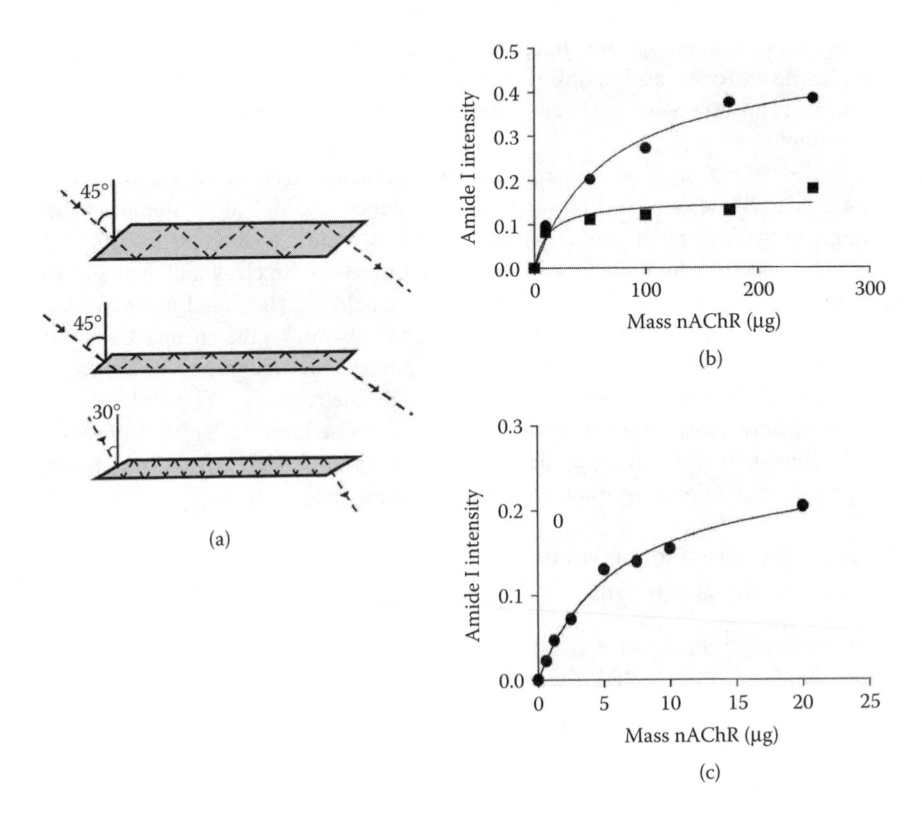

FIGURE 7.5 (a) Thinner IREs with a smaller angle of incidence have more internal reflections and give rise to more intense sample absorbencies than thicker IREs with larger angles of incidence. (b) The absorption intensities observed for dry and hydrated films with varying amounts of nAChR membranes deposited on a 50 × 20 × 2 mm Ge IRE with a 45° angle of incidence. The films were hydrated with ^2H_2O to allow measurement of the amide I band intensity. (c) Absorption intensities observed for dry films with varying amounts of nAChR membranes deposited on a 10 × 5 × 0.5 mm IRE with a 45° angle of incidence.

dimensions to suit customer needs. Flow-through ATR cells (Figure 7.6) are commercially available for all but the smallest IREs. Regardless, Ge IREs can be imbedded in an epoxy resin to create a larger surface area that will fit larger ATR sample cells. Custom-built ATR cells for the smaller crystals are also possible. Note that although the smaller thinner IREs are attractive in that strong absorbance signals can be obtained with limited sample size, the small aperture of the IRE upon which the infrared light is focused limits throughput and leads to high levels of noise. Throughput and thus the level of noise must be optimized with condensing mirrors, which focus the infrared light into the IRE.

Single-bounce hemispherical IREs with very small sampling surfaces represent an alternative to multi-reflection IREs when sample size is limited. While different ATR elements are available, the increased penetration depth of a diamond ATR element leads to absorption signals approaching that obtained with the much larger

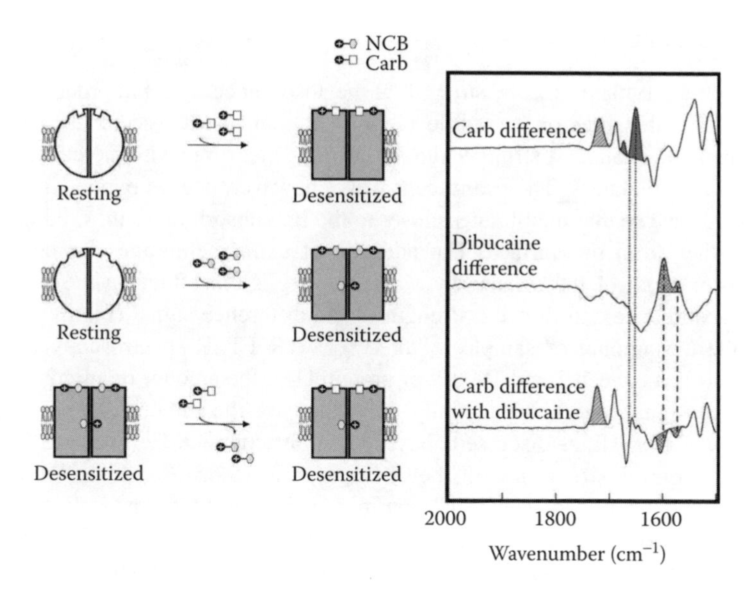

FIGURE 7.6 The difference between FT-IR spectra of the nAChR recorded in the presence and absence of Carb (top schematic on left and top spectrum on right) exhibits features due to nAChR bound Carb (lighter hatched shading) and the Carb-induced R→D conformational change (darker shading). The difference between spectra of the nAChR recorded in the presence and absence of 200 μM dibucaine (middle schematic on left and middle spectrum on right) exhibits features indicative of dibucaine-induced structural change in the nAChR, but these are masked by the strong absorption bands of dibucaine partitioned into the lipid bilayer (light cross-hatched shading) as well as the negative features due to expansion of the nAChR film on the IRE surface. The difference between spectra of the nAChR recorded in the presence and absence of Carb, but while continuously maintaining the nAChR in contact with 200 μM dibucaine (bottom panel on left and bottom spectrum on right), exhibits positive and negative features due to the binding of Carb to and consequent displacement of dibucaine from the neurotransmitter site, respectively. Bands indicative of the R→D conformational change (dark shading) are absent because dibucaine stabilizes the nAChR in a desensitized state prior to the addition of Carb. Note that the Carb vibration near 1720 cm^{-1} (light shading) in both the top and bottom spectrum has an intensity of roughly 7.5×10^{-5} absorbance units while the negative lipid vibration near 1740 cm^{-1} in the dibucaine difference spectrum has a negative intensity of roughly 2.5×10^{-4} absorbance units.

50 mm × 20 mm × 2 mm Ge IREs, even though the sampling area and thus sample requirements are a fraction of that required for the larger IRE. With the proper optics, single bounce accessories have better throughput than the multi-reflection accessories and thus minimal levels of noise. One drawback is that biological membranes do not adhere to diamond surfaces as well they do to germanium. We have found that nAChR membranes dissociate from diamond ATR elements under flow conditions. Careful adjustment of flow rates and acquisition times may, however, allow for the acquisition of accurate difference spectra using diamond IREs. Single bounce accessories with Ge IREs are also available.

7.3.1.3 Sample Size

It is important to maximize the strength of the absorbance signal in order to increase the absolute intensities of the subtle vibrational changes that occur during protein conformational change. A stronger signal is usually measured with increasing sample size and film thickness. The evanescent wave, however, decays exponentially from the IRE surface so that membranes closer to the IRE absorb more infrared light than those farther from the surface. The addition of extra membranes far beyond the penetration depth of the evanescent wave has negligible effect on the strength of sample absorbance and thus the strength of the difference signal (Figure 7.5).

While the amount of sample required for ATR FT-IR spectroscopy is usually reported to be between 1 and 100 μg of protein [15], the amount required to achieve a maximal signal varies substantially depending on the dimensions and physical properties of the IRE, as discussed above. The most common Ge IREs are 50 mm × 20 mm × 2 mm in size with a 45° bevel (angle of incidence $\theta = 45°$). We typically spread roughly 250 μg of our affinity-purified and reconstituted nAChR membranes (lipid to lipid protein molar ratio ~150:1) over a surface area of roughly 4 cm^2 to ensure a maximal signal (Figure 7.5), which corresponds to ~60 μg of protein per cm^2 surface area. Only ~100 μg of protein, or roughly 20 μg protein per cm^2 surface area, are required to achieve a signal ~90% of the maximum. IREs with a proportionally greater d_p require a proportionally greater amount of sample to saturate the absorption signal.

Note that stronger protein absorption signals are achieved with membranes at low lipid–protein ratios. This ratio dictates how many protein molecules within a single bilayer on the IRE surface interact with the evanescent wave of light. Due to varying electrostatic properties, multi-bilayers formed using bilayers with different protein or lipid compositions may have different bilayer spacings in the hydrated state, which will influence the amount of protein required to obtain a maximal absorption [21]. In addition, higher purity samples and smaller proteins give rise to more intense difference signals relative to comparable samples of lower purity or from larger proteins [23].

7.3.2 Membrane Film Deposition

The formation of a stable membrane film on an IRE surface is critical for recording reproducible spectra in the presence of flowing buffer. Stable film formation depends on a number of factors including the method of film deposition. A number of experimental approaches have been utilized for depositing biological membranes on the surface of IREs. While few of the resulting films have been optimized or tested for recording difference spectra in the presence of flowing buffer as described above, these methods represent alternative approaches that may prove optimal for specific applications.

7.3.2.1 Planar Supported Bilayers

Several methods have been developed for preparing single planar lipid bilayers supported on hydrophilic substrates [24]. The relatively weak signals obtained from single supported bilayers are not optimal for measuring ligand-induced spectral differences. Supported bilayers may serve as a starting point for the deposition of additional multi-bilayers as described below.

In most cases, the first step is to deposit a single phospholipid monolayer on the IRE surface by removing the IRE from an aqueous solution through a lipid monolayer located at the air–water interface of a Langmuir trough (phospholipid head groups face toward the IRE) [25]. A second monolayer is then added with a second pass of the IRE through the lipid monolayer, the IRE brought into contact with the air–water interface with its planar surface parallel to the interface. Alternatively, a second layer can be added by equilibrating the monolayer with a solution containing membrane vesicles [26]. In the latter case, membrane vesicles that include integral membrane proteins have been used to incorporate a membrane protein into the supported bilayer. In a similar approach, supported bilayers have been formed on a clean IRE surface upon equilibration with a solution containing vesicles of the biological membrane of interest [27].

A thin film of water (10–20 Å) separates the supported bilayers from the hydrophilic solid supports [25]. Lipids in both monolayers of supported lipid bilayers, both with and without integral membrane proteins, diffuse freely in the membranes. However, integral membrane proteins that extend more than 10–20 Å beyond the bilayer surface may interact with the IRE, slowing their diffusion rates. To circumvent this problem, Tamm and Tatulian used the above noted Langmuir trough method to form supported bilayers that are covalently linked to the surface through thiol linkages and a polyethylene glycol spacer. The spacer increases the distance between the bilayer and the IRE surface allowing rapid diffusion of all components within the inner monolayer [28].

7.3.2.2 Direct Deposition of Biological Membrane Multi-Bilayers

The simplest method of membrane film formation is to spread ~50 μl of a 5 mg/ml integral membrane protein solution in buffer over a clean Ge IRE surface [11]. The bulk aqueous solvent is evaporated with a gentle stream of nitrogen gas to bring the membranes into close contact with the IRE, and the membranes rehydrated with excess buffer. We use Ge IREs because Ge provides a hydrophilic surface similar to glass upon which many biological membranes will adsorb. In addition, the IRE can be oxidized before film deposition to reduce surface tension and enhance the spreading of samples on the IRE surface and thus ultimately the uniformity of film deposition. We immerse the Ge IRE for ~20 minutes in a solution of Chromerge™ and then wash extensively with distilled water (this procedure also removes any remaining biological material). Others clean Ge IREs with a basic detergent followed by a plasma cleaner [15]. Direct film deposition has been used successfully to form multi-bilayers on ZnSe IREs, although this approach with ZnSe is slightly more difficult because of the high surface tension of water on the IRE surface [19,29].

7.3.2.3 Covalent Attachment of Multi-Bilayers

As noted, Tamm and Tatulian formed bilayers on solid supports via covalent thiol linkages [28]. Other methods of covalently linking biological membranes to solid supports are possible, including the use of gold attachment chemistry [30]. Hofer and Fringeli covalently linked the water soluble enzyme, acetylcholinesterase, to silanized Ge surfaces [31]. The high stability of covalently bound films is a potential advantage for recording FT-IR spectra in the presence of flowing buffer and may permit application of the difference technique to water soluble proteins. Further improvements may be required to create covalently linked multi-bilayers (as opposed to single bilayers) that give rise to strong sample absorbances. The formation of a polyacrylamide or agarose matrix on the surface of a deposited membrane film has been suggested as a possible mechanism for improving membrane film stability [20]. Functional immobilized bacteriorhodopsin membranes have been formed in a polyacrylamide gel matrix for orientation measurements [32].

7.3.3 Membrane Film Stability

Initial attempts to stabilize purple membrane from *Halobacterium halobium* on a Ge surface suggested that high concentrations of divalent cations are required for optimal adhesion [20]. Salt concentration influences the stability of nAChR membrane films deposited on Ge IREs. Multi-bilayers composed of nAChR membranes, however, are stable for hours to days in the presence of flowing buffers that are commonly used for *in vitro* studies of the nAChR (250 mM NaCl, 5 mM KCl, 2 mM $MgCl_2$, 3 mM $CaCl_2$, and 5 mM phosphate, pH 7.0). Membrane films of the nAChR that are sufficient, but not optimal, for difference spectroscopy have also been formed in the presence of 20 mM Tris buffer lacking any added salt [33]. High concentrations of salt are not necessarily critical for recording difference spectra in the presence of buffer flow.

The formation of a stable membrane film does not appear to be dependent upon either lipid composition or the presence of integral membrane proteins. Films suitable for difference spectroscopy have been formed on a Ge support from samples containing the nAChR reconstituted into membranes with a variety of different lipid compositions [23,34]. Films of pure DPPC are moderately stable in the presence of flowing buffer (unpublished observations). In contrast, a continuous flow of buffer washes most of the protein found in films of cytosolic proteins, such as myoglobin, from a Ge surface. Although our studies of film adhesion are limited, the above observations suggest that it is perhaps the surface tension between planar membrane multi-bilayers that is essential for the retention of membrane films on the IRE support in the presence of flowing buffer.

Membrane films containing both bacteriorhodopsin and rhodopsin have been deposited on ZnSe ATR crystals [19,29]. In both cases, the films are stable during buffer exchange, although the buffer flow conditions were not as demanding as those described here. NAChR membranes deposited on diamond IRE surfaces are stable in aqueous solution, but sample is lost with rapid buffer flow. The formation of membrane films on a variety of IRE supports that are suitable for difference spectroscopy may be possible with careful adjustment of flow rates and acquisition times.

It is important to avoid changing any experimental condition during a difference experiment that will lead to alterations in membrane film stability or thickness. For example, both the addition of a ligand with an appreciable lipid solubility and a substantial change in salt concentration may lead to swelling and vertical expansion of a membrane film on the IRE surface and thus a reduction in both protein and lipid absorption intensity. Changes in absorbance intensity resulting from membrane film expansion or contraction can be much larger than those that result specifically from protein conformational change, as is shown in Figure 7.6, middle trace [35,36]. To minimize such effects, the trigger and control buffers should be identical except for the added ligand, and the ligand should be used at as low a concentration as possible. Adequate buffering capacity must be present in the flowing solution, particularly in spectra recorded at elevated pH values, to avoid CO_2-dependent changes in pH of the buffers over time. In all cases, appropriate controls must be recorded to ascertain that detected spectral differences result from ligand binding specifically to the protein of interest as opposed to the lipid bilayer. This is recommended instead of phosphate buffer to avoid the precipitation of phosphate salts.

Finally, it is important to note that all membrane multi-bilayers swell upon hydration. Hydration of nAChR membrane multi-bilayers deposited on a Ge IRE leads to a roughly threefold decrease in the absorption intensity of the sample relative to a dry film. It may take 1 to 2 hours in the presence of continuous buffer flow for a film to completely equilibrate. The initial loss of intensity observed upon film hydration must not be mistaken for film instability, when testing the applicability of the technique to new systems.

7.3.4 Data Acquisition Conditions

7.3.4.1 Minimization of Temperature-Dependent Spectral Artifacts

Despite the limited effective sample thickness, which is a consequence of the limited penetration of the evanescent wave into the membrane film, water still absorbs strongly in the 1700–1500 cm^{-1} region of spectra recorded using the ATR technique. Changes in buffer temperature over the course of spectral acquisition may lead to subtle changes in membrane film thickness as well as changes in the width and height of water absorption bands. Maintaining a constant room temperature, as well as a constant temperature for all buffers that come into contact with the membrane film, is critical for the accurate detection of spectral differences resulting from protein conformational change.

To minimize temperature-dependent spectral artifacts, the nAChR sample on the germanium IRE is held in a liquid ATR sample cell that is maintained at a constant temperature by circulating water from a constant temperature water bath/circulator through the body of the IRE cell (this water does not contact the germanium or the sample; Figure 7.4). The trigger and wash buffers that alternatively flow past the surface of the film on the IRE are placed in flasks situated within the working area of the same water bath/circulator. All buffers as well as the germanium IRE itself are thus maintained at as close as possible to a constant temperature.

The temperature inside the spectrometer is usually slightly higher than room temperature. Continuously flowing buffer is used to establish a steady-state temperature intermediate between the temperature of the water bath/circulator and the FT-IR spectrometer. The water bath/circulator is best set close to room temperature because the trigger buffer tubing exposed to air will equilibrate to room temperature while the control buffer is flowing through the ATR cell and vice versa. Even subtle differences between the water bath/circulator and room temperature can lead to baseline distortions in the 1500–1700 cm^{-1} region.

7.3.4.2 Optimization of Signal-to-Noise

The choice of IRE is a critical factor in the optimization of the signal-to-noise ratio, as discussed above. For a given ATR accessory and IRE, the signal-to-noise ratio depends on spectral resolution and length of data acquisition time. The acquisition parameters used in our laboratory were chosen to optimize signal-to-noise while minimizing the appearance of artifacts due to temperature fluctuations and other factors. Better signal-to-noise is achieved with longer acquisition times. Conversely, artifacts due to temperature fluctuations and other factors are less intense with shorter acquisition times.

We have found that artifacts are minimized if spectral differences are measured between two spectra each recorded at 8 cm^{-1} resolution and consisting of 512 scans (roughly 7 minutes of data acquisition time for each spectrum using a DTGS detector). High signal-to-noise is achieved by signal averaging many such differences. Depending on signal strength, 30 to 40 averaged differences generally yield highly reproducible difference spectra with excellent signal-to-noise. Lower spectral resolution reduces acquisition times minimizing spectral artifacts while at the same time improving the signal-to-noise ratio. Studies with the nAChR suggest that spectral information is not lost at 8 cm^{-1} resolution relative to difference spectra recorded at higher resolution.

7.4 LIGAND-INDUCED CONFORMATIONAL CHANGE

The ATR method of recording difference spectra in the presence of flowing buffer has been used primarily to monitor the vibrational changes that occur upon ligand binding to the nicotinic acetylcholine receptor (nAChR) from *Torpedo* (see also Reference 29). In this section, we summarize some of our preliminary studies which show the utility of the technique for both mapping ligand induced conformational states and probing the detailed chemistry of receptor–ligand interactions. We briefly summarize work that has been performed on light activated proteins using the ATR approach. Together these studies demonstrate the enormous potential of the method and illustrate some of the technical problems that can be encountered in ligand binding studies.

The difference between spectra of the nAChR recorded in the presence and absence of the agonist analog Carb (referred to as a *Carb difference spectrum*) exhibits a complex pattern of positive and negative difference bands (Figure 7.6, top difference spectrum). These bands reflect (1) vibrations of nAChR-bound Carb (light hatched shading in Figure 7.6); (2) vibrational changes associated with the formation of physical interactions, such as hydrogen bonds, between Carb and neurotransmitter binding site residues (no shading); and (3) vibrational changes associated with the resting-to-desensitized (R→D) conformation transition (dark shading) (see top schematic in Figure 7.6). Positive difference bands centered near 1663, 1655, 1547, 1430, and 1059 cm^{-1} serve as markers of the Carb-induced transition from the resting to the desensitized state [23,34–38]. For simplicity, we will focus on two of the R→D conformational change marker frequencies centered near 1663 cm^{-1} and 1655 cm^{-1}.

7.4.1 Mapping Conformational States

The conformational states in the nAChR that are stabilized upon the binding of a variety of local anesthetics have been mapped using ATR difference spectroscopy. Pharmacological studies suggest that most local anesthetics bind to the ion channel of the receptor at relatively low concentrations, blocking the flux of cations across the postsynaptic membrane and altering nAChR conformational equilibria. In the absence of bound ligand, the nAChR exists in both an activatable resting and a nonconducting desensitized state, with the equilibrium strongly favoring the resting conformation. Desensitizing local anesthetics shift the equilibrium towards a high-affinity agonist-binding conformation, presumed to be the desensitized state. In contrast, sensitizing local anesthetics stabilize a low-affinity agonist-binding conformation, presumed to be the resting state. At higher concentrations, most local anesthetics also bind to the neurotransmitter sites with undefined effects on nAChR conformational equilibria.

The structural effects of local anesthetic binding were first examined by measuring the differences between spectra of the nAChR recorded in the presence and absence of each local anesthetic (referred to as local anesthetic difference spectra). As shown in the middle trace of Figure 7.6 (see caption), binding of the desensitizing local anesthetic dibucaine to the nAChR membranes leads to intense vibrational changes relative to those observed upon Carb binding to the nAChR. These include positive bands near 1605 cm^{-1} and 1575 cm^{-1} that reflect primarily dibucaine partitioning into the nAChR membranes. Negative lipid and protein bands near 1740, 1650, and 1550 cm^{-1} superimposed on a weak positive broad ^1H_2O band near 1630 cm^{-1} reflect expansion of the nAChR film beyond the penetration depth of the evanescent wave [35,36]. Unfortunately, the vibrational changes associated with dibucaine binding to the lipid bilayer mask the vibrational intensity changes that result specifically from binding to the nAChR.

To monitor the receptor specific vibrational intensity changes, Carb difference spectra were recorded while maintaining the nAChR in continuous contact with dibucaine (see lowest schematic in Figure 7.6). Continuous incubation with dibucaine stabilizes a desensitized conformation of the nAChR prior to Carb addition and thus results in a loss of intensity at each of the two R→D marker frequencies near 1663 cm^{-1} and 1655 cm^{-1} in the corresponding Carb difference spectrum (the former change in intensity is seen more clearly as an increase in the negative band near 1668 cm^{-1}). Surprisingly, the spectral changes indicative of desensitization appear concomitant with both the appearance of negative vibrations that reflect Carb-induced displacement of dibucaine from the neurotransmitter binding site (see lowest schematic in Figure 7.6) and a decrease in intensity of several bands attributed to the formation of physical interactions between Carb and the nAChR [35]. The simultaneous appearance of features indicative of both desensitization and neurotransmitter site binding suggests that dibucaine may influence nAChR conformational equilibria by binding to the neurotransmitter sites, as opposed to the ion channel pore. The overlapping binding affinities of this local anesthetic for the ion channel and neurotransmitter sites prevent an unequivocal interpretation of the conformational effects in terms of the specific site(s) of dibucaine action.

The conformational effects of local anesthetic binding individually to the neurotransmitter and ion channel sites were first examined by recording Carb difference spectra in the presence of the sensitizing local anesthetic tetracaine (Figure 7.7a). At concentrations of tetracaine up to 200 μM, where binding is restricted to the ion channel, Carb difference spectra exhibit an increase (as opposed to a decrease) in the intensities of the R→D marker frequencies consistent with the stabilization of a resting (as opposed to a desensitized) conformation. Note that the increase in intensity near 1663 cm^{-1} is dramatic relative to that observed at 1655 cm^{-1}. Tetracaine binding to the ion channel shifts those conformationally active residues that vibrate near 1663 cm^{-1} from a desensitized to a resting state prior to the addition of Carb, while those that vibrate near 1655 cm^{-1} are essentially unaffected. In other words, ion channel pore binding stabilizes an intermediate between the resting and desensitized conformations that shares structural features in common with both states (Figure 7.8; see also Reference 35).

Carb difference spectra recorded at higher concentrations of tetracaine consistent with both neurotransmitter and ion channel site binding exhibit a completely different pattern of band intensity changes. Specifically, the spectra exhibit the above noted (with dibucaine) spectral features indicative of neurotransmitter site binding. The spectra also show a loss of intensity at both R→D conformational marker frequencies near 1663 cm^{-1} and 1655 cm^{-1}. Neurotransmitter site binding thus reverses the conformational effects that result from binding to the ion channel pore and stabilizes a fully desensitized conformation.

Multiple conformational effects upon local anesthetic binding to the ion channel pore and neurotransmitter sites were also detected in Carb difference spectra recorded in the presence of the desensitizing local anesthetic proadifen (Figure 7.7b). At proadifen concentrations up to 50 μM, where binding is restricted primarily to the ion channel pore, a large decrease in intensity of the R→D marker frequency centered near 1663 cm^{-1} is observed, while intensity near 1655 cm^{-1} is essentially unaffected. As with tetracaine, only the conformationally sensitive residues that vibrate near 1663 cm^{-1} are affected by ion channel pore binding. In contrast to tetracaine, proadifen shifts these residues into a desensitized (as opposed to a resting) conformation. In addition, higher concentrations of proadifen lead to spectral features indicative of both neurotransmitter site binding and the stabilization of a fully desensitized state. Note the baseline distortions that are observed in the Carb difference spectra recorded at high concentrations of proadifen result from instability of the lipid bilayers on the ATR surface [36].

These and other data reveal a much greater complexity of local anesthetic action at the nAChR than was detected using classical pharmacological approaches. Our data led to a revised model of local anesthetic action (Figure 7.8). More importantly, the

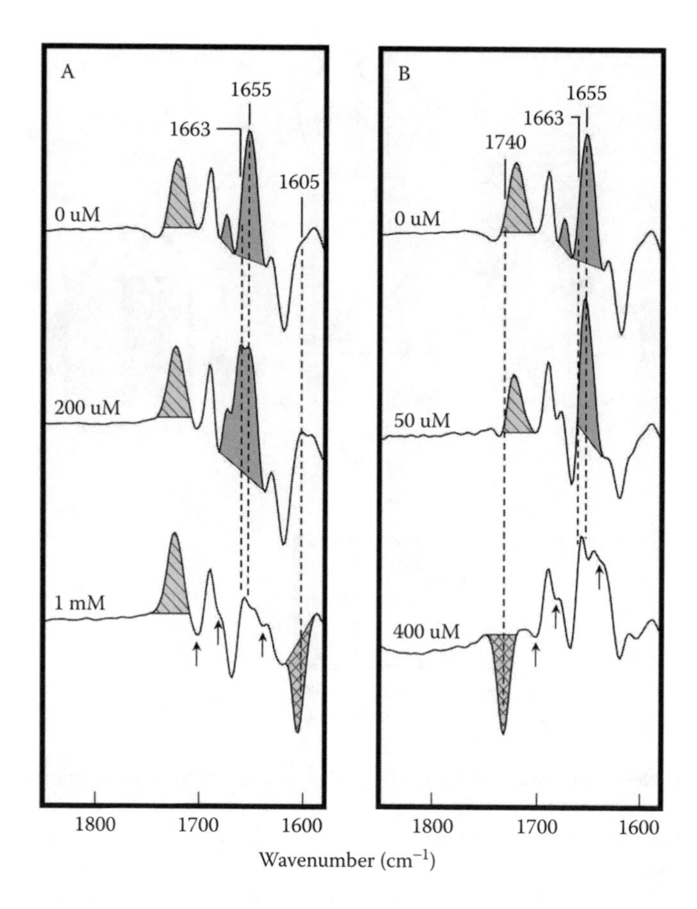

FIGURE 7.7 (a) Selected regions of Carb-difference spectra recorded in the continuous presence of 0, 200 μM, and 1 mM concentrations of the sensitizing local anesthetic tetracaine. At 200 μM, tetracaine binding is restricted to the ion channel pore. At 1 mM, tetracaine binds to both the ion channel and neurotransmitter sites. (b) Selected regions of Carb-difference spectra recorded in the continuous presence of 0, 50 μM, and 400 μM concentrations of the desensitizing local anesthetic proadifen. At 50 μM, proadifen binding is restricted to the ion channel pore. At 400 μM, proadifen binds to both the ion channel and neurotransmitter sites. In both cases, the dark shading at 1663 and 1655 cm^{-1} marks positive intensity that reflects the R→D conformational change, the hatched shading marks bands to nAChR bound Carb, and the cross-hatched shading marks negative features that reflect Carb-induced displacement of the local anesthetics from the neurotransmitter site (see bottom schematic in Figure 7.6). The arrows in (a) show points in the difference spectrum that are normally close to baseline. In (b), relatively large concentrations of proadifen lead to nAChR membrane film instability and thus distortions in the difference spectra [36].

data show that structurally distinct regions of the nAChR interconvert between the resting and desensitized states independently of each other. Difference spectra also detected novel conformational states of the nAChR that are structural intermediates between the resting and desensitized conformations. These novel conformations of the

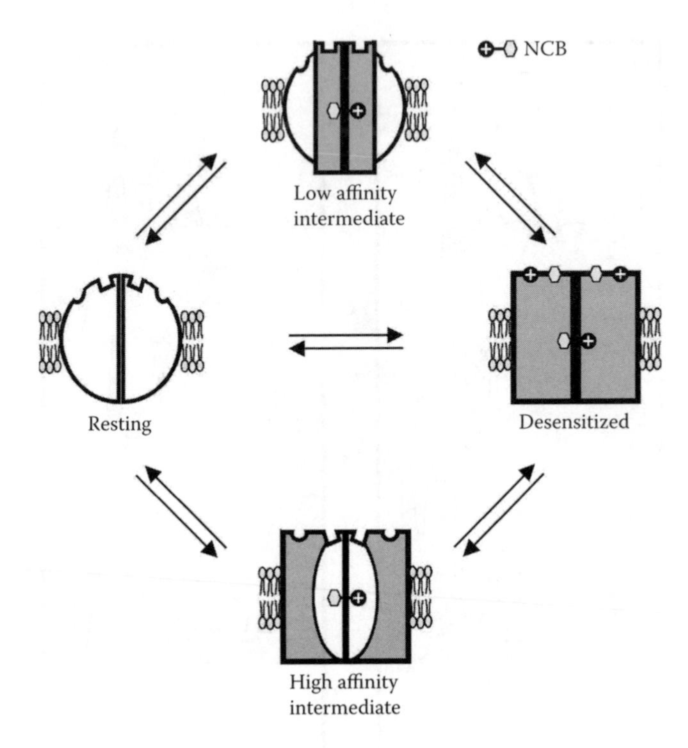

FIGURE 7.8 Schematic model of the conformational effects of local anesthetic binding to the nAChR. In the absence of bound ligand, nAChR exists in an equilibrium between the low-affinity agonist binding resting and high-affinity agonist binding desensitized states. FT-IR studies show that desensitizing local anesthetics bind to the ion channel pore and stabilize a high-affinity conformational intermediate between the resting and desensitized states. Sensitizing local anesthetics bind to the ion channel pore and stabilize a low-affinity intermediate between the resting and desensitized states. The binding of local anesthetics to the neurotransmitter site leads to the formation of a fully desensitized state.

nAChR have been detected in Carb difference spectra recorded from the nAChR reconstituted into lipid bilayers with distinct lipid compositions, and thus may be stabilized under a variety of different physiological conditions [34]. The newly defined conformational states may exist *in vivo* and may play a role in the modulation of a synaptic transmission by endogenous factors, such as receptor phosphorylation. Further work is aimed at defining the nature of the detected structural changes associated with each conformational state.

7.4.2 Chemistry of Receptor-Ligand Interactions

Several bands in the Carb-difference spectra reflect vibrational changes that result from the formation of physical interactions between Carb and neurotransmitter binding site residues and thus provide insight into the chemistry of Carb-nAChR interactions. Our initial goal was to identify protein bands that reflect interactions

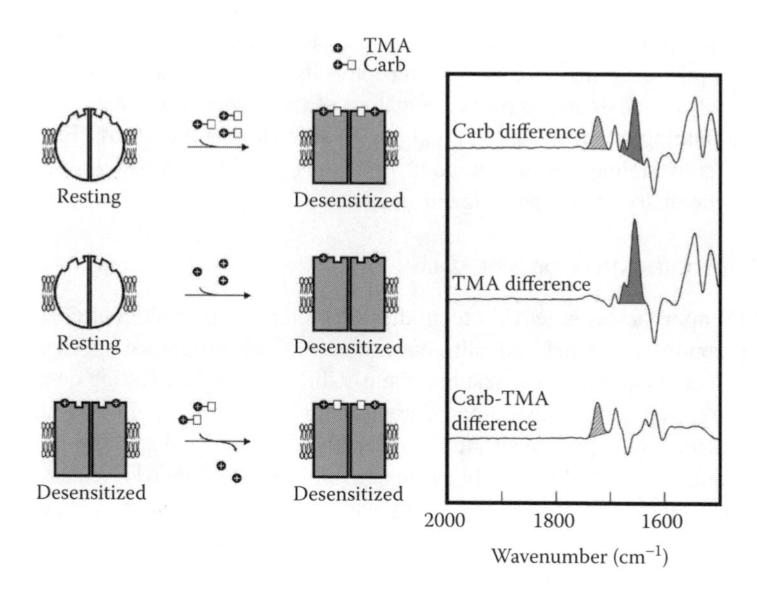

FIGURE 7.9 The difference between FT-IR spectra of the nAChR recorded in the presence and absence of Carb (top schematic on left and top spectrum on right) is compared to the difference between spectra recorded in the presence and absence of TMA (middle schematic on left and middle spectrum on right). The difference between spectra of the nAChR recorded in the with Carb and TMA bound (bottom panel on left and bottom spectrum on right) exhibits positive and negative features due to the physical interactions between the ester carbonyl of Carb and nAChR binding site residues. Shading is as in Figure 7.6.

with specific functional groups on Carb. The difference between spectra recorded in the presence and absence of the agonist analog, tetramethylamine (TMA difference spectra) exhibit all five R→D conformational markers, showing that TMA stabilizes a desensitized conformation of the receptor (Figure 7.9). There are, however, subtle intensity variations between the TMA and Carb difference spectra. These reflect the absence of physical interactions upon TMA binding that normally occur between the ester carbonyl of Carb and the nAChR and are most prominent in the 1650 cm^{-1} to 1700 cm^{-1} region of the difference spectrum. Unfortunately, considerable difference band overlap makes interpretation difficult.

To simplify, Carb difference spectra were recorded from a desensitized nAChR film that was incubated continuously with TMA. The resulting difference spectrum (referred to as a Carb-minus-TMA difference spectrum; see lowest schematic in Figure 7.9) exhibits positive and negative vibrations due to nAChR bound Carb and nAChR bound TMA, respectively. The difference spectrum also exhibits two clear negative/positive couples that reflect shifts in the vibrational frequency of protein vibrations with Carb, versus TMA, binding. Carb-minus-TMA difference spectra recorded in ^2H_2O and at alternative pH values suggest preliminary assignment of the bands and provide insight into the nature of the physical interactions that occur between the ester carbonyl of Carb and the receptor [39,42].

This result shows the versatility of the ATR technique for addressing specific questions regarding the chemistry of receptor–ligand interactions. In this case, the experiment was designed to probe the nature of the physical interactions that occur between binding site residues and a single functional group on Carb. Further work should clearly define these interactions and thus provide detailed insight into the specific chemistry of receptor–ligand interactions.

7.4.3 Modification of Side Chain pK_A

The ATR approach is amenable to studies of protein conformational change under varying conditions of pH and salt concentration. Carb difference spectra recorded at low salt concentrations suggest that the nAChR remains in a resting conformation that undergoes desensitization upon Carb binding [33]. For the light-activated proton pump, bacteriorhodopsin, ATR studies show that both pH and salt alter the rates of conformational change [20,40]. In an elegant application of the ATR technique, time-resolved infrared difference spectroscopy was used to investigate the pH dependence of proton uptake and release from single amino acid side chains in bacteriorhodopsin. The latter study identified transient changes in the pK_As of single residues using the time-resolved ATR difference approach [42].

7.4.4 Changes in Orientation of Specific Functional Groups

Membrane films deposited on planar IREs generally exhibit a strong orientational preference parallel to the IRE surface [15]. Comparison of equivalent light induced difference spectra recorded from bacteriorhodopsin using transmission and ATR reveal subtle band intensity variations that reflect changes in the orientation of individual functional groups [20]. Changes in functional group orientation upon protein conformational change can be probed directly with linear dichroism difference measurements. Linear dichroism difference spectra should be sensitive to changes in both transmembrane α-helix and individual amino acid side chain orientation, and should provide insight into the orientations of bound ligands relative to the bilayer normal. Linear dichroism difference spectroscopy will likely play an important role in the elucidation of membrane receptor conformational change.

7.5 CONCLUSIONS AND FUTURE DIRECTIONS

Our preliminary studies with nAChR illustrate a few important features regarding the use of ATR spectroscopy to monitor ligand-induced membrane protein conformational change. The presented spectra show the extremely high signal-to-noise ratio and reproducibility of the difference spectra obtainable using the ATR approach. The results also demonstrate the versatility of the ATR technique in that the structural effects of a variety of ligands can be monitored despite the encountered difficulties with partitioning of the lipophilic local anesthetics into the lipid bilayers. Experiments can be designed to address specific questions regarding the chemistry of receptor–ligand interactions.

The ATR technique has proven useful both for mapping nAChR conformational states stabilized upon ligand binding and for probing the detailed chemistry of nAChR-ligand interactions. These studies, however, are only in their infancy. The inability to express mutant nAChRs in sufficient quantities for FT-IR has to date prevented the assignment of observed difference bands to structural changes in specific residues. The use of the micro-sampling methods described in the text should bring the spectroscopic requirements in line with current expression capabilities. The ability to combine modern molecular biological approaches with FT-IR difference methods will open up a whole new avenue of investigation and should lead to detailed insight into the nature of localized structural change in the nAChR. In the near future it will be possible to study receptors from neuronal sources, which are of important clinical and pharmacological interest, as well as a host of other integral membrane proteins.

Finally, it is notable that the ATR method of data acquisition is compatible with a caged ligand approach (see, for example, Reference 19). The uniform release of a ligand from its caged precursor with a flash of visible light will allow the kinetics of ligand-induced conformational change to be studied with a very fast time-resolution. The ATR technique with buffer flow will allow signal averaging of flash induced kinetic experiments. Continued technical advances should therefore extend the ATR difference approach to other membrane proteins and lead to more sophisticated studies of the kinetics of ligand-induced membrane protein conformation change.

ACKNOWLEDGMENT

This work was supported by a grant from the Canadian Institutes of Health Research (to J. E. B.).

REFERENCES

1. MS Braiman, KJ Rothschild. Fourier transform infrared techniques for probing membrane protein structure. *Ann. Rev. Biophys. Chem.* 17:541–570, 1988.
2. KJ Rothschild. FTIR difference spectroscopy of bacteriorhodopsin: toward a molecular model. *J. Bioenerg. Biomembr.* 24:147–67, 1992.
3. W Mäntele. Reaction-induced infrared difference spectroscopy for the study of protein function and reaction mechanisms. *Trends Biochem. Sci.* 18:197–202, 1993.
4. F Siebert. Infrared spectroscopy applied to biochemical and biological problems. *Methods Enzymol.* 246:501–526, 1995.
5. A Barth, C Zscherp. Substrate binding and enzyme function investigated by infrared spectroscopy. *FEBS Lett.* 447:151–156, 2000.
6. AJ White, K Drabble, CW Wharton. A stopped-flow apparatus for infrared spectroscopy of aqueous solutions. *Biochem. J.* 306:843–849, 1995.
7. J Backmann, H Fabian, D Naumann. Temperature-jump-induced refolding of ribonuclease A: a time-resolved FTIR spectroscopic study. *FEBS Lett.* 364:175–178, 1995.
8. D Reinstadler, H Fabian, J Backmann, D Naumann. Refolding of thermally and urea-denatured ribonuclease A monitored by time-resolved FTIR spectroscopy. *Biochemistry* 35:15822–15830, 1996.

9. D Moss, E Nabedryk, J Breton, W Mäntele. Redox-linked conformational changes in proteins detected by a combination of infrared spectroscopy and protein electrochemistry. Evaluation of the technique with cytochrome c. *Eur. J. Biochem.* 187:565–572, 1990.
10. M Lubben, K Gerwert. Redox FTIR difference spectroscopy using caged electrons reveals contributions of carboxyl groups to the catalytic mechanism of haem-copper oxidases. *FEBS Lett.* 397:303–307, 1996.
11. JE Baenziger, KW Miller, KJ Rothschild. Incorporation of the nicotinic acetylcholine receptor into planar multilamellar films: characterization by fluorescence and Fourier transform infrared difference spectroscopy. *Biophys. J.* 61:983–992, 1992.
12. JE Baenziger, KW Miller, MP McCarthy, KJ Rothschild. Probing conformational changes in the nicotinic acetylcholine receptor by Fourier transform infrared difference spectroscopy. *Biophys. J.* 62:64–66, 1992.
13. C Zscherp, A Barth. Reaction-induced infrared difference spectroscopy for the study of protein reaction mechanisms. *Biochemistry* 40: 1875–1883, 2001.
14. NJ Harrick. *Internal Reflection Spectroscopy.* New York: Wiley, 1967.
15. E Goormaghtigh, V Raussens, JM Ruysschaert. Attenuated total reflection infrared spectroscopy of proteins and lipids in biological membranes. *Biochim. Biophys. Acta* 1422:105–185, 1999.
16. UP Fringeli, HsH Günthard in *Membrane Spectroscopy* (E Grell, Ed.) pp. 270–332, Berlin: Springer-Verlag, 1981.
17. W Hübner, HH Mantsch. Orientation of specifically $^{13}C{=}O$ labeled phosphatidylcholine multi-bilayers from polarized attenuated total reflection FT-IR spectroscopy. *Biophys. J.* 59:1261–1272, 1991.
18. JE Baenziger, KW Miller, KJ Rothschild. Fourier transform infrared difference spectroscopy of the nicotinic acetylcholine receptor: evidence for specific protein structural changes upon desensitization. *Biochemistry* 32:5448–5454, 1993.
19. J Heberle, C Zscherp. ATR/FT-IR difference spectroscopy of biological matter with microsecond time resolution. *Appl. Spectrosc.* 50:588–596, 1996.
20. H Marrero, KJ Rothschild. Conformational changes in bacteriorhodopsin studied by infrared attenuated total reflection. *Biophys. J.* 52:629–635, 1987.
21. V Kane, JE Baenziger. The optimal size and geometry of attenuated total reflectance infrared accessories for studies of biological membranes. Manuscript in preparation.
22. SE Plunkett, RE Jonas, MS Braiman. Vibrational spectra of individual millimeter-size membrane patches using miniature infrared waveguides. *Biophys. J.* 73:2235–2240, 1997.
23. SE Ryan, CN Demers, JP Chew, JE Baenziger. Structural effects of neutral and anionic lipids on the nicotinic acetylcholine receptor. An infrared difference spectroscopy study. *J. Biol. Chem.* 271:24590–24597, 1996.
24. LK Tamm, SA Tatulian. Infrared spectroscopy of proteins and peptides in lipid bilayers. *Q. Rev. Biophys.* 30:365–429, 1997.
25. LK Tamm, HM McConnell. Supported phospholipid bilayers. *Biophys. J.* 47:105–113, 1985.
26. E Kalb, S Frey, LK Tamm. Formation of supported planar bilayers by fusion of vesicles to supported phospholipid monolayers. *Biochim. Biophys. Acta* 1103:307–316, 1992.
27. AA Brian, HM McConnell. Allogeneic stimulation of cytotoxic T cells by supported planar membranes. *Proc. Nat. Acad. Sci. U.S.A.* 81:6159–6163, 1984.
28. ML Wagner, LK Tamm. Tethered polymer-supported planar lipid bilayers for reconstitution of integral membrane proteins: silane-polyethyleneglycol-lipid as a cushion and covalent linker. *Biophys. J.* 79:1400–1414, 2000.

29. K Fahmy. Binding of transducin and transducin-derived peptides to rhodopsin studied by attenuated total reflection-Fourier transform infrared difference spectroscopy. *Biophys. J.* 75:1306–1318, 1998.
30. M Boncheva, H Vogel. Formation of stable polypeptide monolayers at interfaces: controlling molecular conformation and orientation. *Biophys. J.* 73:1056–1072, 1997.
31. Hofer, UP Fringeli. Structural investigation of biological material in aqueous environment by means of infrared-ATR spectroscopy. *Biophys. Struct. Mech.* 6:67–80, 1979.
32. H Otto, C Zscherp, B Borucki, MP Heyn. Time resolved polarized absorption spectroscopy with isotropically excited oriented purple membranes: The orientation of the electronic transition dipole moment of the chromophore in the O-intermediate of bacteriorhodopsin. *J. Phys. Chem.* 99:3847–3853, 1995.
33. JE Baenziger, BD Ritchie, SE Ryan, MP Blanton, EA McCardy. A Localized Structural Change in the Acetylcholine Receptor at Low Ionic Strength. Manuscript in preparation.
34. JE Baenziger, ML Morris, TE Darsaut, SE Ryan. Effect of membrane lipid composition on the conformational equilibria of the nicotinic acetylcholine receptor. *J. Biol. Chem.* 275:777–784, 2000.
35. SE Ryan, JE Baenziger. A structure-based approach to nicotinic receptor pharmacology. *Mol. Pharmacol.* 55:348–355, 1999.
36. SE Ryan, MP Blanton, JE Baenziger. A conformational intermediate between the resting and desensitized states of the nicotinic acetylcholine receptor. *J. Biol. Chem.* 276:4796–4803, 2001.
37. SE Ryan, HP Nguyen, JE Baenziger. Anesthetic-induced structural changes in the nicotinic acetylcholine receptor. *Toxicol. Lett.* 100–101:179–83, 1998.
38. JE Baenziger, JP Chew. Desensitization of the nicotinic acetylcholine receptor mainly involves a structural change in solvent-accessible regions of the polypeptide backbone. *Biochemistry* 36:3617–3624, 1997.
39. SE Ryan, DG Hill, JE Baenziger. Dissecting the chemistry of nicotinic receptor-ligand interactions with infrared difference spectroscopy. *J. Biol. Chem.* 277:10420–10426, 2002.
40. C Zscherp, J Heberle. Infrared difference spectra of the intermediates L, M, N, and O of the bacteriorhodopsin photoreaction obtained by time-resolved attenuated total reflection spectroscopy. *J. Phys. Chem.* 101:10542–10547, 1997.
41. C Zscherp, R Schlesinger, J Tittor, D Oesterhelt, J Heberle. *In situ* determination of transient pKa changes of internal amino acids of bacteriorhodopsin by using time-resolved attenuated total reflection Fourier-transform infrared spectroscopy. *Proc. Nat. Acad. Sci. U.S.A.* 96:5498–5503, 1999.
42. A Barth. The infrared absorption of amino acid side chains. *Prog. Biophys. Mol. Biol.* 74:141–173, 2000.

8 Step-Scan Time-Resolved FT-IR Spectroscopy of Biopolymers

Mark S. Braiman and YaoWu Xiao

CONTENTS

8.1 INTRODUCTION

The motions of biomolecules and other polymers are often central to what makes them interesting and useful. The general goal of time-resolved vibrational spectroscopy is to examine the dynamics of the chemically bonded structures within materials in a way that helps to explain and manipulate their fascinating behavior.

Biopolymers, specifically the proteins (and a few nucleic acids) that catalyze biochemical processes, have special dynamic properties that stand in stark contrast to most man-made polymers. Enzymes are capable of focusing energy from the environment onto an astoundingly small subset of their internal degrees of freedom, in order to speed up a specific type of dynamic process. Thus, it is not uncommon for the side chain of a single serine (or aspartic acid, or arginine) in a protein of several hundred amino acids to experience a transient change of >6 units in its pK_a, leading to a rapid change from being 100% protonated to 100% deprotonated. Such a large pK_a shift corresponds to a millionfold increase in the rate of deprotonation, compared to the rate for the same amino acid residue in other locations within the same protein. It also corresponds to a change in relative free energy, between the protonated and deprotonated forms of the COOH group, of well over 10 times kT.

Time-resolved FT-IR spectroscopy of proteins has generally focused on detecting changes in the vibrational spectrum arising from such localized changes in structure, against a much larger background of overlapping IR absorption bands from a rather invariant protein scaffold. One of the most interesting results of time-resolved FT-IR spectroscopy on proteins, in fact, is that such changes in pK_a can themselves occur in a very short time. For example, the pK_a of a particular aspartic acid in bacteriorhodopsin (bR) rises from ~2.5 to >11 in under 100 μs after the retinal chromophore is photoisomerized, and drops back to its starting value within the next 10 ms [1]. It is thought that these changes in proton affinity are brought about principally by repositioning of neighboring residues, such that several polar groups which stabilize the negative charge of the aspartate are first brought away from, and then back closer to, this particular residue, within each single-photon-initiated cycle.

The variety of structures assumed by proteins, and especially the focused changes in structure and reactivity that enzymes have evolved, contrast starkly with what chemists have been able to achieve with synthetic polymers. First, few synthetic copolymers involve anywhere near the variety of monomer building blocks — i.e., 20 different amino acids, found in proteins. Second, *in vitro* synthetic methods for even the most complex copolymers have only begun to approach the sequence specificity and accuracy of genetically encoded protein biosynthesis. Finally, neither intentional design of catalytic energy-focusing in polymers, nor human-driven combinatorial synthesis, has yet led to polymer catalysts whose specificity or rate enhancements approach those of enzymes.

Nevertheless, time-resolved FT-IR spectroscopy has also begun to find interesting applications for synthetic polymers. The dynamical properties that have received the most study by time-resolved FT-IR are the structural properties of the average monomer grouping within the polymer, specifically the responses of covalent bonds to external stresses such as application of an electric field [2–5] or rapid stretching [6]. The structural changes that are induced by such external stresses tend to be distributed over a large number of internal degrees of freedom. Even though the resulting small changes in structure tend to cause only small shifts of vibrational frequencies, in bands that are broad to begin with because of heterogeneity, detectability remains as high as in the case of proteins. This is because instead of only a single residue out of hundreds, a very large fraction of the homopolymer subunits are affected similarly by the initiating pulse of mechanical or electrical energy, and they all experience a similar change in their vibrational spectrum.

Over the past decade, commercial step-scan instrumentation and software for controlling time-resolved experiments has become available from at least 3 major instrument manufacturers: Bruker Instruments, ThermoElectron (Nicolet), and BioRad. This has opened up to a wide range of users the possibility of performing experiments with time resolution down into the range of a few ns. However, these experiments must still be customized for each particular type of sample to be investigated. There are many potential pitfalls that have so far limited the publication of TR–FT-IR results to a handful of research groups. The purpose of this chapter is to provide a guide to performing step-scan FT-IR experiments—especially on biological samples, but also with an eye towards polymer applications. The intention is to make this guide sufficiently complete and detailed that, in combination with manufacturer's user manuals, an experienced FT-IR spectroscopist could comfortably venture to perform such an experiment, and would furthermore stand a good chance of getting useful data without requiring additional outside assistance.

8.2 TIME-RESOLVED VIBRATIONAL SPECTROSCOPIC TECHNIQUES

Temporal resolution in vibrational spectroscopy can be obtained by any of a number of techniques for temporally separating three energetic events; e.g., the arrival of a trigger (such as a photolysis laser pulse) at the sample; the subsequent arrival of measuring photons; and the production of an electronic signal in the detector by the photons coming from the sample. The bulk of this chapter will deal in detail with one particular approach to time resolution that has become available with several commercial FT-IR spectrometers within the past few years: step-scan time-resolved FT-IR spectroscopy.

Almost all biological or polymeric materials have very strong IR absorptions. This means that relatively small samples (~1 mg or less) are more than sufficient for most IR spectroscopic measurements. Furthermore, the measuring IR light is photochemically inactive, and therefore nondestructive. However, IR vibrational spectroscopic measurements still present some difficulties. They require detecting the depletion of energy in a beam of IR light, under conditions where all room-temperature objects in the laboratory constantly emit substantial amounts of photons of overlapping wavelengths. On a photon-per-photon basis, this is intrinsically a more difficult process than measuring, e.g., Raman-scattered photons, which are typically detected against a dark background of <1 visible photon per second. Even with a liquid-N_2-cooled detector effectively shielded from ambient IR emission from room-temperature objects, the greatest source of error in an IR measurement is typically random thermal noise (dark background) from the 77K detector element itself.

Fortunately, use of a broadband black body source at ~1500K provides sufficient photons in the mid-IR range to overwhelm the fluctuations in the dark background and make it possible to detect even a tiny sample-induced depletion (<0.01% for a spectral slice of 4 cm^{-1}) within a matter of seconds. Overall performance of IR spectrometers was greatly enhanced a quarter-century ago by the introduction of Fourier-transform instruments. With these it is possible to detect photons over the entire mid-IR range (400–4000 cm^{-1}) simultaneously, by using an interferometer to modulate each IR wavelength at a different frequency. The encoding of the different IR frequencies by the interferometer is subsequently decoded by computation of the Fourier transform of the detected intensity (see Figure 8.1).

8.2.1 Step-Scan Time-Resolved FT-IR Spectroscopy

The use of an FT-IR spectrometer in time-resolved experiments presents a complication, in that the spectral separation device (the interferometer) produces a rapidly time-varying signal all on its own. Within the past 15 years, several technical problems associated with eliminating the problematic time dependence of the interferometer's mirror motion have been solved. The conclusion of these advances is that in step-scan mode, it is now routine to hold constant, to within <1 nm, the relative separation of the fixed and moving mirrors from the interferometer beamsplitter, for periods of several hundred ms or longer. This is long enough to trigger and measure a rapid chemical or biological process of interest.

In order to measure a complete FT-IR spectrum, then, the triggered dynamical process in the sample must be repeated identically at a series of different interferometer positions (see Figure 8.1 and 2). At each of these positions, the transient signal measured at the IR detector is stored as a function of delay time t after the trigger. After all required interferometer positions have been measured, the digitized transient data arrays are re-sorted to form a series of time-resolved interferograms. Each of these corresponds to a specific value of delay time t relative to the initiating trigger, but contains measured intensities at a large number of different values of interferometer optical retardation x. Each such time-dependent interferogram is then Fourier-transformed to give a time-resolved IR intensity spectrum. Such an intensity spectrum can easily be converted into an absorbance (or differential absorbance) spectrum if a reference (static) spectrum of the static (untriggered) sample has also been measured. Commercial step-scan instruments equipped with software and hardware to accomplish these tasks can now be obtained from at least 3 different

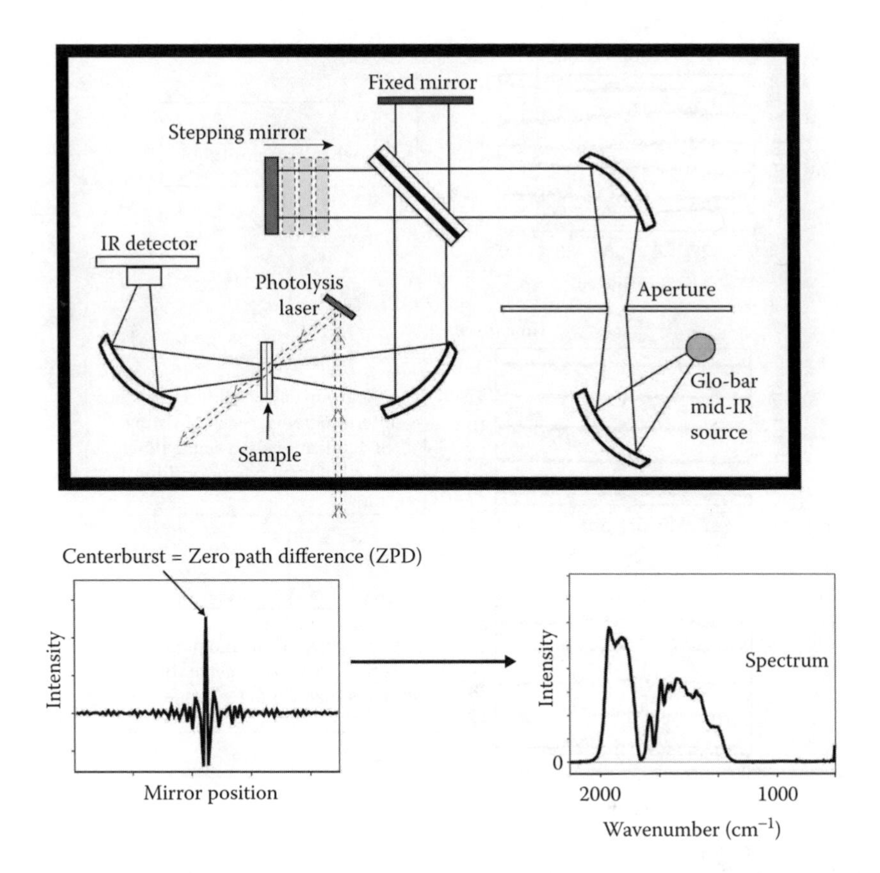

FIGURE 8.1 Schematic of spectral measurement by an FT-IR spectrometer. In the spectrometer (top), collimated broadband IR light is directed into an interferometer, in which a beam splitter divides the light along paths to fixed and moving mirrors. The recombination of light at the beam splitter results in a division of the light energy. One portion of the light is reflected through a sample and then to an IR detector. The other portion is returned in the direction of the source. The energy division between the two output beam directions is determined by the interference of the light reflected from the fixed and moving mirrors. For each wavelength of IR light, a sinusoidal variation in intensity is produced at the detector as a function of the optical retardation, which depends linearly on the moving mirror's position. The optical retardation is the difference in round-trip optical path length from the beam splitter to the fixed and moving mirrors. The interferogram signal — i.e., the measured intensity as a function of optical retardation (bottom left) — represents the superposition of all these sinusoidal patterns reaching the detector, each occurring with a spatial frequency equal to the corresponding IR wavenumber. The spectrum — i.e., the measured intensity as a function of wavenumber (bottom right) — can be recovered from the interferogram by means of a computational Fourier transformation. In a rapid-scan instrument, the interferogram is measured in a fraction of a second while the moving mirror is maintained at constant velocity. In a step-scan instrument, the optical retardation is held constant for individual periods of up to several seconds, then stepped by a fixed amount (typically the 633 nm wavelength of a reference HeNe laser) sequentially through all the positions needed to measure an interferogram.

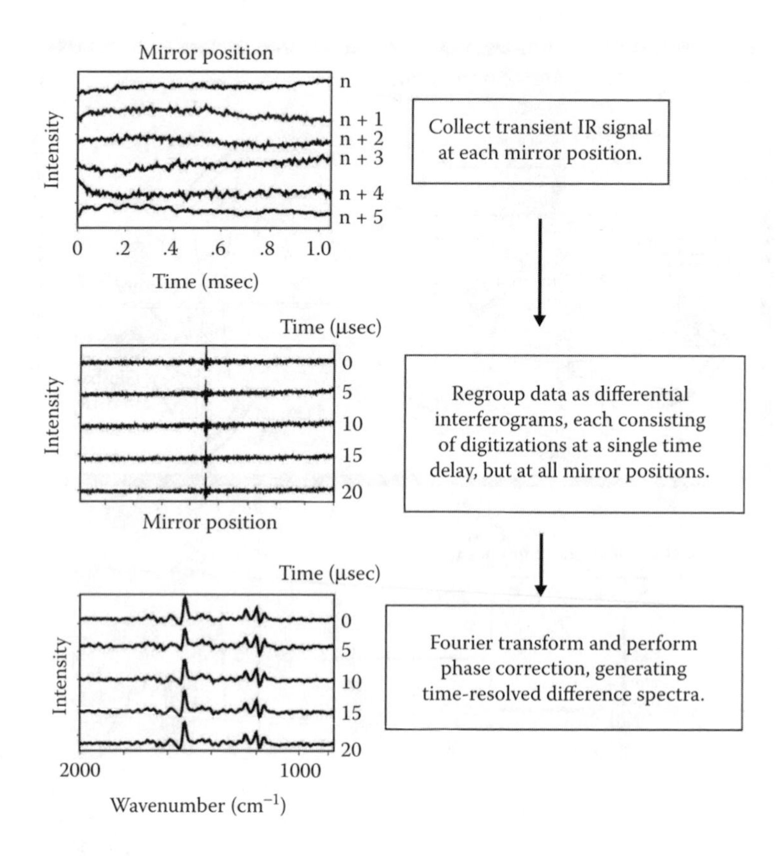

FIGURE 8.2 Measurement of step-scan TR–FT-IR spectra. The data collection itself (top) involves triggering repeatedly the same reaction in the sample, and after each trigger recording a transient IR signal at the detector. The interferometer's moving mirror is stepped sequentially to all required positions for an interferogram — typically 500–2000, depending on the desired spectral bandwidth and resolution. Post-trigger detector signals from only six such mirror positions, measured on an actual photolyzed sample of bacteriorhodopsin, are shown here. The transient signal is amplified, and then may be recorded in either an AC-coupled mode used here, in which the signal value immediately prior to the trigger has settled to 0, or in a DC-coupled mode, where the value before the trigger accurately represents the total light intensity passing through the sample. After data collection is complete, the digitized values from different mirror positions are grouped to form interferograms (middle), each corresponding to a particular time delay after the trigger. The time separation between these interferograms corresponds to the digitization spacing of the A/D converter used to collect the transients. Finally each of the time-resolved interferograms is converted into a time-resolved intensity spectrum by means of Fourier transformation (bottom). If the detector and digitizer are in AC-coupled mode, then these spectra represent difference spectra, as shown here.

manufacturers at prices of $50,000–$100,000, depending somewhat on the range of detector/digitizer time resolution (50 ns or faster) that is sought.

The requirement for hundreds or thousands of repeated triggerings of an identical process means that it is only possible to examine samples that have cyclic dynamical behavior. But there are ways to finesse this requirement for a triggered dynamic cycle if the sample doesn't display it naturally. A suitable triggered dynamic cycle can (at least in theory) be imposed on any sufficiently large uniform sample that is capable of being incrementally introduced into the sample beam and subjected to an initiating pulse of some sort.

8.2.2 Time Resolution Available with Rapid-Scan Instrumentation

While the main focus of this chapter is step-scan time-resolved FT-IR, several other methods need to be mentioned, at least for comparison. Prior to the development of commercial step-scan interferometers, several alternate approaches to time-resolved experiments were developed in which the constant-velocity scanning of the interferometer was maintained. The first class of experiments contains simply those with temporal resolution slower than the time required to obtain a single interferogram on a rapid-scan FT-IR. This time resolution is determined approximately as $\tau_{scan} = 1/(\Delta\bar{\nu} \times \upsilon_{max})$, where $\Delta\bar{\nu}$ is the spectral resolution and υ_{max} is the maximum mirror velocity. For commercially available instruments, τ_{scan} is about 5 ms for a spectrum with 8 cm^{-1} resolution.

This τ_{scan} is faster than the time interval τ_{repeat} for repeat scans, since the latter includes additional time required for mirror turn-around and retrace. Commercial software has been available for several decades that is capable of saving repeated spectral scans at rates of up to ~20 per second for 8 cm^{-1} resolution. This type of rapid-scan technology available for ~50 ms time resolution is generally mature and has been reviewed extensively, and is therefore well beyond the scope of this chapter. However, it should be noted that the recent development of ultra-rapid-scan instrumentation, utilizing rotational rather than linear motion, pushes the possible time resolution of rapid-scanning instruments down by a factor of approximately 50; i.e., into the range of 1 ms per spectrum. At the date of this review, there have been only a few applications of ultra-rapid-scan instruments in development to polymers [7], and applications to biological systems have only been contemplated. When these instruments become commercially available, however, it can be expected that growth in applications to biological systems may be rapid. This may be true especially if the instruments are equipped with stroboscopic software (see below) that allows division of the interferogram into several pieces that can be collected after separate triggers.

As a minor extension of rapid-scanning methods, a somewhat specialized FT-IR approach was developed to allow an improvement in time resolution from the ~50 ms value of τ_{repeat} to the ~5 ms value τ_{scan} for single-interferogram collection [8]. This method was termed rapid-sweep FT-IR spectroscopy, and mainly involved developing software to trigger the sample repeatedly on alternating mirror scans, with reference scans in between. The software then automatically averaged the differential (triggered-minus-reference) as well as the reference interferograms, and used them to calculate a single difference spectrum corresponding to a single ~5 ms window of time after photolysis. The delay between the initiating photolysis event and the 5 ms window could be varied at the start of each data collection. This approach was only applied to bacteriorhodopsin, which has a photocycle time of ~10 ms at room temperature, making it ideally suited for this approach.

8.2.3 Time Resolution Beyond the Rapid-Scan Limit

Since the 1980s and 1990s, several approaches have been developed for rapid-scan (i.e., non-step-scan) instruments that give time resolution shorter than the duration of a single interferogram scan τ_{scan}, a kind of "temporal super-resolution." With several of these approaches, the temporal resolution is given by an integer multiple of the fundamental time unit of the rapid-scan interferometer (τ_d, the interval between digitizations of the interferogram, given by the wavelength of the reference HeNe laser divided by the interferometer's retardation velocity). In particular, the sample is triggered multiple times within each mirror scan, at specific digitization positions of the moving mirror. This requires no modification of the standard interferometer control electronics built into commercial rapid-scan FT-IR instruments. However, τ_d is typically limited to the range 5–100 μs, due to engineering and cost constraints on the acceleration and velocity stabilization of the massive mirror. In theory, all of these methods could be implemented on any standard low-cost microprocessor-controlled rapid-scan FT-IR instrument with some inexpensive digital electronics (most importantly, a programmable pulse counter with a computer interface) as the only additional hardware requirements. In reality, the complexity of writing the requisite additional software correctly has proven to be a daunting task for most individual users. In the absence of a high level of user demand, the software that instrument manufacturers themselves have written to implement this type of time-resolved measurement has also generally been inadequate.

The simplest of the rapid-scan FT-IR approaches for temporal super-resolution is limited to time domains shorter than τ_d, and specifically to samples that are capable of undergoing full relaxation of their triggered dynamic cycle during this time interval. The sample's reaction cycle and a fast external time base are both initiated at intervals of τ_d, and then the detector signal is sampled rapidly using the external time base. With each mirror scan, an entire set of interferogram positions is measured for every desired time delay, so a single mirror scan suffices to obtain a full set of spectra covering all time delays. Repeated scans are required only for improving signal-to-noise ratio. The post-processing is straightforward; i.e., essentially the same as for step-scan measurements (see Figure 8.2). However, relatively few biological processes have a dynamic cycle that can be triggered this quickly, due to practical limitations. For samples covering the typical beam sizes in FT-IR spectrometers (0.25 cm^2), and for the resulting typical energy requirements for each trigger (>5 mJ per pulse), a repetition interval of 100 μs would require a prohibitively large constant average power of 50 W lasting over many hundreds or thousands of identical pulses. Not many lasers are up to this task, and even for nonphotonic triggers there is likely to be a difficult requirement for sample cooling. This form of time-resolved FT-IR experiment with rapid-scanning instrumentation has therefore generally been limited to photoreactions of small molecules, especially those in the gas phase [9]. For these, low background absorption reduces the sensitivity requirement that a large fraction of the sample undergo a reaction after each trigger. This permits the use of much lower trigger pulse energies than 5 mJ. There are only a few measurements on proteins (e.g., changes in $\nu_{C=O}$ upon carbon monoxide dissociation from myoglobin) where a short photorelaxation time and low background absorption over a limited spectral range invite high-repetition-rate experiments of this type.

8.2.4 Stroboscopic Time-Resolved FT-IR Spectroscopy

Most rapid-scan FT-IR experiments that are temporally "super-resolved" are generally categorized as stroboscopic methods. These involve the capture of only a portion of the interferogram from each mirror scan. The sample is triggered at only a select subset of digitization points during each mirror scan. This allows larger intervals between the triggers, to allow the sample to relax fully between them and also to permit the use of affordable high-pulse-energy solid-state laser triggers, such as the frequency-doubled pulsed Nd^{+}-YAG, that can easily provide the required 5 mJ/pulse at rates up to ~30 Hz. However, using this approach, a complete interferogram can be obtained only after multiple scans of the moving mirror. If the time resolution required is only in the 1–10 ms range; i.e., only a small factor down from the time for a single mirror scan, it is possible to keep the number of scans required for a complete interferogram quite low. On the other hand, if the time resolution is in the 10 μs range [10], then hundreds or even thousands of scans are required before an entire time-resolved interferogram is obtained. This in itself is not much of a problem, because such a large number of scans is typically required with biological samples in any case, in order to obtain adequate signal-to-noise ratios.

To permit implementation of stroboscopic experiments with maximal flexibility, simple pulse-counting firmware — along with somewhat complicated software — is needed to keep track of the temporal spacing between the initiating pulse and the digitizations, and to vary the points during each scan where the sample trigger is initiated. Until all required scans are completed, the software must store them separately, and at the end sort all the digitized data points into interferograms. An additional software requirement for biological and polymeric samples comes from the observation that especially with strongly absorbing samples, even a small amount of long-term absorbance baseline drift can result in some rather large artifacts. The only way that has been shown to reduce these artifacts to a manageable level is to interleave reference scans (without a sample trigger) in between scans with a sample trigger, then utilize the reference scans to calculate differential interferograms prior to sorting the points [11]. As a result, interpretable stroboscopic FT-IR experiments have been limited to a few careful practitioners using individually modified commercial software [10,12,13].

No commercial software for stroboscopic FT-IR has ever been available that properly interleaves scans with and without sample triggers as needed to reduce artifacts [11]. This is unfortunate, because with this necessary modification, there are numerous experiments on biological samples where stroboscopic measurements would likely be better than step-scan. In particular, with ~1 ms temporal resolution and covering the spectral range of 0–2000 cm^{-1} with 4–8 cm^{-1} resolution, useful stroboscopic experiments on biological samples can often be completed with only a few dozen sample triggers, whereas step-scan measurements covering the same bandwidth at the same spectral and temporal resolution require at least ~500 triggers. There are biological samples that have sufficient stability and repeatability (cyclicity) only for the smaller number of triggers. Commercial availability of properly written stroboscopic software would likely open up the field of biological time-resolved FT-IR to a somewhat larger variety of samples than can now practically be studied.

8.2.5 Other Approaches to Time-Resolved Vibrational Spectroscopy of Biological Samples

The step-scan and stroboscopic approaches described in the preceding paragraphs represent a class of experiments in which the source IR light is continuous. All time resolution is obtained through the use of a fast-responding detector, amplifier, and digitizer. However, there are at least two alternate methods of obtaining time resolution. One of the simplest is to move the sample past a pair of pump (photolysis) and probe (broadband FT-IR) beams of appropriately small dimension, or to use stopped-flow and/or rapid-mixing techniques if a ligand-gated reaction is being investigated. Due to the general requirement of thin samples, such techniques require carefully-engineered microchannel flow devices, but have been applied successfully to enzymatic reactions [14].

The alternate approach to time resolution is to pulse both the trigger and the measuring IR photons with a known time separation between them. In this case, it is necessary to record only the time-averaged IR signal from the detector in order to obtain time resolution for biological IR experiments down into the ps or even sub-ps range. The leading examples of this approach have been obtained by using IR pulses generated by lasers [15,16]. Detection of the relatively low average intensities of the transmitted ps-pulsed IR light is facilitated by a frequency up-conversion scheme that takes advantage of its high peak intensity. An alternate approach [17] is to use pulsed broadband IR light from a synchrotron at one of the nationally or internationally sponsored user facilities; such pulses are as short as several tens of picoseconds. In order to obtain accurate ps timing, the delay between pump and probe pulses must be generated by an optical delay line rather than electronically. However, the biggest range of delays that can be accurately generated in this fashion is on the order of 1 ns, corresponding to ~0.3 m of optical delay.

All these sub-ns methods, while they can take advantage of standard spectrometers and detectors, have so far required very costly specialized equipment for the source IR light. So far, they have also been limited to samples that can be photolyzed at extremely high repetition rates (thousands or millions of photolysis pulses per second), because only with such repetition rates do the average numbers of photons available start to approach those available from a broadband blackbody source. Physiological insights from sub-ns experiments have naturally tended to focus on local relaxations of chromophoric groups following their photoexcitation. Such ultrafast IR experiments have been reviewed recently in another book from this publisher [18], and are beyond the scope of this chapter to describe in detail.

Time-resolved Raman vibrational spectroscopy, which has generally provided the fastest time resolution of all vibrational techniques, is also outside the realm of this chapter. A disadvantage of Raman spectroscopy for examining protein samples is that the groups most likely to be involved in reaction mechanisms — the acids (glutamic and aspartic, as well as cysteine) and bases (lysine and histidine), and the predominant nucleophiles (cysteine, serine, and threonine) — tend to have much weaker Raman signals than the aromatic residues (tryptophan, phenylalanine, and tyrosine). Useful Raman experiments on proteins therefore tend to be limited to those that involve tyrosine (or rarely the other aromatic residues) in their mechanism; or that bind substrates or cofactors with visible or near-UV absorption bands. Nevertheless, step-scan time-resolved FT-Raman experiments using a continuous near-IR Raman excitation laser are feasible [19]. Those who wish to adapt step-scan instruments for performing such experiments with time resolution of nanoseconds to milliseconds may find some of the information in this chapter applicable.

8.3 APPLICATIONS

For typical biopolymer samples, the expected differential IR absorption signals from any triggered reaction are expected to be a small fraction of the background absorption. As will be seen below in Section 8.4.1, the small size of the differential IR absorption signals from such samples makes it impossible to obtain interpretable absorbance difference spectra with sub-ms time resolution after a single trigger. Instead, it is necessary to average the results from large numbers (typically hundreds or thousands) of identically triggered events. This limits somewhat the variety of samples that have so far been studied with step-scan TR–FT-IR. The vast majority have involved reactions that are triggered by pulsed photolysis lasers, using an intrinsic biological chromophore to absorb the visible light. These will be reviewed first, followed by the smaller class of experiments that involve using a laser pulse to heat the sample or to release a "caged" ligand.

8.3.1 Photocyclic Reactions of Naturally Occuring Chromophores

For reasons explained above, TR–FT-IR measurements typically require a sample that can undergo the same reaction repeatedly many hundreds or thousands of times. The number of times that a system can undergo the same reaction without irreversibly changing is known as the cyclicity. In addition to high cyclicity, another criterion that has determined which protein reactions have been studied most successfully with FT-IR is availability of protein that is both pure and has a relatively low molecular weight. Finally, it is helpful for the biological system to be susceptible to a trigger that is fast, not too costly, and unlikely to perturb the baseline mechanical, electronic, or optical operations of an FT-IR spectrometer. Because they fit these criteria well, biological transducers of solar energy — in particular, the chlorophyll-based photosynthetic reaction centers and the retinal-based microbial rhodopsins— are among the most ideal samples for TR–FT-IR. Likewise, photoactive yellow protein (PYP) has been shown to undergo a reversible photoreaction upon illumination with a laser pulse.

Rounding out these systems, a variety of heme proteins that bind carbon monoxide (CO) quite tightly have been shown to undergo light-triggered release from the heme iron, followed by a multiphase recombination process that sheds some light on the accessibility of these proteins' active sites to the heme group's principal physiological ligand (O_2).

8.3.1.1 Retinal Proteins

In addition to the light-driven ion pumps bacteriorhodopsin (bR), proteorhodopsin (pR), and halorhodopsin (hR), there are two families of sensory rhodopsins (sRI and sRII) that utilize a retinylidene Schiff base chromophore related to vitamin A, and undergo photoreactions with high cyclicity. The highest is obtained with the light-driven proton pump, bR, which was among the first biological samples to be studied with step-scan TR–FT-IR methods [21]. Retinal proteins remain at the forefront of developments that advance the collection and interpretation of TR–FT-IR spectroscopy because of the large number of photons (>300,000) that can be absorbed by the bR chromophore on average before it is bleached or functionality of the protein changes.

8.3.1.1.1 Bacteriorhodopsin (bR)

Many important technical developments in step-scan FT-IR of biological systems were first demonstrated using bR as a sample. This included the first demonstrations of sub-μs time resolution with custom-built [21] and commercial [23] step-scan instruments. Likewise, bR was the first sample to be used successfully with attenuated total reflection (ATR) spectroscopy combined with the laser-induced step-scan time-resolved mode [22], permitting for the first time the use of sample buffers with well-defined compositions.

An example of the richness of the time-dependent spectral information that can be examined in a single ns step-scan TR–FT-IR measurement on bR is shown in Figure 8.3 and 4. Such step-scan experiments have deepened understanding of the linear photocycle of bR, which is now understood to contain at least six intermediates distinguishable by step-scan FT-IR spectroscopy under physiological conditions, according to the reaction scheme bR $\rightarrow K_E \rightarrow K_L \rightarrow L \rightarrow M_1 \rightarrow M_2 \rightarrow N \rightarrow O \rightarrow$ bR. Of these transitions, the $K_E \rightarrow K_L$ transition, which takes place on the ~50 ns time scale at room temperature, was first characterized by step-scan TR–FT-IR [24], and only months later by Raman spectroscopy [25]. Likewise, step-scan TR–FT-IR measurements were crucial for characterizing the $M_1 \rightarrow M_2$ transition [26,27], since the chromophore structure in these two intermediates is evidently too similar to be unambiguously distinguished by either UV-visible or resonance Raman spectroscopic methods.

Earlier stroboscopic and low-temperature time-resolved FT-IR measurements, along with UV-visible spectroscopy, had already clearly established accurate time constants for the other transitions, and had been used in conjunction with extensive isotope labeling and site-directed mutagenesis to assign vibrational difference features, particularly to carboxylic acid groups that undergo transient protonation changes as part of the H^+-translocation mechanism of bR. Dozens of step-scan TR–FT-IR publications on bR and its site-directed mutants have complemented these measurements, and are now too numerous to fit within the scope of this chapter. Selected recent publications representing physiologically important themes have been: (1) The use of ATR measurements with well-buffered systems to define the pK_a values for individual carboxylic acid side chains in the photointermediate states [28,29]; (2) detecting the transient formation of H-bonding networks within the protein along

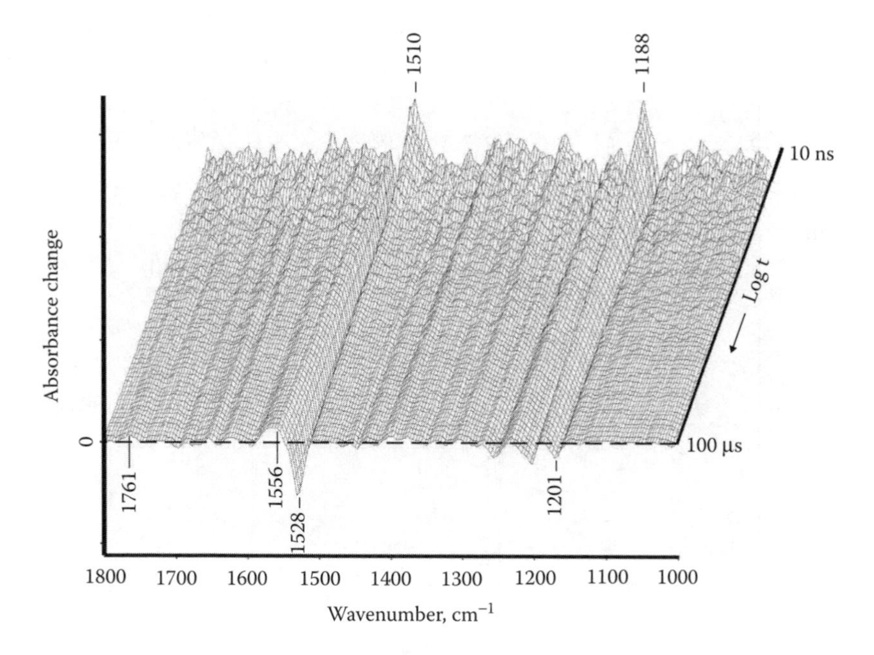

FIGURE 8.3 TR–FT-IR difference spectra of bacteriorhodopsin (bR) from ~8 ns to 100 μs after photolysis, obtained in a single experiment by storing the data with quasi-logarithmic time spacing. AC-coupled step-scan interferograms measured by a Kolmar HgMnTe detector (nominally with 20 MHz bandwidth) were originally digitized with 8 ns spacing, then group-averaged using OMNIC software on the Nicolet 860 spectrometer. Spectral resolution is 8 cm^{-1}. Interferograms obtained with a total of ~100,000 laser flashes were averaged together (S.V. Shilov and M.S. Braiman, unpublished data).

which protons can be conducted [30,31]; and (3) identifying the proton release group; i.e., the chemical group that deprotonates during the bR → M reaction, in addition to the retinylidene Schiff base (which merely transfers its proton internally to its asp85 counterion), and can thereby account for the observed 1:1 stoichiometric proton release from bR to the cytoplasmic medium [32,33,80,].

Several challenges have made the unambiguous identification of a proton release group particularly difficult [28]: First, in the wild type bR this group is almost certainly not an aspartic or glutamic acid, each of which has an easily identified vibrational signature upon deprotonation that is simply not present. Second, it is clear that site-directed mutations can alter the proton release group, with it sometimes ending up as a glutamic or aspartic acid. Finally, the conserved group most strongly implicated in proton release is the guanidino group of arg82, and the strongest characteristic vibrations of protonated and deprotonated arginine happen to overlap strongly with the characteristic amide I and amide II bands of the peptide backbone. Unambiguous assignment of vibrational bands due to arg82 in the bR → M difference spectrum has awaited the development of rather novel isotope labeling methods, described in greater detail in Section 8.4.2.2.

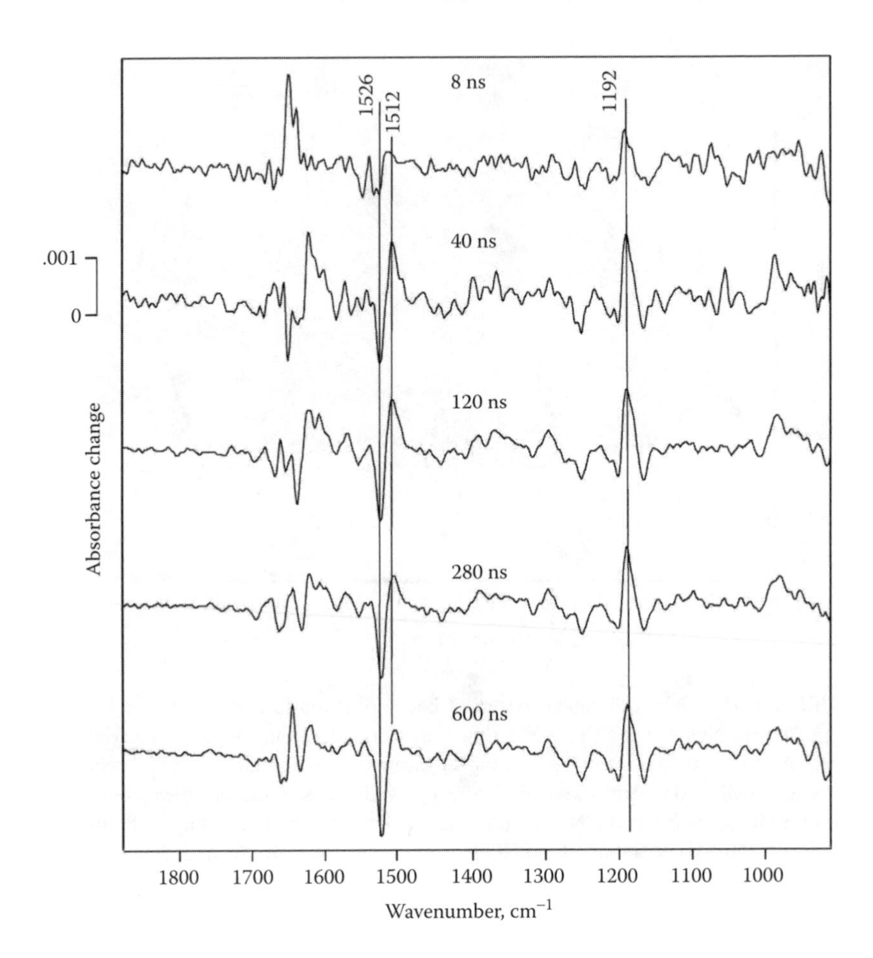

FIGURE 8.4 Several of the spectra at the earliest time points from the data set shown in Figure 8.8, separated to show the SNR and the spectral evolution more clearly. IR difference bands between the unphotolyzed (bR) state and its K photoproduct — e.g., the 1512 cm^{-1} and 1192 cm^{-1} bands characteristic of K — grow in with approximately the response time of the detector (~30 ns). (S. V. Shilov and M. S. Braiman, unpublished data).

8.3.1.1.2 Proteorhodopsin (pR)

Over 780 genes for a widely varied group of proteins named proteorhodopsins — homologs of bR that may also function as proton pumps — have been discovered within the past 5 years in numerous species of oceanic bacteria throughout the world. Step-scan TR–FT-IR difference spectra of pR on the 0.010–100 ms range have been obtained by several groups [34–36]. All of these experimenters were able to obtain step-scan TR–FT-IR difference spectra of an M-like photoproduct that closely resembled analogous spectra from the bR photocycle, but only the spectra of Xiao and Braiman [36] were additionally able to demonstrate the clear presence of an L-like precursor to M, as evidenced by a transient negative vibrational band near 1740 cm^{-1}. Both the observation of the L species and the fast (<100 μs) rise of the M species could be observed only under a particular choice of sample conditions that included a rather high pH value of 9.5 [36]. However, the similarity of the pR photocycle intermediate structures occurring under these conditions to those that are observed in bR at lower pH is intriguing. As pR has only been expressed heterologously in *E. coli*, and has never been isolated from its natural source organism, it is as yet unclear as to whether any of the measurements on it can be directly extrapolated to its true physiological conditions.

8.3.1.1.3 Halorhodopsin (hR)

Like bR, the light-driven chloride pump hR is also found in many species of archaea, and exhibits a photocycle on a similar time scale as that of bR. However, no M-like intermediate with a deprotonated Schiff base chromophore has been identified. The normal chromophore counterion of bR, asp85 (conserved as asp97 in pR), is not present to stabilize the positive charge on the protonated Schiff base. Instead, charge balance is maintained in the active site by the binding of a halide ion, Cl^- under usual conditions, at a location that is accessible only from the external medium. Photoisomerization of the chromophore leads to destabilization of the initial binding site of the halide, resulting in its repositioning at a site accessible only the internal medium. Resetting of the chromophore's isomeric state is accompanied by release of the chloride to the intracellular medium, followed by a slower re-uptake of chloride from the external medium.

Step-scan TR–FT-IR measurements of hR have included sub-μs measurements which demonstrated the presence of an $hK_E \rightarrow hK_L$ transition analogous to the $K_E \rightarrow K_L$ transition in the bR photocycle [37]. More recently, Hutson and Braiman [38] examined the kinetics of the hR photoreaction on the μs time scale. They concluded that the subsequent $hK_L \rightarrow L$ transition is thermally reversible, as evidenced by the persistence of hK_L features in the spectrum for >300 μs, followed by their decay concomitant with the L spectral features (see Figure 8.5).

8.3.1.1.4 Sensory Rhodopsin II (sRII)

In addition to bR and hR, halobacteria have two other retinal proteins, sensory rhodopsins I and II, that mediate photoattractive and photophobic responses, respectively. They exhibit more than tenfold longer photocycles than bR and hR, and are less stable to repeated laser flashes (i.e., they have lower cyclicity). The latter especially makes collection of step-scan time-resolved spectra somewhat challenging. Nevertheless, step-scan TR–FT-IR measurements have recently been reported [39,40], demonstrating the presence of K, L, M, and O-like intermediates in their resemblance to the bR photocycle. More recently, a site-directed mutation of sRII was investigated that eliminates the proton acceptor group and thus blocks the deprotonation of the Schiff base; i.e., of M [41]. These workers showed nevertheless that other protein structural changes were still produced upon photolysis of the mutant, that were similar to those that occur in the wild type M intermediate.

8.3.1.2 Heme Proteins

After bR, hemoglobin was the second protein to be studied with step-scan TR–FT-IR [42]. The toxicity of carbon monoxide is mediated partly through its binding to hemoglobin's prosthetic group. It has been known for some time that the CO is

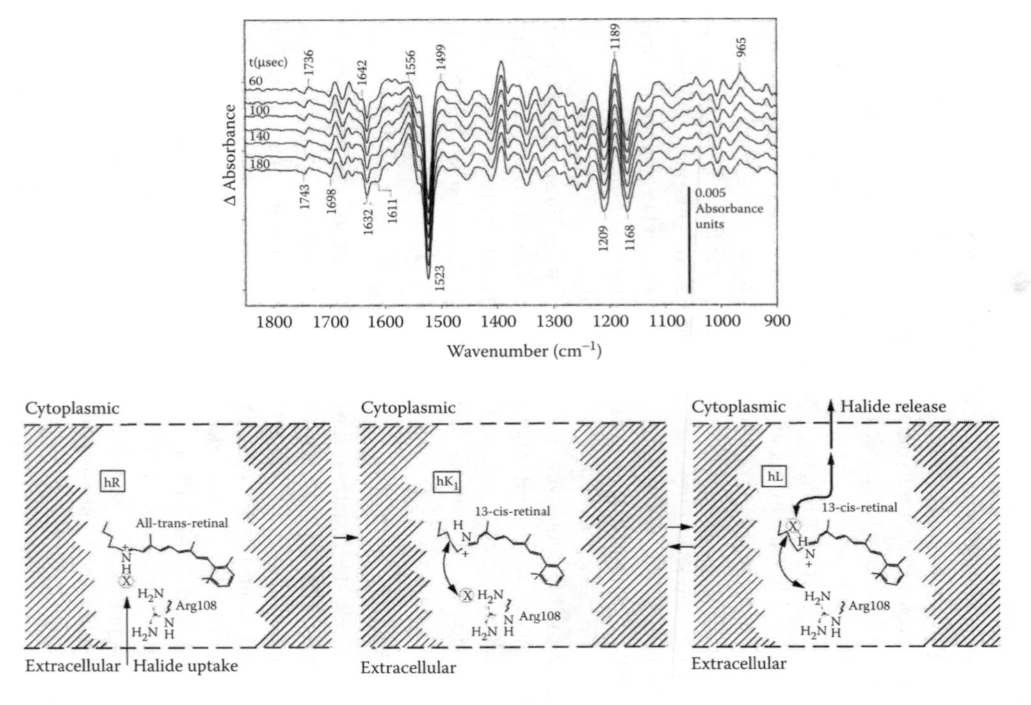

FIGURE 8.5 Upper panel: Microsecond time-resolved FT-IR difference spectra of halorhodopsin (hR), in a time range where its hK_L and hL photoproducts predominate. In each difference spectrum, the baseline near 1800 cm^{-1} represents $\Delta A = 0$. Spectra were originally recorded with a digitizer time resolution of 10 μs, and subsequently were averaged pairwise to give a final time resolution of 20 μs. The band at 965 cm^{-1} is a hydrogen out-of-plane mode that, like the C=C stretch vibration at 1499 cm^{-1}, is characteristic of the twisted retinylidene chromphore of the hK_L photoproduct. The persistence of these characteristic hK_L bands into the time range past 100 μs is a strong indicator of a back-reaction from hL to hK_L, a phenomenon that distinguishes the halide transport mechanism of hR from proton transport by bR. Lower panel: Model for halide movements proposed to account for the observed kinetics. Reproduced from M. S. Hutson, R. Krebs, S. V. Shilov, & M. S. Braiman. *Biophys. J. 80*, 1452–1465 (2001).

released from the heme group when the latter is photolyzed with visible light, and initially remains nearby within the protein binding pocket. The CO then undergoes several possible processes, including rapid rebinding to the heme (termed "geminate" rebinding), or random diffusion out from the protein followed by much slower secondary rebinding. Frei and coworkers showed that it was possible to associate changes in $\nu_{C=O}$ with the CO environment during each of the steps, and to correlate their kinetics with those of heme-group absorption bands [42]. More recently, others have investigated the H93G mutant of myoglobin, in which the native histidine ligand of the heme group is removed, with TR–FT-IR spectroscopy [43]. They demonstrated that in the CO-bound form of this mutant, an alternate histidine is able to substitute as the iron ligand this mutant, but that within ~5 μs after photolysis, a protein conformational change occurs that results in replacement of this substitute internal histidine ligand by a water molecule.

The photolytically driven release of CO has been exploited as a trigger for the more complex biochemical reactions of other heme proteins in numerous step-scan TR–FT-IR measurements. While a few have examined the liver detoxification enzymer P450Cam [44,45], the bulk have been directed at understanding the processes of electron and H^+ transfers in cytochrome *c* oxidase and closely related proteins, collectively known as the heme-copper oxidases. These membrane-bound protein complexes perform the terminal electron-transfer steps in the respiratory chain in mitochondria, bacteria, and archaea; and couple the four-electron reduction of O_2 to H_2O (using four identical one-electron reductant molecules sequentially) to the pumping of protons against an electrochemical gradient. The free energy stored in this process is used by the mitochondria or bacteria to provide the bulk of cellular ATP under aerobic conditions.

The physiological functioning of cytochrome *c* oxidase is quite complicated, consisting of four chemically distinct reactions involving one-electron redox intermediates between O_2 and water, each of which remains heme-bound. Step-scan TR–FT-IR applications to these enzymes (reviewed recently [47,46]) are so far generally limited to investigation of the structure and dynamics of the heme $a3$-Cu_B binuclear center and the coupled protein structural changes following photodissociation of CO from heme and its subsequent binding to and dissociation from Cu_B. While this photochemically triggered process is not itself physiological, it provides physiologically relevant insights into the dynamics of small ligand molecules within the protein active site, and coupling to conformational changes of the protein. A particular advantage of this approach is that CO ligand (unlike the native O_2 of later intermediates that it replaces in the active site) is a very strong IR absorber in a region where no other sample components absorb, so changes in its environment can easily be monitored.

Upon photolysis of the CO adduct of the fully reduced bovine oxidase, picosecond IR measurements shows that the CO moves from heme $a3$ to Cu_B, a distance of about 5 Å, in <1 ps [48,49,50]. The CO photodissociation is directly observed by means of the transient bleach of the Fe^{2+}_{a3}-CO absorption at 1963 cm^{-1}, with concomitant appearance of the Cu^{+}_{B}-CO absorption at near 2065 cm^{-1}. After about 1–2 μs, the CO dissociates from Cu_B and equilibrates with the bulk solution. This process is attributed to a conformational change at heme $a3$ [50]. The CO from the decaying Cu_B^+-CO complex can either diffuse into the external buffer solution, or rebind to heme *a*3. Disappearance of the transient ~2065 cm^{-1} band and appearance of a new one at ~1955 cm^{-1} are indicative of the decay of the Cu_B^{1+}-CO complex and reformation of the Fe_{a3}^{2+}-CO complex, respectively, in *cbb*(3)-type cytochrome *c* oxidase from *Pseudomonas stutzeri* [51]. For other types of heme-copper oxidases from a variety of organisms, formation and disassociation of the transient Cu_B^{1+}-CO complex have also been observed, with somewhat different time scales [52–56]. In several of these cases, specific types of CO ligand motion — i.e., its escape and re-entry through docking sites and channels — have been proposed.

IR investigations within the ν_{CO} frequency range are relatively easy, as a rather thick sample can be used to increase the size of transient absorbance changes because the protein itself as well as the associated aqueous buffer are both almost transparent in this region. In order to expand the frequency range to include characteristic transient difference bands of the protein itself, extreme care must be taken to obtain a concentrated protein sample with a path length <10 μm. With such samples, significant progress has been made to illustrate structural changes within the large cytochrome *c* oxidase protein at early stages following heme-CO photolysis. More specifically, changes happening in or near the three proton transfer pathways, D, K, and Q, have been detected.

For example, changes in COOH groups from aspartic and glutamic acids occur in synchrony with changes in ν_{CO} [57,58,59]. One of the spectroscopic shifts suggests a change in hydrogen bonding of the C=O bond of a conserved glutamic acid coincident with CO dissociation from Cu_B [58]. This highly conserved residue (E242 in bovine mitochondrial cytochrome *c* oxidase), with a $pK_a \geq 9.5$, is situated ~10 Å from the active *a*3-Cu_B binuclear center and ~25 Å from the start of the D-channel. It is thought that protons from the mitochondrial matrix (or bacterial cytoplasm) pass via this glutamic acid to the heme-copper active site as chemical protons (used to make H_2O at the active site) and subsequently to the external medium as pumped protons. Despite the important role of the conserved glutamate, CO photolysis from the fully reduced oxidase surprisingly does not result in changes in this gluresidue's environment in oxidases from *P. denitrificans* [57] or *R. sphaeroides* oxidase [59]. It is therefore hypothesized that in these cases, the functional role of E242 from the bovine mitochondrial cytochrome *c* oxidase is instead played by other groups that have potentially labile protons and occupy a similar beneficial location.

TR–FT-IR also shows evidence for perturbation of a second carboxylic acid side chain at higher frequency [59]. This band is thought to be due to the perturbation of glu62 in cytochrome *c* oxidase (glu89 in cytochrome *bo*3), a residue near the opening of the proton K-channel and required for enzyme function. The third proton pathway (Q) has been identified only in type *ba*3-cytochrome *c* oxidase from *Thermus thermophilus*, and consists of residues that are highly conserved in all structurally known heme-copper oxidases. The Q pathway starts from the cytoplasmic side of the membrane and leads through the propionate of the heme *a*3 pyrrol ring A to asp-372 and then to the water pool. TR–FT-IR spectroscopy indicates that photolysis of CO from heme *a*3 and its transient binding to Cu_B is kinetically linked to spectral features that can be tentatively attributed to a change in the ring A propionate of heme a3 and to deprotonation of asp-372 [60,61]. The ring A propionate of heme *a*3 is therefore hypothesized to be an important proton carrier to the exit or output proton channel as part of the Q-pathway.

8.3.1.3 Photosynthetic Reaction Centers

Photosynthetic reaction centers are generally defined as the smallest isolatable complex that can perform the initial charge separation steps of photosynthesis. Among the three principal types of reaction center, the bacterial reaction center is simplest and is thus somewhat more amenable to TR–FT-IR experiments than photosystem I or photosystem II. (However, very recent step-scan FT-IR measurements on photosystem I from cyanobacteria have been used to investigate the reduction of its A_1 chlorophyll [62].) In the bacterial reaction center from *Rhodobacter sphaeroides*, light excitation induces an intramembrane electron transfer originating at the primary electron donor P (bacteriochlorophyll *a* dimer, also known as the "special pair" of bacteriochlorophylls), proceeding through Q_A (the primary acceptor quinone), and terminating at Q_B (the secondary acceptor quinone), which can dissociate from the reaction center and diffuse throughout the lipid membrane.

Step-scan TR–FT-IR spectroscopy of bacterial reaction centers was applied initially to perform kinetic analysis of $Q_A^-/Q_B \rightarrow Q_A/Q_B^-$ vibrational difference spectra [63], particularly relying on several quinone bands that had previously been partially assigned based on static difference spectroscopy. Bands in the spectral region of the C=O stretching mode of carboxylic acid residues were also resolved at 1730, 1719, and 1704 cm^{-1}. A global fit analysis yielded two exponentials with half times of 150 μs and 1.2 ms, in agreement with previous time-resolved IR studies using tunable diode lasers. More recently, coupling of these reactions to protonation changes of groups in the L subunit (specifically asp210, glu212, his126, and his128) was inferred based on IR bands assignable to the latter groups [64]. Coupling of quinone oxidation to perturbations of the membrane lipid IR bands was also detected with step-scan FT-IR [65,66].

8.3.1.4 Photoactive Yellow Protein

PYP is a cytoplasmic blue-light receptor from purple bacteria, in which it appears to serve as the sensor for negative phototaxis. In its structure, the *p*-hydroxycinnamoyl chromophore is covalently linked by a thioester to the unique cysteine within the protein sequence, and is H-bonded at its phenolate oxygen to the side chain of glu-46. Upon absorption of a photon by the chromophore, PYP undergoes a cyclic sequence of reactions with at least two intermediates. The ground state P (peak absorption at 446 nm) is converted into the intermediate I_1 (peak absorption at 465 nm) in several ns. The intermediate I_1 (also called pR or PYP_L in the literature) is then converted into the long-lived blue-shifted intermediate I_2 (peak absorption of 350 nm, also called pB or PYP_M), which returns to the ground state P. Time-resolved step-scan FT-IR spectroscopy has been applied to investigate the both the structure of I_1 intermediate and the $I_1 \rightarrow I_2$ conversion.

In the first published example [67], the 1-μs difference spectrum after the laser flash shows characteristic bands corresponding to the $P \rightarrow I_1$ conversion. This reaction is mainly characterized by chromophore isomerization, without protein fully responding to this event on the ns time scale. In the $P \rightarrow I_1$ difference spectrum, a band pair at 1739(−)/1730(+) cm^{-1} could be assigned to glu-46, based on E46Q mutagenesis studies. This indicates that glu-46 is protonated in the I_1 state and its H-bonding with chromophore's phenolate oxygen has been preserved and strengthened. The preservation of H-bond between glu-46 and chromophore in I_1 intermediate indicates that the chromophore isomerization is achieved by flipping not the aromatic ring, but the thioester linkage between the chromophore and the protein.

In the $P \rightarrow I_2$ spectrum, corresponding to the difference spectrum measured 10 ms after the laser flash, strong positive bands at 1606 cm^{-1} and 1574 cm^{-1} could be assigned at least in part to coupled C−C and C=C strength vibrations of the chromophore's aromatic ring and vinyl group, indicating that the chromophore is protonated with a cis configuration in I_2. The $I_1 \rightarrow I_2$ conversion is characterized as a biphasic process. The first phase includes two parallel events with a same time constant of 100–200 μs: movement of glu-46 into a hydrophobic environment in about one quarter of the PYP molecules, as indicated by the rise and intensity of the 1759 cm^{-1} band assignable to glu-46; and deprotonation of glu-46 in the main PYP population. The second phase with a time constant of 1–2 ms involves further deprotonation of glu-46 in the minority population. Both steps, but especially the second one described by the 1–2 ms time constant, involve a significant movement of the protein backbone. In fact, the extremely large band pair at 1645(−)/1624(+) cm^{-1} in the $P \rightarrow I_2$ difference spectrum is calculated to attribute ~2.5% of PYP's backbone absorption, larger than any other photoreceptors.

A more recent TR–FT-IR spectral examination of PYP [68] confirms the above two results. Specifically, the H-bond between neutral glu-46 and phenolate oxygen of the chromophore is preserved in the I_1 intermediate; and the $I_1 \rightarrow I_2$ conversion is a biphasic transition. However, there is a small difference in the explanation provided for the biphasic transition. Without observation of the positive 1759 cm^{-1} band during I_2 formation, the first phase predominantly encompasses chromophore protonation and glu-64 ionization, and the second primarily involves large conformational changes in PYP. Deprotonation of glu-46 in the first phase results in a negatively charged carboxylate, which is embedded in an energetically unfavorable hydrophobic environment. The destabilizing electrostatic energy of this single buried COO^- group is sufficiently strong to drive large conformational changes, as happens in the second phase.

8.3.2 Non-Chromophoric Protein Reactions Studied with Step-Scan TR–FT-IR

Several different biological processes that are not intrinsically photochemical in nature have nevertheless been adapted to pulsed laser triggers. One type of trigger that has been exploited considerably is a rapid temperature jump induced by illumination of an aqueous sample by near-infrared light [69]. This trigger was exploited first to study fast protein folding and unfolding reactions. Both helical [70,72] and non-helical [71] proteins have been examined and found to exhibit the first steps of their unfolding reactions on a sub-μs time scale. The pulsed-laser T-jump approach has also been used to study conformational transitions of bR [73].

8.4 EXPERIMENTAL CONSIDERATIONS: A PRACTICAL GUIDE

The next three-part section of this chapter will be focused on a rather detailed discussion of issues that face the experimenter interested in measuring a time-resolved vibrational spectrum using a step-scan FT-IR instrument. The first part focuses on helping the user anticipate the types of noise and other errors that will be present in a typical TR–FT-IR spectrum, because a thorough understanding of such errors is a prerequisite for reducing it to a level that permits accurate interpretations.

The second part deals with selection of instrumental measurement parameters. Commercial software for implementing these experiments requires the selection of a large number of parameters. First-time use of such software, even by someone with a good deal of experience with basic FT-IR measurements, can be bewildering. One of the biggest challenges is that there are a large number of user-selected parameters for which even a small deviation from the optimum can yield an uninterpretable result. The goal is to provide a guide for selection of these parameters, informed by the prior discussion about sources of errors, that may be useful regardless of which manufacturer's software is being used. Finally, the third part deals with non-instrumental considerations, with some advice about sample preparation, photolysis lasers and optical filtering.

8.4.1 Sources of Noise and Other Errors

The intrinsic time resolution of FT-IR spectroscopy on biological and polymeric materials — the time required for IR light energy to be absorbed and converted to random thermal energy — is well below 1 ps, much faster than NMR or EPR spectroscopy. However, this does not mean that it is practical to perform an interpretable sub-ps time-resolved FT-IR experiment on most samples. The principal limitation on time resolution is whether the signal-to-noise ratio will be adequate to answer an interesting question.

Unfortunately the prediction of signal-to-noise ratio in time-resolved FT-IR spectroscopy depends on many factors, ranging from the sample features such as size and cyclicity (stability to repeated reaction cycles), to instrumental parameters such as the bandwidth and spectral resolution needed to observe desired vibrational features. Understanding the tradeoffs available in these parameters is absolutely crucial to being able to set up an experiment that will work correctly. A poor experimental setup that results in a doubling of the noise can be the difference between an experiment that answers a question and one that is merely tantalizing.

What follows is a quick guide to sources of noise and other errors in time-resolved FT-IR experiments. Many of these are similar to sources of error in standard FT-IR experiments, and discussion of these will assume a good background in the general subject, as presented in existing textbooks [74]. Even armed with such general knowledge, however, there are some novel situations and subtle difficulties that should be considered prior to attempting to select parameters for a time-resolved FT-IR measurement.

8.4.1.1 Detector Noise

As stated earlier in this chapter, the single most important source of noise in FT-IR measurements is typically thermal noise in the detector. Within the limited scope of this chapter, we focus on experiments utilizing the broadband sources and semiconductor detector devices found in commercial FT-IR spectrometers. The detectors are typically designed and sold in a modular fashion, each coming with its own matched preamplifier which can be plugged into a standard bus on the spectrometer that provides DC power (typically ±15V) and accepts a standard output of ±12 V. This output signal is then subjected further amplification and filtering, both analog and digital.

Engineering of commercial spectrometers is sufficiently optimized that the principal source of error in normal (rapid-scan, non-time-resolved) spectral measurements is random electronic (dark) noise in the detector/preamplifier itself. In particular this means that the source itself, typically a silicon carbide glower, provides no additional noise due to fluctuations. A simple criterion for testing this is that in a single-beam intensity spectrum, the peak-to-peak noise should be nearly the same regardless of how much of the IR light from the source is reaching the detector; i.e., the intensity noise is the same with an open beam as with 99% of the beam intensity blocked. It is usually a good idea to perform such a test for an instrument both in rapid-scan and in step-scan mode, prior to attempting any time-resolved experiment. (If such a comparison is undertaken, all parameters must of course remain constant, including measurement time and detector gain. Automatic adjustment of such parameters by instrument software must be prevented.)

For step-scan experiments with sub-ms time resolution, detector/preamplifier electronic noise should generally be the ultimate limiting factor in determining the accuracy of the measurement. However, there are other sources of error, discussed below, that can easily exceed the detector noise if the measurement parameters are not properly set. The goal of the time-resolved experimenter should therefore be to reduce these other sources of error to below the detector noise limit. When this is done, the dependence of the residual detector noise on spectral and temporal resolution (see equations below) will end up determining the fastest events that can be detected. It will be seen that the particular characteristics of biopolymers — as well as other polymeric samples — tend to limit the practical time resolution of experiments on these samples to somewhat longer (slower) than the ultimate instrumental time resolution advertised by spectrometer manufacturers.

The general expression for the transmittance noise — the reciprocal of the signal-to-noise ratio (SNR) — in FT-IR spectroscopy is

$$(\delta I_{\bar{\nu}})/I_{\bar{\nu}} = \delta A_{\bar{\nu}} \cdot \ln 10 = (A)^{1/2}/[D^* \cdot \Theta \cdot U_{\bar{\nu}}(T) \cdot \xi \cdot \delta\bar{\nu} \cdot t^{1/2}] \tag{8.1}$$

In this expression, $\delta I_{\bar{\nu}}$ represents the root-mean-square noise level of the light intensity measurement at a particular wavenumber $\bar{\nu}$, while $I_{\bar{\nu}}$ represents the total intensity reaching the detector at that wavenumber. The ratio of these two quantities — i.e., the reciprocal of the SNR in the intensity spectrum — gives the absorbance noise upon division by ln10. On the right-hand side of Equation 8.1 are the experimental parameters that represent detector area A; detectivity D^*; throughput Θ; source intensity $U_{\bar{v}}$ at temperature T; interferometer efficiency ξ; spectral resolution $\delta_{\bar{v}}$; and measurement time t. This expression should be evaluated in order to determine the ultimate feasibility of any particular experiment. In particular, it should be referred to, at least implicitly, in order to choose the right combination of detector, spectral resolution, and amplifier/digitizer time resolution.

8.4.1.1.1 Detector Choice

The first two parameters in Equation 8.1, area A and detectivity D^*, are determined exclusively by the choice of detector. In general, the only commercially available detector types in the mid-IR range (1800–800 cm^{-1}) with a sufficiently high detectivity to consider using for sub-ms step-scan time-resolved experiments in the mid-IR range (1000-2000 cm^{-1}) are HgCdTe or HgMnTe. Whether photovoltaic or photoconductive, the best of these have D^* values exceeding 2×10^{10} cm W^{-1} Hz$^{1/2}$ (2×10^{8} m W^{-1} Hz$^{1/2}$ in SI units) measured at any frequency in the range of 10–100 kHz. The D^* values decrease at higher frequencies, but much more so for the photoconductive than the photovoltaic detectors.

A useful way to think about the meaning of D^* is that its reciprocal (in this case $1/D^* = 5 \times 10^{-9}$ W m^{-1} Hz$^{-1/2}$), when multiplied by the square root of detector area and the square root of electronic bandwidth measured, is equal to the integrated noise-equivalent power (NEP). That is, for a 1 mm × 1 mm detector with the given D^*, if an electronic filter were to select out and amplify the noise only in the frequency range 100,000–100,001 Hz, the root-mean-square (RMS) fluctuations in the detector noise would correspond to 5 pW worth of light energy; i.e., the RMS noise would be the size of the signal change produced by shining 5 pW of IR light on the detector. Increasing the electronic bandwidth that is amplified by the preamplifier increases the integrated NEP nonlinearly, in proportion to the square root of the bandwidth. This is because the different frequency components in the detector noise add incoherently. Determining the actual NEP thus requires summing (integrating) the squared noise over a finite bandwidth, and then taking the square root. Likewise, noise coming from different area elements of the detector (due to thermal excitations of electrons) adds incoherently, so that increasing the detector size increases the amount of the NEP in proportion to the square root of the detector area.

The third parameter above, Θ, is the product of the cross-sectional area and the accepted solid angle of the IR beam. To be more precise, it is the minimum value of this product, considering all locations between the source and detector. In a typical commercial step-scan FT-IR instrument, operating with a sub-ms time-resolved detector, this minimum will occur at the detector. In this case, therefore, $\Theta = A \cdot \Omega$, where Ω is the solid angle of acceptance of the detector (typically ~1 stearadian), and the overall trend of the absorbance noise is to depend inversely the square root of detector area. This expression would therefore generally tend to favor the use of larger-area detectors.

On the other hand, there are three important limitations to the desirability of larger detector areas. The first two apply to both static and time-resolved FT-IR spectroscopy: (1) The area of the detector can reach the maximum image size of either the source or sample; a larger detector will be unfilled with source light. The unfilled detector area is a source of added noise, without producing any additional electronic signal. (2) The detector size should not exceed the maximum throughput possible for a particular spectral resolution and bandwidth. These first two limitations tend to be insignificant for most typical time-resolved experiments on biological systems using commercially available HgCdTe or HgMnTe detectors. Such detectors are nearly all smaller than 2 mm in dimension. A 2 mm detector area is well-matched to the throughput of commercial spectrometers, and works well for resolutions as fine as ~2 cm^{-1} over a bandwidth up to ~4000 cm^{-1}, or ~4 cm^{-1} over a bandwidth up to ~8000 cm^{-1}. Such detectors are optimal for sample diameters of ~2–4 mm, depending on the tightness of the beam focus in the sample compartment. A proportionally smaller detector area would, however, be better if the sample size needed to be ~1 mm or smaller.

The third limitation on detector size, more specifically applicable to time-resolved spectroscopy, is that the rise time for a photoelectric signal from a particular type of detector is proportional to the distance that signal must propagate across the semiconductor material. High sensitivity photovoltaic HgCdTe or HgMnTe detectors, which are optimum time-resolved FT-IR detectors in the mid-IR range, generally require a dimension below ~1 mm for a response time in the 20 ns range.

As can be deduced from the foregoing discussion, there are generally two factors that lead to poorer SNRs with faster detectors. First, the bandwidth of the detector/preamplifier combination is proportional to the reciprocal of the rise time. This predicts that more noise will come from a faster detector/preamplifier combination, in proportion to the square root of the reciprocal of the rise time. An example of this is clear in Figure 8.6, which compares rise times from 2 HgMnTe detectors from Kolmar, nominally with 20 MHz and 50 MHz bandwidths. If a faster detector is used than is truly needed, the higher noise due to the unneeded bandwidth can be removed by applying an external low-pass analog filter, or else by digitizing the data fast enough to capture the entire bandwidth of noise and then signal-averaging in time. However, faster detectors are also generally smaller in area, so that even with the same D^*, they will have higher noise levels than might be obtained with a slower and larger detector. In the example shown here, the 20 MHz detector has an active area 1 × 1 mm, whereas the 50 MHz detector is 0.5 × 0.5 mm. Thus, in choosing between time-resolved detectors with similar D^* but different areas, it is almost always best to choose the largest-area detector that will give adequate time resolution. The only exception would be if all anticipated samples are themselves likely to have a smaller cross-sectional area than the largest detector element being considered.

Commercially available photoconductive HgCdTe detectors are considerably slower than photovoltaics. Nevertheless, a typical 2-mm-square HgCdTe photoconductive detector element is adequate for time resolution of ~2 μs. (It is not generally possible for an end user to measure the response time of a photoconductive element itself, because such a measurement would also reflect the response time of the

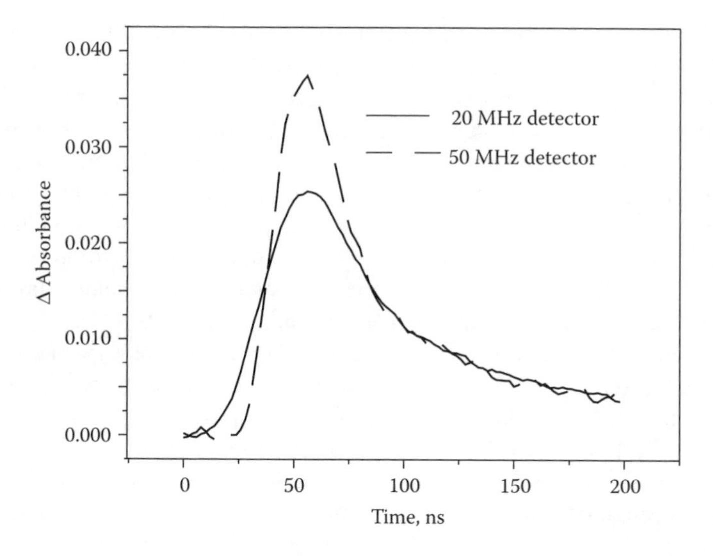

FIGURE 8.6 Comparison of the response times of two different photovoltaic detectors (Kolmar) to a rapid light-triggered event, the intraband transition of sample of CdSe quantum dots. What is plotted is the average absorbance change over a 500 cm^{-1} range centered at the maximum. The actual 0 time in each case is somewhat uncertain. It is known that the transient IR absorbance of the quantum actually reaches its maximum value in well under 1 ns; therefore the rise times of these signals likely reflect the actual rise times of the detectors: ~30 ns for the 20 MHz detector, and ~12 ns for the 50 MHz detector. (S. V. Shilov and M. S. Braiman, unpublished data.)

preamplifier which must be used to provide the bias current. These preamplifiers themselves are typically designed with a response time in the range of 1-5 μs; i.e., well-matched to the response time of the photoconductive elements themselves.) Still, the only good reason to select a photoconductive HgCdTe detector over a photovoltaic is the cost advantage. In fact, for time resolution slower than 10 μs, a photoconductive HgCdTe detector/preamplifier can be obtained (with careful selection from one of the many OEM manufacturers of such detectors) having nearly as good D^* as a photovoltaic limited to operating at the same time resolution, for well under half the price. A 2 mm element would be the recommended size for such a photoconductive detector, unless samples with a cross-sectional dimension of <2 mm are to be examined, in which case a proportionally smaller size (with an equally good D^*) would be advisable in order to improve the SNR.

8.4.1.1.2 Spectral Resolution

The fourth, fifth, and sixth parameters on the right side of Equation 8.1 are $U_{\bar{\nu}}(T)$, ξ, and $\delta_{\bar{\nu}}$ which represent respectively the intensity at wavenumber $\bar{\nu}$ of a black body source at temperature T; the efficiency of the interferometer at modulating the light passing through it; and the width of a spectral resolution element. The value of $U_{\bar{\nu}}(T)$ is fixed by the temperature and emissivity of the source installed in the spectrometer, and the value of ξ is fixed by the interferometer design. Both are therefore generally out of user control; they are within a factor of ~2 for almost all commercial FT-IR spectrometers. The source temperature T is typically 1500K, giving a black body emission intensity (at 1600 cm^{-1}) of 0.14 W m^{-1} stearadian^{-1} in SI units. The efficiency ξ—the ratio of modulated light energy to total light throughput — is ~0.1 for a well-aligned commercial interferometer [74]. Of course, it is crucial to make sure that the source output and spectrometer alignment stay at manufacturer specifications as the spectrometer ages. This can be checked by reference to the standard interferogram that should be obtained at the time of original installation. If they drop significantly below the original performance, a replacement of the source and/or professional realignment of the spectrometer's optical elements is required.

The spectral resolution $\delta_{\bar{v}}$ is a user-adjustable parameter. For a spectral resolution $\delta_{\bar{v}}$ of 4 cm^{-1} (400 m^{-1} in SI units), the modulated light intensity per spectral resolution element reaching the detector through an empty sample compartment is $(0.14 \times 0.1 \times 400) \simeq 6$ Wm^{-1} stearadian^{-1}. If this is multiplied by a typical detector throughput of $\Theta = 10^{-6}$ m^2 stearadian (for a 1-mm^2 time-resolved HgMnTe detector element), the modulated intensity reaching the detector is expected to be about 6 µW, per 4-cm^{-1} spectral resolution element. The magnitude of the modulated intensity — i.e., the signal size measured for each spectral element in the intensity spectrum — is expected to increase linearly with $\delta_{\bar{v}}$, while the noise is unaffected by $\delta_{\bar{v}}$. The SNR ratio in independently measured time-resolved FT-IR spectra does in fact show the expected linear dependence on spectral resolution. This must always be remembered in comparisons of SNR in various spectra; e.g., from different manufacturers demonstrating their instruments. The spectral resolution must be kept constant when making such comparisons.

Sometimes it is desirable to improve the SNR in time-resolved spectra after they have been collected, at the cost of poorer spectral resolution. In this case, however, it must be remembered that recalculating the Fourier transformation by truncating the interferogram *after* the measurement is already completed will also result in a proportional decrease in the effective measurement time t to be used in Equation 8.1. (See also the next section.) This increases the noise in the intensity spectral measurements by a factor of $\sqrt{t}$, corresponding to the square root of the factor by which the resolution has changed. The overall resulting improvement in SNR will be only as $\sqrt{\delta\bar{v}}$, not as $\delta_{\bar{v}}$ itself, as would occur for completely independent measurements at two different spectral resolutions. The same dependence of SNR on $\sqrt{\delta\bar{v}}$, is generally obtained if other mathematical techniques are used to smooth the data after it is collected, as shown in Figure 8.7. No matter what smoothing method is used, the time (and photolysis flashes) spent to obtain the sharper resolution during the original data collection is wasted. This may be a very expensive waste of time, especially if the sample is difficult to prepare and is damaged by repeated photolysis triggers, or is otherwise unrecoverable after the TR–FT-IR experiment.

To sum up this subsection, in order to obtain the maximum improvement in SNR obtainable from broadening the spectral resolution — i.e., an improvement that is linearly dependent on $\delta_{\bar{v}}$, while keeping the total measurement time and number of sample triggers constant — one must commit from the start of the measurement to the broader spectral resolution.

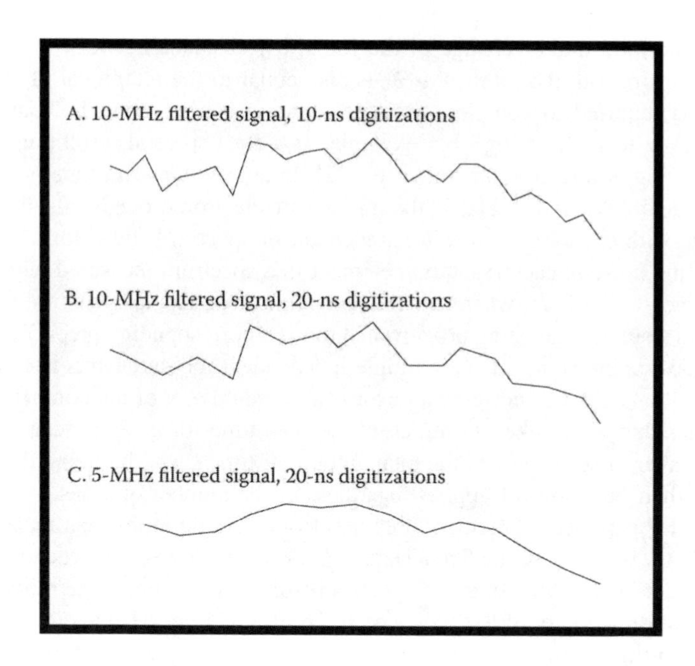

FIGURE 8.7 Illustration of the greater importance of analog bandwidth filter as compared to digitization spacing on the SNR. (a) 30 ns worth of a hypothetical transient detector signal, measured with a time resolution determined by the 10 MHz bandwidth of the analog filter: $\tau = 1/(2\pi f) = 16$ ns. The 10 ns digitizations are assumed to be performed by a digitizer with a substantially higher electronic bandwidth than 10 MHz, so that they accurately reflect the "instantaneous" value of the analog signal at each time point. (b) The same signal as it would be measured with the digitizer spacing set by the software to 20 ns. Note that there is no improvement in the SNR over (a) despite the degraded time resolution of the stored data. This is because the raw signal (i.e., the analog data) was not filtered to blur its temporal resolution prior to digitization. (c) The same signal showing the effect of a twofold reduction in analog bandwidth prior to digitization. In this case, there is clearly a twofold slowing of the time resolution, as well as a $\sqrt{2}$-fold improvement in the SNR.

8.4.1.1.3 Measurement Time

The last experimental parameter to be included in determining the noise level, and therefore the SNR, is the measurement time *t*. This is not equal to the start-to-finish duration of the experiment, but rather the time actually spent on digitization of a rapid-scan detector signal [74]. Because of time not spent digitizing (e.g., during the moving mirror turnarounds and retrace), *t* is typically only ~50% of the total duration of the measurement; and a much smaller fraction for a step-scan measurement.

As stated above, our typical FT-IR detector has an NEP, expressed per unit of electronic bandwidth, of 5 pW Hz$^{-1/2}$ at 100 kHz. To convert this to a NEP expressed per unit of spectral bandwidth, remember that each element of spectral bandwidth $\delta_{\bar{v}}$ represents an electronic bandwidth of $\upsilon \cdot \delta_{\bar{v}}$. Here υ is the interferometer's retardation velocity, the time rate of change of the round-trip distance between the fixed and moving mirrors. But the product $\upsilon \cdot \delta_{\bar{\nu}}$ is also equal to the reciprocal of the mirror travel time required to complete a single scan (measured from the center of the interferogram to its far wing). For example, at 4-cm^{-1} spectral resolution, with the mirror moving at a retardation velocity of 31.25 cm s^{-1}, the scan time is 8 ms, and the reciprocal of this, 125 Hz, is the amount of electronic bandwidth that can be associated with each 4 cm^{-1} resolution element of spectral bandwidth.

Thus the noise in each resolution element in a spectrum measured with a single scan is given by NEP/$\sqrt{t}$, where t is the scan time. This results in the transmittance noise from a single scan being proportional to $1/\sqrt{t}$, corresponding precisely to Equation 8.1. Averaging signals from multiple independent measurements is expected to reduce the transmittance noise by a factor of the square root of the number of scans. This means that if we take t to represent the scan time for a single scan multiplied by the number of scans — i.e., the total amount of time spent digitizing the detector signal — then Equation 8.1 applies regardless of the number of scans.

The NEP per spectral element obtained from a single rapid-scan measurement lasting 8 ms would thus be $5\,\text{pW Hz}^{-1/2}/\sqrt{.008\,\text{s}}$, or 55 pW. In order to make a measurement lasting this long at a spectral resolution of 4 cm^{-1}, the mirror moves at a retardation velocity of (0.008 s × 4 cm^{-1})$^{-1}$ = 31.25 cm s^{-1}. This mirror motion would modulate light at 1600 cm^{-1} with a frequency of (1,600 × 31.25 = 50,000) Hz. That is, the spectral data point centered at 1600 cm^{-1} represents the peak-to-peak amplitude of the detector signal modulated at ~50,000 Hz. In this case, the modulated light intensity from the source reaching the detector (previously calculated as 6 μW per 4-cm^{-1} spectral resolution element) is ~1 × 10^{5} times the 80-pW noise level after just 8 ms of measurement time. This is the SNR, and its reciprocal, 1 × 10^{-5} (or 0.001% on a percent scale) represents the expected transmittance noise. The expected absorbance noise is just this transmittance noise divided by ln10, or about 4 × 10^{-6} absorbance units.

In fact, it is generally not possible to obtain a large-bandwidth spectrum with the stated SNR with a single 8 ms scan. Besides the decrease in black-body emission intensity in regions of the mid-IR spectrum away from the maximum for a temperature of 1500K (about 2000 cm^{-1}), there is no high-D* detector that avoids saturation (i.e., that remains linear) when the integrated intensity of this entire mid-IR spectral range is focused onto it. Linear response of the detector is a crucial assumption in FT-IR spectroscopy. In practice, therefore, the open-beam intensity must be reduced by a factor of approximately 10 to avoid saturation of the detector, and this is predicted to increase the transmittance noise tenfold. When this is taken into account, the transmittance noise between two single scans when using the stated measurement parameters is indeed close to the 0.01% value predicted from the detector's specifications and Equation 8.1 (spectrum not shown).

8.4.1.1.4 Modifications for Step-Scan Time-Resolved FT-IR Experiments

During a TR–FT-IR experiment, only a small fraction of the total measurement time is typically used to digitize data. It is furthermore illogical that the total time duration of signal digitization after each sample trigger could affect the SNR ratio of every time-resolved spectrum collected, because (for example) the spectra corresponding to shortest times are calculated only from digitizations occurring immediately after the trigger.

Hence it is immediately clear that Equation 8.1 cannot be used directly to predict SNR in time-resolved experiments — at least not with t still representing the total digitization time of the entire experiment. Yet a logical consideration of the right side of Equation 8.1 suggests that even in a time-resolved experiment, SNR should depend on all of the other terms besides t in the same way as in a rapid-scan measurement. This means there should be an appropriate value of t to substitute in Equation 8.1. What is it?

In fact, the correct value of t to use in 1 for the case of a step-scan TR–FT-IR experiment is simply $M \times N \times \delta t$ (S. V. Shilov and M. S. Braiman, unpublished observations). Here δt is the time resolution of the digitizations, M is the number of points in each interferogram, and N is the number of interferograms averaged together to give each final time-resolved spectrum. This product represents, roughly speaking, the combined duration of all the digitizations that are actually used in the calculation of each individual time-resolved spectrum.

Note, however, that it is important to use for δt the time resolution present in the analog signal being fed into the digitizer, not the actual digitization spacing (see Figure 8.7). That is, if the detector has a rise time of 50 ns, but its signal is being digitized by a 500 MHz digitizer (i.e., with 2 ns spacing), then the appropriate value to use for δt is 50 ns, not 2 ns. Likewise, if the same detector signal is filtered through an f_{max} = 11-kHz low-pass filter prior to digitization with 2 ns spacing, then the appropriate value to use for δt is $1/(2\pi f_{max})$ = 14 μs. The formula thus accurately predicts the improvement in SNR that can be obtained by putting a low-pass filter on the signal prior to digitization.

Furthermore, the number N to use in the formula above is the number of *independently* measured interferograms that are averaged together. The reason, in brief, is that the SNR of a signal that has already been filtered with a time constant of 50 ns, and digitized every 50 ns, cannot be improved fivefold by averaging 25 digitizations taken every 2 ns. The 25 digitizations taken during the rise time of the detector are not truly independent, and therefore their noise is correlated. This correlation disappears smoothly and exponentially as digitizations are measured farther and farther apart, with an expected exponential decay time constant of 50 ns in the autocorrelation function of the noise.

The foregoing definition of δt as the response time of the analog signal coming from the detector assumes, of course, that no new noise at frequencies above the 20 MHz detector/preamplifier bandwidth is introduced by subsequent amplifiers or by the digitizer. However, matching the detector/preamplifier bandwidth to the bandwidths of the subsequent amplifying/digitizing electronics is a crucial part of the design of any time-resolved experiment. This rule must be observed strictly in order to obtain optimal results (see Section 8.4.1.3). All further discussion here is based on the assumption that this rule is observed carefully when setting up each time-resolved experiment.

With the foregoing interpretations, the stated substitution of $M \times N \times \delta t$ for t in Equation 8.1 can be seen to apply perfectly well for the case of a rapid-scan measurement, as long as the time resolution of the individual digitizations is set to $\delta t = t/(MN)$. That is, δt is equal to the temporal spacing between successive digitizations in the rapid-scan measurement. In fact, the use of an amplification time resolution (i.e., inverse bandwidth) very close to the digitizer spacing is one of the assumptions explicity made in the derivation of Equation 8.1 for rapid-scan FT-IR measurements [74]. Under normal conditions, most rapid-scan spectrometers impose a low-pass filter on the analog signal prior to digitization. If this low-pass filter is properly optimized, it imposes an analog time resolution that matches quite precisely the digitizer spacing.

The suitability of Equation 8.1 with the substitution $t = M \times N \times \delta t$ has been tested for step-scan time-resolved interferograms collected with an empty sample chamber (S. V. Shilov and M. S. Braiman, unpublished). The SNR of a 0-Absorbance baseline, collected using an AC-coupled detector, was found to show the predicted proportionality to $M^{-1/2}$, $N^{-1/2}$, and $\delta t^{-1/2}$ for both 20-MHz and 50-MHz photovoltaic detectors available from Nicolet Instruments. Furthermore, the SNR of each step-scan time-resolved measurement corresponded closely with that of a measurement in rapid-scan mode having a total measurement time equal to the value of $M \times N \times \delta t$ from the step-scan measurement.

8.4.1.1.5 A Simple Rule of Thumb for Predicting Detector-Limited Absorbance Noise in Time-Resolved Experiments

There is a very simple insight to be drawn from the foregoing discussion, which is particularly useful for predicting the SNR of a planned time-resolved spectral measurement. This is based on the realization that $M \times N$ is equal to the number of trigger events that must be applied to the sample in order to obtain the signal-averaged time-resolved spectrum. Therefore, the absorbance noise in an FT-IR spectrum with time resolution δt is expected to be equal to that which could be obtained in a rapid-scan experiment on the same sample and at the same spectral resolution, in which the total data collection time equals δt times the total number of triggering events from the time-resolved experiment.

When a sample is in place, there will typically be a substantial reduction in energy reaching the detector relative to the open beam, at least in certain spectral regions. This will correspondingly reduce the SNR below that predicted for an open sample compartment. These bandwidth-specific increases in noise in the time-resolved spectrum can be determined quickly and empirically determined in advance of the measurement. One must simply measure the zero-absorbance baseline in two successive rapid-scan measurements, using one of the two spectra as the background for the other. The total measurement time should be set equal to $\delta t \times M \times N$; i.e., to the planned temporal resolution of the time-resolved measurement multiplied by the number of trigger events that the sample will be subjected to.

Thus, to estimate the anticipated absorbance noise in spectra with a time resolution of 50 ns and 8-cm^{-1} resolution, measured with 100,000 flashes, as in the earliest time points in Figure 8.3 and 4, one would measure the zero-absorbance baseline measured with (100,00 $\times$ 50 ns) = 5 ms measurement time in the sample and in the background. Such an experiment requires one scan for each of these two measurements, using the fastest mirror velocity available with commercial spectrometers. As mentioned above, for the 20 MHz photovoltaic detector used to measure these spectra, the single-scan noise is about 0.01% when the IR beam is attenuated tenfold (data not shown). A wet biological sample between two windows typically produces roughly the same overall tenfold attenuation; i.e., with the sample in place it is possible to open the aperture to its maximum and still not saturate the detector. The transmittance noise obtained from two successive single-scan measurements with a typical wet protein sample is therefore close to 0.01%. This corresponds to absorbance noise of $1 \times 10^{-4}/(\ln 10) = 4 \times 10^{-5}$. This is not a perfect prediction, but is only worse by about a factor of two than the actual absorbance noise observed in spectra obtained with 30 ns time resolution and 100,000 flashes (Figure 8.4).

This useful rule of thumb only takes into account detector noise, and therefore works fairly well for biological TR–FT-IR experiments in which other sources of noise (see below) are reduced to a level of insignificance. It certainly does not make sense to attempt an experiment for which this rule of thumb predicts a noise level higher than the predicted differential absorption signals.

8.4.1.2 Mirror Position Error

The IR signal at the detector is expected to vary strongly with relative positions of the fixed and moving mirrors, so even a tiny fluctuation in the position of the moving mirror can result in substantial noise in the spectrum. The moving mirror in an interferometer is specifically designed to glide nearly without friction along a bearing, usually an air bearing, during a rapid-scan measurement. During rapid-scan measurements, there is also a control mechanism that incorporates a feedback loop and an electromagnetic force generator to maintain a constant velocity. In order to adapt a rapid-scan instrument to step-scanning, the feedback loop must be altered to keep the mirror motionless at each step of the interferogram. Furthermore, extra resistance of the mirror to external force fluctuations (such as acoustic noise in the environment) should also generally be engineered in.

Rödig and Siebert [75] made a systematic study of the effects of mirror motion on noise in time-resolved FT-IR difference spectra. They detected peak-to-peak fluctuations in the mirror position of 1.3 nm using an air-bearing-equipped Bruker step-scan instrument specially modified with a home-built vacuum housing. Fluctuations of this size typically do not produce noise that is significant compared to detector dark noise. Other workers using Bruker's commercially available vacuum bench accessory with step-scan FT-IR have reported similar results.

Our own experience with a non-vacuum air bearing bench by the same manufacturer indicates that the fastest time-resolved measurements (in the range of 0.01–10 μs) also do not show much sensitivity to mirror motion. The frequency range of the acoustic noise that is transmitted into the nonevacuated bench appears to be limited predominantly to below several kHz. For slower time-resolved spectra, however, the acoustic noise can easily become the predominant source of noise in the spectrum. Anyone considering the purchase of a step-scan instrument for the purpose of making measurements in the time range above several μs should take this concern into account, and consider a vacuum accessory. For sub-μs measurements, however, the cost of a vacuum bench is likely to be unjustified.

The graphite bearing of the Nicolet 860 spectrometer is an alternative to an air bearing that appears to work extremely well with step-scan instruments. This instrument's moving mirror has just enough residual friction to allow it to remain motionless after it reaches its desired position and the solenoid is completely de-energized. This appears to be true even in the presence of typical laboratory acoustical noise. Thus we have not been able to detect any noise resulting from mirror position fluctuations in our time-resolved FT-IR spectra measured on this system.

8.4.1.3 Amplifier/Digitizer Errors

Detector dark noise, even with the very best detectors available, generally sets the limit of useful time resolution with biological samples, for which signals are expected in general to be small compared to the background absorbance. However, as implied above, it is easy to establish wrong settings for the amplifier and digitizer that will cheat the user out of attaining this optimum.

The foremost consideration should be that regardless of the desired experimental time resolution, the temporal spacing between digitizations should be less than the fastest significant noise component present in the analog signal; i.e., less than the shortest correlation time of the electronic noise reaching the digitizer. In order to meet this requirement, it may be necessary to take one of two steps: (1) introduce a low-pass analog filter on the signal prior to digitization, in order to reduce its bandwidth; or (2) set the spacing of the digitized points closer, as required to match the noise correlation times (reciprocal of the electronic bandwidth) of the combination of existing detector/amplifier(s)/filter(s).

The latter choice may require rewiring the signals to permit selection of a digitizer with faster maximum rate. Commercial time-resolved FT-IR spectrometers frequently come with two or three different digitizers. However, the bit depth of faster digitizers tends to be more limited, which may end up requiring that the time-resolved spectrum be measured with an AC-coupled signal in order to reduce the dynamic range. Note that use of an AC-coupled signal has the potential for introducing phasing errors into the final spectrum. (See below.)

If appropriate low-pass analog filters are not available, then it may be necessary to digitize the data more finely in time — and to generate a larger number of spectra — than is desirable for final data presentation. If this happens, the proper approach to reduce the number of time-resolved spectra is to group-average them. Unfortunately, it is possible with most current versions of commercial time-resolved software to select a subset of the digitizer measurements actually to store in the computer, without group-averaging them. Thus, for example, with a standard 500 MHz digitizer, it is possible to select a final spectral spacing of 2, 4, or 8 ns. Built-in software limitations on the total number of time-resolved spectra collected create a temptation to use a spacing larger than 2 ns, in order to include data out to the 100-μs range or beyond. However, succumbing to this temptation can lead to a substantial degradation of SNR. While the final spectra will be spaced more widely in time than 2 ns, and therefore appear to have a slower time resolution, the value of δt that determines noise in the formulas above will not have increased as it would if the points spaced at 2 ns are co-added to give a wider final spacing (see Figure 8.7).

It is also tempting to assume that the electronic bandwidth of any commercial detector/preamplifier combination that comes with a spectrometer will always be properly matched to the temporal spacing of the digitizers provided, but this is not a safe assumption. In general, it is best to run a triggerless experiment — even one without a sample in place — in order to determine the optimum digitization spacing. Start with a slightly faster digitization than one expects will be needed, and compare SNR as the spacing between digitizations is increased, with all other parameters held constant. When the SNR starts to get worse, the spacing has gotten too large.

8.4.1.4 Sample/Trigger Drift

One of the assumptions of TR–FT-IR is that the same sample changes occur identically after repeated excitation triggers. Several types of spectral errors result from imperfections in this assumption. When the sample composition changes slowly with time, the expected time-resolved spectra from the sample also vary slowly in time. Substantial mathematical analysis may be required in order to extrapolate the spectra to what they would be if the sample composition remained constant.

One of the potential worst-case scenarios is when the step-scan spectra are measured with DC coupling over a long period of time, and no pretrigger interferograms are included in the time-resolved data collection. In this case, it might seem possible to ratio the DC-coupled spectra to single-beam spectra from the same sample collected a few minutes or hours away in time. However, the likelihood is that even minor sample drifts will produce baseline shifts that may be accompanied by drifts in the sizes of water vapor and carbon dioxide bands, and that can easily overwhelm the very small differential absorbance signals expected from triggered biological samples, as in Figure 8.8.

Most such problems can be corrected to first order simply by collecting an AC-coupled differential interferogram, or alternatively making sure that DC-coupled interferograms include a pretrigger sample that can be used as a background spectrum. Even when such care is taken, however, there remain problems associated with sample drift. The most general solution is to find a way to minimize the amount of sample change over the course of the entire experiment. One possibility is to introduce multiple fresh samples of identical composition into the FT-IR sample during the course of the experiment, either between collections of entire interferograms, or even within the time required to collect a single interferogram.

In a particular subset of cases, the sample changes that occur can be described simply as a long-term decrease in the amount of the dynamically active starting material, with the only new components being introduced corresponding to inactive species. An example would be a bleaching of a chromophore by repeated pulsed laser triggers that results in the formation only of nonabsorbing or photochemically inactive species. In this case, the time-resolved difference spectra obtained from the sample are expected to remain invariant in time, except for a decrease in their amplitude. A similar effect occurs if the intensity of the sample trigger drops slowly with time

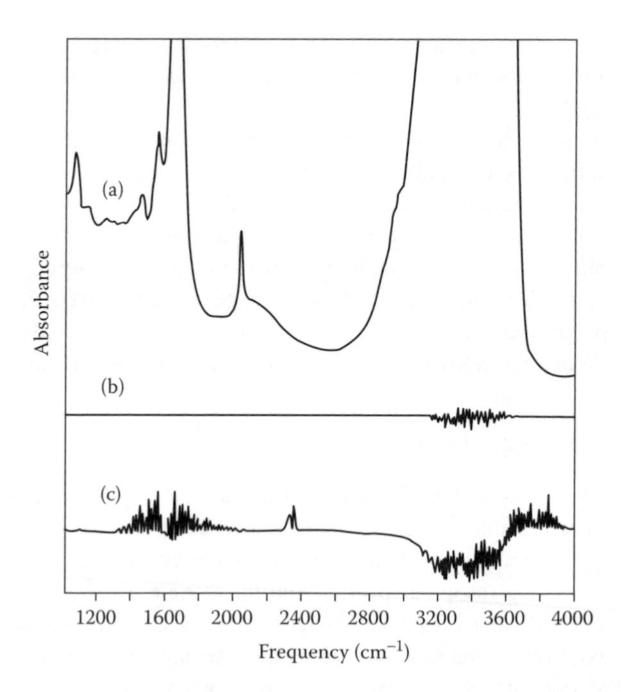

FIGURE 8.8 A demonstration of the importance of correcting non-differential (DC-coupled) time-resolved spectra by referencing them to spectra of the unphotolyzed sample that are collected as contemporaneously as possible. In (a) the static absorbance spectrum of a typical biomembrane sample (in this case, photosynthetic reaction centers) is presented. In (b) a time-resolved difference spectrum measured on this same sample is presented on the same vertical scale as (a) (but with an offset). The time-resolved DC signal measured by the detector at each mirror position was corrected by digitally subtracting the DC level measured immediately before the trigger. This spectrum was measured without an actual sample trigger, so that it should represent a flat zero-absorbance line. Deviations from zero absorbance represent measurement errors, which are clearly small in this case after repeated signal averaging over several hours. In (c) the DC signal measured over the same time period as in (b) was corrected instead by ratioing to a background spectrum that was collected for several minutes prior to the commencement of the time-resolved signal collection, which itself lasted for several hours. Drifts in the sample's liquid H_2O content and in the spectrometer's CO_2 and H_2O vapor content resulted in large artifactual errors in the baseline. Reproduced from M. S. Braiman & K. J. Wilson. *Proc. SPIE 1145*, 397–399 (1989).

(e.g., due to a slow misalignment of optical components directing a photolysis laser onto the sample), producing a similar slow drift in the amount of sample proceeding through the dynamical cycle. If the amplitude of the reaction cycle varies by no more than ~10% within the collection period of each complete step-scanned interferogram, then it is generally possible to collect and store multiple intensity difference spectra, and to convert them to absorbance difference format according to the formula

$$\Delta A(\bar{v}, t) = -\log\{1 + [\Delta I(\bar{v}, t)/I_o(\bar{v})]\} \qquad (8.2)$$

prior to averaging them together. (Here $\Delta I(\bar{\nu}, t)$ represents the time-resolved intensity difference spectra as a function of time t after the trigger, and $I_o(\bar{\nu})$ represents the background intensity spectrum of the sample measured in the absence of the trigger.) The cycling amplitude changes produce uniform rescaling of the absorbance spectra at all values of t. This means that averaged data can still be used accurately for kinetic analysis; i.e., for determination of time constants and spectral shapes.

In fact, if (as is typical for biological samples) $\Delta I << I_o$ at all spectral frequencies being examined regardless of the shifting amplitude of the dynamical cycle, and I_o itself does not drift significantly over long time periods, then it is not really necessary to convert time-resolved differential intensity spectra obtained from different stepped mirror scans to absorbance difference spectra prior to averaging (or co-adding) them. The absorbance difference spectra can be calculated afterwards. Correspondingly, since the Fourier transform process is linear, it is also possible to co-add time-resolved differential interferograms from different stepped interferometer scans, even if the sample's cycling amplitude changes somewhat (up to ~10%) between repeated stepped scans. This statement is only correct, however, if the interferogram does not change its phase between repeated scans.

On the other hand, if the amplitude of the reaction cycle varies by more than ~10% within the collection period of a single interferogram scan, then errors may be introduced that seriously distort the shapes of absorption difference bands. It is possible to analyze such effects by conceptualizing them as resulting from multiplication of the time-resolved differential interferograms by an apodization function [8]. This function is given by the varying amplitude of the triggered reaction. Fourier transformation of the product of the apodization function with the interferogram is expected to result in the convolution of the correct time-resolved differential spectra with a bandshape function given by the Fourier transform of the apodization function. For example, if the amplitude of the triggered reaction decreases exponentially over the course of each stepped interferometer scan, then the resulting bandshape function is approximately a Lorentzian, which represents the Fourier transform of an exponentially decaying function. The faster the exponential decrease in cycling amplitude occurs, the greater the broadening effect on the spectral difference bands.

From the foregoing discussion it should be clear that, for biological samples, sample drift errors that might distort band shapes can be minimized by selecting a strategy that completes the collection of individual step-scanned interferograms as quickly as possible, and with the application of as few triggering pulses as possible. That is, if extensive signal-averaging is required in order to obtain adequate SNR, it is generally valuable to scan through the entire interferogram multiple times, rather than to repeat the sample trigger multiple times with the mirror held stationary. For typical biological samples, it is also possible (and often very desirable, in order to minimize data storage requirements) to automatically co-add in memory a number of repeated interferometer scans prior to storing the multiple time-resolved differential interferograms to disk. This latter approach was initially implemented in Thermo-Nicolet's Omnic step-scan software, and was only more recently introduced as an option in the Bruker Opus step-scan software.

8.4.1.5 Excitation Trigger Noise

If changes in the amplitude of the sample's dynamic cycle occur randomly rather than as a steady drift, the effects on the final spectra also will look more like random noise. This is most likely to occur if the excitation trigger varies randomly in energy. This variation in the sample cycling amplitude can sometimes result in a contribution to noise that exceeds that produced by thermal noise in the detector. This noise is termed multiplicative noise [75], because the random variations in cycling amplitude get multiplied by the size of the interferogram signal that is measured at the detector.

It was shown by Rödig and Siebert [75] that the biggest problems occur when sample absorbance changes produced by the trigger affect not only the interferometrically modulated IR intensity, but also the steady (interferometrically unmodulated) IR intensity, which is even larger than the modulated intensity (see Figure 8.9). The unmodulated intensity, which covers a broad spectral bandwidth, is usually hidden from the user in rapid-scan experiments because the detector is AC-coupled. The same is generally true in step-scan TR–FT-IR experiments on samples that exhibit balanced positive and negative absorption changes at all times after being triggered. In such a case, the

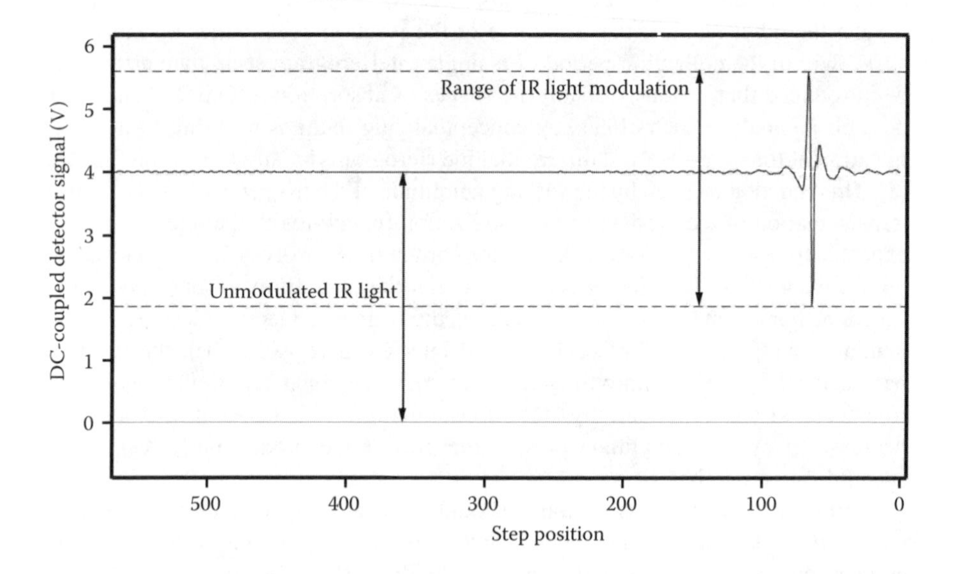

FIGURE 8.9 Comparison of the sizes of modulated and unmodulated intensities in the DC-coupled interferogram from an open-beam sample compartment. In a Nicolet 860 spectrometer, the IR beam was blocked and the offset of the DC-coupled detector carefully zeroed; then the interferogram was measured with an open sample compartment. The unmodulated DC light level is normally insignificant to FT-IR spectroscopists, because it is removed by the Fourier transformation process and it contributes insignificantly to the noise in a rapid-scan spectrum. In a time-resolved FT-IR spectrum, where there may be "noise" in the sample modulation (e.g., in the strength of the photolysis pulse), the unmodulated DC light level can become modulated by triggered changes in the sample absorbance. Random variations in the amplitude of the triggered reaction can contribute significant multiplicative noise to the spectrum, as explained in the text.

broadband IR intensity that is interferometrically unmodulated cannot produce measurable transient signals in an AC-coupled detector, because sample-induced intensity increases at some wavelengths are cancelled out by decreases at nearby wavelengths, regardless of the amounts of photoproducts that are formed.

However, this is not always what occurs. In fact, it is quite possible for a photoreactive sample (such as CdSe quantum dots) to experience a large transient mid-IR absorbance increase that is not balanced by any absorbance decrease in a nearby spectral region. In such cases, the photolysis laser pulse produces a modulation in the AC-coupled detector signal that is due not only to changes in the interfermetrically modulated component of the IR beam, but also due to a transient change (in this case a decrease) of the interferometrically unmodulated component. In the case of the CdSe quantum dots, such a transient signal can be visualized easily on an oscilloscope display of the detector signal (S. V. Shilov, M. Shim, P. Guyot-Sionnest, and M. S. Braiman; unpublished observations). Its direction (corresponding to a transient decrease in IR intensity reaching the detector) marks it as distinct from IR light emitted upon transient heating of the sample by the photolysis laser. The transient signal shows no systematic variation in magnitude as the interferometer is stepped through the entire interferogram. However, a pulse-to-pulse variation in photolysis laser intensity (~2%) results in a distinct increase in the random noise present in the baseline of the differential time-resolved interferograms. The random noise remains after Fourier transformation (see Figure 8.10). These spectra show surprisingly large absorbance noise, compared with time-resolved difference spectra measured from biological samples (e.g., bacteriorhodopsin as shown in Figure 8.4). The source of the added noise from the quantum dots is unambiguously clear, based on kinetic analysis. The large noise level is not present in difference spectra measured prior to arrival of the photolysis laser pulse, and after photolysis the unusually large noise decreases with the same kinetics as the observed fast transient mid-IR absorption band due to the CdSe quantum dots (see Figure 8.10). This type of noise is obviously multiplicative noise, since it is largest in the difference spectra that have the largest magnitude peaks.

As implied above, it is expected that biological samples and polymers will be somewhat less susceptible to such noise than are CdSe quantum dots, because the former tend to exhibit a balance of absorbance increases and decreases over the broad spectral ranges measured. However, photolysis of biological samples by a laser with random intensity variations can still result in measurable added noise, simply due to the variation in the amplitude of the interferometrically modulated component of the IR beam [75].

Several strategies have been investigated to reduce the latter noise. One is to use such a high photolysis laser intensity that the sample's photoreaction is always saturated. However, in this case the detector signal from sample heating becomes more important, relative to the signal due to the photoreaction, and the heating signal is also susceptible to random pulse-to-pulse intensity variations [75]. It was shown [75] that a better approach is to measure the variation in the energy of each visible laser pulse precisely with a visible photodiode, and then to normalize each transient IR intensity change to the visible pulse energy. This approach has been implemented in the Bruker Opus software, which utilizes one of the spectrometer's digitizer channels to keep track of the amplitude of the trigger pulse (e.g., from a visible photodiode

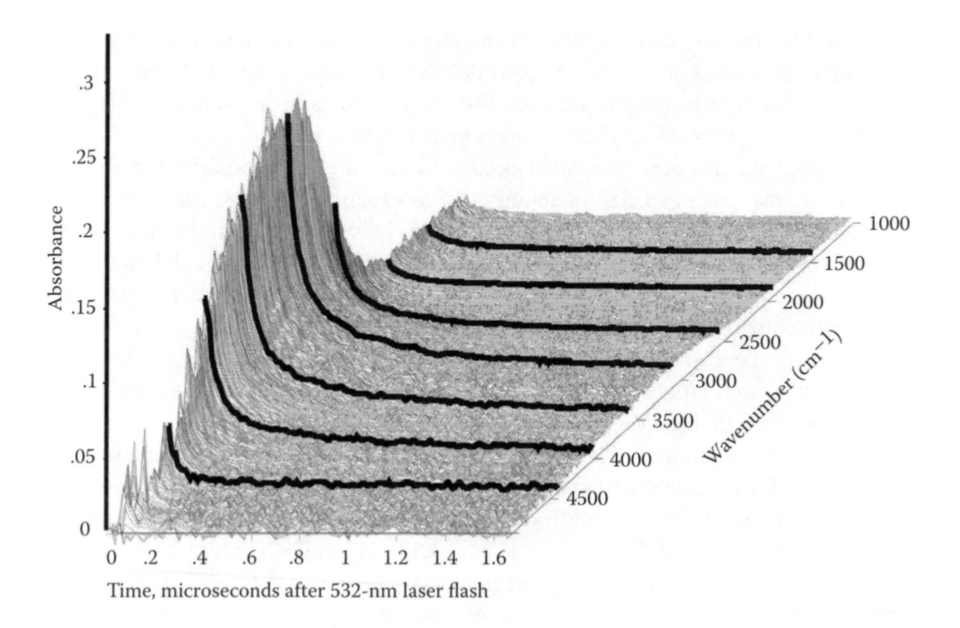

FIGURE 8.10 Sub-μs spectra of the transient charge-separated state of ~3 nm thiocresol-capped CdSe quantum dots, capped with thiocresol and measured in CCl_4 colloidal suspension. The large, broad, transient spectral absorption band corresponds to the intraband transition, which results from excitation of an electron from the lowest excited state within the conduction band to the next-higher state. The relatively high noise level of the spectra results from ~9% flash-to-flash variations in intensity of the frequency-doubled Nd^+-YAG photolysis laser, which produces random modulation of the sample's transient absorbance, in proportion to the amount of photoproduct that is produced by the flash. This produces multiplicative noise (see text), which shows up in the spectra as noise features that remain at constant in position as the photoproduct decays. This is because the photoproduct concentration at any time delay is still proportional to the amplitude of the flash. Thus the amplitude of the spectral noise features remains proportional to the overall size of the transient absorption band. As a result, the absorbance noise in the kinetic trace at any individual spectral frequency (see, for example, the bold traces) is significantly less than the absorbance noise within the initial time-slice spectrum at far left. (Details of measurements described in Reference 77.)

onto which a small fraction of that laser pulse energy is directed). However, since the software can generally keep track of only two digitizers in a single experiment, this eliminates the possibility of collecting both AC-coupled and DC-coupled interferograms in a single experiment, and reduces the options for correcting phase errors (see below).

8.4.1.6 Phase Errors

A well-known feature of FT-IR optical interferometers is that they do not produce symmetrical interferograms. This is because there is no single position of the fixed and moving mirrors that can be defined as producing a zero-path difference simultaneously for all wavelengths of light. For rapid-scan instruments, this is often attributed to a frequency-dependent phase delay, or "chirp," introduced by the analog electronics that transmit the signal from the detector to the digitizer. However, if the analog electronics were truly the only source of the frequency-dependent phase of rapid-scan interferograms, then a step-scan interferogram should show no chirp whatsoever; i.e., it should be perfectly symmetric. In fact, even for a step-scan interferogram there is an optical dispersion present in the beam-splitter materials that produces a slow wavelength-dependent shift in phase. This means that, as with rapid-scan interferograms, Fourier transformation of step-scan interferograms results in a non-zero imaginary component.

Consequently, step-scan interferograms (like rapid-scan interferograms) must be phase-corrected as part of the Fourier processing. However, the standard algorithms for phase correcting interferograms do not automatically work correctly with step-scan time-resolved spectra. The biggest problem occurs when the detector signal is AC-coupled, so that only a transient differential interferogram is collected. Such AC coupling of the detector is often required, especially at the fastest time resolutions, in order to ensure that the fast (but low bit resolution) digitizer accurately captures the tiny intensity changes produced at the detector by the sample's small absorbance changes. Not only are these changes small at individual wavelengths, but positive and negative absorbance changes at different wavelengths throughout the broad bandwidth tend to cancel each other out, resulting in even smaller overall transient signals at the detector.

When an AC-coupled transient interferogram is Fourier transformed, one of the key assumptions of standard phase correction algorithms is no longer valid. This assumption is that the phase-corrected Fourier-transformed spectrum should be everywhere positive; i.e., the amplitude of the IR signal measured by the detector should be ≥ 0 at every wavelength. In fact, if the sample absorbance increases transiently at a particular wavelength, then the IR intensity associated with that wavelength transiently drops, corresponding to a negative AC-coupled signal from the detector. This produces a flipping of the sinusoidal variation in intensity generated by the interferometer, with a Fourier amplitude that is -1 times what it would be if the sample absorbance had decreased, instead of increased, at that particular IR wavelength.

The presence of both negative and positive IR intensity components in a single digitized signal confuses all commercially implemented phase correction routines. The Mertz and Mertz-signed phase correction algorithms, for example, are both likely to flip the sign or reduce the intensity of spectral difference bands in some spectral regions, as shown in Figure 8.11. Note that the breadth of the spectral regions that are actually flipped is determined by the spectral resolution of the phase correction. Going to broader resolution can reduce the number of completely flipped regions. However, this also tends to reduce the overall accuracy of the magnitude of the computed intensity difference bands, in the process changing relative intensities of many bands, moving some intensity at each frequency to the computed imaginary part of the Fourier transform [78].

The most general solution to the phase problem associated with time-resolved differential interferograms is to collect a DC-coupled interferogram along with the

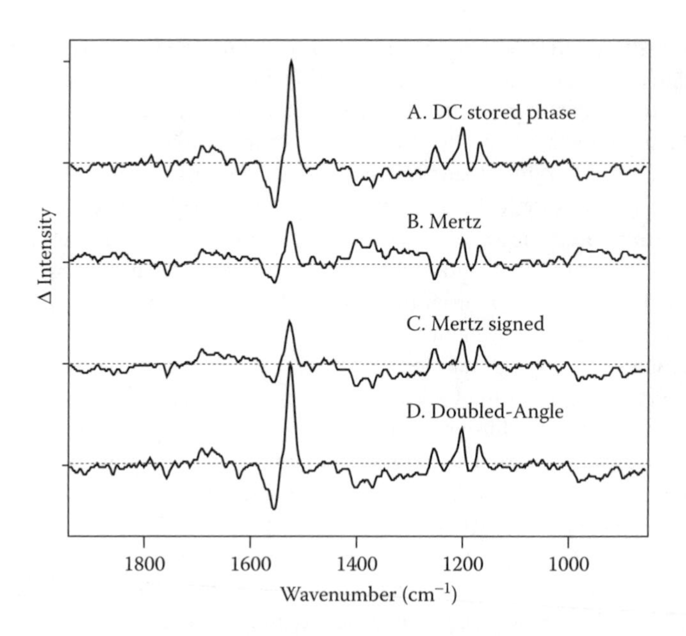

FIGURE 8.11 Results of different phase correction routines applied to the same time-resolved differential interferogram. (a) Using stored phase from DC-coupled interferogram measured separately, but within minutes of the AC-coupled interferograms. (b) Mertz phase correction. (c) Mertz signed phase correction. (d) Doubled-angle phase correction as proposed by Hutson and Braiman [78,79]. The stored DC phase is considered to be the most accurate, as long as the DC measurement can be made within several minutes of the AC-coupled measurement. The doubled-angle method gives nearly identical results. However, with the Mertz and the Mertz signed methods, there is missing band intensity which shows up in the imaginary part of the Fourier transformation (not shown).

AC-coupled interferograms. A Fourier transform of the DC-coupled interferogram is then used to calculate a phase array that can be transferred to the AC-coupled interferograms. A key question to consider is how close in time the AC- and DC-coupled measurements need to be, in order for the transferred phase to be reliably accurate. In our experience, there is a significant risk of thermally and mechanically induced optical phase drift in commercial FT-IR spectrometers over a period of about 1 hour. This is typically less than the time required to complete a single step-scan measurement. Thus, it is a risky practice to take a DC-coupled measurement either before or after the AC-coupled step-scan interferogram is completely collected, even though this is the only practical way if user intervention is required to switch the detector's coupling from AC to DC. Instead, it is most reliable and accurate to have dual detector outputs, dual digitizers, and software capable of storing both AC- and DC-coupled outputs of the detector at every stepped mirror position. This approach has become available recently with at least some commercially available combinations of step-scan instrument and fast detector/preamplifier, in particular the Nicolet 860.

In the absence of such excellent hardware and software solutions, a few specialized phase correction methods have been proposed for differential interferograms, such as those obtained with step-scan TR–FT-IR. These have been discussed and compared in detail for a variety of broadband time-resolved measurements at varying spectral resolution [78,79]. The most important conclusion reached is that for TR–FT-IR, it is inadvisable to rely on the simple Mertz Signed method that restricts the phase angle to a range of $\pm\pi/2$. This does not reliably eliminate the phase-flipping errors over a large-bandwidth spectrum. For a spectrum covering a large bandwidth, such errors can be corrected better by the doubled-angle method of Hutson and Braiman [78]. However, the latter method requires collection of a two-sided interferogram, and is more likely to fail completely if the SNR of the collected interferograms is low.

8.4.1.7 Spectral Aliasing (Interferogram Undersampling) Errors

As discussed further below, it is possible to reduce the total amount of data collection time in a step-scan measurement by enlarging the size of the steps, thereby covering the same interferogram travel distance with a smaller number of digitizations. Typically the steps are taken in integral multiples of the reference HeNe laser wavelength, 632.8 nm. Instrument manufacturers' manuals generally make clear that doubling the step size, for example, results in halving of the bandwidth of measurement, but do not always make clear what types of errors can result from undersampling of the interferogram in this fashion.

In fact, reducing the sampling of the interferogram results in a calculated spectrum that covers a smaller Nyquist bandwidth, but that may still contain extraneous signals, referred to as aliased signals, that are due to detected wavelengths outside that Nyquist bandwidth. For example, if the interferogram is sampled with 4× the normal spacing, the Nyquist bandwidth is only $1/(8\lambda_{HeNe})$, or 1975 cm^{-1}. This does not mean, however, that this automatically limits the spectral information to the range 0–1975 cm^{-1}. In fact, if there is a transient IR intensity increase at (for example) 2900 cm^{-1} in the C–H stretching region, it will show up as an aliased transient increase in the Fourier-transformed spectrum at $2 \times 1975 - 2900 = 1050$ cm^{-1}, where it would look like a transient increase in a C-C stretch vibration intensity, or could even cancel out a transient decrease in such a C-C vibrational band. Improper interpretation of the superimposed aliased signals would be an embarrassing error. The only reliable way to avoid aliased signals in step-scan measurements is with appropriate optical filtering. That is, one must use optical filters to ensure that no interferometrically modulated light outside the Nyquist bandwidth reaches the detector.

Since Globar sources put out light over a bandwidth exceeding the 7900 cm^{-1} Nyquist frequency at the normal interferometer sample spacing ($1 \times \lambda_{HeNe}$), it is important with step-scan measurements to ensure that even in this situation there is appropriate optical filtering in place. A Ge-coated KBr beamsplitter cuts out most of the light at higher frequencies than 7900 cm^{-1}. However, there is a hole in the Ge thin-film coating to permit the reference laser to pass through, which allows a small amount of aliased short-wavelength light through in step-scan measurements. By contrast, in a rapid-scan experiment with the same sample spacing, this aliased signal is further reduced to insignificance by the use of low-pass electronic filters applied to the analog signal. More details on the methods used to correct this and other sources of errors and artifacts are given in the subsequent sections.

8.4.1.8 Selection of Software Parameters for TR–FT-IR Experiments

For a new user of commercial step-scan time-resolved spectroscopy software, the number of spectral measurement parameters that must be specified can seem daunting. With the likelihood that a poor choice for any single parameter can ruin a measurement, there can be a painful learning curve to make such experiments work. The purpose of this section is to provide a useful guide for novices to make successful step-scan time-resolved measurements.

8.4.1.8.1 Interferogram Parameters

The following set of parameters determine the number of points M in each time-resolved interferogram, and in turn the value of $t = \delta t \times M \times N$, which helps to determine the SNR of the final data (see Section 8.4.1.1 for definitions of t, δt, M, and N). However, each can contribute different additional errors. Considerable thought must therefore be given to optimizing these parameters.

8.4.1.8.2 Spectral Resolution

The effect of spectral resolution on absorbance noise was discussed in detail in Section 8.4.1.1. For a fixed total effective measurement time $t = \delta t \times M \times N$, the noise is expected to double for each halving of the spectral resolution. However, halving the spectral resolution parameter, without making any other changes in measurement parameters, will end up doubling M and thus t, which will reduce the noise by a factor of $\sqrt{2}$. (Currently available TR–FT-IR software does not currently allow a value of total measurement time t to be specified directly).

The overall result is that noise typically depends on the square root of the spectral resolution specified in time-resolved software. As an example, a change in spectral resolution from 8 cm^{-1} to 4 cm^{-1} is expected roughly to double the amount of time and the number photolysis flashes required for the measurement, while resulting in a $\sqrt{2}$ worsening of the SNR.

In general, therefore, it is important to specify the coarsest spectral resolution that will permit answering the desired question. For most biological samples (as well as other polymers), the typical homogeneous bandwidth of bands is in the range of 5–10 cm^{-1}, so there is rarely any value in making a measurement with resolution finer than 4 cm^{-1}. An initial experiment at 8 or 16 cm^{-1} resolution is faster and has better SNR, and is therefore strongly advised in order to determine whether any transient differential signal is detectable at all. However, if the expected transient difference spectra consist only of closely spaced alternating positive and negative bands, such low-resolution measurements may obliterate all signals.

8.4.1.8.3 Spectral Bandwidth (Sample Spacing)

Besides spectral resolution, the other parameter that has the greatest direct effect on M is sample spacing. The default sample spacing of the points in the interferogram is one point for each HeNe laser wavelength ($\Delta x = \lambda_{HeNe} = 632.8$ nm). Here, Δx is defined with respect to the optical retardation, which is the round-trip distance from the beam splitter to the mirror; the distance the mirror itself travels between digitizations is actually $\lambda_{HeNe}/2$. The Nyquist sampling theorem requires the bandwidth of the light being sampled to be less than half of the reciprocal of the sample spacing; i.e., $\Delta\overline{\nu} \leq 1/(2\Delta x)$. For the default spacing, then, the maximum optical bandwidth is 7900 cm^{-1}. A doubling of the sample spacing will reduce the bandwidth by a factor of 2, to 3950 cm^{-1}. A quadrupling of the sample spacing will reduce the bandwidth by an additional factor of 2, to 1975 cm^{-1}.

If these reduced bandwidth values are used, a long-pass optical filter should be placed in the IR beam, to block the intensity at shorter wavelengths that could be aliased into the measured band pass. A benefit of this optical filtering, however, is the reduction of the integrated intensity at the detector, bringing the response back down into the linear range for high-throughput conditions and allowing the aperture to be opened more fully than might otherwise be possible. Measurement of an uninteresting spectral region can thus be sacrificed to result in an improved SNR in the region of greatest interest.

Most spectrometers also allow a half-normal spacing, which will increase the measurable bandwidth to 15,800 cm^{-1}. This is useful for measurement of near-IR absorption spectra, that can extend even a bit into the visible spectral range. This setting is also typically used for obtaining rapid-scan FT-Raman spectra. There has not yet been much work on step-scan time-resolved measurements in the near-IR spectral region. It remains unclear whether time-resolved measurements in this range can ever approach the performance of dispersive spectrometers equipped with relatively inexpensive gated multichannel detectors.

8.4.1.8.4 Single- or Double-Sided Interferogram

A double-sided interferogram includes equal numbers of points on both sides of the zero-path difference (ZPD). A typical interferogram from an FT-IR spectrometer is close to symmetric about the center-burst. Therefore, to some degree the information from the two sides is redundant. In particular, the spectral resolution is not increased measurably by collecting a double-sided rather than single-sided interferogram. The total amount of mirror travel (and correspondingly the total number of points M) required to collect a double-sided interferogram is double that of the corresponding single-sided interferogram, with no gain in resolution. Most commercial spectrometers therefore permit only single-sided interferograms for resolutions of less than 4 cm^{-1}.

As with rapid-scan measurements, it is typically best in TR–FT-IR measurements to collect only a single-sided interferogram. However, selection of double-sided interferograms is not generally expected to lead to any significant disadvantages. The near-doubling of the digitizations M required to measure a double-sided interferogram contributes to a $\sqrt{2}$ -fold improvement in the SNR, just as if 2 separate single-sided interferograms were being measured. For typical biological samples, obtaining adequate SNR almost always requires signal-averaging at least 2 interferograms, so no additional time is required to measure the time-resolved spectra by using double-sided interferograms. The one possible disadvantage of double-sided interferograms is that twice as much sample drift (e.g., due to irreversible photolysis by a triggering laser) is expected during the collection of each interferogram.

Note that there is one important exception to the general recommendation of collecting single-sided interferograms. If an AC-coupled measurement is to be made, and there is no possibility of measuring the DC-coupled interferogram within ~1 hour of the AC measurement, then the doubled-angle phase correction method for AC-coupled interferograms should be used [78] (see also Section 8.4.1.6 and Section 8.4.1.8.22). In this case, storage of a complete double-sided interferogram is a requirement for proper phase correction.

8.4.1.8.5 Points before Interferogram Peak

This parameter needs to be specified only if single-sided interferograms are selected. This need arises because the symmetry between the two sides of the interferogram is imperfect, reflecting a phase chirp. This is a slow variation in the moving mirror's ZPD position, as a function of IR wavelength. The chirp results from the presence in the beam splitter of materials (e.g., KBr, Ge) with a wavelength-dependent refractive index. Proper calculation of the spectrum from the interferogram requires a phase correction procedure that determines the magnitude of the phase chirp. All phase correction procedures useful for time-resolved spectroscopy require that a double-sided interferogram be measured at least to low resolution. Therefore, even when single-sided interferograms are selected, there is an opportunity to specify the amount of the second side to be measured, through the parameter defining the number of points before the interferogram peak.

In most cases, only enough prepeak points are required to give a resolution of 64 cm^{-1} in the calculated phase spectrum. This corresponds to 256 points before the interferogram peak if the default sample spacing (spectral bandwidth) is used. If the sample spacing is doubled (to ~3900 cm^{-1}), then 128 points before the peak are recommended; and if it is quadrupled (to give a bandwidth of ~1950 cm^{-1}), then 64 points are recommended.

8.4.1.8.6 Timing and Averaging Parameters

The following set of 10 timing parameters should typically be specified at the start of each time-resolved measurement. However, not all of these 10 are user-selectable in all versions of software from the different manufacturers of commercial step-scan instrumentation. This is probably the area of greatest distinction between the different step-scan time-resolved instruments on the market. In some cases, the lack of availability of user choice for a particular parameter may make a particular spectrometer (or at least a particular version of software) completely unsuitable for a certain desired measurement. It is therefore strongly recommended to try to understand this subsection in particular before purchasing an instrument for step-scan time-resolved measurements.

8.4.1.8.7 Settling Time

This is set to the time needed for feedback-induced oscillation in the mirror position following each step to die down. It should typically be set to a value close to 100 ms, in order to allow the resulting signal fluctuations at the detector to drop below the detector noise.

However, if no static interferogram is being collected, and the post-trigger delay is large (see below), then it makes sense to reduce the settling time so that the sum of the post-trigger delay and settling time equals 100 ms. This will allow less total time to collect the data, while still affording at least 100 ms in between the mirror's step and the first digitization thereafter.

8.4.1.8.8 Static (DC-Coupled) Pre-Trigger Average Time

This is the time interval during which the static interferogram is collected, if the option of collecting one has been selected. (See Section 8.4.1.8.21 for more information about making this selection.) The static interferogram is supposed to correspond to the state of the unperturbed sample immediately prior to the trigger. Accurate digitization of the static value cannot begin until the settling time is over; and even longer, if this is required to allow the trigger interval (see below) to have passed since the preceding trigger. Once the static interferogram digitizations begin, they last for the static average time. This should ideally be set to no more than about 10% of the trigger interval, but no less than two to three times the longest time-resolved interval that is to be measured. The former condition ensures that the static spectrum contributes insignificantly to the total measurement time; whereas the latter insures that the static spectrum contributes insignificantly to the noise in any time-slice spectrum.

The static DC-coupled interferogram can (and should) typically be collected with much better SNR than the time-resolved measurements, at a cost of only a small percentage increase in the total duration of an experiment. This is because the multiple time-resolved digitizations will typically involve a large number of points at very narrowly spaced intervals. If the static measurement is averaged for a duration corresponding (for example) to only 10 of these intervals, then it will still contribute a negligible amount of noise to each absorbance spectrum calculated from an individual post-trigger time point, compared to the noise coming from the time-resolved signal itself.

In the case of a logarithmic time base, the last time-resolved interval to be measured may be many times (even millions of times) longer than the initial dwell. If the static average time is set to be too short, then the noise in the spectrum corresponding to this last interval will result as much from noise in the static spectrum as from noise in the time-resolved measurement. A good rule of thumb is to set the static average time to about 10% of the total duration of the time between repeated sample triggers.

8.4.1.8.9 Post-Trigger Delay

This represents the amount of time during which a signal is digitized, but then discarded rather than stored. Unless quasi-logarithmic data averaging is selected (see below), it is generally best to set this delay equal to 0. The analog-to-digital (A/D) converter starts working when the external triggering device (e.g., the photolysis laser) sends a synchronizing transistor–transistor electronic pulse to the A/D board. In most cases, it is best not to discard data once this synchronization pulse arrives at the A/D board. In fact, it is generally desirable to have the external device send out the synchronization pulse slightly in advance of when the actual sample trigger will occur. This allows measurement of spectra both before and during the actual triggering event. This is extremely useful for documenting that the timing of the experiment actually occurred as desired. For sub-μs spectra especially, it can be very difficult to time the A/D converter to start its conversions at exactly the same instant the trigger arrives at the sample. It is possible to determine this timing after the fact if the A/D signal obtained during the trigger event has been saved.

The main reason for setting a non-zero value for the post-trigger delay is that current versions of software typically have a limit to the number of samples (that is, the number of time points) that can be measured. If one wants to measure a large number of time points corresponding to times significantly delayed from the trigger (which is typically the case for quasi-logarithmic data averaging), then it may be necessary to sacrifice some digitizations that are measured close to or before the trigger.

One situation where a significant post-trigger delay may be desirable is when a Q-switched laser is used in a mode where its flash lamp and the spectrometer internal digitizer are slaved to the same spectrometer-generated timing pulse. (This situation arises in particular with the Nicolet-OMNIC software when measuring time-resolved spectra with the internal digitizer.) After the flashlamp is triggered, there is typically a Q-switch delay of about 200–300 μs before the Q-switch is turned on and photons leave the laser. If the digitizer time spacing is set to 5 μs, for example, there will be a good 40–50 digitizations of nothing but prephotolysis baseline. When the total number of saved digitizations is limited to 100–200, this can constitute a substantial waste of resources. In this case, the post-trigger delay can be set to ~200 μs, and one can still obtain a recording of the sample before, during, and after the photolysis.

8.4.1.8.10 Initial Dwell

The initial dwell is the spacing between the first two stored time points. If a linear time base is chosen, it will of course also be equal to the spacing between all successive time points. It is important to set this initial dwell equal to, or slightly faster than, the time resolution of the analog signal. If this spacing is set to a larger value, then at least some of the interspersed digitizations are usually lost. This will result in a poorer SNR than that predicted by Equation 8.1.

8.4.1.8.11 Number of Samples (Number of Time Points)

The number of samples, or number of time points, is just the number of time-resolved interferograms (or spectra) in the final multifile saved to disk. It is strongly recommended to set this to the largest value allowed by the software, unless that value exceeds by a factor of 10 or more the longest intrinsic time constant for the sample response to the trigger. More typically with current software, the number of time points will be limited by either the digitizer's on-board memory (in the case of a logarithmic time base), or by the space reserved in RAM for the data collection (for a linear time base).

The latter limit is typically 100–200 time points with current commercial software. However, it can be seen that with time-resolved measurements, the size of a multifile stored in memory with $M = 4100$ points per interferogram and 200 time points will be only ~2 Mbyte (assuming 16-bit digitization). It can be expected with continuing decreases in the cost of computer RAM and processor speed, the time and space required to signal-average and to store even much larger multifiles should continually become less constraining.

8.4.1.8.12 Linear/Logarithmic Time Base

With a linear time base selected, equally spaced time points will be collected. In this case, the SNR at each time point is expected to be the same. A limit in the number of samples of 100–200 with current commercial software tends to limit the total temporal duration of the processes that can be followed with a particular time resolution.

However, if a logarithmic time base is chosen, the spacing between subsequent time points increases. In this case, roughly equal numbers of time points will be collected for each decade (factor of 10) in time that passes. This provides the advantage of allowing for observation of spectral evolution over many orders of magnitude in time. It is recommended that ~10 time points be collected per decade of time, which may be expressed alternatively as a factor of ~1.25 between successive time values. These choices allow the coverage of 7 decades of time with only ~70 time points stored.

8.4.1.8.13 Trigger Interval

This must at a minimum be greater than the initial dwell τ multiplied by the number of samples S. If a nonlinear time base is used, the trigger interval must be set to an even larger number than this: typically, at least $10^{S/D}\tau$, where D is the number of points per decade of time.

Finally, regardless of the preceding instrumental limitations on the trigger interval, it should in no case be set to less than 5–10 times the longest relaxation time required to return the sample to its starting state. If this rule is violated, then the sample will likely accumulate a significant steady-state concentration of one or more reaction products. This can lead to multiple-photon events — i.e., the trigger event can induce a further reaction in the already-created intermediate state. This is likely to lead to significant artifacts in the spectrum, especially if the secondary reaction is not reversible. The resulting drifts in the sample can get to be quite large over the course of the measurement.

8.4.1.8.14 In-Scan Co-additions

This represents the number of times the trigger will be repeated before the mirror moves to its next position. It is recommended generally to leave this at 1, except possibly for survey scans. The reason is that with multiple triggers per mirror position, there is an increased likelihood of sample drift over the course of measuring a single interferogram. Such drift can result in significant errors during the Fourier-transform process, that are likely to be made smaller by instead making multiple scans of the interferogram (see section 8.4.1.4).

8.4.1.8.15 Step Interval

This is the minimum time between interferometer steps. If it is set to less than the trigger interval multiplied by the number of in-scan co-additions, it will have no effect, since this product represents the amount of time required to complete all the sample triggers at each mirror position. It is generally recommended to set this step interval to a small value, so it has no effect on the measurement. However, under some circumstances, it may be desirable to set the step interval to a larger value.

8.4.1.8.16 Number of Scans Co-added

Signal-averaging is required for improvement of the SNR. The value of $t = M \times N \times \delta t$ in Equation 8.1 above is proportional to N, which is simply the number of scans co-added, multiplied by the number of in-scan co-additions. It is important therefore to set the number of scans co-added to a large value. However, if the measurement conditions are untrustworthy for any reason for a period of more than a few hours — e.g., the sample is unstable; the detector will not be refilled reliably with liquid N_2; or the electrical power is subject to outages — it is extremely important to limit the number of scans attempted in any single measurement. This is because, in general, if the measurement is interrupted or degraded at any time prior to the completion of the total number of scans specified, the entire data set will be lost.

8.4.1.8.17 Analog Signal Parameters

These are parameters that can affect the amplification of the analog detector signal. They are generally software-selectable only for slower (microsecond, rather than nanosecond) time resolution; i.e., those that utilize the same 16- or 20-bit digitizer that is utilized in the collection of rapid-scan spectra. Such digitizers are usually installed directly in the spectrometer bench, and are therefore referred to as internal digitizers, as opposed to the external 8- or 12-bit digitizers that are used for sub-μs time resolution. The latter are typically mounted in the external computer that is used to run the spectrometer. For such external digitizers, it is usually necessary to adjust the following parameters directly by use of an external amplifier/ noise filter.

It should be remembered that the optimal settings of all the following parameters depend somewhat on whether AC- or DC-coupled interferograms are to be measured. If the former, then there is by definition a high-pass filter built into the AC coupling of the detector preamplifier, corresponding to a characteristic instrumental decay time. For typical AC-coupled detector/preamp combinations, this is near 1 ms, which may significantly limit measurement of the tail end of triggered biological decay processes. On the other hand, DC-coupled measurements typically preclude accurate measurement of the fastest (sub-μs) processes in biological samples, because the fast digitizers needed for such measurements are not available with adequate bit resolution to accurately capture the tiny changes in the interferogram produced by the biological reaction. For a broadband FT-IR spectrum, the dynamic range of the interferogram digitizer must be, typically, tenfold greater than dynamic range in the intensity spectrum itself (see Figure 8.12). This is because the transient changes in the interferogram result from a combination of closely spaced absorbance increases and decreases, which mostly cancel each other out for typical biopolymer reactions. It is crucial to digitize the residual transient signals accurately, even though they are typically more than 1000 times smaller than the maximum value of the DC-coupled interferogram signal. Since the software-implemented digitizers capable of operating at more than 100 MHz are typically limited to 8-bit resolution, it is simply impossible to make DC-coupled measurements of biological samples with sub-μs time resolution.

8.4.1.8.18 Internal Electronic Bandwidth Filter(s)

If a user-selectable low-pass bandwidth filter is provided, it should be set to the lowest frequency available that does not provide a response time constant greater

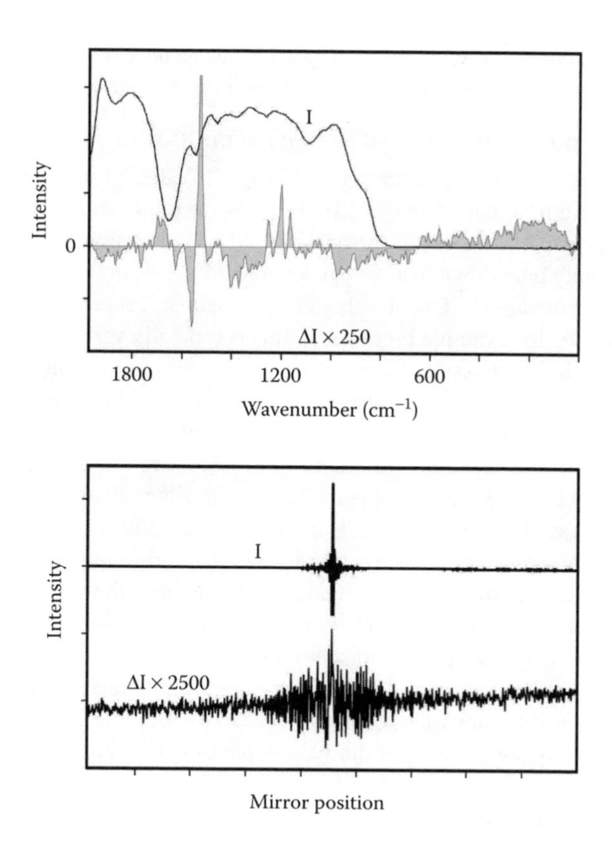

FIGURE 8.12 Comparison of the sizes of the maximum transient differential IR intensities expected from protein samples, in this case bR. A presentation of this comparison in the spectral domain (top) shows that decreases in IR intensity at some wavelengths tend to be balanced by decreases at other nearby wavelengths. This situation is expected to prevail for most polymeric materials. In most ways, the alternating positive and negative bands in TR–FT-IR spectra obtained from biopolymer samples pose a great challenge. The presence of negative as well as positive bands in neighboring spectral regions makes it more difficult to calculate phase spectra, as described above. This also requires substantial amplification of the tiny differential interferogram signal, which has peak-to-peak voltages that are typically less than 0.1% of those in a static interferogram measured on the same sample (bottom). On the other hand, the presence of alternating bands tends to reduce the transient changes in the spectrally integrated IR intensity reaching the detector; and this in turn reduces the size of multiplicative noise arising from modulation of the interferometrically unmodulated broadband IR light, shown in Figure 8.8.

than the desired time resolution. It should be remembered that the time constant provided by an electronic bandwidth filter of frequency f is usually expected to be $1/2\pi f$, although this may be off by a small factor depending on what convention is being used to describe the cutoff point of the filter (e.g., the 50% or 3-decibel roll-off point; a $1/e$ point; or a 10% or10-decibel point).

A safe choice is usually to use the highest-frequency low-pass internal filter available (typically 20–50 KHz), along with the shortest time interval available with the internal A/D converter, typically 5–10 μs. That is because these options are usually designed by the manufacturer to reach the same limit of fastest time resolution.

8.4.1.8.19 Input Range, Internal Amplification, Gain

Standard detector preamplifiers are powered by ±15 V DC power supplies, and will output a maximum signal of about ±12 V, or 24 V total range. The internal A/D converter, typically having 16-bit (or possibly 20-bit) precision, will thus be able to digitize accurately only down to a voltage level of 24×2^{-16}, or 0.36 mV (or 0.023 mV for the 20-bit converter). Due to the large dynamic range of the DC-coupled interferogram (see for example Figure 8.9), this is typically very close to the detector noise level. While it is possible to perform signal-averaging as long as even one bit of the digitizer's dynamic range varies due to the noise, the signal averaging will be more efficient if several bits of noise are digitized.

Therefore, if the dynamic range of the step-scan interferogram has been decreased — e.g., by the use of an optical bandwidth filter, an optically thick sample, or AC-coupling (see Figure 8.12) — it is important to amplify the detector signal prior to digitization. As with rapid-scan measurements, it is advisable to select the largest electronic gain (i.e., the smallest range) that does not risk overflowing the A/D converter at any time during the measurement. For AC-coupled measurements on biological samples, this should typically be a gain factor of 100–200. If the spectrometer's internal digitizer is used, an internal amplifier can usually be programmed via the time-resolved software to produce a gain value of up to ~128. This may be expressed either as a gain fact, or as a dynamic range of the AC-coupled interferograms in volts.

It may be expected, however, that there will occasionally be samples that produce a transient AC-coupled signal larger than ±120 mV, especially from non-biological materials. The most likely reason for this is that the sample converts a substantial amount of the photolysis trigger power into a transient IR emission, although a sample that has a very strong transient IR absorbance increase or decrease may also cause a comparably large intensity change at the detector. A specific example is CdSe quantum dots (see Figure 8.10). With such samples, a gain of 100–200 will be too large, and the A/D converter will be overfilled for at least part of the measurement. This will be obvious from a glance at the time-resolved interferograms, especially at the points measured very shortly after the trigger and near the centerburst (zero-path difference) of the interferogram. These points will show clipping at the maximum or minimum levels of the digitizer. In such cases, the gain setting must be reduced until no clipping is observed. A possible alternative to gain reduction is reduction in the strength of the trigger — e.g., reducing the intensity of a photolysis laser pulse. This is usually not advisable, however, since it will result in a proportional decrease in signal size. For DC-coupled measurements, the gain can be selected by first performing rapid-scan measurement to determine the value that comes closest to filling the A/D converter without overflowing it. This same gain level (or corresponding internal range level) should then be used for the time-resolved measurement.

8.4.1.8.20 Post-Processing Parameters

These parameters control what the operating software does with the interferograms stored in memory once the data collection is finished.

8.4.1.8.21 Result(s) Stored

If it is specified that the interferograms will be converted to spectra prior to data storage, then the parameters for phase correction must be chosen with great care. In currently available time-resolved software on commercial spectrometers, it is usually possible to specify storage of either interferograms or spectra, but not both. Unfortunately this means that if spectra are stored, phase correction performed at the time of original data collection is irreversible. That is, no phase errors can be corrected subsequently by reprocessing the data using better-selected phase correction parameters. This is a significant limitation. There is a substantial advantage to being able to go back and test for errors in phasing well after the measurement, and to correct them if they have occurred. Likewise, the spectral resolution set at the time of original data collection is fixed; it is not possible afterwards to reprocess the data at lower resolution, if that would help reduce noise.

Unfortunately, if a large number of interferograms are stored (especially AC-coupled interferograms), there are limited software routines available afterwards for automatically Fourier-processing them. For example, in neither the Nicolet (OMNIC) nor Bruker (OPUS) software available as of this publication is there a built-in command which allows utilization of the phase array from a stored DC interferogram to phase correct an entire set of time-resolved interferograms that have been stored to disk. By contrast, it is easy to select such a phase correction at the time of original data collection. It is of course still possible to perform the desired phase correction after extracting individual files from the original multifile data. However, manually entering the repeated processing commands is laborious.

Nevertheless, until the user is confident of the parameters to be used for phase correction, it is generally recommended that the result to be stored should consist of interferograms only. Recourse should be made to additional external programs, such as GRAMS from Thermo-Galactic Industries (Keene, NH) to perform multifile operations such as phase correction.

Ideally, there is an option in the data collection software that selects for storage of a static DC-coupled interferogram, obtained from data measured immediately prior to each sample trigger. If available, this option should almost always be selected, since it imposes almost no additional data collection time. This interferogram generally provides the best way to phase-correct the time-resolved interferograms (which will generally have poorer SNR than the static interferogram, even if they are DC-coupled as well). Furthermore, it provides the best background spectrum, which is needed to convert the time-resolved single-beam spectra into absorbance difference spectra. For reasons stated above in Section 8.4.1.4, it is important to calculate the background spectra from data digitized as contemporaneously as possible with the time-resolved signals.

Despite this recommendation, there are several reasons why it may be difficult or impossible to collect a static DC interferogram contemporaneously with a set of AC-coupled time-resolved interferograms. Not all fast IR detectors are equipped with preamplifiers that can provide a DC-coupled output. This is especially true of the most economical photoconductive detector/preamplifier combinations. Furthermore, in early versions of the Bruker IFS66 bench and OPUS time-resolved software, use of the internal digitizer has precluded recording of both an AC-coupled and DC-coupled interferogram, even when both signals are available from the preamplifier. (That is, only the external digitization option allowed contemporaneous measurement of AC- and DC-coupled signals.)

Recording of high-quality time-resolved spectra is still possible in these cases. One option is to store only AC-coupled differential interferograms. In this case, the most accurate phase correction will be obtained by the doubled-angle method (see below) [78]. Subsequent noncontemporaneous measurement of the static background, using the same spectrometer and sample in rapid-scan mode, can be used for calculating absorbance spectra. This approach yields only minimal errors in the final data, since most of the spectral distortions that can arise from drift are eliminated by use of an AC-coupled signal. However, use of a rapid-scan measurement for phase correction is far from ideal, due to the likelihood of phase drift which is not at all corrected by use of an AC-coupled signal in the step-scan time-resolved interferograms.

The other option that has been used very successfully is to store only DC-coupled time-resolved interferograms [28,29]. In this case, it is generally easy to set other timing parameters in the software so that the time-resolved interferograms actually begin several μs prior to the sample trigger. After the data collection is completed, these can be used both for phase correction, and as a background for calculation of absorbance difference spectra. This approach generally requires digitization with at least 16-bit depth, so is suitable only for slower time resolution (greater than 5 μs digitization spacing).

One alternative that is made available in OMNIC software is to AC-couple the detector and digitizer, but to add the digitized AC value to the DC-coupled signal level that is digitized just prior to the triggering of the sample. The addition takes advantage of the large bit-resolution of the computer's main memory to store the resultant interferogram with a large dynamic range. There is no particular disadvantage to doing this as long as the DC-coupled pre-trigger interferogram is also stored, so that it can be ratioed out (effectively subtracted away) afterwards, and this in effect allows the storage of phase information from the DC interferogram with each subfile present in the time-resolved measurement.

8.4.1.9.22 Phase Calculation

As mentioned above, the best options for phase correction are as follows:

(1) If the time-resolved interferograms are DC-coupled, then they can be phase-corrected using standard routines, since the spectra collected from them can be assumed to contain only positive intensities. It is best to use a single phase array to phase-correct all of the time points in a data measurement. This is especially true if only a small number of scans have been performed, since the phase arrays are then likely to include a significant amount of noise. The noise can be reduced by co-adding together all of the time points measured before calculating the phase array.

    Routines for automatically applying the exact same phase array to a large series of time-resolved files may not be present — e.g., in older versions of Nicolet OMNIC or Bruker OPUS. The only exception is in the case where the DC output from the detector is divided (e.g., with a BNC "T" connector); the pre-trigger DC level measured at the second digitizer channel can be saved as the static DC interferogram, and then used to phase correct all the other interferograms. If it is desired to use the phase array based on the co-added DC interferograms from all time delays to phase correct each time point individually, then this must be done by hand or else a special macro must be written.

(2) If the time-resolved interferograms are AC-coupled, then the stored static DC interferogram should be used to calculate the phase array. This phase array should be applied in the phase correction of all of the time-resolved interferograms. All these steps are typically accomplished just by checking one or two boxes in the parameter selection window.

As stated several paragraphs above, it may be possible to apply these phase correction routines automatically to multiple time-resolved files only if selected at the time of original data collection; this may require that only the final spectra (and not the interferograms) be stored to disk. Additionally, limitations present in some versions of commercial time-resolved software and hardware may make it impossible to use option #2 above. If such a limitation is encountered, and no DC interferogram will be available for phase correction of an AC-coupled data set, then it is best to choose from the following alternatives:

(1) Measure and save double-sided AC-coupled interferograms. At any time after the measurement is complete, one can export the data into GRAMS™ format, and Fourier-process the data with a standard phase correction procedure (e.g., the Mertz phase correction method). While the Mertz procedure is likely to introduce phase errors (see above), it can be used to determine if there are bands in the difference spectrum that stand out well above the noise. If so, then the Fourier processing should be repeated with the doubled-angle phase correction method [78,79], which is likely to give the most accurate result.

 The GRAMS™/Array BASIC source code that implements the doubled-angle phase correction method is included in an appendix to Reference 78. However, practical usage of this will likely require the separate purchase of a license for GRAMS™, not available from the authors. Note that use of the doubled-angle phase correction method absolutely requires the measurement of double-sided interferograms.

 For this method to work, difference bands need to stand out above the noise only in the time-averaged result, not in the individual spectra time slices. This is because the doubled-angle method as implemented by the authors can utilize a stored phase array that is obtained from the time-averaged interferograms.

(2) Immediately before the AC-coupled time-resolved measurement (or possibly immediately after it), switch the output of the detector from AC-coupled to DC-coupled mode, and measure a step-scan interferogram with exactly the same parameters as the time-resolved measurement. Spectral

resolution, sample spacing, and (especially) the number of prepeak points must all be the same. Use this interferogram as the source of a phase array that can be used to phase-correct all of the time-resolved interferograms.

8.4.2 Other Experimental Considerations

8.4.2.1 Sample Preparation

In general, similar sample preparation considerations apply to step-scan time-resolved FT-IR methods as to normal rapid-scan measurements on biological samples. The most important is a limitation on the effective sampling thickness of ~10 μm, which is generally observed in all biological FT-IR spectroscopy. This limitation avoids excessively high background absorption in regions where water and protein both tend to absorb strongly, specifically near 1650 cm^{-1}. It must be remembered that a background absorbance of 2 would correspond to a 100-fold reduction in intensity of the measuring beam. The effect of this on the SNR is the same as a 100-fold reduction in throughput Θ, except that it is limited to a specific spectral region. That is, it is expected to result in a 100-fold reduction of the SNR in the region with a background absorbance of 2, relative to spectral regions where the background absorbance is 0 (see Equation 8.1).

Regardless of the sample thickness, differential absorbance signals from biological samples are a small fraction (<2%) of the background absorption of the sample. It is therefore crucial that the largest possible fraction of the sample within the sampling region undergo the triggered reaction cycle that produces the differential signals. This imposes a requirement for a high level of sample purity. For a photoreactive protein sample, for example, it is equally important to avoid the presence of inactivated protein (denatured or lacking chromophore), of other proteins, and of water, since these all contribute nearly equally to the 1650 cm^{-1} spectral region that usually has the highest background absorbance. It is only slightly less important to reduce the presence of lipids and detergents to the minimum needed for sample stability. These molecules by themselves consist largely of polymethylene units, which absorb IR light rather weakly and in band regions where they tend not to interfere with sample interpretations. However, lipid and detergent head groups bind water rather tightly, so that excessive amounts will raise the minimum water content that must be present in the sample in order to maintain biological functionality.

Of course, based on these considerations, it should also be clear that larger proteins are expected to exhibit poorer SNRs, since the IR signals resulting from structural changes of individual amino acid side chains will constitute a smaller fraction of the total background IR absorption in the sample. However, it is also of interest to note that there seems to be a lower size limit in nature for chromophoric proteins to undergo triggered photoreaction cycles with a high level of cyclicity and specificity.

In fact, the best SNRs for signals due to individual amino acid side chains in TR–FT-IR have been obtained from PYP (MW = 14,000) and from bacteriorhodopsin in the form of purple membranes (MW = 26,000). Purple membrane samples, which have a dry weight content of 75% protein and 25% native halobacterial lipids,

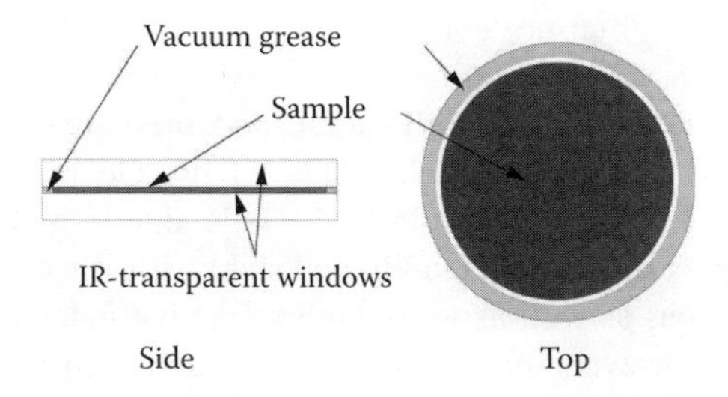

FIGURE 8.13 Sample holder for concentrated membrane protein samples. While an annular spacer of ≤10 μm thickness can be used to define the sample thickness, it can be difficult to keep the sample from losing water with such spacers. Here the sample thickness is instead controlled mostly by the high viscosity of the protein suspension and of the vacuum grease (Dow Corning) that is deposited from concentrated hexane solution or suspension as a film around the rim. Once the sample is compressed by the windows (typically CaF_2 or BaF_2) to a thickness of 5–10 μm, it becomes very resistant to further thinning.

exhibit photochemical reaction cycles that remain nearly unchanged down to a water content (by weight) of ~50%. Such samples are prepared most easily by pelleting the sample in a high-speed centrifuge (200,000 × g), followed by squeezing the sample between two windows, typically 2-mm-thick CaF_2 or BaF_2 disks that are 19 mm, 25 mm, or 37 mm in diameter (see Figure 8.13). Lower-speed centrifugation followed by partial air drying also works, as does ultrafiltration. However, it is more difficult to specify pH and concentrations of other species in the sample when such techniques are used.

The edges of such samples can be sealed either with a gasket of defined thickness, or with a thin coating of vacuum grease spread around the periphery of the window surface prior to application of the sample (see Figure 8.13). Once the windows' edges both contact the gasket or vacuum grease, the windows are mounted into a commercial sample holder that maintains continuous pressure to keep them sealed against water loss. In our experience, samples sealed with vacuum grease hold their water for up to several weeks — somewhat longer than with a gasket alone.

For samples squeezed between 2.5-cm-diameter windows, it is usually important to mask off all but a defined area of the sample with ~5 mm diameter. The hole in the mask serves as a target for the photolysis laser. A simple but effective mask consists of a disk of aluminum foil laminated against black electrical tape. Both tape and foil are punched out with a 5-mm-diameter hole punch or cork borer, and then the mask is placed against the window that will be facing the photolysis laser beam prior to clamping the sample and windows in the sample holder. The sample holder is aligned in the IR beam so as to maximize the throughput through the hole in the mask, and then a small mirror mounted in the sample compartment is used to direct the photolysis laser beam through the same hole, at a ~30° angle away from the IR beam. If the sample window thickness is greater than 2 mm, then the lack of co-linearity in the laser and IR beams can make their degree of overlap at the sample inadequate. In this case, it is advisable to make the mask hole larger than 5 mm and to expand the photolysis laser beam diameter while increasing the pulse energy correspondingly, so as to ensure complete coverage of the IR-illuminated area of the sample by the photolysis beam.

Attenuated total reflection is a well-established alternative sampling technique for biological FT-IR samples, including TR–FT-IR. Ge is an optimal internal reflection element (IRE) for most biological FT-IR, due to its low reactivity with biomembrane samples that can be readily and stably adsorbed to its surface in the presence of bulk water, as well as to its high refractive index and resulting short penetration depth. However, for photochemically activated biological reactions, Ge is generally unsuitable due to a long-lived (~0.5 ms) isotropic emission of broadband IR light that occurs when it is illuminated by an actinic visible or UV light pulse. This can be avoided by using an IRE made from a material whose electronic band gap is sufficiently large as not to absorb the wavelength of the actinic light pulse [28]. ZnSe and diamond IREs have both been used to measure biomembrane samples adsorbed to their surfaces and kept in contact with a suitable buffer solution.

8.4.2.2 Isotope Labeling for Vibrational Assignments

In general, it is important to base vibrational assignments on frequency shifts induced by specific labeling with stable isotopes. It can be a great challenge to find suitable conditions for introducing labels with sufficient specificity to make definitive assignments to particular biopolymer subunits involved in a reaction. For proteins expressed in cell culture, it is often possible to obtain specific labeling of one of the 20 amino acid types, but finding the appropriate conditions for selective labeling may involve considerable effort.

As an example, one recent challenge has been to locate vibrational bands due to an active-site arginine (arg82) in bR. Despite knowledge for over 15 years that mutation of arg82 generally eliminates or greatly reduces light-induced transient proton release that occurs during the bR $\rightarrow$ M photoreaction, until recently no one had been able to identify any perturbed arginine-side-chain IR absorption bands in the bR $\rightarrow$ M difference spectrum. The absence of arginine-specific labeling contrasts with numerous other amino acid types that had previously been specifically labeled for assignment of bR $\rightarrow$ M difference bands, including aspartic acid, tyrosine, lysine, threonine, valine, proline, and tryptophan.

The long-time absence of arginine isotope labels, despite clear evidence for the role of this type of residue in the photocycle, was partly due to the fact that arginine's strongest characteristic vibrational bands, due to guanidino C-N stretch vibrations, lie in the spectrally congested 1620–1680 cm^{-1} region. Furthermore, most of the isotopic substitutions (at H, N, or C) that would be expected to shift the strong arginine bands could also shift other types of nearby vibrational bands — e.g., the water H-O-H bend and the amide I backbone vibrations. As with arg82, these other groups have long been known to participate in the structural changes in the bR$\rightarrow$M reaction. Additionally, the strongest vibrational difference band that would accompany arginine deprotonation is at ~1555 cm^{-1}, nearly coincident with the expected frequency of amide II band changes that might be expected to accompany any conformational rearrangement of the peptide backbone. The amide II group would be expected to show isotope sensitivity similar to that of the deprotonated guanidino group, so that an arginine deprotonation signal near 1555 cm^{-1} could be difficult to distinguish from a mere backbone rearrangement.

Nevertheless, recently it has become possible to label arginines selectively in bR with ^{15}N, by growing halobacteria in media containing $^{15}N_{\eta 2}$-arginine, with no significant scrambling of the label to the amide backbone or other amino acid side chains. This has permitted the unambiguous assignment of a negative band near 1660 cm^{-1} and a significant portion of the positive intensity near 1555 cm^{-1} in bR $\rightarrow$ M difference spectra to arginine. (See Figure 8.14.) Confidence in the assignment of a portion of the complex band at 1555 cm^{-1} to arginine is bolstered by the accuracy of the TR–FT-IR measurement, and in particular by the temporal evolution of the isotope-sensitive band at 1555 cm^{-1}. At the earliest measurement times (tens of μs), when only the photointermediate preceding M is expected to be present, the bandshape in this region is identical to a very high degree in ^{15}N-arg and ^{14}N-arg (natural isotope abundance) samples. The difference between the two isotopic variants of the sample rises with the expected rise time of the M intermediate at the 0°C measurement temperature. Quantitative measurement of the area of the arginine-assigned 1555 cm^{-1} band, along with comparisons to guanidine model compounds, permits the conclusion that an arginine side chain within the protein undergoes nearly stoichiometric deprotonation between the L and M states [32].

For more detailed assignment of vibrational bands within proteins, it is desirable to be able to obtain even more site-specific isotope labeling than in the foregoing example. There are of course multiple arginines within the bR primary sequence, so how (for example) could one pin down the vibrational assignment to a single arginine residue?

In fact, a variety of approaches involving a combination of site-directed mutagenesis and chemical labeling have been attempted to obtain single-site specificity for isotope labeling. For small proteins that can be refolded *in vitro*, total synthesis is an option. For larger proteins, synthesis of a segment, then linking it to the rest of the protein, has been attempted as well.

For assignment of arginine vibrations in the bR $\rightarrow$ M difference spectrum, Hutson et al. demonstrated the feasibility of performing TR–FT-IR measurements on samples containing an isotope-labeled "pseudoarginine" residue at a single position within the sequence [80]. The ψ-arg side chain was synthesized (see Figure 8.15) by substituting the natural arg82 with a cysteine residue, followed by reaction of the cysteine sulfhydryl group with thioethylguanidinium bromide. The latter molecule can easily be synthesized with either ^{14}N or ^{15}N at its two terminal nitrogens. It was shown that TR–FT-IR spectra obtained from the resulting ψ-arg-bR samples were very similar to those obtained from the wild-type protein (see Figure 8.16). However, the isotope substitution at only 2 atoms in the entire 26,000 MW protein resulted in a clear shift of a negative band near 1,670 cm^{-1}, as well as a positive band near 1,560 cm^{-1}. This establishes that the arginine undergoing deprotonation between the unphotolyzed (bR) and M states is indeed the highly conserved arg82.

Such site-specific labeling may be facilitated in other proteins by recent approaches to the incorporation of unnatural amino acids into proteins, which might include unique reactive sites besides the sulfhydryl group of cysteine. Alternatively,

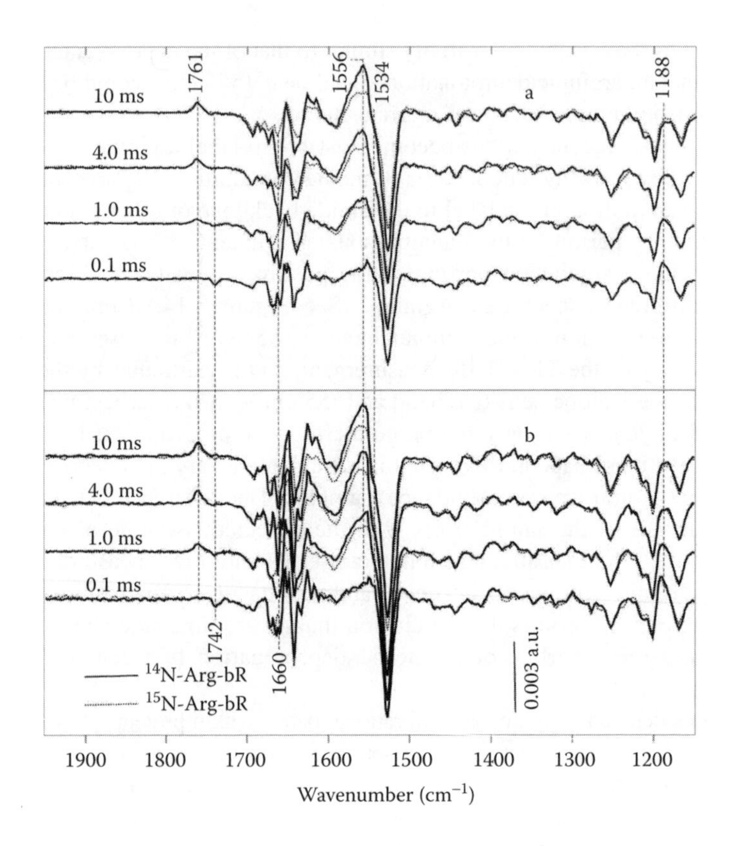

FIGURE 8.14 Step-scan time-resolved FT-IR spectra of bR collected at 0°C demonstrate the use of specific isotope labels to assign transiently appearing vibrational bands to amino acid side chains. Data from two pairs of independently measured samples are shown in the upper and lower panels to show reproducibility of the isotope-induced changes. Spectra were from samples containing either natural-abundance isotopes (^{14}N-arg-bR, dotted lines) or were selectively labeled with η-$^{15}N_2$-arginine — i.e., with the label on the two terminal nitrogens of the guanidino group — at all 11 of bR's arginines (^{15}N-arg-bR, solid lines). Spectra in each panel are stacked in ascending order from 0.1 ms to 10 ms after photolysis. The results show that a substantial portion of the positive intensity near 1556 cm^{-1} at the later time points is assignable to the guanidino group of an arginine in the M photoproduct state, with corresponding negative intensity due to an arginine at 1660 cm^{-1}. However, no difference bands due to arginine are present in the L-bR difference spectra (corresponding to the earliest time point). The isotope-sensitive portions of the 1556 cm^{-1} and 1660 cm^{-1} bands rise with the same kinetics as the L → M transition at this temperature. It is characteristic of deprotonated arginine, and represents approximately as much integrated intensity as the band due to the 1761-cm^{-1} carbonyl stretch vibration of transiently protonated asp85. These changes are attributed to the nearly stoichiometric deprotonation of the active-site arginine (arg82) during the L → M transition, with a net transfer of this proton to the external medium. (Taken from Y. Xiao, M. S. Hutson, M. Belenky, J. Herzfeld, & M. S. Braiman. *Biochemistry 43*, 12809–12818 (2004).)

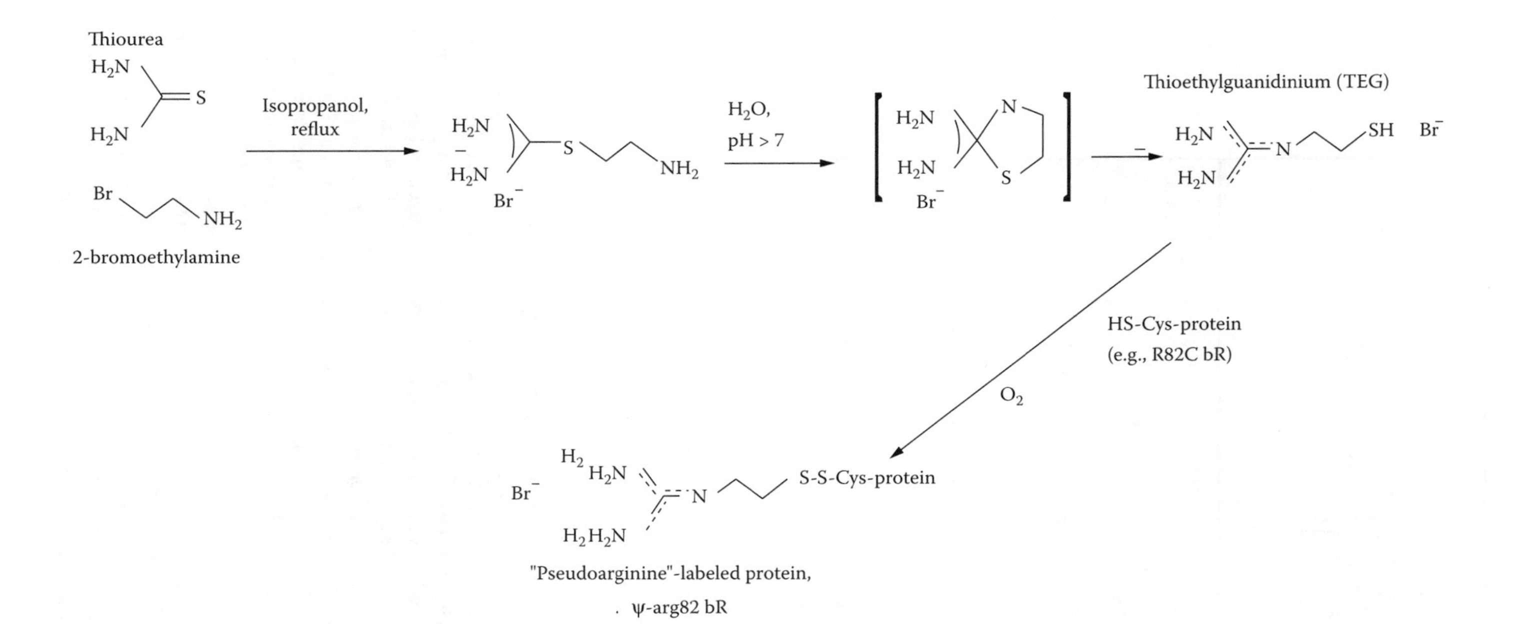

FIGURE 8.15 Combination of site-directed mutagenesis and chemical labeling used to introduce an isotope-labeled "pseudoarginine" residue into a protein, for site-specific vibrational assignment. The pseudoarginine side chain is formed from the disulfide-linked combination of cysteine and thioethylguanidinium. The latter can easily be synthesized from the commercially available ingredients thiourea and 2-bromoethylamine hydrobromide [80]. When the thiourea is obtained with ^{15}N labels at its two nitrogens, the label ends up exclusively at the two terminal nitrogens of the pseudoarginine residue. To apply this approach to a protein, it should either contain no natural cysteines (as in the case of bR), or else the native cysteines should be mutated to nonreactive but otherwise similar residue, such as serine. The latter approach has been utilized to create stable cys-less mutants of hR and pR as well as the further-mutated form with the conserved active-site arginine mutated to cysteine. (R. A. Krebs, M. S. Hutson, R. Parthasarathy, and M. Braiman, unpublished data.)

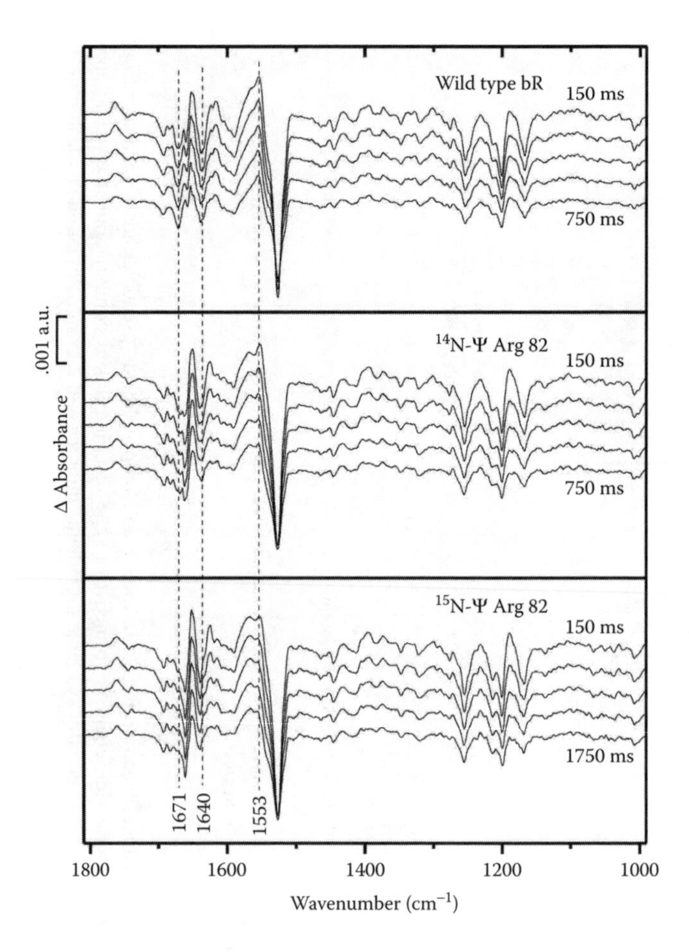

FIGURE 8.16 Sub-ms time-resolved FT-IR spectra of bR containing site-specific pseudoarginine, used to assign vibrational frequencies of the guanidino group of arg82 in the M photoproduct state. The top panel was collected from a sample of wild-type bR. The bottom two panels were collected from a sample of R82C bR that had been reconstituted with thioethylguanidinium and reacted to form a disulfide-linked pseudoarginine-82 (Ψ-arg82) residue. For the middle panel, the Ψ-arg82 contained the natural-abundance nitrogen isotope ^{14}N; for the bottom panel, the two terminal nitrogens in the guanidino group were ^{15}N-substituted. The sensitivity to the isotope substitution of a portion of the differential absorbance at 1671 cm^{-1} and 1553 cm^{-1} confirms that it is the highly conserved active-site arginine (arg82) that is undergoing deprotonation to give rise to these bands. In each panel of 5 spectra, data from a time-resolved experiment were measured with 10 μs digitizer spacing, then group-averaged to give a final time resolution of 150 μs. Spectra represent data accumulated from a total of 30,000 photolysis flashes. The indicated times represent the middle of the averaged time range. Thus the top spectrum in each of the 3 panels includes data points between 80 and 220 μs after photolysis. The detailed time evolution of the data are presented to show that the relative mixes of L, M, and N species, while not identical, are too similar to account for the observed differences between the samples. Figure reproduced from M. S. Hutson, U. Alexiev, S. V. Shilov, K. J. Wise, & M. S. Braiman. *Biochemistry 39*, 13189–13200 (2000).

direct incorporation of fluorinated amino acid side chains — e.g., the guanidino group of arginine — could help to shift their characteristic vibrational frequencies into frequency ranges that conflict less with those of the protein backbone.

8.4.2.3 Optical Filtering

Sharp cutoff long-pass interference filters with cutoffs near 2.5 μm (4000 cm^{-1}) and 5 μm (2000 cm^{-1}) are generally suitable for 2× and 4× undersampled interferograms, respectively, although the imperfect coincidence of their band passes with the Nyquist bandwidth means that there will be a small residual region where aliased signals may occur in the spectrum. As a less expensive and more readily available alternative, antireflection-coated Ge, with a cutoff near 5200 cm^{-1}, works well for 2× undersampled spectra in which the spectral region of interest does not extend beyond 7900 – 5200 = 2700 cm^{-1}. The latter is the lower limit of the spectral region where signals between the Nyquist frequency (3900 cm^{-1}) and the optical low-pass cutoff (5200 cm^{-1}) could appear as aliased bands.

The common use of Ge substrates for such long-pass IR filters affords an additional benefit, in that it also blocks typical visible and UV laser wavelengths that are commonly employed as photolysis triggers for TR–FT-IR experiments. Mid-IR detectors can respond to such visible wavelengths, so it would seem to make the most sense to place the long-pass IR filter right in front of the detector. This location, close to a focus of the IR beam, also allows for use of the smallest possible diameter optical filter, which can be mounted very inexpensively against the detector surface with a taped-on sleeve made of cardboard or plastic.

The Bruker IFS66 has an internal filter wheel accessory that allows for software selection from among multiple 1-inch-diameter band pass filters. Unfortunately, this wheel is located between the IR source and the interferometer, where the filters can provide no protection for the IR detector (or the HeNe reference laser detector) from scattered or reflected light from a photolysis laser. This generally means that another long-pass filter or Ge window must still be placed in the sample compartment or against the front of the detector.

8.4.3 Triggering of Photolysis Lasers

All time-resolved experiments require careful consideration of the triggering method for initiation of the reaction cycle. For most of the biological samples studied to date with TR–FT-IR, pulsed laser photolysis has been used to trigger the reaction. The characteristics of short pulse duration (as low as 5 ns), high pulse energy (typically 5–20 mJ per pulse) and pulse-to-pulse stability over the course of many thousands of flashes makes Q-switched pulsed solid-state lasers — e.g., Nd-YAG with or without frequency-doubling — the most frequently described source of actinic light for TR–FT-IR experiments.

One feature of Q-switched solid-state photolysis lasers that must not be forgotten is the ~250 μs delay between triggering of the flash lamps and triggering of the Q-switched output. The need for a very constant value of this delay time, coupled with the fact that the step-scan FT-IR instrument requires a longer delay (several ms) after each interferometer step for mirror vibrations to die out, establishes a conflict as to which instrument, the spectrometer or the laser, will provide the master clock for the experiment.

In our experience, letting the spectrometer serve as the master minimizes the amount of time it takes to collect the entire spectrum, and also reduces the number of sample-damaging photolysis pulses that occur at times when the spectrometer is not ready to collect data. But in order for the spectrometer to serve as an overall master, at least for sub-μs time-resolved experiments, it must be able to perform a certain amount of hand-shaking with the laser. The spectrometer needs to issue a flash lamp trigger pulse ~250 μs prior to the time it expects to initiate digitization of the detector signal. The laser then generates the internal Q-switch delay internally, using a delay value that is optimized by the user. In order to minimize the effects of uncertainty in the value of this Q-switch delay time, the fast external digitizer card is triggered by a Q-switch advance trigger pulse from the laser. This advance trigger pulse is set to occur ~100 ns prior to the actual opening of the Q-switch, in order to allow collection of a few digitizations immediately prior to the photolysis pulse reaching the sample.

For slower time-resolved experiments in which the spectrometer's 200 KHz internal digitizer is used, it is generally not possible to trigger the digitizer from the Q-switch advance trigger. In this case, it is desirable for the spectrometer software to have a feature that allows an internal delay of up to ~250 μs between sending the trigger pulse to the laser flash lamps and initiating digitization. This is especially important if the software is to be used to perform quasi-logarithmic averaging of data from different time points following the photolysis flash. The zero time point in this case should be measured from the pulsed laser output, not from the earlier firing of the flash lamp.

8.5 FUTURE PERSPECTIVES

In order to extend step-scan FT-IR methods to other non-photochemical enzyme reactions, it will likely be necessary to find ways to increase the concentration of the substrate rapidly by means of UV-laser-induced release of a "caged" precursor compound. This would trigger substrate binding to the enzyme and initiate the subsequent reaction. The caged substrate is not thermally returned to the caged state, and this reaction is therefore not intrinsically cyclic. However, by means of a moving sample stage, fresh regions of sample can be introduced into the path of the IR and photolysis beams. This approach has been outlined [81,82], and a demonstration of step-scan TR–FT-IR spectra have been obtained from the photolytic release of caged ATP, without any protein present. In order to attain the goal of studying of the submillisecond enzyme reaction steps for which the step-scan method is best suited, the proposed approach is clearly more challenging than use of an intrinsic chromophore. This is mainly because the photolytically uncaged ligand concentration must be quite high, in order to avoid having diffusion and binding being the only observable steps on these fast time scales — at least within the viscous concentrated protein solutions that must be used for IR studies. Nevertheless, this approach points the way to more general time-resolved FT-IR methods that should be applicable to a wider range of enzymes lacking intrinsic photochemistry.

Additionally, applications of time-resolved vibrational spectroscopy to synthetic polymers and biomolecules can be expected each to enhance progress in the other. Polymer scientists have been accumulating useful experience obtaining TR–FT-IR spectra from reactions triggered by fast mechanical (stretching, compressing, twisting) [6] or electrical pulses [2,4,5]. Such non-photochemical triggers should be useful for initiating all sorts of interesting biochemical processes *in situ* in FT-IR spectrometers, including cross-bridge forming and breaking in linear motor proteins (myosin, tubulin); rotation-induced proton transport in mitochondrial ATP synthase and the flagellar rotor; and voltage-gated triggers of membrane proteins (K^+ and Na^+-specific channels of nerves).

Conversely, design of polymers with useful photoelectric and electromechanical properties has advanced rapidly in the past decade; and other types of dynamic polymers (photomechanical, photochemical) are likely to follow. The ability to synthesize combinatorially large numbers of polymers of defined complex composition is likely to lead eventually to synthetic molecules that have some of the interesting properties of proteins, in particular the ability to focus energy (whether from a light pulse, an electrical pulse, a temperature change, or elsewhere) in order to promote a chemical change at a single defined bond out of many thousands that form an invariant scaffold. As these developments take place, it is expected that time-resolved FT-IR experiments on polymers will increasingly resemble those on proteins, in particular with respect to their utility for detecting structural changes at individual bonds within very large macromolecular assemblies. Such experiments should aid in evaluating the utility of such polymeric materials, as well as in guiding their design.

ACKNOWLEDGEMENTS

We thank Richard A. Krebs, Sergey Shilov, and Rangadorai Parthasarathy for assistance in preparing figures, and Peter Lassen for critical reading of the manuscript. We also appreciate extensive technical assistance from spectrometer and manufacturers (Bruker Instruments, ThermoElectron Corporation and Kolmar Technologies).

REFERENCES

1. M. S. Braiman, A. Dioumaev, and J. R. Lewis. *Biophys. J. 70*, 939–947 (1996).
2. S. Okretic, M. A. Czarnecki, and H. W. Siesler. *Polym. Mater. Sci. Eng. 75*, 37–38 (1996).
3. S. V. Shilov, S. Okretic, H. W. Siesler, and M. A. Czarnecki. *Appl. Spectrosc. Rev. 31*, 125–165 (1996).
4. S. V. Shilov, H. Skupin, F. Kremer, E. Gebhard, E., and R. Zentel. *Liq. Cryst. 22*, 203–210 (1997).
5. M. A. Czarnecki, S. Okretic, and H. W. Siesler. *Vibrat. Spectrosc. 18*, 17–23 (1998).
6. H. Wang, R. A. Palmer, C. J. Manning, Christopher J., and J. R. Schoonover. *AIP Conf. Proc. 430*, 555–559 (1998).

7. P. R. Griffiths, B. L. Hirsche, and C. J. Manning. *Vibrat. Spectrosc. 19*, 165–176 (1999).
8. M. S. Braiman, P. L. Ahl, and K. J. Rothschild. *Proc. Nat. Acad. Sci. U.S.A. 84*: 5221–5225 (1987).
9. P. M. Aker and J. J. Sloan. *Springer Proc. Phys. 4 (Time-Resolved Vib. Spectrosc.)*, 6–10 (1985).
10. M. S. Braiman, O. Bousché, and K. J. Rothschild. *Proc. Nat. Acad. Sci. U.S.A. 88*, 2388–2392 (1991).
11. M. S. Braiman and K. J. Wilson. *Proc. SPIE 1145*, 397–399 (1989).
12. G. Souvignier and K. Gerwert, Klaus. *Biophys. J. 63*, 1393–1405 (1992).
13. T. J. Walter and M. S. Braiman. *Biochemistry 33*, 1724–1733 (1994).
14. R. Masuch and D. A. Moss. *Appl. Spectrosc. 57*, 1407–1418 (2003).
15. R. Diller, M. Iannone, R. Bogomolni, and R. M. Hochstrasser. *Biophys. J. 60*, 286–289 (1991).
16. R. Diller, M. Iannone, B. R. Cowen, S. Maiti, R. A. Bogomolni, and R. M. Hochstrasser. *Biochemistry 31*, 5567–5572 (1992).
17. R. A. Palmer, G. D. Smith, and P. Chen. *Vibrat. Spectrosc. 19*, 131–141 (1999).
18. P. Hamm and R. M. Hochstrasser, in *Ultrafast Infrared and Raman Spectroscopy* (Vol. 26), Ed. M. D. Fayer, New York: Marcel Dekker, pp. 273–347 (2001).
19. G. S. Jas, C. Wan, and C. K. Johnson. *Spectrochimica Acta 50A*, 1825–1831 (1994).
20. K. Gerwert, in *Infrared and Raman Spectroscopy of Biological Materials* (Vol. 24), ed. H.-U. Gramlich and B. Yan, New York: Marcel Dekker, pp. 193–230 (2001).
21. O. Weidlich and F. Siebert, *Appl. Spectrosc. 47*, 1394–1399 (1993).
22. J. Heberle and C. Zscherp. *Appl. Spectrosc. 50*, 588–596 (1996).
23. W. Hage, M. Kim, H. Frei, and R. A. Mathies. *J. Phys. Chem. 100*, 16026–16033 (1996).
24. A. K. Dioumaev and M. S. Braiman. *J. Phys.Chem. B 101*, 1655–1662 (1997).
25. O. Weidlich, L. Ujj, F. Jager, and G. H. Atkinson. *Biophys. J. 72*, 2329–2341 (1997).
26. B. Hessling, J. Herbst, R. Rammelsberg, and K. Gerwert. *Biophys. J. 73*, 2071–2080 (1997).
27. C. Rödig, I. Chizhov, O. Weidlich, and F. Siebert. *Biophys. J. 76*, 2687–2701 (1999).
28. C. Zscherp, R. Schlesinger, J. Tittor, D. Oesterhelt, and J. Heberle. *Proc. Nat. Acad. Sci. U.S.A. 96*, 5498–5503 (1999).
29. C. Zscherp, R. Schlesinger, and J. Heberle. *Biochem. Biophys. Res. Commun. 283*, 57–63 (2001).
30. R. Rammelsberg, G. Huhn, M. Lubben, and K. Gerwert. *Biochemistry 37*, 5001–5009 (1998).
31. J. Wang and M. A. El-Sayed. *J. Phys. Chem. A 104*, 4333–4337 (2000).
32. Y. Xiao, M. S. Hutson, M. Belenky, J. Herzfeld, and M. S. Braiman. *Biochemistry 43*, 12809–12818 (2004).
33. F. Garczarek, L. S. Brown, J. K. Lanyi, and K. Gerwert. *Proc. Nat. Acad. Sci. U.S.A. 102*, 3633–3638 (2005).
34. T. Friedrich, S. Geibel, R. Kalmbach, I. Chizhov, K. Ataka, J. Heberle, M. Engelhard, and E. Bamberg. *J. Mol. Biol. 321*, 821–838 (2002).
35. A. K. Dioumaev, L. S. Brown, J. Shih, E. N. Spudich, J. L. Spudich, and J. K. Lanyi. *Biochemistry 41*, 5348–5358 (2002).
36. Y. Xiao, R. Partha, and M. S. Braiman, *J. Phys. Chem. B 109*, 634–641 (2005).
37. A. K. Dioumaev and M. S. Braiman. *Photochem. Photobiol. 66*, 755–763 (1997).
38. M. S. Hutson, R. Krebs, S. V. Shilov, and M. S. Braiman. *Biophys. J. 80*, 1452–1465 (2001).

39. C. Hackmann, J. Guijarro, I. Chizhov, M. Engelhard, C. Rödig, and F. Siebert. *Biophys J. 81*, 394–406 (2001).
40. M. Hein, A. A. Wegener, M. Engelhard, and F. Siebert. *Biophys J. 84*, 1208–1217 (2003).
41. M. Hein, I. Radu, J. P. Klare, M. Engelhard, and F. Siebert. *Biochemistry 43*, 995–1002 (2004).
42. X. Hu, H. Frei, and T. G. Spiro. *Biochemistry 35*, 13001–13005 (1996).
43. S. Franzen, J. Bailey, R. B. Dyer, W. H. Woodruff, R. B. Hu, M. R. Thomas, and S. G. Boxer. *Biochemistry 40*, 5299–5305 (2001).
44. J. Contzen and C. Jung. *Biochemistry 37*, 4317–4324 (1998).
45. F. Meilleur, J. Contzen, D. A. Myles, and C. Jung. *Biochemistry 43,* 8744–53 (2004).
46. C. Koutsoupakis, E. Pinakoulaki, S. Stavrakis, V. Daskalakis, and C. Varotsis. *Biochim. Biophys. Acta 1655*, 347–352 (2004).
47. B. H. McMahon, M. Fabian, F. Tomson, T. P. Causgrove, J. A. Bailey, F. N. Rein, R. B. Dyer, G. Palmer, R. B. Gennis, and W. H. Woodruff. *Biochim. Biophys. Acta 1655*, 321–331 (2004).
48. R. B. Dyer, K. A. Peterson, P. O. Stoutland and W. H. Woodruff, *J. Am. Chem. Soc. 113*, 6276–6277 (1991).
49. W. H. Woodruff. *J. Bioenerg. Biomemb. 25*, 177–188 (1993).
50. O. Einarsdottir, R. B. Dyer, D. D. Lemon, P. M. Killough, S. M. Hubig, S. J. Atherton, J. J. Lopez-Garriga, G. Palmer, and W. H. Woodruff. *Biochemistry 32*, 12013–12024 (1993).
51. S. Stavrakis, K. Koutsoupakis, E. Pinakoulaki, A. Urbani, M. Saraste, and C. Varotsis. *J. Am. Chem. Soc. 124*, 3814–3815 (2002).
52. E. Pinakoulaki, T. Soulimane, and C. Varotsis. *J. Biol. Chem. 277*, 32867–32874 (2002).
53. E. Pinakoulaki and C. Varotsis. *Biochemistry 42*, 14856–14861 (2003).
54. K. Koutsoupakis, S. Stavrakis, T. Soulimane, and C. Varotsis. *J. Biol. Chem. 278*, 14893–14896 (2003).
55. C. Koutsoupakis, T. Soulimane, and C. Varotsis. *J. Am. Chem. Soc. 125*, 14728–14732 (2003).
56. C. Koutsoupakis, T. Soulimane, and C. Varotsis. *J. Biol. Chem. 278*, 36806–36809 (2003).
57. B. Rost, J. Behr, P. Hellwig, O. M. Richter, B. Ludwig, H. Michel, and W. Mäntele. *Biochemistry 38*, 7565–7571 (1999).
58. D. Heitbrink, H. Sigurdson, C. Bolwien, P. Brzezinski, and J. Heberle. *Biophys. J. 82*, 1–10 (2002).
59. J. A. Bailey, F. L. Tomson, S. L. Mecklenburg, G. M. MacDonald, A. Katsonouri, A. Puustinen, R. B. Gennis, W. H. Woodruff, and R. B. Dyer. *Biochemistry 41*, 2675–2683 (2002).
60. K. Koutsoupakis, S. Stavrakis, E. Pinakoulaki, T. Soulimane, and C. Varotsis. *J. Biol. Chem. 277*, 32860–32866 (2002).
61. C. Koutsoupakis, T. Soulimane, and C. Varotsis. *Biophys. J. 86*, 2438–44 (2004).
62. V. Sivakumar, R. Wang, and G. Hastings. *Biochemistry 44*, 1880–1893 (2005).
63. R. Brudler and K. Gerwert, *Photosynth. Res. 55*, 261–266 (1998).
64. A. Remy and K. Gerwert, *Nature Structural Biology 10*, 637–644 (2003).
65. A. Mezzetti, W. Leibl, J. Breton, and E. R. Nabedryk. *FEBS Lett. 537*, 161–165 (2003).
66. A. Mezzetti, D. Seo, W. Leibl, H. Sakurai, and J. Breton, *Photosyn. Res. 75*, 161–169 (2003).
67. R. Brudler, R. Rammelsberg, T. Woo, E. D. Getzoff, and K. Gerwert. *Natur. Struct. Biol. 8*, 265–270 (2001).

68. A. Xie, L. keleman, J. Hendrixs, B.J. white, K.J. Hellingwerf, W.D. Hoff. *Biochemistry 40,* 1510–1517 (2001).
69. R. H. Callender, R. B. Dyer, R. Gilmanshin, and W. H. Woodruff. *Ann. Rev. Phys. Chem. 49*, 173–202 (1998).
70. S. Williams, T. P. Causgrove, R. Gilmanshin, K. S. Fang, R. H. Callender, W. H. Woodruff, and R. B. Dyer. *Biochemistry 35*, 691–697 (1996).
71. S. J. Maness, S. Franzen, A. C. Gibbs, T. P. Causgrove, and R. B. Dyer. *Biophys. J. 84*, 3874–3882 (2003).
72. D. M. Vu, J. K. Myers, T. G. Oas, and R. B. Dyer. *Biochemistry 43*, 3582–3589 (2004).
73. J.-P. Wang and M. A. El-Sayed. *Biophys. J. 76*, 2777–2783 (1999).
74. P. R. Griffiths and J. deHaseth, *Fourier Transform Infrared Spectrometry*, New York: John Wiley and Sons (1986).
75. C. Rödig and F. Siebert. *Appl. Spectrosc. 53*, 893–901 (1999).
76. M. S. Braiman and K. J. Wilson. *Proc. SPIE 1145*, 397–399 (1989).
77. M. Shim, S. V. Shilov, M. S. Braiman, and P. Guyot-Sionnest. *J. Phys. Chem. B 104*, 1494–1496 (2000).
78. M. S. Hutson and M. S. Braiman. *Appl. Spectrosc. 52,* 974–984 (1998).
79. M. S. Hutson and M. S. Braiman. *Vibrat. Spectrosc. 19*, 379–383 (1999).
80. M. S. Hutson, U. Alexiev, S. V. Shilov, K. J. Wise, and M. S. Braiman. *Biochemistry 39*, 13189–13200 (2000).
81. C. Rödig and F. Siebert. *Vibrat. Spectrosc. 19*, 271–276 (1999).
82. R. Rammelsberg, S. Boulas, H. Chorongiewski, and K. Gerwert. *Vibrat. Spectrosc. 19*, 143–149 (1999).

Index

C

D

E

M

Q

R

T

U

V

W

X

Y

Z